GLOBAL MOBILE SATELLITE COMMUNICATIONS

Global Mobile Satellite Communications

For Maritime, Land and Aeronautical Applications

by

STOJČE DIMOV ILČEV

University of Durban — Westville,
South Africa

 Springer

A C.I.P. Catalogue record for this book is available from the Library of Congress.

ISBN 1-4020-7767-X (HB)
ISBN 1-4020-2784-2 (e-book)

Published by Springer,
P.O. Box 17, 3300 AA Dordrecht, The Netherlands.

Sold and distributed in North, Central and South America
by Springer,
101 Philip Drive, Norwell, MA 02061, U.S.A.

In all other countries, sold and distributed
by Springer,
P.O. Box 322, 3300 AH Dordrecht, The Netherlands.

Printed on acid-free paper

Printed in the Netherlands.

Dedicated to the memory of my beloved father

Prof. Dimo Stoev Ilčev

CONTENTS

FOREWORD

Global Mobile Satellite Communications (GMSC) are specific satellite communication systems for maritime, land and aeronautical applications, which will enable connections between Mobile Earth Stations (MES), such as ships, land vehicles and aircraft on the one hand and telecommunications subscribers on shore on the other, through the medium of space segment configuration (satellite constellation), Land Earth Stations (LES) and PTT, Terrestrial Telecommunications Network (TTN) or other landline providers.

This book is very important for modern shipping, road, rail and aeronautical concerns, because GMSC are providing more effective business, trade and prosperity in the new millennium, in the first place for safety and security matters and secondly, for commercial communications. Thus, the most considerable marketing and technical point of this book is due to a great deficiency of suitable manuals on the present international book market, which completely describe GMSC fundamentals, space segments, ground segments (MES and LES) and Global Mobile Personal Satellite Communications (GMPSC).

The book discusses hot topics in telecommunications techniques and technology, which will be useful for technical staff on board vessels, in land vehicles and aircraft, on offshore constructions and for those possessing satellite handset phones. Included are sea, land and air transport sets with many other requirements for more effective trade, which need development, design, utility, implementation, usage and knowledge about GMSC for safety and commercial applications. Otherwise, GMSC solutions are very important to all modern transportation companies, dispatchers, agencies, brokers and the successful management, commerce, carriage tracking and logistics of their fleet.

In general, this book may become the manual for a broad range of readers with different levels of technical education and knowledge, for professional staff involved in GMSC and their technical managers, engineers, professors, students, instructors and participants in GMDSS courses, consultants and supervisors of MES and for military officers and cadets. This book could find an important place in libraries, universities and institutions.

Mobile satellite systems have today become very considerable reading matter for students in many maritime and transportation universities, faculties of electrical engineering and for all modern transportation companies, GMSC providers, operators and manufacturers and for their management staff. Everyone involved in Mobile Satellite Communications (MSC) has to know something about these technology and transmission systems.

In writing this book the author has used the expertise, prospects, literature and manuals of numerous experts, specialists and institutions mentioned at the end of this handbook as well as information from Internet WebPages.

The author has been a professional expert in maritime radiocommunications since 1969, as a radio officer on board ocean-going cargo ships using Morse MF/HF radiotelegraphy and MF/HF/VHF radiotelephony and later as deck officer and Electronics/GMDSS Operator with Inmarsat Ship Earth Stations (SES). In addition, for over 15 years he has managed the IS Marine Radio Company for research, service, installation and engineering of MSC and satellite navigation systems and equipment on board ships and integration with modern IT systems. The author has also used his Doctoral dissertation, Master's theses, technical manuscripts, papers and practical experience with radiocommunications, navigation and GMSC systems.

For basic and principal technical information he has drawn heavily mostly on the following sources:

- "Mobile Satellite Communication Networks", written by R. Sheriff and Y. F. Hu; and "Satellite Communications Systems", written by G. Maral and M. Bousquet. Both books were published by Wiley in 2001 and 1994, respectively.
- "Mobile Satellite Communications - Principles & Trends", written by Madhavendra Richharia and published by Addison-Wesley in 2001.
- "Mobile Antenna Systems Handbook", written by K. Fujimoto and J. R. James; "Mobile Satellite Communications", written by S. Ohmori, H. Wakana and S. Kawase; and "Low Earth Orbital Satellites for Personal Communication Networks", written by A. Jamalipour. All three books were published by Artech House, in 1994, 1998 and 1998, respectively.
- "Satellite Communications: Principles and Applications" and "Electronic Aids to Navigation: Position Fixing". Both books written by L. Tetley and D. Calcutt were published by Edward Arnold, in 1994 and 1991, respectively.
- "An Introduction to Satellite Communications", written by D. I. Dalgleish; and "Satellite Communication Systems" edited by B. Evans. Both books were published by IEE, in 1991 and 1993, respectively.
- "Never Beyond Reach", edited by B. Gallagher and published by Inmarsat, in 1989.
- "Спутниковая связь на море", written by L. Novik, I. Morozov and V. Solovev; and "Международная спутниковая система морской связи – Инмарсат", written by V. Zhilin. Both books were published by Sudostroenie, Leningrad, in 1987 and 1988, respectively.
- "Telekomunikacije satelitima", written by R. Galić, Školska Knjiga, Zagreb, 1983.
- "Radio wave Propagation Information for Predictions for Earth-to-Space Path Communications", edited by C. Wilson and D. Rogers, ITU, Geneva.

Readers will find that this book has been written using up-to-date systems, techniques and technology in satellite communications. The material has been systematized in such a way as to cover satellite development, systematization and definition of all nomenclature and synonyms of mobile satellite communications systems and services, new kinds of launcher systems and the presentation of all types of satellite orbit constellations and spacecraft, the newest concepts of transmission models and accesses including IP networking, a complete introduction to mobile antenna systems and propagation, Inmarsat, Cospas-Sarsat, Big LEO, Little LEO, navigation and tracking systems, including forthcoming augmentation satellite system for Communications, Navigation and Surveillance (CNS) mobile solutions, stratospheric platforms as communications systems, and new coming DVB-RCS.

Furthermore, new concepts and innovations in GMSC, such as Inmarsat GAN, BGAN, F77/F55/F33 Fleet solutions, Swift64 and integration of GMSC systems with new Personal Videophone Technology and Mobile Videophone over IP (VPoIP) will be discussed. Finally, the historic moment is approaching when we can use MES terminals and say: **"Hallo, can you see me, over"?**

PREFACE

This book was written in order to form a bridge between potential readers and current GMSC trends, mobile system concepts and network architecture by using a very simple style with easily comprehensible many technical information, characteristics, graph icons, figures, illustrations and mathematic equations.

The GMSC for augmented maritime, land and aeronautical communications, navigation and surveillance applications are new solutions for modern transportation concerns and their fleets at sea, on land and in the air for the enhancement of commercial and distress communications and tracking solutions.

Therefore, MSC are greatly important for all transportation companies, their successful commerce, carriage and management of vessels, land vehicles and aircraft. Furthermore, modern, innovative techniques and technology in MSC are needed for newly developed GMPSC, broadband and multimedia communications and for IT and Global Navigation Satellite Systems (GNSS).

This handbook consists in 10 chapters on the following particular subjects:

Chapter 1: INTRODUCTION gives a short background to the development of Radio and MSC, overview, concept and applications of GMSC for maritime, land and aeronautical applications, including international organizations, operators and new GMPSC systems.

Chapter 2: SPACE SEGMENT discusses the fundamentals of space platforms and orbital parameters, lows of satellite motions, new types of launching systems, satellite orbits and geometric relations, spacecraft configuration, payload structure and satellite link design.

Chapter 3: TRANSMISSION TECHNIQUES introduces an essential basic knowledge of modulation and demodulation, coding and error corrections, multiple access, Mobile Internet and Broadband protocols and forthcoming mobile DVB-CNS solutions.

Chapter 4: MOBILE ANTENNAS FOR SATELLITE APPLICATIONS provides considerable information about mobile antennas for ships/rigs, land vehicles and aircraft; electromechanical characteristics, with basic relations and classifications of antennas, antenna platforms and tracking systems.

Chapter 5: PROPAGATION AND INTERFERENCE CONSIDERATION comprises all the particulars about propagation loss effects, path depolarization causes, trans-ionospheric contribution and propagation effects that are specific and important for GMSC maritime, land and aeronautical environments and requirements.

Chapter 6: INMARSAT GEO GMSC SYSTEMS describes the Inmarsat system, space segment, ground segment, mobile standards and services, including commercial GMSC, SCADA (M2M), business and emergency solutions.

Chapter 7: COSPAS-SARSAT LEO AND GEO GMSC SYSTEM presents distress and safety satellite systems, emergency satellite beacons for all mobile applications, LEOSAR and GEOSAR space and ground segments, integration and types of coverage.

Chapter 8: NON-GEO GMSC SYSTEMS comments particularly upon modern GMPSC as a part of GMSC systems, such as Globalstar and Iridium of Big LEO and Leo One and Orbcomm of the Little LEO satellite constellation. Accordingly, this book does not present other regional GEO and Non-GEO infrastructures, as they are not GMSC systems, as are regional GEO, regional navigation and tracking MSS OmniTRACS and EutelTRACS and Global Navigation Satellite Systems (GPS, GLONASS and forthcoming Galileo).

Chapter 9: GLOBAL SATELLITE AUGMENTATION SYSTEMS (GSAS) gives a retrospective of determination and navigation satellite systems and makes clear the very new WAAS, MTSAT and EGNOS systems for aeronautical CNS, enhanced air and surface flight control, maritime coastal navigation and vehicle tracking.

Chapter 10: GLOBAL STRATOSPHERIC PLATFORM SYSTEMS (GSPS) are new wireless systems still under development, which will use constellations of aircraft and airships equipped with transponders and large antenna systems, which will provide global or regional multimedia mobile and fixed communications.

Finally, there are References, Facts about the Author, Acronyms and an Index presented.

The author has been engaged for more than 30 years in mobile maritime radio and satellite communications, research, development, education and technical fields. He believes that this book will provide better an understanding of GMSC for everyone with an interest in such modern satellite communications systems.

The Senior Editor of Artech House, Mr Mark Walsh (MWalsh@ArtechHouse.com) and his technical reviewer Mr Bruce R. Elbert have expressed some opposing concerns about this project. Namely, they were most concerned that the book would contain obsolete or no longer relevant material, such as the ICO system, which is not in service and is on hold for an extended period, while both Globalstar and Iridium are soon to encounter bankruptcy. Mr Walsh and Mr Elbert seemingly forgot that Iridium and Globalstar are new systems, while ICO will soon start operations and they have their doubts about matters in other chapters, including hot topics about Inmarsat GAN, BGAN, Fleet F33/55/77 and Swift64 solutions; Cospas-Sarsat; Satellite Augmentation and Stratospheric Platforms Systems. They are also concerned that according to the author, these systems will solve the roaming problem with GSM. This is not entirely true, because, on the contrary, the author wrote that cellular systems are solving the roaming problem by integration with mobile satellite or stratospheric platform systems. They also expected more technical descriptions of the systems that are included. However, future readers of this book will have the chance to evaluate and conclude how many new and hot technical topics are included.

ACKNOWLEDGEMENTS

Above all, the author would like to express his special appreciation to **Nera SatCom AS** manufacturer of MSS, satellite and microwave broadband equipment from Norway and **South African Airways**, for their generous contribution as sponsors of this book. Their unselfish financial and moral helped to complete this project, while other companies such as Inmarsat, Thrane & Thrane and others refused to contribute financially, although their systems and products are for the most part introduced in this book.

Technical information, drawings and photos of the latest MSC systems and products are included in this book thanks to the very kind collaboration of many manufacturers of MES and mobile antenna systems, such as Nera, Jotron, Thrane & Thrane, Skanti, S. P. Radio, JRC, Kyocera, Cal Corporation, Ball Aerospace, Canadian Marconi, Rockwell-Collins, Telit, Dassault Electronique, Tecom Industries, Omnipless and others.

The author has also included some important information for which he is grateful to the writers of various manuals, brochures and magazines published by IMO, ICAO, Inmarsat, Cospas-Sarsat, ITU, Globalstar, Iridium, ICO, Orbcomm and other regulatory bodies and operational companies.

Therefore, he wishes to thank all the above-mentioned esteemed authors, equipment manufacturers and satellite providers for their precious assistance. He also expresses his very sincere respect and appreciation to all old colleagues and present GMDSS operators on ships; to all mariners who have navigated with him; to Prof. Dr Borislav Ivosević, Dean; Prof. Dr Jovo Tauzović; and to all his ex-students from the Maritime Faculty at Kotor; to all his ex-pupils from the Nautical School at Kotor, Montenegro (former Yugoslavia) and he wishes them calm seas.

The author also expresses his special gratitude to his mentor Prof. Dr Boris Spasenovski, Prof. Dr Tatiana Ulčar-Stavrova and Prof. Dr Ljuben Janev for boundless support to end his postgraduate studies for a Masters Degree at the Faculty of Electrical Engineering at Skopie, Macedonia; also to his mentor, Prof. Dr Zoran R. Petrović and to Prof. Dr Djordje Paunović and Prof. Dr Miroslav Dukić of the University and Faculty of Electrical Engineering "Nikola Tesla" at Belgrade, Serbia (former Yugoslavia) for kind assistance in completing his doctoral dissertation.

He is also thankful to his ex-employer Mr Andre (father), President and owner and the unforgettable Captain Von Arx, being with him on m/v Nyon during 1985 and the management, staff and all crewmembers of the shipping company Suisse-Atlantique SA of Lausanne, Switzerland.

The author is very grateful to his ex-employer Mr Gianluigi Aponte, President and owner of MSC at Geneva; Capt. Salvatore Sarno, Chairman at MSC S.A. in Durban, his wife Mrs. Sandra Sarno; Capt. Antonio Maresca, Manager of MSC at Piano di Sorrento in Italy; Mr Paolo Bolla, Manager of MSC in Baltimore; to all ex-crew members, especially to good friends Capt. Salvatore Ambrosino, Ch/Eng Mitrano Raffaele and Ch/Eng Jakić Igor, of the unforgettable m/v MSC Sandra, sailing in the period from 26 June 1995 until 10 February 2000, and to all his friends, ex-pupils and students within the Mediterranean Shipping Company (MSC) for unreproachful consideration and encouragement.

The author acknowledges the key contributions of prospects, brochures and manuals made by the following companies and their responsible persons:

- **Alcatel Telecommunications Review**, Marcoussis, France, Ms Catherine Camus
- **ARINC**, Annapolis, MD, USA, Ms Elizabeth M. Leger
- **Canadian Marconi Co**, Kanata, Canada, Ms Cathy Bain
- **Cospas-Sarsat**, London, UK, Mr B. Ruark, Mr W. Carney and Ms D. Hacker
- **ESA**, Paris, France, Mr A. de Agostini, Ms D. Detain and Ms C. Troudart
- **ESA**, Bruxelles, Belgium, Mr Alexandre Steciw, Head of Galileo Structure
- **Eutelsat,** Paris, France, Ms Nathalie Bavière
- **Globalstar**, San Jose, USA, Ms Bernadette M. Coronado
- **GT & T**, Louvain-La-Neuve, Belgium, Ms Marjorie Perot
- **IMO**, Publication Department, London, UK
- **Inmarsat,** London, UK, Mrs S. Taylor, Ms C. Butler and Miss C. Habib
- **IPM International**, Geneva, Switzerland, Mr D. Zumkeller
- **Iridium LLC,** Tempe, USA, Mr R.W. Smith
- **ITU,** Publication Department, Geneva, Switzerland
- **Jotron**, Tjodalyng, Norway, Mr W. R. Gilbert
- **JRC**, Tokyo, Japan, Mr Toshi Anemiya
- **Kyocera**, Yokohama, Japan, Mr Kenichiro Kawazoe
- **Nera ASA,** Billingstad, Norway, Ms Ulrica Risberg
- **Omnipless**, Cape Town, South Africa, Mr Marius de Plessis
- **Orbcomm LCC**, Dulles, USA, Mr J. J. Stolte and Mr G. Flessate
- **Qualcomm Inc**, San Diego, USA, Mr Brian Jones
- **Radio Holland**, Durban, South Africa, Mr Bruce Dunn
- **Rockwell Collins**, Reading, UK, Mr Derek Berry
- **Sea Launch**, Long Beach, USA
- **SITA**, Vienna, USA, Mrs Elizabeth L. Young
- **Tecom Industries**, Erie, USA, Mr D. Radawicz
- **Telit,** Roma, Italy, Mr Luigi Ferri
- **Thrane & Thrane**, Lyngby, Denmark, Ms Tina Gade Nielsen

The author is very frustrated in not being able to obtain details of certain systems from a very few Satellite Operators, Service Providers and Equipment Manufacturers, who shall not be named at this time.

Much appreciation from the author is extended to his great supporter Mr Brian Jones from Qualcomm Inc.; to Mr Asim G. Kiyani, Key Account Manager of Nera SatCom AS; and to the technical reviewers:

- **Sheriff E. Ray**, PhD, University of Bradford, UK
- **Gultchev Stoytcho**, PhD, University of Surrey, UK

Finally, he would like to express very heartfelt appreciation to his family for their help and understanding while the manuscript was being written, especially to his dear children, son Marijan (3rd Deck Officer for the Shipping Company Azalea) and daughter Tatijana (final year student of English and Italian languages at Niksic University in Montenegro) and to his sister Prof. Tatijana Ilčeva and niece Ivana.

1

INTRODUCTION

The safety of navigation through all past ages has been a primary preoccupation for all seamen and shipping owners. Distress and disasters at sea caused by the blind forces of Mother Nature or by human factors have occurred during the course of many centuries on ships and in the life of seafarers. For many centuries, seafarers sailed without incoming information about trip, navigation and weather conditions at sea. At that time, only audio and visual transfers of information from point to point were used. However, no earlier than the end of the 19th century, were developed new disciplines, such as the transmission of news and information via wire initially, then by radio (wireless) and latterly, via modern satellite and stratospheric platform communications systems.

The facts about airplanes and land vehicles are well determined and clear because these transport mediums have more reliable environments and routes than ships. After a disaster with airplane, train, truck or bus it is much easier to find out their positions and to provide alert, search and assistance. With the exception of safety demands, an important question is the utilization and development of new mobile radio and satellite communications and navigation systems for commercial and social utilization at sea, on land and in the air.

1.1. Abstract

Communication satellites provide the bridges for a number of new, specialized markets in commercial and private telecommunications and make ties between nations. In the course of more than 40 years they have obtained global links in the public and private Terrestrial Telecommunication Network (TTN). Soon after Mobile Satellite Communications (MSC) and Navigation came to serve navy, ground and air forces worldwide and for of economic reasons, they also provided commercial MSC. For 30 years, MSC was used, particularly because ocean-going vessels have become dependent for their commercial and safety communications on Mobile Satellite Services (MSS). Although, aircraft and land vehicles started before ships, due to many unsuccessful experiments and projects they have had to follow the evident lead of Inmarsat maritime MSC service and engineering. Thus, the modified ship's Mobile Earth Stations (MES) are today implemented on land (road or railway) vehicles and aircraft for all civil and military applications, including remote or rural locations and industrial onshore and offshore installations.

The GPS, DGPS, GLONASS and other new satellite navigation and determination systems provide precise positioning data for vessels and aircraft and also serve land navigation and fleet vehicle management. Because of the need for new, enhanced services, these systems will be augmented with satellite communications and ground surveillance facilities. At the end of this race a new mobile satellite revolution is coming, whereby anyone can carry a personal handheld telephone using simultaneously satellite or cellular/dual systems at sea, in the car, in the air, on the street, in rural areas, in the desert, that is to say everywhere and in all positions. These integrated systems will soon be implemented, with new stratospheric platform wireless systems using aircraft or airships.

1.2. Overview

The word **communications** is derived from the Latin phrase "**communication**", which stands for the social process of information exchange and covers the human need for direct contact and mutual understanding. The word "**telecommunications**" means to convey and exchange information at a distance (**tele**) by the medium of electrical signals.

In general, telecommunications are the conveyance of intelligence in some form of signal, sign, sound or electronic means from one point to a distant second point. In ancient times, that intelligence was communicated with the aid of audible callings, fire and visible vapor or smokes and image signals. We have come a long way since the first human audio and visual communications, in case you had forgotten, used during many millennia. In the meantime primitive kinds of communications between individuals or groups of people were invented. Hence, as impressive as this achievement was, the development of more reliable communications and so, wire and radio, had to wait a couple of centuries more.

The invention of the telegraph in 1844 and the telephone in 1876 harnessed the forces of electricity to allow the voice to be heard beyond shouting distance for the first time. The British physicist M. Faraday and the Russian academic E. H. Lenz made experiments with electric and magnetic phenomenon and formulized a theory of electromagnetic (EM) induction at the same time. The British physicist James C. Maxwell published in 1873 his classical theory of electromagnetic radiation, proving mathematically that electromagnetic waves travel through space with a speed precisely equal to that of light. The German physicist Heinrich Rudolf Hertz during 1886 experimentally proved Maxwell's theoretical equations. He demonstrated that HF oscillations produce a resonant effect at a very small distance away from the source and that this phenomenon was the result of electromagnetic waves. Thus, it is after Hertz that the new discipline of Radio technology is sourced and after whom the frequency and its measuring unit (Hz) are named.

An English physician, Sir Oliver J. Lodge using the ideas of others, realized that the EM resonator was very insensitive and he invented a "coherer". A much better coherer was built and devised by a Parisian professor, Edouard Branly, in 1890. He put metal filings (shut in a glass tubule) between two electrodes and so a great number of fine contacts were created. This coherer suffered from one disadvantage, it needed to be "Shaken before use". Owing to imperceptible electric discharges, it always got "baked" and blocked.

The Russian professor of physics Aleksandar S. Popov in 1894 successfully realized the first practical experiments with EM waves for the transmission of radio signals. In the same year, he succeeded in making a reliable generator of EM waves, when the receiving or detecting systems in common use still were not at all satisfactory. Accordingly, using the inventions of his predecessors and on the basis of proper experiments, Popov elaborated the construction of the world's first radio receiver with a wire-shaped antenna system in the air attached to a balloon, see **Figure 1.1. (A).** In 1895, Popov improved Branly's receiver by the insertion of choke coils on each side of the relay to protect the coherer and also by replacing the spark gap with a vertical antenna insulated at its upper end and connected to the ground through the coherer. He then mounted a small bell in serial connection with the coherer's relay anchor, whose ringing effected automatic destabilization and successive unblocked function of the receiver system. On 7 May 1895, he demonstrated his new apparatus to the members of the Russian Physic-Chemical Society: a lightning conductor as an antenna, a metal filings coherer and detector element with telegraph relay and a bell. The relay was used to activate the bell, which is announcing the occurrence of transmitting

Figure 1.1. Popov's Radio Receiver (A) and Ship's Morse Sender (B)

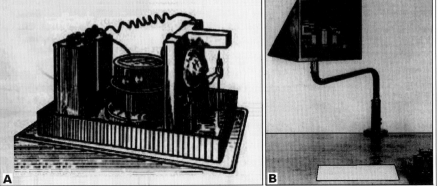

Courtesy of Webpage: "The Radio I" from Internet & Book: "Never Beyond Reach" by B. Gallagher

signals and in this way serving as a decoherer (tapper) to prepare the receiver to detect the next signals. This was the first telegraph station in the world, which could work without any wires. In May the same year, he reported sending and receiving radio signals across a 550 m distance. Soon, instead of a bell he contrived to use a clock mechanism to realize direct, fast destabilization of metal filings in the coherer upon receipt of the signals. In December 1895, he announced the success of a regular radio connection and on 21 March, 1896 at the St. Petersburg University, he demonstrated it in public. Finally, on 24 May 1896 Popov installed a pencil instead of the bell and sent the first wireless message in the world the distance of 250 m between two buildings, conveying the name "GENRICH GERZ" (the name of Hertz in Russian) by Morse code using his homemade transmitter and receiver. In March, 1897, Popov equipped a coastal radio station at Kronstadt and the Russian Navy cruiser Africa with his wireless apparatus and in summer, 1897, Popov started to make experiments at sea, using radios on board ships before the entire world. In 1898, he succeeded in relaying information at a distance of 9 km and in 1899, a distance of 45 km between the island of Gogland and the city of Kotka in Finland. With all his inventions Popov made advances on the discoveries of Hertz and Branly and created the groundwork for the development of maritime radio.

In 1895, a few months later than Popov, a young Italian experimenter, Guglielmo Marconi, started to use radio and was the first to put EM theory into practical application. By the next year, he had sent Morse code messages at a distance of 2 miles. Moving to England to obtain patents on his equipment, he demonstrated radio reception over 8 miles. In 1897, he exhibited the use of radio between ship and shore and, according to Western literature practically started the use of maritime radio. By the same year, Marconi succeeded in getting his wireless telegraphy transmissions officially patented for the first time in the world. The owners of the Dublin Daily Express in Ireland invited Marconi to conduct wireless reports of the Kingston Regatta of July, 1898 from the steamer Flying Huntress, the first ship equipped with a commercial wireless system. Using an antenna hung from a kite to increase the effective height of his masts and a LF of 313 kHz at 10 kW of power, on 12 December 1901 Marconi crackled out the first wireless message to span an ocean in the form of Morse code; three dots forming the letter "s". This telegraph signal was sent from Newfoundland in Canada and received in Cornwall on the west coast of England.

Unlike Popov, Marconi was a good businessman and was able to turn his research into a financial and manufacturing empire. In the **Figure 1.1. (B)** is illustrated Sailor first used Morse Sender for transmitting distress messages and later for commercial ship to shore and vice versa direction at the turn of the 20 century.

In 1900, R. Fessenden made the first transmission of voice via radio in the USA, Fleming in 1904 discovered the diode valve, while their countryman and pioneer Lee de Forest developed and used a triode valve, which made it possible to use radio not only for radiotelegraphy but for voice communications. As early as 1907, he installed a triode valve mobile radio on a ferryboat operating on the Hudson River near New York City.

1.2.1. Development of Mobile Radiocommunications

The very impressive development of mobile radio for maritime use at first and later on for aero applications, initiated mobile distress and safety radio. Once the principles of radio were understood, mobile radio has been a matter of the steady development of technology to extend communications accessibility, coverage and reliability by reducing the size, cost and power consumption of equipment and improving efficiency. With further innovations an age-old barrier between ships and shore was eliminated and possibility to communicate with mobile radios independent of space and time were created. These early radio devices were primitive by today's standards, incorporating spark transmitters, which blasted their signals across almost the entire radio spectrum. It is supposed that the first vessel to have a Ship Radio Station (SRS) was the American liner St. Paul, equipped in 1899. The next one, early in the following year was the German vessel s/s Kaiser Wilhelm der Grosse.

Thereafter, mobile radio spread rapidly throughout the shipping and safety business. By 1899, however, A. S. Popov had been the first in the world who successfully carried out demonstrations of mobile wireless telegraphy communications at a distance of 20 miles between warships of the Black Sea fleet. The first recorded use of radio for saving life at sea occurred early in March, 1899. The lightship on the Goodwin Sands near Dover on the south coast of England was fitted with one of the first seaborne Marconi SRS and used it to report to the Coastal Radio Station (CRS) that the German steamer Elbe had run aground.

The first distress signal was CQD (Come Quick Distress), used from 1904 only on British ships equipped with Marconi devices. After the collision of two passenger ships, the British s/s Republic and the Italian s/s Florida, running in thick fog in the early hours of 23 January 1909, the radio officer on board the s/s Republic sent for the first time in history the distress signal: "CQD MKC (call sign of s/s Republic), CQD MKC, CQD MKC" and text: "Republic rammed by unknown steamship, 26 miles southwest of Nantucket, badly in need of assistance". After the catastrophe of the White Star liner s/s Titanic in the early hours of 15 April 1912, the UK proposal for a distress signal was the already established CQD, the USA proposed the NC of International signal codex and Germany preferred SOE.

By 1912, there were 327 established CRS and 1,924 SRS available for public, commercial and safety use. Use of radio at sea became very attractive and indispensable, creating an immediate need for enforcement of the rules and regulations under international radio coordination. Because of this, the first Preliminary Radio Conference was held in Berlin in 1903, where some of the basic principles for the use of radiotelegraphy at sea were established. In the subsequent Berlin Conference of 1906, two radio frequencies, 500 and 1000 kHz were earmarked for correspondence. This conference also established a standard for International distress signals in radiotelegraphy, SOS.

The radiotelephony distress signal MAYDAY was adopted in 1927 at Hanover. The name of this signal derives from the French phrase "M'aidez", which means, "Help me". The first international SOLAS (Safety of Life at Sea) Convention was adopted during 1914 in London, partly as a result of the Titanic disaster. It stipulated Morse telegraphy on 500 kHz and battery-operated backup radio unit. Ships carrying more than 50 passengers were required to carry radio devices with a range of at least 100 Nm and larger ships had to maintain continuous radio watch with a minimum of 3 radio officers. At the Conference in Washington in 1927 the Regulations were established as a supplement to bring into force three safety calls in radiotelegraphy for distress, urgency and security, SOS, XXX and TTT, respectively. These three signals were obligatory only on 500 kHz, with a silence period of 3 minutes after every 15th and 45th minute. At that time was also introduced the use of radiotelephony at sea and, soon after that, the first radiotelephone communications between s/s America and coastal radio station Deal Beach, New Jersey in the USA, was realized. Then, at the Conference in Madrid radio stations call signs and frequencies were determined, the International Telecommunications Union (ITU) was established and Radio Regulations (RR) adopted. At the Conference in Atlantic City in 1947 a supplementary ITU RR was adopted and a new radiotelephony distress frequency on 2,182 kHz accepted, instead of the old one on 1,650 kHz, with silence periods of 3 minutes after every 00th and 30th minute. Three telephone safety calls were previously used for distress, urgency and security such as: MAYDAY, PANPAN and SECURITE, respectively, on 2,182 kHz and more recently, on 156.8 MHz (16 VHF channel).

Finally, the new era of transistors commenced and later on the period of revolutionary integrated circuits started after 1957. In the meantime, was the change to frequency instead amplitude modulation with a new ARQ system for use in maritime radio telex services.

1.2.2. Evolution of Satellite Communications

The first known annotation about devices resembling rockets is said to have been used by Archytus of Tarentumin, who invented in 426 B.C. a steam-driven reaction jet rocket engine that flew a wooden pigeon around his room. Devices similar to rockets were also used in China during the year 1232. In the meantime, human space travel had to wait almost a millennium, until Sir Isaac Newton's time, when we understood gravity and how a projectile launched at the right speed could go into Earth orbit.

Finally, the twentieth century came with its great progress and the historical age of space communications began to unfold. Russian scientist Konstantin Tsiolkovsky (1857–1935) published a scientific book on virtually every aspect of space rocketing. He propounded the theoretical basis of liquid propelled rockets, put forward ideas for multi-stage launchers and manned space vehicles, space walks by astronauts and a large platform system that could be assembled in space for normal human habitation. A little later, the American Robert H. Goddard launched in 1926 the first liquid propelled engine rocket.

At the same time, between the two World wars, many Russian and former USSR scientists and military constructors used the great experience of Tsiolkovsky to design many models of rockets and to build the first reactive weapons, particularly rockets called "Katyusha", which one Soviet Red Army used against German troops at the beginning of the Homeland War (Second World War). Thus, towards the end of the Second World War, many military constructors in Germany started with experiments to use their series V1 and V2 rockets to attack targets in England. After that, in October 1945, the British radar expert and writer of

Figure 1.2. Sputnik I & Explorer I

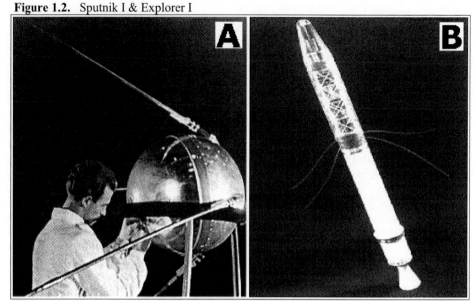

Courtesy of Book: "Never Beyond Reach" (Sputnik I) by B. Gallagher & Webpage: "Explorer I" from the Internet

science fiction books Arthur C. Clarke proposed that only three communications satellites in Geostationary Earth Orbit (GEO) could provide global coverage for TV broadcasting.

The work on rocket techniques in Russia and the former USSR was much extended after the Homeland War. The satellite era began when the Soviet Union shocked the globe with the launch of the first artificial satellite, Sputnik I, on 4 October 1957, shown in **Figure 1.2. (A).** This launch marked the beginning of the use of artificial Earth satellites to extend and enhance the horizon for radiocommunications, navigation, weather monitoring and remote sensing and signified the announcement of the great space race and the development of satellite communications. That was soon followed on 31 January 1958 by the launch of US satellite, Explorer I, illustrated in **Figure 1.2. (B)** and so, the development of satellite communications and navigation and the space race began. The most significant progress in space technology was on 12 April 1961, when Yuri Gagarin, officer of the former USSR Air Forces lifted off aboard the Vostok I spaceship from Bailout Cosmodrome and made the first historical manned orbital flight in space.

1.2.3. Experiments with Active Communications Satellites

After the launch of Sputnik I, a sustained effort by the USA to catch up with the USSR was started. This was reflected in the first active communications satellite named SCORE launched on 18 December 1958 by the US Air Force. The second satellite, Courier, was launched on 4 October 1960 in High-inclined Elliptical Orbit (HEO) with its perigee at about 900 km and its apogee at about 1,350 km using solar cells and a frequency of 2 GHz. The maximum emission length was between 10 and 15 min for every successive passage. The third such satellite was Telstar I designed by Bell Telephone Laboratories experts and launched by NASA on the 10 July 1962 in HEO configuration with its perigee at about 100

Fgure 1.3. Telstar I (A) and Intelsat I (B)

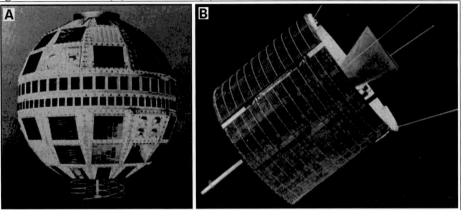

Courtesy of Book: "Satellite Communications" (Telstar I) by R. M. Gagliardi & Webpage: "Explorer I" from Internet

km and apogee at about 6,000 km, see **Figure 1.3. (A).** The plane of the orbit was inclined at about 45° to the equator and the duration of the orbit was about 2.5 h. Because of the rotation of the Earth, the track of the satellite as seen from the Earth stations appeared to be different on every successive orbit. Thus, over the next two years, Telstar I was joined by Relay I, Telstar II and Relay II. All of these satellites had the same problem, they were visible to widely separated LES for only a few short daily periods, so numbers of LES were needed to provide full-time service.

On the other hand, GEO satellites can be seen 24 hours a day from approximately 40% of the Earth's surface, providing direct and continuous links between large numbers of widely separated locations. The World's first GEO satellite Syncom I was launched by NASA on 14 February 1963, which presented a prerequisite for the development of MSC systems. This satellite failed during launch but Syncom II and III were successfully placed in orbit on 26 July 1963 and 19 July 1964, respectively. Both satellites used the military band of 7.360 GHz for the uplink and 1.815 GHz for the downlink. Using FM or PSK mode, the transponder could support two carriers at a time for full duplex operation. Syncom II was used for direct TV transmission from the Tokyo Olympic Games in August 1964. These spacecraft continued successfully in service until some time after 1965 and they marked the end of the experimental period.

Technically, all these satellites were being used primarily for Fixed Satellite Service (FSS) experimental communications, which were used only to relay signals from Fixed Earth Stations (FES) at several locations around the world. Hence, one FES was actually located aboard large transport vessel the USNS Kingsport, home-ported in Honolulu, Hawaii. The ship had been modified by the US Navy to carry a 9.1 m parabolic antenna for tracking the Syncom satellites. The antenna dish was protected, like present mobile antennas, from the marine environment by an inflatable Dacron radome, requiring access to the 3-axis antenna through an air lock within the ship. Otherwise, the Kingsport ship terminal was the world's first true MES and could be considered the first Ship Earth Station (SES). The ITU authorized special frequencies for Syncom communication experiments at around 1.8 GHz for the downlink (space to Earth) and around 7.3 GHz for the uplink (Earth to space). This project and trial was an unqualified success proving only the practically of the GEO system for satellite communications but, because of the large size of the Kingsport SES antenna, some experts in the 1960s concluded that MSC at sea would never really be practical. However, it was clear that the potential to provide a high quality line-of-sight path from any ship to the land and vice versa via the satellite communications transponder existed at this time.

Intelsat was founded in August 1964 as a global FSS operator. The first commercial GEO satellite was Early Bird (renamed as Intelsat I) developed by Comsat for Intelsat, see **Figure 1.3. (B)**. It was launched on 6 April 1965 and remained active until 1969. Routing operations between the US and Europe began on 28 June 1965, a date that should be recognized as the birthday of commercial FSS. The satellite had 2 x 25 MHz width transponder bands, the first with 2 Rx uplinks (centered at 6.301 GHz for Europe and 6.390 GHz for the USA) and the second 2 Tx downlinks (centered at 4.081 GHz for Europe and 4.161 GHz for the USA), with maximum transmission power of 10 W for each Tx. This GEO system used several LES located within the USA and Europe and so, the modern era of satellite communications had begun.

In the meantime, considerable progress in satellite communications had been made by the former USSR, the first of which the Molniya I (Lightning) satellite was launched at the same time as Intelsat I on 25 April 1965. These satellites were put into an HEO, very different to those used by the early experiments and were used for voice, Fax and video transmission from central FES near Moscow to a large number of relatively small receive only stations.

In other words, that time became the era of development of the international and regional FSS with the launch of many communications spacecraft in the USSR, USA, UK, France, Italy, China, Japan, Canada and other countries. At first all satellites were put in GEO but later HEO and Polar Earth Orbits (PEO) were proposed, because such orbits would be particularly suitable for use with MES at high latitudes. The next step was the development of MSC for maritime and later for land and aeronautical applications. The last step has to be the development of the Non-GEO systems of Little and Big Low Earth Orbits (LEO), HEO and other GEO constellations for new MSS for personal and other applications.

1.2.4. Early Progress in Mobile Satellite Communications and Navigation

The first successful experiments were carried out in aeronautical MSC. The Pan Am airlines and NASA program in 1964 succeeded in achieving aeronautical satellite links using the Syncom III GEO spacecraft. The frequencies used for experiments were the VHF band (117.9 to 136 MHz), which had been allocated for Aeronautical MSC (AMSC).

The first satellite navigation system, called Transit was developed by the US Navy and become operational in 1964. The great majority of the satellite navigation receivers has worked with this system since 1967 and has already attracted about 100,000 mobile and fixed users worldwide. The former USSR equivalent of the Transit was the Cicada system developed almost at the same time.

Following the first AMSC experiments, the Radiocommunications Subcommittee of the Intergovernmental Maritime Consultative Organization (IMCO), as early as 1966 discussed the applicability of a MSC system to improve maritime radiocommunications. This led to further discussions at the 1967 ITU WARC for the Maritime MSC (MMSC), where it was recommended that detailed plan and study be undertaken of the operational requirements and technical aspects of systems by the IMCO and CCIR administrations.

A little bit later, the International Civil Aviation Organization (ICAO) performed a similar role to that of IMCO (described earlier) by the fostering of interest in AMSS for Air Traffic Control (ATC) purposes. The majority of early work was carried out by the Applications of Space Technology to the Requirement of Aviation (Astra) technical panel. This panel considered the operational requirements for and the design of, suitable systems and much time was spent considering the choice of frequency band. At the 1971 WARC, 2 x 14 MHz of spectrum, contiguous with the MMSC spectrum, was allocated at L-band for safety use. Hence, the work of the Astra panel led to the definition of the Aerosat project, which aimed to provide an independent and near global AMSC, navigation and surveillance system for ATC and Airline Operational Control (AOC) purposes. The Aerosat project unfortunately failed because, whereas both the ICAO authority and world airlines of the International Air Transport Association (IATA) agreed on the operational benefits to be provided by such a system, there was total disagreement concerning the scale, the form and potential cost to the airlines. Finally, around 1969, the project failed for economic reasons.

The first experiment with Land MSC (LMSC) started in 1970 with the MUSAT regional satellite program in Canada for the North American continent. However, in the meantime, it appeared that the costs would be too high for individual countries and that some sort of international cooperation was necessary to make MSS globally available.

In 1971, the ICAO recommended an international program of research, development and system evaluation. Before all, the L-band was allocated for Distress and Safety satellite communications and 2 x 4 MHz of frequency spectrum for MMSS and AMSS needs, by the WARC held in 1971. According to the recommendations, Canada, FAA of USA and ESA signed a memorandum of understanding in 1974 to develop the Aerosat system, which would be operated in the VHF and L-bands. Although, Aerosat was scheduled to be launched in 1979, the program was cancelled in 1982 because of financial problems.

The first truly global MSC system was begun with the launch of the three Marisat satellites in 1976 by Comsat General. Marisat was a GEO spacecraft containing a hybrid payload: one transponder for US Navy ship's terminals operating on a government UHF frequency band and another one for commercial merchant fleets utilizing newly-allocated MMSC frequencies. The first official mobile satellite telephone call in the world was established between vessel-oil platform "Deep Sea Explorer", which was operated close to the coast of Madagascar and the Phillips Petroleum Company in Bartlesville, Oklahoma, USA on 9 July 1976, using AOR CES and GEO of the Marisat system.

The IMCO convened an international conference in 1973 to consider the establishment of an international organization to operate the MMSC system. The International Conference met in London two years later to set up the structure of the International Maritime Satellite (Inmarsat) organization. The Inmarsat Convention and operating agreements were finalized in 1976 and opened for signature by states wishing to participate. On 16 July 1979 these agreements entered into force and were signed by 29 countries. The Inmarsat officially went into operation on 1 February 1982 with worldwide maritime services in the Pacific, Atlantic and Indian Ocean regions, using only Inmarsat-A SES at first. Moreover, the Marecs-1 B2A satellite was developed by nine European states in 1984 and launched for the experimental MCS system Prodat, serving all the mobile applications.

In 1985, the Cospas-Sarsat satellite SAR system was declared operational. Three years later the international Cospas-Sarsat Program Agreement was signed by Canada, France, USA and the former USSR. In 1992, the Global Maritime Distress and Safety System (GMDSS), developed by the International Maritime Organization (IMO), began its operational phase.

Hence, in February 1999, the GMDSS became fully operational as an integration of Radio MF/HF/VHF (DSC), Inmarsat and Cospas-Sarsat LEOSAR and GEOSAR systems.

The Transit system was switched off in 1996 to 2000 after more than 30 years of reliable service. By then, the US Department of Defense was fully converted to the new Global Positioning System (GPS). However, the GPS service could not have the market to itself, the ex-Soviet Union developed a similar system called Global Navigation Satellite System (GLONASS) in 1988. While both, the Transit or Cicada system provides intermittent two-dimensional (latitude and longitude when altitude is known) position fixes every 90 minutes on average and was best suited to marine navigation, the GPS or GLONASS system provides continuous position and speed in all three dimensions, equally effective for navigation and tracking at sea, on land and in the air.

The USA Federal Communications Commission (FCC) is reasonably encouraging toward private development of the Radio Determination Satellite System (RDSS), which would combine positioning fixing with short messaging. Thus, in 1985, Inmarsat developed the Standard-C system and later examined the feasibility of adding navigational capability.

Although, the ESA satellite navigation concept, called Navsat, dates back to the 1980s, the proposed project has received relatively little attention and even less financial support. Since 1988, the US-based Company Qualcomm has established the OmniTRACS service for mobile messaging and tracking. Soon after, Eutelsat promoted a very similar system named EuroTRACS integrated with GPS and the Emsat communications system.

At the beginning of this millennium two Satellite Augmentation Systems were developed for Aeronautical Communications, Navigation and Surveillance (CNS): the American WAAS and Japanese MTSAT. The next similar project, EGNOS, is Europe's first real venture into global satellite navigation systems. Hence, it will augment the two military satellite navigation systems now operating, the US GPS and the Russian GLONASS and make them suitable for safety critical applications such as flying aircraft or navigating ships through narrow channels and port approaches. This system is a joint project of the ESA, the European Commission (EC) and Eurocontrol (the European Organization for the Safety of Air Navigation) and will become fully operational for commercial usage in 2004. The European Union contribution is the Global Navigation Satellite System (GNSS) as a precursor to a new system known as Galileo. This full GNSS, under development in Europe, is a joint initiative of the EC and the ESA in order to reduce dependency on the GPS service. The target of new the Galileo project is to start with operations by 2005 and to become completely operational by 2008.

In the meantime several very interesting projects are developing in Europe, Japan and the USA for new mobile and fixed multimedia Stratospheric Platform Communication Systems powered by fuel or the Sun's energy and manned or unmanned aircraft or airships equipped with transponders and antenna systems at an altitude of about 20–25 km.

1.3. Development of Global Mobile Satellite Systems (GMSS)

Once the principles of radio were understood, mobile radiocommunications have been a matter of steadily developing and perfecting the radio technologies, extending accessibility and the possibility of radio networks, enhancing range, extending coverage and reliability, reducing the size, cost and power consumption of radio devices and improving efficiency. With further MSC innovations an age long barrier was eliminated between vessels and shore, vehicles and dispatch centres, aircraft and airports and thus, facilities were created to provide mobile offices in ships, land vehicles and aircraft and to communicate with LES independently of space, place and time. So, the world is going to reduce communications barriers and move people across borders for business, social, safety, economic, technical and prosperity purposes. Therefore, the new mobile satellite industry must ensure that mobile communications and navigation services will be responsive to these extraordinary changes and globalization trends.

The MSC systems and technology also offer other benefits and perspectives. In many developing countries telephone density is still at a low level in urban and non-urban areas, because the cost of upgrading such facilities through wireless or TTN means is prohibitive for much of the world areas. Remote, rural and mobile service sectors in many regions are outside the reach of communications facilities, so the new MSS technology, with its instant ubiquitous coverage, may provide cost-effective solutions for developing countries.

1.3.1. Definition of Global Mobile Satellite Communications (GMSC)

The GMSC are GEO or Non-GEO satellite systems, which refers to all communications solutions that provide global MSC service directly to end users from a satellite segment, ground satellite network and TTN landline and/or radio infrastructures. The term GMSC means not only global coverage but also includes local or regional MSS solutions as an integral part of the worldwide telecommunications village. Namely, some of the regional or local MSS can be afterwards integrated to establish a Global MSC network. **Table 1.1.** gives an overview of telecommunication systems showing the respective satellite fit.

The GMSC solution is a modern mobile communication structure systems, which began providing communication links to vessels initially in the 1970s and later to aircraft and all kinds of land vehicles. It must be noted that GMSC providing global and regional coverage represents a new technology era in which wire terrestrial and wireless cellular voice, image,

Table 1.1. GMSC Position within the Telecommunication Structure

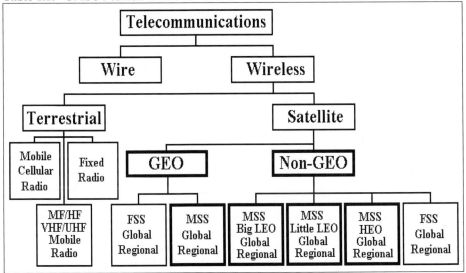

Courtesy of Book: "GMPCS Reference Book" by ITU

video and data systems are combined with MSC applications to provide communication services available anytime and anywhere. Additionally, new satellite technologies, such as GMPSC and Very Small Aperture Terminals (VSAT), have also allowed global personal and commercial mobility. In fact, some of the new GEO or Non-GEO GMPSC systems have entered the field of MSC solutions, which for the past 20 years has been occupied predominately by Intergovernmental Satellite Organizations. In recent years, a growing number of private entities have been prepared to develop and invest in satellite technology, such as Iridium, Globalstar, Teledesic, Ellipso, Orbcomm, Leo One, etc.

At the same time, satellite technology continues to advance; so satellite mobile terminals have become smaller, better and cheaper. Some GMSC systems now being developed are the initiative of the private sector or consortiums. This implies that there should be changes in policy, particularly in countries that do not foresee sufficient private participation in the telecommunication sectors to allow these systems to thrive and to realize their potential.

As mentioned, GMSC systems can provide global or regional coverage. This capability has raised questions about national sovereignty, integrity and security of a country covered by a particular GMSC network. Generally speaking, communication networks in the concerned country must always comply with national regulations that govern integrity and assistance to law enforcement and security agencies. These typically have requirements for national routing, location determination, call monitoring and legal interception. Therefore, seven categories of players in the GMSC community can be identified:

1. National Regulatory Authorities – The international community has to recognize the sovereign right of each country to adopt its telecommunication regulations and that the authority acts in the name of and on behalf of a certain state. It is the responsibility of the National Regulatory Authorities, according to their national laws, regulations and policies, to grant the appropriate authorization to allow GMSC services in a country.

2. GMSC System Operators – These are the owners or operators of the space segment, who have assumed all the financial, technical and commercial risks of developing a GMSC system applications and seek the harmonization of procedures governing the provision of GMSC services to avoid a proliferation of administrative impediments liable to constrain the development of the market.

3. GMSC Gateway/LES Operators – Gateways are LES links between the space segment and TTN, from which, as main sources, GMSC terminal traffic is drawn. The LES in some cases, depending on the business structure of the GMSC system, can be considered as a part of the space segment and can be managed by the GMSC satellite network operator.

4. PSTN Operators – The traditional PSTN operators provide most telecommunication services and networks, both wired and wireless, in a certain country. Furthermore, they are indispensable business partners and responsible for interconnection with terrestrial landline telecommunication networks.

5. Local and/or Regional Service Providers – Local service providers are responsible for the local or regional provision of GMSC services, distributing GMSC terminals and billing GMSC customers. Otherwise, GMSC system or Gateway operators could also be local or regional service providers.

6. GMSC Terminal Manufacturers – These are companies that manufacture MSC and semi-fixed terminals for mobiles, including GEO and Non-GEO satellite networks.

7. GMSC Terminal Users – These are the customers whom all the other players are called upon to serve. They should receive good quality service at the best possible price, within the strict confines of the laws and regulations of the host countries.

1.3.2. Definition of Global Navigation Satellite Systems (GNSS)

The first generation GNSS, as defined by the experts of the ICAO/GNSS panel, plans for some system augmentations in addition to the basic GPS and GLONASS constellations in order to achieve the level of performance suitable for augmented civil aviation applications in oceanic flight and also for enhanced maritime routing applications worldwide, especially in narrow passages, coastal navigation and approaching ports.

The GPS network is a satellite-based all-weather, full jam resistant, continuous operation radio navigation system, which utilizes precise range measurements from the GPS satellites to determine exact position and time anywhere in the world. This system provides military, civil and commercial maritime, land and aeronautical users with highly accurate worldwide three-dimensional, common-grid, position/location data, as well as velocity and precision timing to accuracies that have not previously been easily attainable. The GPS service is based on the concept of triangulation from known points similar to the technique of "resection" used with a map and compass, except that it is done with radio signals transmitted by satellites. The GPS receiver must determine when a signal is sent from selected GPS satellites and the time it is received. Nothing except a GPS receiver is needed to use the system, which does not transmit any signals; therefore they are not electronically detectable. Because they only receive RF satellite signals, there is no limit to the number of simultaneous GPS users.

The Russian Federation (former USSR) provides the GLONASS service from space for accurate determination of position, velocity and time for mobile or fixed users worldwide and in all-weather conditions anywhere. Therefore, three-dimensional position and velocity determinations are based upon the measurement of transit time and Doppler Shift of RF signals transmitted by GLONASS satellites.

The GNSS consists in many players with similar GMSC systems and three major segments:
1. The Space Segment has 24 satellites (21 functioning satellites and 3 on-orbit spares) and is controlled by a proprietary satellite operator or service provider.
2. The Control Segment is operated by Master Control and Monitor Stations.
3. The User Segment is represented by the military/civil authorities for maritime, land and aeronautical users located worldwide. This segment offers Standard Positioning Service (SPS) and Precise Positioning Service (PPS), available to all users around the world. Access to the SPS does not require approval by a certain service provider but PPS is only available to authorized users via the service provider administration.

1.3.3. Network Architecture of GMSC

The increased availability of MSS and GNSS solutions means that many mobiles and individuals will have radio connections and determinations at their disposal whenever and wherever they are traveling, including worldwide shipping, long-distance road and railway haulage, transcontinental flights and universal personal handheld terminals. The new MSC technology is very attractive to mariners, drivers and aviators alike. As discussed, the MSS began in the 1970s for vessels and ocean rigs; land vehicles have been served at a modest level since the late 1980s but a stage of rapid development began in the mid-1990s; service to aircraft has started to grow faster than for land vehicles since 1992. The MSC system consists in space and ground segments, represented by communications satellites and the ground segment comprises different types of MES and Networks, respectively.

1.3.3.1. Space Segment and Configuration of MCS Links

The space segment provides the connection between the subscribers on shore and mobile users via LES or Gateways. It consists in one or more operational or spare spacecraft in a corresponding constellation. The satellite constellation can be formed by a particular type of orbit, such as GEO, Non-GEO (LEO, PEO and HEO) or combinations of these orbits. The satellites can be independent or connected with each other through Inter-Satellite Link (ISL) or Inter-Orbit Link (IOL). The space segment can be shared among different radio networks in different areas in both time and space.

There are also constellations of multipurpose satellites, which platform can serve more than two payloads, such as a combination of meteorological and navigational payloads, etc.

An MSC link is an RF connection between an LES or Gateway stations and MES via GEO or Non-GEO satellites. The part of the MSC link between the LES and satellite is called the feeder link, while the link between the MES and the satellite is called the service link. Both feeder and service links consist in an uplink from the ground towards a satellite and a downlink in the opposite direction (satellite to ground), and both falls under the category of MSS. The feeder link of the LES is categorized as a fixed utilization in MSS and not at all as a part of FSS as is stated in the MSS Handbook published by ITU, page 78.Therefore, the location of LES in MSS is fixed and is not a part of FSS, although some LES in MSS can also provide FSS.

1.3.3.2. Ground Segment and Networks

The ground segment consists in two major network elements: user mobile or portable terminals and ground support stations. The user network comprises four main categories of user terminals whose characteristics are highly related to its applications and operational environments as follows:

a) Mobile Earth Stations (MES) – The MES group of terminals are designed for group usage and installation on board collective transport systems such as: Ship Earth Stations (SES) mounted on ships and other floating objects at sea, on lakes and rivers; Vehicle Earth Stations (VES) mounted on road or railway vehicles and Aircraft Earth Stations (AES) mounted on airplanes, helicopters and other flying objects. Otherwise, the MES terminals are composed of in-mobile and out-mobile and outdoor equipment such as: Above Deck Equipment (ADE) and Below Deck Equipment (BDE) for ships; Above Haul Device (AHD) and Below Haul Device (BHD) for land vehicles and Above Cockpit Unit (ACU) and Below Cockpit Unit (BCU) for aircraft.

b) Personal Earth Stations (PES) – The PES often refer to handheld or palmtop devices for personal utility and carriage in hand, pocket or bag. The transceiver and antenna are integrated into one unit together with an additional antenna for cellular roaming. Other PES categories include those situated on board a mobile platform, such as small ships, cars and airplanes, or they can be transformed as a public payphones in rural areas, ships, oilrigs and airplanes or can be installed as a fixed office unit with outdoor antenna.

c) Transportable Earth Stations (TES) – The TES terminals are typically similar in dimensions to that of a briefcase or laptop computer. As the name implies, these terminals can be transported from one remote or rural site to another; however, operation while mobile will not normally be supported. Every TES contains the transceiver modem and antenna units, and can also serve as indoor equipment.

d) Fixed Earth Stations (FES) – The FES terminals are similar to public urban payphones or fixed office units. These units with antenna can be mounted in some rural areas for public service, while office units can be mounted like remote household or business sites.

The Ground Network consists in six main network elements, which support, maintain and control the space segment and user network as follows:

1. Land Earth Stations (LES) – The LES infrastructure is actually a Gateway station for MSC service and provides an interface to the satellite access network and existing TTN, such as PSTN/PLMN/ISDN via local exchanges. A single LES can be associated with a particular spot-beam or global beam or can access the FSS network. Similarly, LES could provide access to more than one spot or global beam in cases where the coverage of beams overlaps. The main structure of LES can be an external antenna system and the internal sets of RF/IF components, Traffic Channel Equipment (TCE) and Gateway subsystems. The synonym LES is usually used for Inmarsat MSS in general, for all types of Gateways. The Coast Earth Stations (CES) are for Inmarsat Maritime and Land Applications, while Ground Earth Stations (GES) are for Aeronautical Application only.

2. Network Control Centres (NCC) – The NCC, also known as the Network Management Station (NMS), is connected to the Customer Information Management System (CIMS) to coordinate access to the satellite resource and perform the logical functions associated with Network Management (NMF) and Control Functions (NCF). In fact, the NMF can include: development of call traffic profiles; congestion control; system resource management and network synchronization; provision of support in user terminal commissioning; Operation and Maintenance (OAM) functions and the management of interstation signaling links.

The NCF service includes: common channel signaling functions; definition of Gateway configurations and Gateway selection for any mobile station origination. The CIMS is responsible for maintaining Gateway configuration data; performing system billing and accounting and processing call detail records. Thus, the third element of the entire GMSC network can be the control segment, consisting in NCC, SCC and CIMS.

3. Satellite Control Centres (SCC) – The SCC monitors the current performance of a certain space segment and controls the satellite's position in orbit. The spacecraft control functions include the following elements: generation and distribution of satellite ephemera; generation and transmission of commands for inclined orbit operations, payload and bus; reception and processing of telemetry, transmission of beam pointing commands and performance of range calibration. The call control function is associated to SCC, which includes the provision of real-time switching for mobile-to-mobile calls.

4. Network Coordination Stations (NCS) – One NCS serves one ocean region to monitor and control communication traffic within four Inmarsat ocean regions: AORW, AORE, IOR and POR. Each NCS communicates with LES in its own ocean region using special interstation signaling links, with other NCS cites and with NCC located in Inmarsat Headquarters, making possible the transfer of information throughout the system. The NCS is involved in setting up calls to and from MES and in assessing the channel to which both the MES and LES are to tune for the call. For these functions to be performed, each MES Rx must initially be synchronized to the NCS common channel and logged-in to the NCS for its ocean region.

5. Local User Terminal (LUT) – The LUT is a special ground Rx station that receives alert data from Cospas-Sarsat satellites of LEOSAT and GEOSAT constellations, derives the position of the beacons (EPIRB, PLB and ELT), retrieves/checks coded information and forwards the resultant information to MCC.

6. Mission Control Centre (MCC) – The MCC is a ground main infrastructure of the Cospas-Sarsat system which receives alert data from its LUT stations and distributes that information to affiliate SAR points of contact or forwards it to other MCC, or vice versa.

7. Rescue Coordination Centres (RCC) – The RCC operates a system responsible for promoting the efficient organization of SAR service under the Cospas-Sarsat system and for coordinating the conduct of distress and SAR operations with other MES within a certain on-scene region.

8. Terrestrial Telecommunications Network (TTN) – The TTN element is a local ground exchange service, which provides an interface between subscribers ashore and Gateways, i.e., mobile users.

1.4. GMSC Applications

The present GMSC systems are in use for maritime, land and aeronautical applications. Recently, several Personal mobile multipurpose applications using GEO and Non-GEO satellites have been developed and introduced. The lately developed regional networks using stratospheric aircraft and airships will be introduced as very low satellite systems.

1.4.1. Maritime Mobile Satellite Communications (MMSC)

The commercial MMSC systems are designated for very large and medium ocean-going vessels, passenger cruisers, small coastal and river ships, fishing boats, pleasure yachts and rescue boats. These systems are also available for navy vessels, offshore rigs and platforms, including any kind of off/onshore infrastructures. The MMSC system is a successor to the Conventional Maritime Radiocommunications system, which for almost a century was very successful on the commercial and distress scene at sea. In fact, the biggest MSS operator for MMSC is Inmarsat, while other global, regional or local GEO or Non-GEO systems providing MMSC are Iridium, Globalstar, Optus, Emsat, Thuraya, MSAT, AMSC, N-Star, Orbcomm, Leo One and others who have introduced their own MMSS.

1.4.1.1. Maritime Transportation Augmentation System (MTAS)

The development of the MTAS was to identify the possible applications for enhancement of global DCS MF/HF/VHF radio and satellite communication, navigation, surveillance and safety systems including transport security and control of vessels and freight at sea, on lakes and rivers and the security of passengers on board cruisers and hovercrafts. These enhancements include many applications for the better management and operation of vessels and they are needed more than ever because of world merchant fleet expansion. Just the top 20 world ships registers have about 40,000 units under their national flags. Above all, the biggest problem today is that merchant ships and their crews are targets of the types of crime traditionally associated with the maritime industries, such as piracy, robbery and recently, a target for terrorist attacks. Thus, IMO and flag states will have a vital role in developing International Ship and Port Security (ISPS). The best way to implement ISPS is to design a Port Control System by special code augmentation satellite tracking, monitoring and surveillance of all vehicle circulation in and out of the port area. The establishment of MTAS will meet most of these requirements and will complement the services already provided by marine radio beacons.

1.4.1.2. Service for MMSC Users

The first-class two-way MMSC will be essential for mariners to contact and constantly exchange information between vessels, owners, agents, shippers, port authorities, families and friends, or to deal with emergencies, distress and rescue situations at sea. Navy ships can use these facilities for fleet defense, tactical, emergency and information purposes. Therefore, shippers will be nearer to their fleet units, using not only commercial MSC but also reliable distress and intership communications and will also have important 24-hour Maritime Safety Information (MSI), such as Weather (WX) and Navigation Warning (NX).

1.4.2. Land Mobile Satellite Communications (LMSC)

The development of LMSC application for vehicles first started with the unsuccessful Canadian MUSAT regional program in 1970. The earliest experiment was carried out in the USA, where it was realized that the cellular system could not economically provide coverage of vast rural areas, as could MSC. After the cellular system was allocated the spectrum of 810–960 MHz, the ITU WARC-97 allocated sections of this band to the LMSS in Regions 2 and 3 only. This allocation stimulated much research in North America with the initiation of Canadian MSAT and NASA/JPL MSAT-X projects. The MSAT LMSS would be operated by TMI (Telesat Mobile Inc.), a joint venture between Telesat, Canadian Pacific and C Itoh.

In the meantime, the US-based Marisat network was developed and the next ESA Prosat program, which started in 1982 with the initial phase, involved a number of propagation experiments via the Marecs AOR spacecraft. Furthermore, the American Mobile Satellite Consortium (AMSC) in the USA and TMI in Canada, were both inaugurated in 1988 and started to collaborate on a low-bit rate messages MSC system in 1991, in the same way as Inmarsat-C, by leasing Marisat and Inmarsat transponders. The first AMSC and MSAT spacecraft were launched in 1995 and 1996, respectively. Both are establishing the interim use of Inmarsat-C for LMSS. However, this is perhaps one of the major reasons why Inmarsat has decided to reconfigure AOR into two regions, AORE and AORW, in order to improve coverage of the North American land mass.

In Europe, a less ambitious scheme has been initiated such as the unsuccessful Swedish Trucksat project. The geographic nature of Europe is very attractive for LMSS because of the predominant high mountain localities and many small countries, which is not at all convenient for cellular roaming. Therefore, the ESA has been considering, together with Eutelsat, the possibility of including an L-band transponder on the Eutelsat II spacecraft to provide a space segment for a new MSC solution known as the European Land Mobile Satellite System (ELMSS). The discussions about this project were unsuccessful and ESA is now considering the use of Italsat as a host satellite.

In addition, Australia has become a new domestic MSS provider via the Australian MSC operator Optus (former Aussat), who decided to include an L-band transponder on their second generation B1-series spacecraft and to allow the provision of LMSS service in rural areas. They introduced the new MSS Optus in 1992, compatible with both AMSC and MSAT systems, launched for the commercial Australian market in 1994 and providing as well as the AMSC service. Similarly, Japan commenced with MSS implementation within the large Experimental Mobile Satellite System (EMSS) program funded by the Japanese Ministry of Post and Telecommunications. In effect, this project has included the launch of the ETS V satellite, the development, experiments and trial of various speech and data-only LMSC terminals.

Inmarsat also started in 1988 with LMSC service, proposing its second standard Inmarsat-C two-way data-only system. This standard was initially developed for ships but was later adapted for land and air services. For such reasons, several trials were carried out by ESA throughout Europe to establish its performances. After some time, the MOBSAT Group carried out comparative trials of Inmarsat-C and the ESA Prodat and concluded in early 1989 that while Prodat provides a higher throughput under realistic operating conditions, the Inmarsat-C was preferable due to its more flexible network configuration. Soon after, Inmarsat proposed a low-cost MSC telephony system, known as Inmarsat-M, compatible with the service proposed by AMSC, MSAT and Optus. The next standard suitable for LMSC is Inmarsat mini-M, designed to exploit the spot-beam power of the new Inmarsat-3 satellites, as the smallest, lightest and most cost-effective MSC unit ever made.

All systems discussed above have been developed to use L-band frequency spectrum via the ITU WARK MOB-97. The new initiative Qualcomm OmniTRACS system has been developed to use the secondary LMSC location at Ku-band (11/14 GH). This US-based project started in 1986 and trials were carried out two years later. The OmniTRACS system has also been trialed in Europe, using Eutelsat I-F1 spacecraft during the summer of 1989. However, soon after, Eutelsat announced the intention to launch their own service known as EutelTRACS, based on the OmniTRACS CDMA system.

1.4.2.1. Land Transportation Augmentation System (LTAS)

The LTAS has been set up to identify the possible applications for global radio and satellite communication, navigation, tracking and determination; safety systems; transport security and control of all vehicles and freight on roads and railways, including the security of passengers in buses and trains. Namely, these enhancements will comprehend the local and regional road and railway transportation, including a drastically density of vehicles on the roads and railways, which in the future will need some regulation and control using augmentation satellite communication, positioning and tracking solutions. Therefore, this potential benefit will assist vehicle tracking and control to cope with increased traffic and to improve the safety and control of track lines and signaling.

1.4.2.2. Service for LMSC Users

On a worldwide scale, millions of medium to large trucks, buses, cars, trains and other types of land vehicles lack any form of in-cab communications with their dispatch bases, owners, agents, families and friends, or the capability to deal with emergencies such as damage to cargo, engine breakdowns, collisions and rescue situations on the roads. Now transport companies can locate their vehicle fleet and stay in touch with them, no matter where and when they roam. In the same spirit, bus and railway companies can always be in contact with their rolling stock and coach and train personnel and passengers will have the possibility to make private phone calls. The best LMSC solutions on board road and railway vehicles will be utilization of MSC service by two-way low-rate data and message facilities in combination with voice only services such as Inmarsat-C and Inmarsat mini-M standards, respectively. Besides, there are similar solutions like Iridium, Globalstar, Optus, AMSC, MSAT, OmniTRACS and other systems also suited for Army ground forces.

1.4.3. Aeronautical Mobile Satellite Communications (AMSC)

Commercial AMSS is being implemented for aircraft and helicopters, though the different service varies widely depending upon the group or aircraft concerned. The general aviation communities with large aircraft flying worldwide are potential users of AMSS, while light airplanes flying short routes are not now considered to be potential users of MSS for economic reasons. The Army air forces are also potential users of AMSC terminals.

1.4.3.1. Development of AMSC

As discussed, the MUSAT system was the first LMSS in the VHF 200–400 MHz band, developed by Canadian specialists and started in 1970, also providing AMSC experiments, mainly for military applications, using the ATS-6 spacecraft. This program was changed to the new MSAT for all three applications in 1980 because the 800 MHz band was allocated to MSS by the WARC-79. The next program was designed by Japanese experts to develop an AMSC and MMSC and started with exploitation in 1975. The ESA started with the Prosat system in 1982 to contribute to the Inmarsat system. The ICAO embarked on a study into the Future Air-Navigation System (FANS) and new concept of ADSS using the MSC system in 1985. The ADSS was developed and finally accepted as the future prime means of aircraft surveillance and control over oceanic and remote regions.

The Oasis study carried out and completed a research project in 1982 by Stanford Research Laboratories, which confirmed the benefits of MSS in terms of improving ATC in ocean areas and recommended that "satellites of opportunity" be considered as a way of easing the financial burden of introducing such a system. Thus, the Inmarsat organization noted this recommendation and commissioned the Oasis study to investigate the applicability of the Inmarsat system to the AMSS. This study was started by Racal-Decca and confirmed that a 200 b/s data service could be provided to aircraft using 0 dBi antennas and the existing Inmarsat space constellation, which was considered for ATC and AOC, including the provision of data service to passengers. The Oasis study was practically validated by Racal at the 1984 Farnborough International Air Show using aircraft-to-ground data links via Inmarsat AOR satellite and BT Goonhilly GES. Following the successful events at this air show, the Skyphone consortium was formed by Racal, BT and British Airways (BA) to design and develop a system capable of supporting preoperational trials of a public telephony service for passengers on a few BA long-haul airplanes.

A similar experiment was performed by Comsat Labs, in conjunction with Mitre, Collins and Ball Aerospace, in 1985. Simultaneously, ESA was working on the Prosat program that at first included propagation measurements of the aeronautical channels. Furthermore, in the second phase the ESA program will be renamed Prodat in consideration of the design and prolonged demonstration of the two-way low rate data transmission system for all GMSC applications.

The USA-based program, developed by the ARINC organization, commenced in 1985. This project proposed the AvSat system to provide integrated voice and data AMSS between air and ground services throughout the world. AvSat developed a system in 1986 based on six satellites with the advanced TDMA scheme. However, this program collapsed because of the excessive costs of deploying a six-satellite constellation, leaving Inmarsat as the only provider of L-band links, so in 1986 Inmarsat confirmed its intention to set up a versatile AMSS and invitations were issued to tender for the development of the avionics.

The world's first aeronautical telephone call from a commercial jetliner over transoceanic flight routes was successfully carried out from a Japan Air Lines (JAL) aircraft Boeing 747 in October 1987 using the Inmarsat space and ground satellite segments. Otherwise, this experiment was jointly carried out by KDD, JAL and CRL of the Ministry of Post and Telecommunications in Japan. It was based on ETS-5/EMSS research and development program of CRL, which was reorganized from AMES in 1984. The other significant AMSS voice trials were conducted by the Inmarsat, BT, BA and Racal Decca companies and conducted a Skyphone service in May 1988, using a fully avionics-compliant package located in the bay of BA Boeing 747 aircraft, not in the passenger cabin.

The AMSS L-band frequencies, which had been exclusively allocated for ATC and ACC, were opened to APC at WARC-87. With these trends, L-band became the most suitable spectrum for integrated GMSC, which provide worldwide service to ships, land vehicles and aircraft. Since 1990 Inmarsat has carried out a program of AMSC tests, demonstrations and trials and later started to lead an AMSC service. Accordingly, large airways companies worldwide can now use Inmarsat for reliable corporate, social and safety AMSC service.

The Inmarsat system is not ideal for AMSC, in particular because some scheduled flights are using lines far over the North Pole, beyond the coverage of GEO. Meanwhile, this problem can be solved similarly to MMSC, by using integrated Radio and Satellite Mobile Communications under GMDSS and establishing a Global Aeronautical Distress and Safety System (GADSS). On the other hand, by using a combination of GEO and PEO or HEO constellations, the coverage problem for both polar zones will be solved.

1.4.3.2. Present Status of Aeronautical Communications

Business or corporate airlines have for several decades used HF communication for long-range voice and Tlx communications during intercontinental flights. For short distances, nearby airports during approach or departures, all aircraft have used the well-known VHF/UHF radio. Data communications are since recently also in use, primarily for flight plan and worldwide weather (WX) reporting in a form similar to the VHF/UHF Airborne Flight Information Service (AFIS) system. Apart from data service for the aircraft cockpit and cabin crew, cabin voice solutions and passenger telephony have also been developed. On the other hand, some airlines servicing major transcontinental routes have provided free telephone services to passengers as marketing ploys, whereas others want the service to generate additional revenue.

The airline data service requirements are divided into two main areas: AAC-AOC (Airline Administrative and Airline Operational Control) and ATC services. Efficient operation of modern aircraft requires that automatic engine airframe health monitoring and control is available wherever the aircraft is flying. Worldwide access to remote databases while in flight is required for optimum flight profile planning. Therefore, the ARINC (Aeronautical Radio Inc.) program named ARINC Communication and Reporting System (ACARS) and the SITA (Societé Internationale de Télécomommunications Aeronautique) system known as AIRCOM still provide these facilities via VHF radio, although they have started to use AMSC to make these services available worldwide and more effectively.

The success of Inmarsat MMSC and LMSC systems together with the rapid growth of VHF air-to-ground radiocommunications has encouraged the development of the AMSC system. Avionics companies have the best solution to organize MSC links and within their own fleets and also to enable reliable voice, Fax, video, image, data and Internet service for their passengers, crewmembers and corporate purposes. At present, Inmarsat is the only global operator providing the AMSC service, while other GEO and Non-GEO operators also provide AMSC, such as Globalstar, Iridium, Optus, AMSC and MSAT.

As with Inmarsat's MMSC system, the AMSC service is provided to users via the Inmarsat signatories, who are predominantly national PTT or Telecommunications organizations. In the meantime, several competing service provider agreements have been announced with: US-based Arinc, Japanese Avicom, Satellite Aircom consortium of France Telecom, SITA, Teleglobe Canada and Telstra Australia, Skyphone consortium of BT, Singapore's SingTel and Telenor and the multinational Skyways Alliance led by Comsat.

1.4.3.3. Aeronautical Transportation Augmentation System (ATAS)

The ATAS has been set up to identify the possible applications for global radio and satellite communication, navigation, surveillance and safety systems including security and control of aircraft, freight and passengers and SAR service in accordance with ICAO regulations and recommendations. The world's commercial airline fleet is expected to double in the next 20 years, which will result in crowded routes, leading to fuel wastage and delays, which could cost millions of dollars annually. In this sense, the new augmentation system for aeronautical satellite CNS/ATM is designed to assist navigation both en-route as well as during landing and in airports. In fact, the potential benefits will assist ATC to cope with increased air traffic as well as improving safety and reducing the infrastructure needed on the ground.

When planning aircraft routes and landing schedules at busy airports, it is essential to ensure that aircraft are always a safe distance from each other. The trouble is that it is not always possible to know where the planes are. Thus, it is necessary to leave a very large safety margin but when it is possible to know precisely where the planes are, it will be easy to reduce the margins safely and increase the numbers of planes in each corridor. At any rate, the new WAAS, EGNOS, MTSAT and forthcoming Galileo GNSS will provide a guaranteed service with sufficient accuracy to allow airlines and pilots to know their current position and safety margins reliably and precisely enough to make substantial efficiency savings. The GNSS can also help pilots to land planes safely, especially in poor weather and dense fog, in which only with DGPS or the augmented satellite CNS system is reliable. Unfortunately, small airports are unlikely to invest in this system but they can use local augmented system or when Galileo becomes operational, the need for a differential antenna will reduce costs. Galileo's guaranteed service and use of dual frequencies will increase accuracy and reliability to such an extent that planes will be able to use its navigational signals for guidance with their on-board technology alone.

1.4.3.4. Service for AMSC Users

New aeronautical services will provide mile-high opportunities to bring AMSC to general aviation planes, commercial airlines, helicopters, their crewmembers and passengers. They can use air-to-ground AMSC facilities to optimize commercial and safety communications, fuel, maintenance and revenue demands. All the communication features found in the office are made available in the airborne environment, with direct dialing and continuous global coverage. The AMSC service provides passengers, pilots and crew to communicate critical business decisions or social, private and confidential cockpit and seat phone calls.

The airborne equipment enables direct dial phone calls, data and messaging services, HSD, Internet, E-mail, E-commerce, videoconferencing, databases and Aeronautical Information Services (AIS), such as news highlights, WX reports and stock market information. The GMSC system benefits long, medium and short haul airline customers by facilitating accurate positional information for better aircraft utilization and more productivity, good air traffic management and they are very important for increasing airline revenues. In this way, except for Crew and Passenger Service, Aircraft Operation Management, ATC and HSD/ISDN services through the Swift64 Inmarsat system integrated with GNSS solutions, will offer two very important services:

1. Aircraft Security Control – All planes can be provided with the service of Aeronautical Safety Information (ASI) via satellites, such as Turbulence Warnings, Weather (WX) and Ace Reports to assure safe flights. Anti hijacking actions are important for the safety of lives and property on board aircraft, as hijacking is usually violent, causing injury or death among crew and passengers or the loss of valuable cargo. All airports have security control of passengers on all gates but it will also be important to find out how the AMSC system can provide automatic surveillance and monitoring of aircraft and cargo before departure.

2. Remote Troubleshooting System (RTS) – For almost one year, Dassault Aviation has been using the Inmarsat-ISDN solution for a new remote analysis and technical assistance system called Telemaque. It enables the technical support specialists at Dassault Aviation to carry out all diagnoses or repair operations in real time with customer sites located anywhere in the world. Using an audio and video system connected to an Inmarsat-ISDN transmitting (Tx) antenna, the aircraft sends images in real time to the technical support team at Dassault, enabling them to identify the fault precisely. Besides, secure software for application sharing and file transfer enables the two teams to connect up and work on the same documents, such as databases and plans of the aircraft located at the Dassault office. Dassault Aviation started with Global Telemaque Service (GTS) to all its customers in 2001, which offered video camera, a digital camera, an Inmarsat antenna, a PC, an earpiece and remote maintenance software.

1.4.4. Global Mobile Personal Satellite Communications (GMPSC)

At the beginning of the 1990s, several private US firms proposed new concepts for MSC known as GMPSC, using a group of LEO or MEO spacecraft. Typical systems were Iridium, Globalstar and Odyssey. In 1991, Inmarsat also proposed the new Project 21, named Inmarsat-P, which would provide GMSC for personal applications using Non-GEO systems. After feasibility studies, Project 21 is now going to use the ICO, which is the same as MEO and belongs to ICO Global Communications. With regard to this, in 1992, WARC-92 responded to these activities and allocated the L-band (1626.5 to 631.5 MHz) and S-band (2483.5 to 2500 MHz) for MMS using LEO or MEO satellites.

Those people living and working far from the reach of cellular wireless networks, for example, roaming construction engineers, exploration workers, medical staff, journalists, rangers, farmers, fishing boat or yacht crew members and small planes or helicopters, can also have access to the GMSC network. Namely, every customer can choose to use satellite and/or a convenient cellular network when they are in urban roaming areas, because mobile phones will be dual-mode and triple-mode handsets. Therefore, it is possible to produce dual-mode satellite/GSM handheld phones (the waterproof boat kit solution is optional) and triple-mode satellite/cellular models for D-AMPS/AMPS or for CDMA/AMPS systems.

The GMPSC handsets are lightweight, easy to carry, simple to use and, similarly to current mobile phones, are able to provide a GMSC service with both satellite and cellular network access. These phones can offer voice, Fax, data, SMS and voice mail. The dimensions of unit are about 54 x 26 x 145 mm and about 240 g in weight. The graphic display is with fixed icons and keypad consists in 6 menu-driven soft keys. This unit can support European Community, East European, Chinese, English and other languages. The delivery features include voice, data and Fax messaging in both satellite and cellular modes. The handset has manual or automatic satellite/GSM mode selection, using a small extending antenna.

The special, rugged GMPSC equipment with external antenna can be fitted on board ships, land vehicles or aircraft similar to the ship borne, vehicle borne and airborne Inmarsat solutions. Otherwise, similar equipment is designed for use in Rural and Remote (R&R) environments as well. Three billion people living in remote villages and households have no access to a cellular or landline network using semi-fixed installations but providing this telephone service to remote rural areas is often very slow, complex and quite an expensive project. The GMPSC system can extend the existing telephone utilization to any location in the world, with more products and solutions and with much cheaper equipment and usage charges providing the following services:

1. Rural Villages and Remote Communities – The GMPSC model of rural payphone will offer the ability to communicate with any location in the world using voice facilities and emergency numbers for access to medical, firefighting or security services. The terminal will be similar to a typical city payphone operated by phone cards, ruggedly constructed to withstand vandalism, simple to use and able to indicate the amount of money remaining on the card and will be powered using mains network supply or solar panels and batteries. Additional services include remote fault diagnosis, local support to ensure rapid repair and high availability. This payphone can be used in remote suburban and village environments, in desert and forest areas, on cruise ships, on board aircraft, sea platforms and oilrigs. Otherwise, they are easy and cost-effective to install and operate, using prepaid chip cards. Local authorities can use payphones for emergency calls, using a special access card.

2. Phones for Remote Households – The GMPSC model of an indoor phone offers every remote household connection to TTN via a standard phone set connected to an interface box and external outdoor antenna mounted on the roof. This box can be linked to standard Tel/Fax or ISDN/Internet lines, using mains power, 12V battery or solar cell energy.

3. Remote Business Sites – The GMPSC system is giving an advantage to every business based in suburban and R&R areas, providing in the developing and developed world the following services:

a) Basic GMPSC remote service is suitable for companies requiring voice, Fax, LSD and SMS facilities via a compact system, which includes a small interface box and a compact rod antenna. The system will utilize existing TTN devices easily attached to a company's Tel, Fax or PC and is ideal for remote post offices, credit card companies, administrative authorities, small private companies and companies requiring telemetry and SCADA.

b) Advanced remote service is specified for companies requiring full telecommunication services such as voice, Fax, HSD and efficient data transmissions and SMS facilities, using a GMPSC interface box and an A4-sized antenna. The interface box can easily be fixed to Tel, PABX, Fax or PC, while outdoor antenna can be mounted onto an appropriate on-site building. This GMPSC service is ideal for geological, mining, agriculture, manufacturing, construction, energy and gas exploration companies, local government administration and other organizations requiring high volumes of computer data service.

c) Portable GMPSC remote service has been determined for the early stage of exploration and survey work, or for journalists using a portable laptop-sized terminal, which can easily be connected to a telephone handset or PC configuration. This service is ideal for the immediate transmission of data from a remote site to a headquarters or research location, for transmitting urgent copies of visual images or photos and for any customer requiring advanced telecommunications with easy installation in remote locations.

4. SCADA (Supervisory Control and Data Acquisition) – The Inmarsat D+, C, GAN, OmniTRACS or Orbcomm and other SCADA (M2M) equipment or systems may be used in the point-to-multipoint broadcast of Automatic Remote Monitoring and Messaging Data (ARMMD), which installations are fitted with automatic sensors that regularly report back to a control centre via satellite. With its greatly reduced power consumption and by being remotely controlled, this device is a very effective way of remotely collecting basic environmental and industrial data via messages of up to 2000 bits, together with one of four alert signals, which may be sent as a single message. Therefore, the setting up of theses M2M systems and devices are all remote industrial and manufacturing installations, oil and gas pipeline pumping stations, offshore platforms, meteorological observatories, water treatment and energy stations, maritime lighthouses and buoys signaling and other remote systems to run unattended is fairly easy. Keeping this equipment working efficiently, gathering management data and guarding against remote catastrophic failure, all without overspending on expensive technical labor and travel, is a tougher challenge.

5. Government Services – The GMPSC system network will also provide possibilities to organizations involved in fast-moving emergencies and interventions such as disaster relief agencies and SAR organizations, military and international peace-keeping forces, police squads, medical teams and civil authorities. This service offers assured communications by PC data, voice, Fax, SMS and encrypted messages. For personal use it will offer dual mode functions portable handsets for both satellite and cellular connection. On the other hand, for group use the system will offer mobile handset terminals for fitting to ships, vehicles, aircraft and temporary accommodation at camps or bases. Therefore, Inmarsat, Iridium, Globalstar and other current GMSC systems have developed both common and special military mobile and portable satellite tactical and defense communications equipment.

1.5. International Coordination Organizations and Regulatory Procedures

International coordination in MSC has been carried out by the International Coordination Organizations, which include ITU, IMO, ICAO, IHO, WMO and MSUA.

1.5.1. International Telecommunications Union (ITU) and Radio Regulations

The ITU organization of the UN with all member governments has carried out the entire international coordination and regulation of mobile radio and satellite communications by the ITU RR. The ITU was inaugurated in 1932 and reorganized in 1992, with head office, all committees and departments located in Geneva, Switzerland. Numerous provisions of the telecommunication services applicable or useful to all stations have been defined and introduced by the general RR articles and manual on mobile radio service and in a special manual for use by the Maritime Mobile and Maritime Mobile-Satellite Services. Otherwise, the ITU also publishes many additional lists of recommendations concerning RR, systems, radiocommunications, etc. The administrative structures established by the ITU Convention comprises a Secretariat headed by the Secretary General, an Administrative Council, Registration Board for RF and Consultative Committees for radio and telecommunications. The entire terminology, definition of radio and satellite services, technical standards and frequency allocations are defined in the RR and drafted by the World Administrative Radio Conference (WARC) of the ITU. In effect, the WARC was reorganized in 1992 as a World Radio Conference (WRC) of the ITU. The basic concept of the present RR relevant to satellite communications was proposed by the WARC-ST (WARC-satellite) in 1971 and by all other conferences held during later years. The WARC-92 covers the spectrum needs of numerous telecommunications services, from HF/VHF Broadcasting to personal satellite communications networks. Thus, the requirements for MSC systems below 1 GHz, above 1 GHz and in the 20/30 GHz bands have dominated the agenda of the conference. After Spain, the USA heads off to the WRC-93 in November armed with a clear mandate to seek more spectrums for MSS. Thus, the USA position for the Conference, to be held from 15 to 19 November in Geneva, is to establish MSS on the agenda for the 1995 and 1997 WRC.

The task of the International Radio Consultative Committee (CCIR) is to form study groups to consider and report on the operational and technical issues relating to the use of radio and satellite communications. Hence, the International Telecommunications Consultative Committee (CCIT) offers the same telecommunication services. The study groups produce new recommendations on all aspects of radiocommunications, which are considered by the Plenary Assembly of the CCIR and if accepted, become incorporated into the RR and published separately. Besides, another subgroup of the ITU is the International Frequency Registration Board (IFRB), which considers operating frequencies, transmitter sites and the location of satellites in orbit. Consequently, satellite communication systems have been internationally authorized by the IFRB and the present ITU-R (ITU-Radio).

The administration of a country that intends to establish satellite communication systems, operated in GEO, HEO or LEO configuration, must send to the ITU-R, not earlier than five years before the date of introducing the system, information on each satellite network of the planned service, information such as the frequency bands to be used, the modulation type of signals and the radiation characteristics of antennas, satellite constellations and Earth stations. The ITU-R publishes the information in a weekly circular distributed to all relevant administrators around the world. If any administration is of the opinion that interference will be caused to its existing or planned space radiocommunication services, it can, within four months after the publication of the relevant weekly circular, send its comments to the administration concerned. If no such comments are received from anyone within the period mentioned above, it might safely be assumed that no administration has any objections to the planned satellite network, for the system in question, on which details have been published. Accordingly, an organization set up to receive comments about registration endeavors to resolve any difficulties that may arise and therefore provides any additional information necessary. Soon after completing all coordinations, the planned satellite system will be internationally authorized by recording it in the Master International Frequency Register (MIFR) of frequency assignments.

1.5.2. International Maritime Organization (IMO) and Regulations

The service and regulations reaching development, studies and agreements of the Maritime Safety Committee regarding distress and safety at sea are providing by the IMO. It consists in an Assembly, a Council and 4 main Committees: Maritime Safety; Marine Environment

Protection; Legal and Technical Cooperation. There is also the Facilitation Committee and a number of sub-committees, which support the work of the main technical committees. Since its establishment in 1959, IMO and all its member governments have striven to enhance the International Convention for the Safety of Life at Sea (SOLAS – 1974). In 1972 IMO, with the assistance of CCIR, commenced a study of satellite communications systems, which resulted in the establishment, in 1979, of the Inmarsat Organization. With the continuing support of CCIR, ITU, WMO, IHO, Cospas-Sarsat and Inmarsat, IMO has developed the GMDSS system, which entered very late into force in early 1999, following almost 30 years of careful preparation.

Shipping is perhaps the most international of all the world's great industries and one of the most dangerous. It has always been recognized that the best way of improving safety at sea is by developing international regulations that are followed by all shipping nations and from the mid-19th century onwards a number of such treaties were adopted. In 1948 an international conference in Geneva adopted an Intergovernmental Convention, formally establishing a new international organization named IMCO (the original name of this organization was changed in 1982 to IMO). The IMO Convention entered into force in 1958 and the new organization met for the first time the following year. The purposes of the organization are to provide the machinery for cooperation among governments in the field of governmental regulation and practices relating to technical matters of all kinds affecting shipping engaged in international trade; to encourage and facilitate the general adoption of the highest practicable standards and rules in matters concerning maritime safety; efficiency of navigation and prevention and control of marine pollution from ships.

The first IMO task was to adopt a new version of the SOLAS International Convention, the most important of all treaties dealing with maritime safety. This was achieved in 1960 and the IMO then turned its attention to such matters. In 1974 the International Convention for the SOLAS adopted the SOLAS 74 Convention based on the requirement that certain classes of ships, when at sea, keep continuous radio watch on the international distress frequencies assigned in accordance with the ITU RR and carry corresponding radio equipment capable of transmitting over a minimum specified range. Shipping, like all of modern life, has seen many technological innovations and changes. Some of these have presented challenges for the organization and others have presented opportunities. The enormous strides and efforts made in new radiocommunications technology for example, have made it possible for the IMO to introduce major improvements to the maritime distress and safety system. In this sense, in the 1970s a global SAR system was initiated worldwide. The 1970s also saw the establishment of the Inmarsat at London, which has greatly improved the provision of commercial and distress radio and other messages to ships. A further advance was made in 1992, when the GMDSS began to be phased in. In February 1999, the GMDSS became fully operational, so that now any ship in distress anywhere in the world can be virtually guaranteed assistance because even if the crew does not have time, the message will be sent automatically.

Other measures introduced by the IMO have concerned the safety of container units, bulk cargoes, gas tankers and other ship types. Special attention has been paid to crew standards and adoption of a special convention on standards of training, certification and sea watch keeping. The challenge also facing the IMO and its 162 member states is how to maintain this success at a time when shipping is changing more rapidly than ever before. The revised technical Annex of the SAR Convention clarifies the responsibility of all governments and puts greater emphasis on the regional or local approach and coordination between maritime and aeronautical SAR operations. In fact, both parties are also required to ensure the closest practicable coordination between mutual service. Each RCC and Rescue Sub Centre (RSC) should have up-to-date information on the SAR facilities and communications in the area and should have detailed plans for the conduct of SAR operations.

In addition, concurrently with the revision of the SAR Convention, the IMO and ICAO technical experts jointly developed the International Aeronautical and Maritime Search and Rescue Manual (IAMSAR), published in three separate volumes covering: Organization and Management, Mission Coordination and Mobile Facilities. Therefore, the IAMSAR Manual revises and replaces both the IMO Merchant Ship Search and Rescue Manual (MERSAR), first published in 1971 and the IMO Search and Rescue Manual (IMOSAR), first published in 1978. The MERSAR Manual was the first step towards developing the 1979 SAR Convention and it provided guidance for those who, during emergencies at sea, may require assistance from others or who may be able to provide assistance themselves. In particular, it was designed to aid the master of any vessel who might be called upon to conduct SAR operations at sea for persons in distress. The second IMOSAR Manual was adopted in 1978. It was designed to help governments to implement the SAR Convention and provide guidelines rather than requirements for a common maritime search and rescue policy, encouraging all coastal states to develop their organizations on similar lines and enabling adjacent states to cooperate and obtain mutual assistance. It was also updated in 1992, with the amendments entering into force in 1993. This manual was aligned as closely as possible with the ICAO Search and Rescue Manual to ensure a common policy and to facilitate consultation of the two manuals for administrative or operational reasons. In such a way, MERSAR was also aligned, where appropriate, with IMOSAR.

1.5.3. International Civil Aviation Organization (ICAO) and Regulations

The consequence of the studies initiated by the USA and subsequent consultations between the major allies was that the US government extended an invitation to 55 member states to attend, in November 1944, an International Civil Aviation Conference in Chicago. In fact, 54 states attended this conference at the end of which a Convention on International Civil Aviation was signed by 32 states, setting up the permanent ICAO policy and task as a means to secure international cooperation and the highest possible degree of uniformity in regulations and standards, procedures and organization regarding civil aviation matters.

One of the primary objectives of the ICAO with its corresponding expert committees and departments is to promote the safety and security of civil aviation worldwide. With 188 contracting states and its active involvement in global aviation safety issues, the ICAO is well-positioned to assume a coordinating role for safety initiatives for the reduction of aviation accidents. Recognizing this, in 1997 the Air Navigation Commission proposed an ICAO Global Aviation Safety Plan (GASP) to the ICAO Council. The GASP serves to focus the safety-related activities within ICAO on those safety initiatives, either planned or in progress, that offer the best safety dividend in terms of reducing accident numbers and rates worldwide. For that reason is necessary to develop new GADSS similar to GMDSS.

The main objective of ICAO's technical aviation security program is to assure the safety of passengers, aircraft crew, ground personnel and other members of the general public by first attempting to deny offenders access to aircraft. In accordance with the Chicago Convention (Annex 17), the ICAO Council adopts Standards and Recommended Practices (SARP) for the safeguarding of international civil aviation security. The Security Manual for Safeguarding Civil Aviation Against Acts of Unlawful Interference contains guidance material on the interpretation and implementation of the SARP of Annex 17. This Manual and Annex 17 are kept under constant review and amended in the light of the new trends and recent technical developments which have a bearing on the effectiveness of preventive measures against acts of unlawful interference. In order to assist any state to implement the aviation security standards contained in Annex 17, the ICAO has developed the Training Program for Aviation Security, currently comprising seven important Aviation Security Training Packages (ASTP). Through its global network of ten Aviation Security Training Centres, ICAO promotes regional cooperation with regard to AVSEC training, the main objectives being to improve the quality of training of aviation security personnel worldwide and to assist technically and financially every active member state with their own national AVSEC training programs. The major objectives of the ICAO GASP are to:

1) Reduce the number of accidents and fatalities irrespective of the volume of air traffic, achieving a significant decrease in worldwide accident rates, placing emphasis on regions;

2) Urge all contracting states to provide the needed support of the ICAO GASP;

3) Endorse the concept of concentrating the security and safety-related activities of ICAO on those initiatives, planned or currently under way, that offer the best safety and safety dividend in terms of reducing the accident rate;

4) Instruct the Council and Secretary General to participate in efforts by states to improve existing safety/security database systems and the exchange of safety-related information and to participate in activities aimed at the development of a comprehensive data analysis and information dissemination network, taking into account the need to adequately protect privileged information and its sources;

5) Encourage the free communication of security and safety-related information amongst users of the aviation system, including the reporting of accident and incident data by states to the ICAO Accident/Incident Data Reporting (ADREP) system.

6) Commence a study of MSC systems to enhance air traffic control, to improve the flow of management data, meteorological, security and safety information and to participate in the development of the Global Satellite Augmentation System for Aeronautical CNS.

The ICAO and IMO have jointly collaborated to study, research and develop an important IAMSAR Manual. With the revision of the SAR Convention and previous manuals they published the IAMSAR Manual, which covers all particulars about SAR programs and procedures for maritime and aeronautical SAR applications. This manual was aligned as closely as possible with the previous ICAO Aeronautical Search and Rescue Manual.

1.5.4. International Hydrographic Organization (IHO)

The IHO is an international intergovernmental consultative and technical organization that was established in 1921 to support safety in navigation and the protection of the marine environment. The main IHO activity is standardization of nautical charts and documents, bathymetry and ocean mapping and related publications, technical assistance and training. The special activity of IHO related to MSS is radio navigational warnings implemented by the GMDSS as an integrated Radio and Satellite communication system of Inmarsat and Cospas-Sarsat systems. The GMDSS system has improved the dissemination of Maritime Safety Information (MSI), taking advantage of modern communications technology. The far offshore navigation warnings are broadcast via Inmarsat-C Enhanced Group Call (EGC) SafetyNET Service, whilst polar areas warnings are covered by DCS HF radio services and coastal area warnings are transmitted via DCS MF/VHF radio and NAVTEX on a single frequency, timeshared and automatic broadcast. The NAVTEX and DSC radio services are already in operation in most parts of the world in framework of the GMDSS mission.

The advent of digital data transmission, computers and video display systems is having a considerable impact on hydrographic and navigation technology. This has made possible the development of ECDIS (Electronic Chart Display and Information Systems), which has become a major focus of activity. Several working groups have been established by the IHO to coordinate these new developments and ensure the standardization of systems and specifications. In 1992, the IHO adopted standards for the formatting of ECDIS data so that it can be readily exchanged between hydrographic offices. At this point, the Committee on ECDIS (COE) examines both the database requirements and the standardization and overall parameters of such systems.

1.5.5. World Meteorological Organization (WMO)

The WMO coordinates all global scientific activity to allow prompt and accurate weather information, WX and tropical storm forecasting and other services for public, private and commercial use, including the international shipping and airline industries, sent by GMSC media. More exactly, the major part of WMO's activities contribute to the safety of life and property, the socioeconomic development of nations and the protection of the entire Earth environment, including the Global Ozone Observing System.

The World Meteorological Convention, by which body the WMO was created and adopted at the 12th Conference of Directors of the International Meteorological Organization, met in the USA capital Washington in 1947. Although the Convention itself came into force in 1950, the WMO commenced operations as its successor in 1951 and, later that year, was established as an UN's specialized agency by agreement between the UN and the WMO.

The purposes of the WMO are to facilitate international cooperation in the establishment of networks of stations for making meteorological, hydrological and other observations and to promote the rapid exchange of meteorological information by means of meteorological radio or satellite communications, the standardization of meteorological observations and the uniform publication of observations and statistics. The WMO also furthers the specific application of meteorology to maritime, aviation, agriculture and other human activities, climatology, atmospheric sciences, hydrology and instruments and methods of observation. Each of them meets every four years, in an effort to promote operational hydrology and to encourage research and training in meteorology.

The main WMO task is to organize, control, collect and offer up-to-the-minute worldwide meteorological observations and weather information through many member-operated observation systems and telecommunication links with four PEO and five GEO satellites, about 10,000 land observation stations, 7,000 ship stations and 300 moored and drifting buoys carrying automatic weather stations. Namely, each day, high-speed links transmit over 15 million data characters and 2,000 weather charts through 3 World, 35 Regional and 183 National Meteorological Centres cooperating with each other in preparing weather analyses and forecasts in an elaborately engineered fashion. Thus, transoceanic ships and airplanes, research scientists on sea/air pollution or global climate change, the media and the general public are given a constant supply of timely data, which are very important for safe navigation and flight. Moreover, it is through the WMO that the complex agreements on standards, codes, measurements and communications are internationally established.

Data from all over the world are needed to provide weather (WX) forecasts and warnings. In other words, an aircraft does not take off, nor does a ship leave a port, without a WX received via agent or by means of radio and/or satellite communication systems for safety utilization. Combining facilities and services provided by all members, the program's primary purpose of the system is to make available meteorological and related geophysical and environmental information enabling countries to maintain efficient meteorological services. Otherwise, facilities in regions outside any national territory (outer space, wide ocean areas and Antarctica) are maintained by members on a voluntary basis. Accordingly, the World Weather Watch System comprises the Global Observing System, the Global Data Processing System, the Global Telecommunication System, Data Management and System Support Activities.

The Hydrology and Water Resources Program concentrates on promoting worldwide cooperation in the evaluation of water resources and the development of hydrological networks and services, including data collection and processing, hydrological forecasting and warnings very important for the safe navigation of ships and the flight of aircraft and the supply of meteorological and hydrological data for design purposes.

1.5.6. Mobile Satellite Users Association (MSUA)

The MSUA was established in 1992 as a non-profit association to promote the interests of users of MSS worldwide. It fosters effective communication among MSS users, suppliers of equipment and services, operators of the satellite systems and the various governmental entities that may affect the future of the entire industry. Membership is not limited to American entities; it is open to organizations worldwide engaged in any of these activities.

The MSUA has created a website that is the premier source of MSS information. Namely, the Association maintains a website on the World Wide Web: – (www.msua.org). The Website will publicize MSUA activities and provide, to members only, information and analyses that enhance their business opportunities. It is possible to find a description of the Association, a great list of links (service providers, manufacturers, consultants, etc) as well as links to all the major operational systems such as Inmarsat, Iridium, Globalstar, Leo One, Orbcomm and many others. Three classes of membership are open to applicants: Corporate, Associate and Personal. In pursuing its mission, the MSUA media informs its membership on issues that affect them and conveys all members' views on such issues to appropriate authorities, specifically, the Association reviews, reports and analyzes the activities of all members and MSS providers and reports on those executive and legislative activities of governments that affect MSS and navigation services, etc.

1.6. Satellite Communications Organizations and Operators

Government and intergovernmental satellite communications organizations and operators are divided into international, offering almost global service, regional, usually covering few countries or entire continents and domestic, for local service only.

The GMSC system operators are the only entity responsible for the operation of the GMSC space and ground network configurations providing global, regional or domestic coverage. However, the GMSC service providers are any entity commissioned by a GMSC system operator to provide GMSC services to the public within a region or country and which may require an authorization to do so under the applicable legislation of the country concerned.

1.6.1. International Satellite Communications Organization

International satellite communications organizations and operators are Intelsat, Inmarsat, Intersputnik, Eutelsat, ESA and other global, multinational or intergovernmental operators serving outside domestic and regional boundaries.

1.6.1.1. Intelsat

The FSS were the first to develop and there was rapid recognition that these new global possibilities necessitated the creation of some kind of international organization. Thus, this led to the creation of the Intelsat international organization based in Washington. On 20 August 1964, 11 countries signed a charter agreement creating the Intelsat, the first open worldwide satellite communications network and appointed the Comsat Corporation as its first manager. However, Comsat placed a contract for a GEO and the Intelsat fixed system started to offer transatlantic satellite services in 1965 after the successful deployment of Intelsat I (formerly Early Bird), the world's first GEO spacecraft. Today, more than a fifth generation of Intelsat birds provides international and domestic satellite communication service on behalf of over 112 member nations. Since 1973, Intelsat has operated with an organizational structure that has only four tiers: the Assembly of Parties, the Meeting of Signatories, the Board of Governors and the Executive Organ.

During the 1990s, users were served by the following satellite constellations: Intelsat K and the Intelsat V to VIII series of spacecraft. The first Intelsat VIII was launched in 1996 on board an Ariane 4 booster. Moreover, the first satellite in the Intelsat IX series, Intelsat 901, was scheduled to be launched in 2000 to the 60°E GEO location. The next series of Intelsat GEO spacecraft 902, 903 and 904 were scheduled in the period between 2000 and 2002.

Figure 1.4. Intelsat V MCS

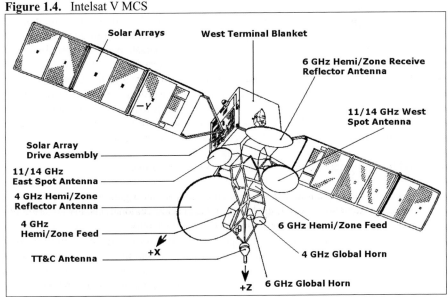

Courtesy of Book: "Telekomunikacije Satelitima" by R. Galic

The Intelsat provided as well as MSS using payloads carried by its GEO spacecraft Intelsat V MCS series: A (F5), B (G6), C (F7) and D (F8). The Inmarsat organization leased capacity on Intelsat V MCS flights F5 through F8 for MMSS, shown in **Figure 1.4.** These four satellites used portions of the L-band (from 0.5 to 1.7 GHz) and C-band assigned for such purposes by the ITU. The Intelsat V MCS A (F5) was launched on 28 September 1982 over IOR in orbital location at 63°E; The Intelsat V MCS B (F6) was launched on 19 May 1983 over AOR in orbital location at 18.5°W. The Intelsat V MCS C (F7) was launched on 19 October 1983 as a spare spacecraft over IOR in orbital location at 66°E and the Intelsat MCS V D (F8) spacecraft was launched on 3 March 1984 over POR in orbital location at 53°E longitude and at the end of 1985 was relocated to the assignment of 179/180°. Thus, MCS uses radio frequencies in the top C-band transponder and an additional L-band spectrum to link CES with ships as part of the Inmarsat network. Because of the additional power requirements by the MCS packages on Intelsat V F5 – F8 satellites, the Ku-band operations on these four spacecraft must be limited or curtailed when they are activated for Inmarsat use. However, when the L-band signal amplifier is operating in the high power mode, the 11 GHz Ku-band capacity of these spacecraft is switched off. These satellites after many years of service are deployed and not used by the Inmarsat system any more.

1.6.1.2. Inmarsat

The Inmarsat organization is the only international and nonmilitary MSC provider in the world. The Inmarsat MSC network provides Tel; Tlx; Fax; slow, medium and high-speed data transfer; image and video; videoconferencing; videophone; new mobile MPDS and ISDN; LAN and IP service; and in the framework of the GMDSS system, enables distress and safety service for maritime, land and aeronautical MES, including government, rural, military and offshore infrastructures.

In the early 1970s the IMO, then known as the IMCO, began to consider the possibility of using MSS to improve maritime communications, not least for safety purposes. Towards the end of 1973, IMCO convened a conference to decide on the principle of establishing an international maritime safety system and to conclude the necessary agreements. The work of this conference culminated in September 1976 in the adoption of what became the Inmarsat Convention and its complementary Operating Agreement, requiring always acts exclusively for peaceful purposes. Inmarsat was established on 16 July 1979 by the major maritime nations to finance this project, which is to investigate using satellites to form links with vessels and offshore oilrigs at sea. Inmarsat also owes much of its success to the foresight and commitment of the IMO, which played a crucial role in its creation.

On 1 February 1982, Inmarsat officially took control of satellites previously operated by three Marisat spacecraft (series F1, F2 and F3), a joint venture among nations begun in the early 1970s through the efforts of the Comsat General. Inmarsat has since expanded its space segment by leasing additional capacity from Intelsat and ESA, Intelsat V MCS series of four spacecraft and two Marecs spacecraft (B and B2A), respectively. However, today an additional Inmarsat second, third and forthcoming fourth generation of GEO, Inmarsat-2, Inmarsat-3 and Inmarsat-4 spacecraft, respectively, are being exploited.

In many ways, the Inmarsat organization is patterned after the Intelsat system and almost all countries comprise the Inmarsat Assembly as members, each country casting one vote. The Assembly meets once every two years to formulate general policy, long-term strategy and objectives. Besides, each government selects a representative Signatory to the Inmarsat

Operating Agreement from the public or private sector. The Assembly has also primarily to establish financial, technical and operational standards. The Inmarsat Council, similar to a corporate board of directors, meets three times a year to implement all policy decisions of the Assembly. The Director General of Inmarsat oversees the day-to-day management of the organization, with an executive staff headquartered in London.

Inmarsat is financed in two ways: signatories must pay fees based on their use of the network, or can purchase investment shares in Inmarsat that reduce the user fees in proportion to the investment and signatories earn a return of 14% per annum on their investment in Inmarsat. During 1987, the largest numbers of investment shares were held by the USA, UK, Norway and Japan.

1.6.1.3. Intersputnik

The Russian Federation, that is to say, the former the Soviet Union, is not a member of Intelsat. Instead, in 1971 it created a similar multinational organization named Intersputnik, which provides FSS for its 14 member states and a number of other associated countries. This system uses the various families of former Soviet communications satellites, such as Molniya, Raduga and Gorizont, using GEO, HEO and PEO satellite coverage.

The Russian satellite system conceivable can beam radio broadcasting, TV programs, voice and data traffic to almost any location on Earth, and only Intelsat can supply more global FSS links than Intersputnik. This Organization is an open international intergovernmental that any sovereign state can join. The fundamental structure of Intersputnik was determined by the Cooperative Agreement on the Establishment of the Organization. A representative for each member nation serves on the Board of the Members, which is the main governing body of Intersputnik organization. Besides, this Board selects a Director General to chair the Intersputnik Directorate based in Moscow.

The Organization also provides three MSS for all three applications, using payloads carried by its GEO spacecraft Gorizont (Horizon), Raduga (Rainbow) and Morya (Seamen):

1. The Volna (Wave) Network – The Volna MSC system served to connect maritime and aeronautical MES terminals via space segment constellation to LES and ground-based telecommunication facilities for former USSR ships and aircraft. This MSS consists in communications payloads carried by the spacecraft Gorizont and Raduga. The Volna Network provided radio and TV service for mobile stations on UHF frequency bands between 335–399/240–328 MHz. On the other hand, the Volna Network provided service uplink/downlink on L-band between 1636–1644/1535–1542 MHz for MMSS and also on L-band 1645–1660/1543/158 MHz for AMSS applications, while feeder link used 6/3 GHz uplink/downlink bands for both MSS applications.

2. The Morya Network – In 1989 the former Soviet Union expanded its MMSS with the Morya MSC Network, using existing Soviet satellites series Morya for carrying the MSC payload. Namely, the Morya Network provided MMSS on two 2.5 MHz wide frequency uplink/downlink bands centered on 1637.25/1535.75 MHz (service link) and 6084.0/3758.3 MHz (feeder link).

3. The Gals Network – Piggybacked on the former Soviet Union's Raduga spacecraft, named Gals (Tack), a special telecommunications payload was serving as satellite links for the former USSR (today Russia) military forces. Thus, this Network will operate using the X-band spectrum for up linking (7.9 to 8.4 GHz) and down linking (7.25 to 7.75 GHz), for defense maritime, land and aeronautical applications.

1.6.1.4. ESA

The idea of creating an independent space organization in Europe goes back to the early 1960s when six European countries: Belgium, France, Germany, Italy, the Netherlands and the UK, associated with Australia, to develop and build a heavy launcher called Europa. In 1975, a Convention was endorsed at the political intergovernmental level to set up the European Space Agency (ESA). Finally, the Convention entered into force on 31 October 1980. Since then the founding ESA members have been joined by 4 new members from Europe, while some other European countries have expressed their interest to join ESA. In addition, the Cooperation agreements have also been signed to allow Canada to participate in certain ESA programs and to sit on the ESA Council.

This makes the dreams of Europe's space scientists come true, by creating and operating new scientific spacecraft for the ESA. The project management interacts with the aerospace and instruments industry in the 15 member states and oversees the construction, launch and operation of the spacecraft. There is close liaison with ESA's European Space Operations Centre (ESOC) in Darmstadt, Germany, with Arianespace and other cooperative agencies providing launchers and with the GES needed for communication with the spacecraft.

The ESA Research and Scientific Support Department was restructured in 2000/2001 into Divisions. In such a way, each Division is the home of the project scientists for the study, projects development and operation of the missions, serves the space scientists of Europe, under the supervision of the Science Program Committee, on which all member states are represented. Any scientist or group within the member states can propose a space science mission. Before a final selection, some rival proposals are selected for detailed study of their scientific, technological and budgetary implications. Under a science communications initiative, ESA is intensifying its efforts to keep the press, public and schoolteachers well informed about the science program and the progress of its various missions.

Apart from construction, launch and operation of the spacecraft, ESA has developed Artemis MSS, derived from former Prodat MSS for all three applications. At present ESA members are developing GSNS named Galileo and satellite augmentation systems EGNOS for maritime, land and aeronautical satellite CNS.

1.6.1.5. Eutelsat

The European Telecommunications Satellite (Eutelsat) Organization was provisionally founded in 1977 by representatives of 17 members of the European Conference of PTT and telecommunication Administrations, with headquarters based in Paris. Its major mandate was to establish and run the European satellite communications system as a regional operator but because it is currently offering global FSS service on an international basis, it can be classified into the group of global and international satellite organizations.

The constitution and financing of Eutelsat are modeled on those of Intelsat. Although it did not formally come into existence until 1984, Eutelsat started work in 1977 with ESA on the exploitation of the Orbital Test Satellite (OTS) experimental communication birds and on the design of the operational European Communications Satellite (ECS) series. Finally, the latter were in due course taken over and operated by Eutelsat, which is now procuring its second generation of satellites. At this point, Eutelsat passed from an interim organizational structure to definitive operational status on 1 September 1985 and all shares were divided at that time among about 25 member nations. During the early stages of planning, Eutelsat designers thought their spacecraft primarily would carry voice and high-speed data service. Although Eutelsat was set up to handle long distance traffic, a significant part of its revenue now comes from the relaying of satellite TV programs for distribution through the terrestrial telecommunications network or into the cable system. Throughout its 25 years experience, Eutelsat has placed innovation at the centre of its development. The Company has distinguished itself, notably by being the first in Europe to distribute satellite TV and pioneering the use of the Digital Video Broadcasting (DVB) standard for the transmission of digital TV channels and data.

Eutelsat is one of the world's leading operators of multipurpose satellite infrastructures. It provides capacity on 23 satellites that offer a broad portfolio of services, which include direct TV and radio broadcasting for the consumer public, professional video broadcasts, corporate networks, Internet services and MSC systems. Therefore, Eutelsat is providing regional MSS for all three applications with the current Emsat MSC system for maritime and land applications and the EutelTRACS system for mobile tracking and messaging. The latter system is developed by technical cooperation with Qualcomm, a US-based company, to establish communication network and equipment infrastructures.

1.6.2. Former International MSS Operators

Former international global and regional MSS organizations and operators were Marisat, developed by the US-based Comsat Company Marecs, which was formed by European nations and Prodat was a project of ESA.

1.6.2.1. Marisat MSS

The World's first maritime MSC system as a new application of the GEO system was unveiled in 1976 with only three satellites and ocean networks that are providing MMSC services in the Atlantic, Pacific and Indian Oceans. The Hughes Aircraft Company, known today as Boeing Satellite Systems Inc, under contract to Comsat General Corp, built three multifrequency communications spacecraft called Marisat (Maritime Satellite), for the space segment of the world's first MMSC operator. The Comsat General was developed a Marisat system for MSS at first for only maritime applications. In 1971, frequency bands around 1.6 GHz were allocated for satellite communications connections with ships and aircraft. The Marisat satellites were designed initially for US Navy vessels and they had a UHF transponder on board in a band from 240 to 400 MHz. Because there was sufficient margin for additional payload, L and C-band transponders were installed on the Marisat satellite to provide commercial MSC traffic for maritime applications.

These satellites had a dual role at that time: to provide space segment facilities that were leased to the US Navy for military communications with naval ships and it also enabled the use of transponders for Comsat General itself to operate MSS for traffic with merchant ships, virtually worldwide. All three Marisat satellites were launched with the same type of USA rocket "McDonnell Douglas 2914 Delta" during 1976, on 19 February, 9 June and 14 October, for the needs of the company Comsat General. Marisat F1, F2 and F3 satellites were placed in GEO planes at 15°W, 72.5°E and 176.5°E longitude, respectively.

All satellites have been leased from Comsat, in effect; Marisat F1 spacecraft served as an in-orbit spare for the Marecs A spacecraft in the Marisat AOR region at a position 15°W. Then this satellite was leased as a spare in Inmarsat AORE region and removed to 106°W.

Figure 1.5. Marisat Space and Ground Segments

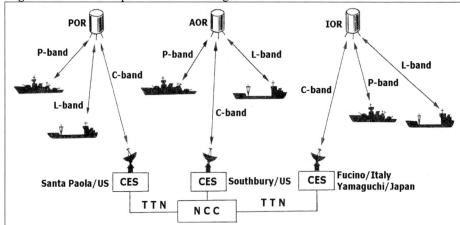

Courtesy of Book: "Sputnikovaya Svyaz na More" by L. I. Novik, I. D. Morozov and V. I. Solobev

Marisat F2 spacecraft served as a spare in IOR region at a position 73°E and lately, for the Inmarsat space segment. The Marisat F3 spacecraft served at a position 176.5°E in the POR region, and afterwards reprogrammed as a spare for the Marecs B2 satellite. This satellite was finally relocated as in-orbit spare to 182.5°E. The service at that time was welcomed by merchant shipping and by 1982 around one thousand vessels were equipped to use the Marisat system. All three Marisat satellites also served as an emergency backup, one in each of the three ocean regions: AOR, IOR and POR. Accordingly, after many years, these satellites are no longer in exploitation either by the Comsat or Inmarsat systems.

The Navy fleet used P-band frequencies (Tx = 248–260 MHz/Rx = 1300–312 MHz) for the MMSS link, while the merchant-shipping element of the Marisat payload used the newly allocated L-band frequencies (Tx = 1537–1541 MHz/Rx = 1638.5–1642.5 MHz) for its MMSS links and C-band (Tx = 6174.5–6424 MHz/Rx= 3945.5–4199 MHz) for feeder links. Fixed CES for mobile service were located at Santa Paola for POR, at Southbury for AOR and at Fucino and Yamaguchi for IOR, see **Figure 1.5.** The system provided access to the satellites, linking ships at sea through the PSTN with the TTN subscribers ashore for Tel, Tlx, Fax, data and HSD transmissions. The Marisat system was controlled by Network Control Centre (NCC) located at Washington. Satellite Tracking, Telemetry and Command (TT&C) are also conducted over C-band frequencies.

However, the governments of many other countries were not quite content for control of MSC with their ships to rest with a foreign commercial corporation. Owing to this problem, in 1976 under the aegis of IMCO an agreement was drawn up for the establishment of an Inmarsat organization, initially for maritime service only.

1.6.2.2. Marecs MSS

The ESA organization's Maritime European Communications Satellite (ECS) or Marecs project covered the study, development, launch and in-orbit operations of communication spacecraft to be integrated in a global MMSC. Development began in 1973 with funding from Belgium, France, Italy, UK, Spain and Germany, then later joined by the Netherlands,

Norway and Sweden. In effect, the program started at first as the experimental Maritime Orbital Test Satellite (Marots) but was subsequently changed to an operational system resulting in a name change, satellite redesign and delayed development.

The Marecs satellite was part of the GMSC system configured to provide high quality full duplex, reliable real-time voice, Fax, Tlx and data services between SES and CES with automatic connection to the TTN. The Marecs satellites operated by ESA were members of Inmarsat first generation MMSC network. The Marecs-1 B2A spacecraft was successfully launched via an Ariane rocket from Kourou, French Guiana on 9 November 1984 at AOR region location 26°W. It was later relocated to the assignment of AORW at 55.5°W as an operational satellite of the Inmarsat network. The next Marecs A satellite was successfully launched via an Ariane rocket from Kourou, French Guiana on 20 December, 1981 on a POR assignment at a position of 177.5°E and was later relocated to POR at 178.5°E, as a spare satellite leased by Inmarsat.

The Marecs satellite consists in two modules: a service module, which is a derivative of the ECS bus and a payload module. The spacecraft has a design life of 7 years, a 3-axis altitude control and TT&C that uses VHF spectrum during transfer orbit and C-band through the communications subsystems on station. Thus, the payload is capable of operating without continuous ground control and it consists in a C to L-band forward transponder and an L to C-band return transponder, incorporating SAR channels. The Marecs spacecraft was based on the British Aerospace ECS 3-axis stabilized platform, with two sun-tracking solar arrays providing 955 W (BOL) with 2 NiCd batteries for eclipse power supply. Payload had three repeaters: shore-to-ship with 5 MHz bandwidth, ship-to-shore with 6 MHz bandwidth and shore-to-shore with 0.5 MHz bandwidth. These provide 35 two-way voice channels plus search and rescue capabilities. Hence, one 2 m diameter L-Band antenna and 2 horns (one transmit, one receive) for 4/6 GHz channels provide almost 1/3 of Earth coverage.

1.6.2.3. Prodat MSS

The ESA organization promoted a special Prosat program, which included propagation measurements of the aeronautical channels. The second phase of the Prosat MSS project primarily consisted in the design, development and prolonged demonstration of a two-way low rate data system. The latter service is referred to as Promar, while the low cost digital data only MSC service included in the configuration is known as Prodat, to be available for maritime, land and aeronautical mobile operations, on 1.6 GHz band to the satellite, which in turn relays the signals to ground using a 6 GHz carrier.

The Prodat program conducted successful field trials with low data terminals using the Marecs satellite. In parallel, two L-band MSC payloads are being procured by ESA to promote European MSS: the European Mobile System (EMS) payload on the Italsat 1-F2 satellite and the L-band Land Mobile (LLM) payload on the Artemis satellite. The EMS allows voice and data communications with a capacity of 300 channels. The available capacities are partially being used to demonstrate and evaluate the emerging MSC system. The European MSS operational phase started with the EMS payload in orbit during 1996 and is continuing with the LLM payload, which was planned to be launched in 1997.

The Prodat MSS was operated through ESA ground stations and Inmarsat communications satellites and provided a low speed, low power and narrow bandwidth service between the terrestrial Public Data Network (PDN) and a population of maritime, land and aeronautical mobile terminals.

1.7. Frequency Designations and Classification of Services

The assignment of a radiocommunications frequency, band or channels is performed by an authorized administration for radiocommunications via platform, satellite or space stations to use a radio frequency (RF) spectrum or frequency channels under specified conditions. In this sense, the allotment of a RF or frequency channel comprises the entry of a designed frequency channel in an agreed plan, adopted by a competent conference for use by one or more administrations for a terrestrial or space radiocommunications service in one or more identified countries or geographical areas and under specified conditions. The allocation of a RF band makes possible its entry in the Table of Frequency Allocations of a given RF band for the purpose of its use by one or more terrestrial or space radiocommunications service or the radio astronomy service under designated and specified conditions. The radio spectrum shall be subdivided into nine RF bands and designated by progressive whole numbers, in accordance with **Table 1.2.** At any rate, in satellite communication fields the frequency bands are often denoted with alphabetical symbols such as L to Ka-bands. Frequency band numbers and names are defined by the ITU RR, ITU Tables of Frequency Allocations in general or for a particular band and the mentioned alphabetic symbols by the IEEE Standard Radar Definitions.

Table 1.2. Frequency Bands Designation

Band No.	Abbreviation	Band name	Name	Symbol	Frequency
4	VLF	Very Low Frequency	Myria m		3-30 kHz
5	LF	Low Frequency	Km		30-300 kHz
6	MF	Medium Frequency	Hm		300-3000 kHz
7	HF	High Frequency	Dam		3-30 MHz
8	VHF	Very High Frequency	m		30-300 MHz
9	UHF	Ultra High Frequency	dm		300-3000 MHz
				L-band	1-2 GHz
				S-band	2-4 GHz
10	SHF	Super High Frequency	cm		3-30 GHz
				C-band	4-8 GHz
				X-band	8-12 GHz
				Ku-band	12-18 GHz
				K-band	18-27 GHz
				Ka-band	27-40 GHz
11	EHF	Extremely High Frequency	mm		30-300 GHz
12	VEHF	Very Extremely High Frequency	deci mm		300-3000 GHz

Frequency designations for MSS are used by a number of different administrations for their national or international MSS networks and can be systematized into two main categories:
1. Frequency Allocations for Service Links – The frequency allocation for service links in current commercial use are in L-band at 1.5 GHz (downlinks) and at 1.6 GHz (uplinks). In addition to the L-band designations, there are other MSS frequency band allocations for service links between MES and spacecraft, shown in **Table 1.3.** New frequency bands below 3 GHz were allocated for MSS at WARC-92 and WRC-95. The new allocations below band of 1 GHz are very narrowly designated for only LEO configurations, while GEO satellites are excluded from most of the terms. Very soon, it will be clear which of the new bands above 1 GHz should be used for MSS via GEO satellites and the rest will be available for Big LEO systems and satellites in other Non-GEO constellations. Moreover, there are also LMSS allocations for land vehicles at 14.0–14.5 GHz.

Table 1.3. Frequency Spectrum for MSS Service Links

Uplink (MHz) - Tx Earth (MES)-Spacecraft	Downlink (MHz) - Rx Spacecraft-Earth (MES)	Comments
128 and 240 except		Distress and safety band allocated for former SAMSARS and DRCS system, respectively
121.5 and 406.025 243 except	1544.5	Distress/safety bands allocated for Cospas-Sarsat system
1645.5-1646.5	1544-1545	Commercial/Distress bands allocated for Inmarsat system
	2483.5-2500 (Globalstar)	Principal current frequency allocations for MSS at L-band using by MMSS, MLSS and AMSS applications
1610-1621,35 1626.5-1660.5 except 1645.5-1646	1525-1559 except 1544-1545	
		New MSS frequency allocations made in WARC-92, WRC-95 and WRC-97 for GEO and Non-GEO MSS and RDSS applications; last allocation will take effect from January 2005
148-149.9 148-150.05 312-315 399.9-400.05 & 406-406.1 454-460 1610-1626.5	400.505-400.,645 (Faisat) 137-138 (Leo One) 387-390 400.15-401 400.505-400.645	
2660.2-2690 1980-2010	(Faisat) 2483.5-2500 2500-2535 (N-Star)	Satellite component of IMT-2000 worldwide
27500-30000 1215-1260	2170-2200	Frequency allocations for GNSS applications
	17700-20200	Teledesic

2. Frequency Allocations for Feeder Links – The feeder links between LES and satellites are illustrated for most MSS in **Table 1.4.**

In general, all frequency allocations can be designated for global and regional coverage. The subcategories of frequency allocation are divided into three regions and among three mobile applications. All frequencies used between satellite and MES are part of MSS and can be classified into three main allocations: MMSS for ships, LMSS for land vehicles and AMSS for aircraft. Some allocations differentiate among the MMSS, the LMSS and the AMSS scheme and are somewhat complex in detail. The regional frequency allocations for corresponding countries are systemized into Regions 1, 2 and 3, respectively. Thus, there are also allocations for other services within these frequency bands in some countries but most of these shared allocations are of secondary status. Finally, new Ka-band allocations are being studied by NASA for commercial MSS and may present a feasible opportunity.

The ITU has defined many communication services within its RR that can be carried out by satellite systems and has developed many rules for worldwide RR of the services in order to maximize the peaceful use of outer space. In a more general sense, depending on the specific purpose and services, satellite communications can be classified into several services, as explained later. In effect, many of these services have been provided under the auspices of special satellite organizations set up to develop, operate and market the service. As already mentioned, classification of an MSC service can be achieved by considering the role played by the geographical extent of network coverage on international, regional and local bases. All these organizations set up to handle these classes of satellite systems have a similar structure, only specific where the service spans national boundaries. Otherwise, the main classification of the satellite service in connection with types of users has been realized on fixed and mobile units.

Table 1.4. Frequency Spectrum for MSS Feeder Links

Uplink (GHz) - Tx Earth-Spacecraft	Downlink (GHz) - Rx Spacecraft-Earth	Comments
0.148-0.14855	0.1370725-0.1379275	E-Sat
0.148-0.15005	0.400150-0.401000	Leo One
0.150-0.15005	0.400505-0.400645	Faisat (WARC-92)
6.345-6.425	4.120-4.200	N-Sat
6.425-6.725	3.400-3.629	Inmarsat, Thuraya
5.091-5.250	6.875-7.075 5.091-5.150 temporary 5.150-5.216 from 2010	New MSS frequency allocations made in WRC-95 for GEO and Non-GEO MSS and RDSS applications (ICO, Globalstar)
15.45-15.65	15.40-15.70	
13-13.25	10.75-10.95	MSAT
14.236-14.250	12.736-12.750	Emsat
14 & 30	11 & 20	MTSAT
29.1-29.3	19.4-19.6	Iridium

1.7.1. Fixed Satellite Service (FSS)

The FSS enables a radiocommunications link between two or more Fixed Earth Stations (FES) at given positions, when one or more satellites are used, see **Figure 1.6.** Thus, the given position may be a specified point or any fixed point within a particular area. In some cases this particular service includes satellite-to-satellite links, which may also be operated in certain inter-satellite services. Moreover, the FSS may also include feeder links for other space communication services, including MSS or MSS LES can provide service for FES.

Figure 1.6. Fixed Satellite Service

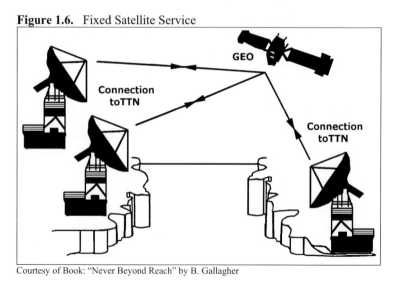

Courtesy of Book: "Never Beyond Reach" by B. Gallagher

The FSS signals are relayed between many FES, which are relatively large, complex and expensive systems. The FES terminals are connected to the conventional TTN and the service is intended for long distance voice, video and data communications. According to the WARC-85/88 principle plan the FSS shares frequency bands with terrestrial networks in the 4/6, 12/14 and 20/30 GHz, which guarantees every country equal access to the GEO. A typical example of FSS is the Intelsat, one of the pioneers in satellite communications. The first generation of Intelsat system operated in the C band (4/6 GHz). At present, many regional systems such as Optus in Australia and JCSAT in Japan operate in the Ku band (30/14 GHz), such as Olympus and CS, which provide coverage throughout most of Europe and Japan, respectively. This service may include Satellite Voice and VSAT networks.

1.7.1.1. Satellite Voice Network

Voice service is inherently interactive communication in nature providing global telephone infrastructures. In fact, telephony system represents first form of two way wire or wireless communications on distance. Voice band channels are useful for relatively low data rate applications such as Fax, E-mail and low speed Internet access on less than 64 Kb/s. The digital standard for PSTN access at the subscriber level is ISDN service which provides 144 Kb/s of active data subdivided into two 64 Kb/s circuit switched bearer channels plus one data channel (2B+D). The FSS voice network uses GEO satellite configuration and a bandwidth per channel from 8 to 64 Kb/s with FDMA and TDMA satellite transmissions. The system enables the international telephone trunks to extend the global coverage, thin route and to improve rural services in developing regions.

1.7.1.2. VSAT Network

The VSAT (Very Small Aperture Terminals) devices are quite similar to Inmarsat MES. This equipment is small Earth Stations capable of receiving only from and transmitting and receiving (transceivers) to or from spacecraft. Therefore, VSAT devices are classified as a communications media with either one-way or two-way facilities.

In the broadest sense, the term one-way VSAT includes the data terminals designed for the reception only of DSB transmission and conventional TVRO, using PAL and similar TV system. This device can also receive data using modulated sub carriers. Two-way VSAT can transmit and receive signals at rates to approximately 64 Kb/s. The low directivity of VSAT antennas limits the power and hence the boresight EIRP, which may be transmitted. The transceivers are usually completely solid state and can be highly integrated.

This equipment represents an important addition to the telecommunications world, because they can provide a service directly to their customers at virtually any geographic location covered by suitable satellite beam. They do not require any support from a local TTN and can even be run from portable or alternative power supply. This system is useful for private data network within countries and regions to promote business needs. Examples of such networks are Equatorial, Intelnet, Intelsat and other systems.

The VSAT system main applications are serving for data and documents distributions, rural communities, business utilizations and for disaster area communications. Data distribution can be in two-way between Central Hub stations for archive and data processing and all VSAT users. Documents distribution by VSAT can be only one-way satellite transmission from Hub library stations to all users. The users can be in touch with Hub via TTN.

Figure 1.7. VSAT Equipment

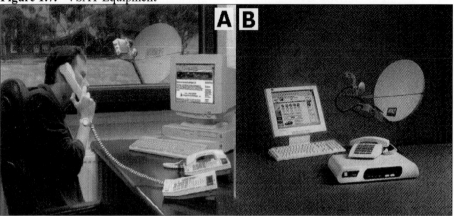

Courtesy of Prospect: "VSAT" by GT&T

Rural application of VSAT devices is very important for improving capability to transmit and receive much information from rural areas to central locations and vice versa. Rural communities would also like to provide speech service via VSAT systems, because of very limited telephone lines facilities, if they are available at all. Voice VSAT system operating at 4.8 Kb/s is already in use. High quality speech can be carried on satellite channels with a data rate of 9.6 Kb/s using modern voice encodes and modems capable of operating at low C/N values, something like Inmarsat aeronautical voice system. A voice circuits set up between two VSAT devices include a double hop and accordingly a delay of about 0,5 sec. A Hub located in a city can provide both voice and data services to many villages.

Disaster areas and in generally for emergency needs VSAT devices for alert and security communications, almost regarding the terrain, locations and safety of life. The reason is similar that many ships, vehicles and aircraft now carry satellite distress beacons.

Business VSAT applications are of an essential interest because of their large potential to provide persons and companies with the competitive edge. In generally, business wants to establish private networks to link their all locations and move their information in a safe manner and for the lowest possible cost. The service can be numerous like airlines and bus reservations, car rental conference facilities, insurance and newsgathering.

Although, the term VSAT equipment is generally used in connection with very small and fixed location terminals for business use, there are comparable developments in related fields should be considered. For example, very low cost TVRO, which can receive MAC signals from high power satellites, could offer VSAT type service in higher data rates and less cost.

The VSAT technology brings all of the features and benefits of unit or bi-directional FSS down to an extremely economical and usable for business data transmission. The system can provide also very efficient bypass with TTN for voice, data and video services using sophisticated digital technology and advanced communication network protocols.

The VSAT system enables use one or more 56 Kb/s data channels, each of which can be subdivided or applied directly. Voice communications is also possible using 16 or 32 Kb/s, depending on the compression algorithm. The VSAT network uses GEO configuration, a channel bandwidth from 64 to 512 Kb/s with FDMA and TDMA transmissions.

In **Figure 1.7. (A)** is shown GT&T Faraway SES100 VSAT cost-effective state-of-the-art satellite solution for companies having high communication flow to their clients, agents or branch offices located abroad or for remote and rural areas where communication services are unavailable, unreliable or too expensive. This equipment is using V-sat communication network with African C-Band and European, African and Middle East Ku-Band coverage. The system is providing on demand High-quality voice up to 8 telephone lines, Group III Fax, Data transfer compatible via external Hayes compatible modem with rate from 4.8 up to 64 or 128 Kb/s, IP/X25/X400, Internet (E-Mail and Websites), Videoconference and Broadcasting (TV and Radio) services. The Main unit with high capacity chassis allows several telephones (up to 8 interfaces), Fax and PCs to be connected simultaneously via parabolic antenna. The antenna is available with diameter of 65, 98, 120, 180 and 240 cm dishes depending on position in coverage area and distance from subsatellite point. This configuration employs a full-mesh, DAMA and PAMA network architecture that maximize the use of available space and ground-based resources.

Mobile satellite users can also be included in this category of satellite communications. Most probably in the near future VSAT system will offer full MSS similar to Inmarsat-C or mini-M systems. With an additional mobile antenna VSAT can offer MSS for maritime, land and aeronautical applications. Thus, another very last VSAT model is GT&T IPsky2 of V-Sat two-way transceiver system together with Internet modem and router lunched in January 2002 can be easily adapted in mobile unit with slight transformation of antenna system only, see **Figure 1.7. (B)**.

This equipment is a low cost and highly compact Internet dedicated DVB-RCS-MPEG2 solution that also offers prepaid VoIP and Fax by satellite. In reality, customer will have an ultra-fast asymmetric Internet interface that is directly connected to the Internet backbone of 155 Mb and 34 Mb fibre using an E-Mai, FTP, TCP and Web Access service available on 24-hours basis. The maximum available download (outbound) speed of service via GEO is 2 Mb/s including very high level of transmission and reception coding for high speed and reliability. Return path or inbound speed is 33, 76.8 or 153.6 Kb/s depending on antenna size of 75, 96 or 1.2 m, respectively. Several PCs (from 5 to 13 or more if necessary) and Ethernet LAN can be connected to the IPsky2 modem/router through a Proxy server, public IP network, private FTP, VPN, etc.

1.7.2. Mobile Satellite Service (MSS)

The MSS consists in three types of GMSC services: maritime, land and aeronautical and may include the space segment, Mobile and Land Earth stations (MES and LES) with all applications (SES, VES, AES, TES, PES), Coordination and Control stations and centres (NCC, SCC, NCS, RCC, LUT) and TTN interface with subscribers, see **Figure 1.8.**

The MSS enables satellite linkage between MES and one or more space stations or between space stations used for this service, or between two or more MES by means of one or more space stations. This service may also include service and feeder links necessary for its operations and in such a way MES can be connected via service link to the satellite and from the satellite via feeder link to the LES.

The LES in MSS network can be Coast Earth Station (CES) for both maritime and land mobile applications and Ground Earth Station (GES), for aeronautical mobile application only. In a more general sense, this modern classification can be a provisional proposal but for the moment gives the reasonable ideas for future establishment some practical universal

Figure 1.8. Mobile Satellite Service

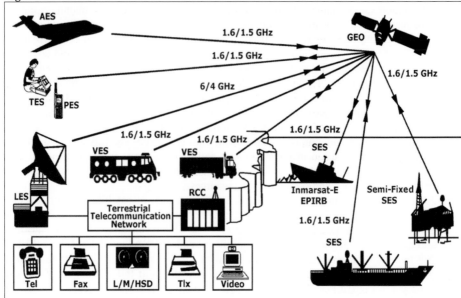

Courtesy of Book: "Never Beyond Reach" by B. Gallagher

nomenclature of GMSC's terms. The LES is a special Earth Station in MSC located at a specified fixed point or within a specified area on land, to provide a feeder link for MSS. The Base Earth Station (BES) or Gateway is an Earth station in mobile or fixed satellite service located at a specified fixed point, or within a specified area on land, to provide a feeder link for the MSS, especially for Non-GEO systems.

Satellite radio beacons for indicating distress emergency position is a special Earth station in the MSS, the emissions of which are intended to facilitate SAR operation for distress in maritime, land and aeronautical applications, EPIRB, PLB and ELT, respectively.

The service link is a connection between MES and satellite, while the feeder link means of duplex connection between CES or GES in the given location and spacecraft of the various MSC services operate in frequency bands allocated to the FSS. Thus, the given location of MSS may be at a specified fixed point or at any point within specified areas. Therefore, a satellite link comprises one uplink and one downlink performing radio linkage between a transmitting and receiving Earth station.

1.7.2.1. Maritime Mobile Satellite Service (MMSS)

The MMSS is a service in which MES is located on board merchant or military ships, other floating objects, rigs or offshore constructions, hovercrafts and/or survival craft stations providing commercial, logistic, tactical, defense and safety communications. In addition, the special maritime Emergency Position Indicating Radio Beacon (EPIRB) terminal, either portable or fixed stations on board ships may also participate in this service. The EPIRB is a special Earth station in the MMSS, the emission of which is intended to facilitate urgent SAR operation for vessels in distress for maritime applications.

The MMSS service enables mobile satellite links between CES and Ship Earth Station (SES), between two or more SES and/or between associated ships and other satellite communications stations in all positions at sea or in ports. The SES is a mobile Earth station in the MMSS capable of surface movement at sea within the geographical limits of a country or continent. In distinction from conventional maritime communications, a ship fitted with SES in or near a port may operate with CES or other SES in cases of distress and commercial operations. The CES is a maritime Earth station located at a specified fixed point on the coast to provide a feeder link for MMSS. The SES is a maritime Earth station fixed on board ships or other floating objects, which can provide communications links with subscribers onshore via CES and communications spacecraft.

The ship on scene radiocommunications and alert service performs a distress and safety service in the MMSS between one or more SES and CES, or between two or more nearby SES, or between SES and RCC, or between portable or floating EPIRB and LUT stations in which alert messages are useful to those concerned with the movement and position of ships and of ships in distress.

1.7.2.2. Land Mobile Satellite Service (LMSS)

The LMSS is a service in which MES is located on different types of cars, trucks, buses, trains and other civil or military vehicles, providing logistics and business communications, or it can be a Transportable Earth Station (TES). Besides, the land emergency Personal Locator Beacon (PLB) terminal may also participate in this service as a special Earth station in the LMSS, the emissions of which are intended to facilitate SAR operation of vehicles and/or personal distress and emergency on the ground for land applications. This unit is very suitable for military applications, for use in polar expeditions and for desert, remote and rural environments.

The LMSS enables mobile satellite links between CES and Vehicle Earth Station (VES), between two or more VES and/or between associated MSC stations. The VES is a mobile Earth station in the LMSS capable of surface movement on dry roads or railways on land within the geographical limits of a certain country or continent. The CES is a maritime Earth station also used for LMSS located at a specified fixed point on the coast to provide a feeder link for MMSS. The VES is a land mobile Earth station fixed on board road or rail land vehicles, providing communications links with terrestrial subscribers via CES and communications spacecraft.

The land vehicle or persons on scene radiocommunications and alert service performs a distress and safety service in the LMSS between one or more VES and CES, between two or more nearby MES, between VES and RCC, or between PLB and LUT stations in which messages are useful to those concerned with the position of vehicles or persons in distress.

1.7.2.3. Aeronautical Mobile Satellite Service (AMSS)

The AMSS is a service in which MES is located on different types of airplanes, helicopters and other civil or military aircraft providing logistics, flight regulations, air traffic control, safety, business and private communications, primarily along national or international civil air routes. This service is also intended for all kinds of satellite communications, including those relating to flight margin coordination in corridors and air traffic control, primarily outside national and international civil aeronautical routes. The special Emergency Locator

Transmitter (ELT) terminal either portable or fixed on board aircraft stations may also participate in this service as a special Earth station in the AMSS, the emission of which is intended to facilitate the SAR operations for aircraft in distress and emergencies at sea, on the ground and for aeronautical applications.

The AMSS enables mobile satellite links between GES and Aircraft Earth Station (AES), between two or more AES and/or between associated airplanes/aircraft and other satellite communications stations. The AES is a mobile Earth station in the AMSS capable of flying and maneuvering in the air over the geographical limits of a country or continent. The GES is an aeronautical Earth station located at a specified fixed point on land to provide a feeder link for AMSS. The AES is an aeronautical Earth station fixed on board aircraft, which can provide communications links with subscribers on land via GES and spacecraft.

The aircraft on scene radiocommunications and alert service performs a distress and safety service in the AMSS between one or more AES and GES, between two or more nearby MES, between AES and RCC, or between ELT and LUT stations in which messages are useful to those concerned with the movements, flight locations and/or positions of aircraft on land, in distress.

1.7.3. Personal Mobile Satellite Service (PMSS)

The PMSS is a service in which MES is handled by individuals and serving everyone in wherever position possessing handset satellite phones. The system can serve for both their GMSC and Global Mobile Personal Satellite Communications (GMPSC). The land mobile emergency PLB terminal may also participate in this service or actually may need some coordination with regard to the Cospas-Sarsat system.

This service enables mobile satellite links between BES (Gateways) and Personal Earth Station (PES), between two or more PES and/or between other satellite communication stations using the same satellite providers or frequency spectrum. The PES is a mobile Earth station in the PMSS carried handheld by individuals for personal and/or business facilities or capable of surface movement at sea, on land and in the air for professional utilization. As is evident, this new system can also provide three types of GMSC services: maritime, land and aeronautical.

The BES or Gateway is an Earth station located at a specified fixed point to provide a feeder link for PMSS. The PES is a personal Earth station or handheld terminal carried by individuals or fixed on board ships, in vehicles or aircraft, which can provide two-way communications links with subscribers anywhere on Earth via satellites and Gateways.

The PES or persons on scene radiocommunications and alert service performs a distress and safety service in the PMSS between one or more PES and BES, between two or more nearby MES or PES, between PES and RCC, or between PLB and LUT stations in which messages are useful to those concerned the position of mobiles or persons in distress.

1.7.4. Radio Navigation Satellite Service (RNSS)

The RNSS can be used for the purposes of safety navigation and secure sailing or flight, including obstruction warnings. The RNSS is an on-directional system, in which only mobile stations can know its own position, others cannot. This service may include:

1. Maritime Radio Navigation Satellite Service (MRNSS) – This is a special service in which Earth Station or satellite navigation equipment is located on board ships.

Figure 1.9. Broadcast Fixed and Mobile Satellite Service

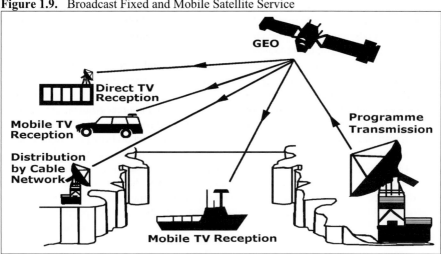

Courtesy of Book: "Never Beyond Reach" by B. Gallagher

2. Land Radio Navigation Satellite Service (LRNSS) – This is a special service in which Earth stations or satellite air navigation equipment is located on board land vehicle.

3. Aeronautical Radio Navigation Satellite Service (ARNSS) – This is a special service in which Earth stations or satellite navigation equipment is located on board aircraft.

1.7.5. Radio Determination Satellite Service (RDSS)

The RDSS is a special service for the purpose of radio determination involving the use of one or more space stations and may include feeder links necessary for its own operations. This service provides the satellite radio determination of the position, velocity and/or other characteristics for each moving object and all mobile applications, or to obtain information relating to these parameters by means of the propagation properties of radio waves.

A typical RDSS is Iridium, which provides two-way voice communications and positioning services in a self-consistent manner. Using this system, a mobile station has the possibility to determinate its own position, when other user stations can also know that position, for example, handheld, fixed and other Earth stations or a station installed in an office or onboard vehicle. In distinction from a radio navigation one-directional system, the RDSS system is a bi-directional system.

1.7.6. Broadcast Mobile Satellite Service (BCMSS)

The BCMSS is a one-way radiocommunication solution in which signals transmitted by Earth stations or retransmitted by space stations are intended for direct reception by the general public via satellite receivers and suitable antennas. Direct Broadcasting Satellite (DBS) signals are transmitted from fixed positions on the ground to the satellite and then back to all MES, community distribution by cable network, or direct individual receivers. This service requires the generation of high RF power from the satellite to permit reception by small antennas onboard mobiles or on the ground, shown in **Figure 1.9.**

Developments in digital audio/video broadcasting are, of course, not solely limited to the terrestrial domain of the home market. Digital services on offer by satellite are now making significant in-roads into the mobile entertainment market, which especially are maritime and aeronautical. New digital broadcasting service of CD-quality music programmes and DVD-quality video transmissions via satellites to many customers in mobiles have recently been attracting the attention of many people in transportation industries.

The term "direct reception" in the BCMSS shall encompass both individual fixed or mobile reception and community reception. Individual reception is any reception of the emissions from space stations in the BCMSS by simple domestic installations at home or on board mobiles and in particular those processing small antenna, intended for private or common use on board mobiles. Community reception is any reception of the emissions from a space station in the BCMSS by professional receiving equipment, which in some cases may be complex and can have antenna larger than those used for individual reception. This service is intended to be used by a group of the general public at one fixed or mobile location or through a distribution system covering a limited area.

The present BCMSS operates at 12 GHz band and is designed for community reception, equipped with fixed satellite terminals and large antennas. Besides, if a satellite has enough power to transmit signals to be received by small antennas suitable for individual reception equipped with satellite Very Small Aperture Terminals (VSAT), or very small aperture antennas, the system is called a DBS system. Although the present system is designed for fixed service and terminals, new systems are also serving all three mobile applications equipped with auto tracking antennas. Some transportation means such as large ocean-going ships, cruisers, airplanes flying on intercontinental airlines, trains and buses traveling on international or interregional routes can receive TV programs from direct Direct Video Broadcasting (DVB) and Direct Audio Broadcasting (DAB) stations. These mobile stations can be equipped with special tracking receiving only antenna systems as a separate unit or in combination with two-way communications satellite antennas. Therefore, this service may include audio (sound), video (television) and data transmission.

1. Audio Broadcasting – Audio broadcasting is a special radio emission via satellite for BCMSS, in which signals transmitted by Earth stations or retransmitted by space stations are intended for direct reception of audio signals by the general public or mobile units via satellite receivers and corresponding antennas. In Europe, USA and Japan the satellite DAB systems in the frequency L and S-bands have been investigated to allow the direct transmission of high quality programs, comparable to CD, to be developed. Usually, road vehicles are equipped with such equipment.

2. Video Broadcasting – Video broadcasting is a special TV emission via satellite for BCMSS, in which signals transmitted by Earth stations or retransmitted by space stations are intended for direct reception of DVB video and audio signals by the general public or mobile units via satellite TV receivers and corresponding antennas. All mobile units at sea, on land and in the air can be equipped with suitable video and tracking antenna equipment. The Inmarsat system also provides audio and videoconferencing via HSD MPDS or mobile ISDN interface using Fleet F77 service for ships and Swift64 for aircraft.

3. Data Broadcasting – Data broadcasting, using a different speed rate, is a special data transmission via satellite for BCMSS, in which signals transmitted by Earth stations or retransmitted by space stations are intended for direct reception of data signals by the general public or MES terminal via special HSD satellite data receivers and corresponding omnidirectional or tracking antennas.

1.7.7. Broadband Mobile Satellite Service (BBMSS)

The BBMSS is a multimedia one or two-way radiocommunication service in which signals transmitted by mobile or fixed Earth stations or retransmitted by space stations are intended for direct reception by the general public or mobile units through satellite receivers and corresponding antennas. This new system, similar to the optical-fibre terrestrial network, will provide various advanced interactive multimedia services operating at higher bit rates, such as voice (audio), video, different speeds of data, teletext, videoconferencing, distance learning, high resolution graphics, HiFi audio, high definition TV, color Fax and imaging, mobile Internet and PC communications. Such advanced networks will soon cover urban, fixed and mobile environments and most populations living in rural/remote areas, which cannot be covered by landline and optical-fibre or wireless TTN.

These signals are transmitted from LES to the satellite and then back to all mobile or fixed applications via optical-fibre cable TTN. This service requires the generation of high RF power from the satellite to permit reception by receivers and small antennas on the ground.

The BBMSS has been developed by broadband MSC operators and will start to offer a multimedia service for personal and all three mobile applications, which allows any person or mobile to communicate anytime and anyplace. Thus, it will provide two categories of high-speed wireless access communications. The first will be serviced both outdoors and indoors, which can enable a high-speed rate up to 30 Mb/s and the second will provide ultra-high-speed indoors only, which can transmit high-speed signals up to 600 Mb/s. The second system cannot provide wide coverage areas or services in mobile environments, so the main application is limited to a "hot spot" of indoor premises.

The next example of BBMSS is DVB-RCS (Digital Video Broadcasting-Return Channel). This system ties: Terrestrial Networks (Broadband, Broadcasting, Internet, UMTS/GPRS Cellular, Private and Public) via satellite HUB (LES) with antenna and C, Ku or Ka-band satellites, with satellite Router Terminals for Remote Internet, VoIP, Videoconference, all E-services, Interactive TV/radio, Broadband LAN/WAN, Multicasting, Intranet/VPN, etc.

The newly designed Intelligent Satellite Transport System (ISTS) will be also part of the BBMSS infrastructure, which comprises an advanced information and telecommunications network for users, roads and vehicles. The ISTS is expected to greatly contribute to solving problems such as traffic accidents and congestion. Not only solving such problems, ISTS will provide multimedia services for vehicle drivers and passengers. Otherwise, the ISTS consists in several development areas, including technical advances in MSC and navigation systems, control and tracking system, electronic toll collection system, assistance for safe driving and so forth, which is appreciated as one of the most promising mobile satellite multimedia businesses.

Digital networks used to operate 64 Kb/s channels with the switching (transmission) based on the so-called Synchronous Transfer Mode (STM). This mode was developed mainly for transmission; thus, switching functions are difficult to handle at different bit rates. At this point, in the interest of more flexible broadband switching, therefore, a new Asynchronous Transfer Mode (ATM) has been developed that can handle traffic relating to services that require widely differing bit rates. In ATM basically the information is put in fixed-length cells that are switched and transported to the broadband network and at the point of destination reconstituted in its original synchronous form. Typical new services requiring broadband switching equipment operating at higher speeds (bit rates) include: Desktop publishing; Multimedia service; Videoconferencing; Color Fax; HiFi music/HDTV, etc.

1.7.8. Meteorological Mobile Satellite Service (MLMSS)

The MLMSS is a special infrastructure for obtaining meteorological data transmission via satellite, in which data signals transmitted by Earth stations or retransmitted by space stations are intended for direct reception by the meteorological centres. In fact, this service is used for one-way transmission of meteorological data only from meteorological centres through the MSC system, including hydrological, observations and exploration particular to ships, oilrigs and aircraft mobile stations. In such a way, this meteorological information, bulletins and weather warnings are very useful for the safe navigation and flight of ships and aircraft, respectively.

In more general sense, the meteorological satellites provide mobile customers at sea and in air with a unique and long-sought opportunity to look at Earth from space. These special spacecraft enable observation and measurement the many forces of nature, which converge on our Planet. Certainly, from sophisticated weather (WX) satellites is possible to obtain daily meteorological data and climatological information, situation, measurements, images and forecast. The current need for quick and easy access to marine and aeronautical WX data and forecast information has become increasingly important. The growing demand of new sources of energy has led the offshore oil exploration into more remote and hostile seas. Environmental constraints have narrowed the maneuvering margins of many at-sea operations, so there is a need for an operational WX/forecast system, which can quickly disseminate maritime and aeronautical WX to the users.

Much of this information is transmitted from WX satellites via direct readout to receiving ground stations where it can be displayed, analyzed and prepared for customers. These Direct Readout Service (DRS) were developed and operated by US NOAA, Russian (CIS), the ESA, Japan, China, India, and etc. The most popular of these WX services are Wefax transmitted by the US Geostationary Operational Environmental Satellites (GOES), and the Automatic Picture Transmission (APT) from PEO NOAA satellites. The GOES-8, 9 and 10 spacecraft series are currying as well as payloads with GEOSAR transponders of GMSC Cospas-Sarsat system. Therefore, remotely sensed meteorological data are transmitted directly from GEO or PEO satellites in "real time" to many Ground Forecasting Centre (GFC) and GES within signal range of the WX satellite. The WX satellite images are designed with a format so that they could be received, processed and reproduced by relatively expensive GES equipment, and retransmitted free of charge to anyone with the appropriate satellite receiving and display equipment.

The former USSR has been attempting for decades to develop an effective GEO program for WX applications. Namely, due in part to the very high latitude of their launch sites, it has been difficult for Russia to design, build and launch a dependable GEO WX satellite system. After all, Russia had launched in 1994 a Geostationary Operational Meteorological Satellite (GOMS) called Elektro-1 that is placed in GEO at $76°$ E. During 1996 and 1997 the GOMS Wefax service has been operating erratically, and there been problems with the imaging sensors. An Elektro-2 satellite was scheduled for launch in 2000 and to improve GOMAS Wefax service. This satellite is also projected to be part of Cospas-Sarsat system, because is currying as well as GEOSAR transponder.

The ESA organization also operates a series of GEO WX satellites called Meteorological Satellite (Meteosat), which provide low resolution DRS similar to GOES Wefax, and called Secondary Users Data Station (SUDS). Thus, the Meteosat-6 was replaced by Meteosat-7 in 1988, and new series of MSG satellites were scheduled to be launched in 2001.

Figure 1.10. WEFAX Coverage Map

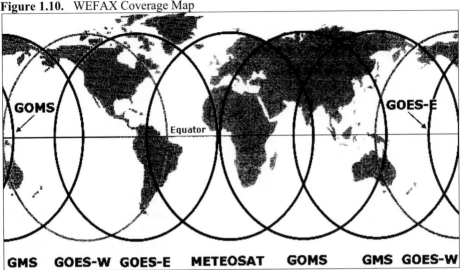

Courtesy of Book: "Maritime Safety Information Broadcast" by ALRS

Japan has launched a series of Geostationary Meteorological Satellites (GMS). The last one GMS-5 was launched in 1994 at position of 140° E, which transmits both: a primary data stream and Wefax transmission on 1691 MHz. A new, advanced series of GEO MTSAT satellites for WX and Aeronautical Augmentation services will start with operation in 2003 and replace GMS-5 spacecraft, which is described in Chapter 10.

The Indian INSAT-2A, 2B and 3 multipurpose satellites have several payloads; among the rest they carry both WX and GEOSAR (Cospas-Sarsat) transponder. The Chinese GEO Wefax program began with the launch of the FY-2 (Feng Yun) satellite in 1997.

The basic footprint for the current operational GEO satellites transmitting WEFAX data is presented in **Figure 1.10.** There are many other MLMSC obtaining WX for mariners like AWT and others, and also for avionics such as METAR.

1.7.8.1. WEFAX System

The WEFAX (WX Fax) satellite images are designed with the format for the DRS provided by the US GOES system. Otherwise, similar services are transmitted from the European Meteosat, Russian GOMS, Japanese GMS and satellites from other countries. The Wefax system retransmits data consists of processed images produced by the primary imager on the GOES above mentioned GEO as well as other meteorological data and images relayed from PEO satellites. This System was first incorporated into the GOES satellites in 1975.

The format of the WEFAX signal was designed to be received and reproduced by low cost GES. This satellite delivered WEFAX signal should not be confused with the HF WX radio Fax transmission from Coast Radio Station (CRS), which some of service is still using. Namely, the WEFAX system is a line-of-sight satellite transmission, with different contents and uses specific receiving equipment. The GOES WEFAX data is transmitted as an analog signal at 1,691 MHz and 240 lines per minute. A large amount and variety of data can than be obtained. In the current WEFAX schedule over 100 images can be received in a 24 hour

Figure 1.11. GOES WDCP Meteorological Satellite System

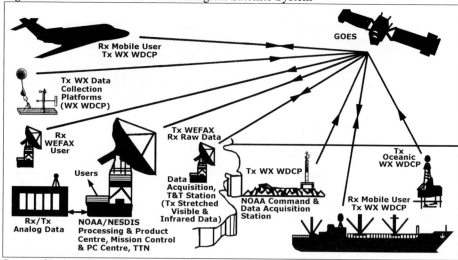

Courtesy of Book: "Maritime Safety Information Broadcast" by ALRS

period. These consist of scheduled data transmissions of quadrants of the full Earth disc and equatorial regions in visible and infrared spectra, composite images from PEO, WX and ice charts, and operational messages. The WEFAX image is initially formatted by the special Ground Control Station (GCS), using high resolution GOES image data polar orbiter mosaics and weather charts, and rebroadcast through the GOES satellite back to DRS of GES. In fact, it is a delayed a DRS product or near real time. The basic structure of the GOES network of Tx (transmitted) from Weather Data Collection Platforms (WDCP) and Rx (received) weather information by mobiles is shown in **Figure 1.11.**

In order to receive the Wefax transmissions most stations use a parabolic or dish antenna with stabilized and tracking platform for large ship MES. Thus, continuous tracking of the antenna is not required for a fixed GES. Namely, once properly aligned to the satellite downlink signal, the antenna is locked into permanent position and rarely requires any further adjustment. Except the antenna, a basic WEFAX RDS station is typically comprised of the following electronic components: preamplifier, receiver, demodulator card to decode the satellite signals, PC for display the satellite imagery, a storage system (hard disc, CD writer or ZIP drive) to memorize and archive the satellite imagery, and computer software applications to manipulate the imagery (image processing and enhancement). The basic WEFAX receiving configuration system can be purchased for about 800/1500 US$ without PC configuration. The commercial DRS now available have been designed for a variety of computers like IBM compatible, Apple Macintosh and even some UNIX systems.

While most of the details in the preceding article have emphasized the US NOAA GOES satellites and their WEFAX system, it is important to keep in mind that the WEFAX service is global in nature. Otherwise, the Wefax transmission of the Russian, Japanese and European GEO are nearly identical in their technical characteristics. The small differences between these systems are normally easily resolved in the computer software that comes with a commercial purchased receiving system, and such system will be usable even if moved to different locations.

1.7.8.2. Automatic Picture Transmission (APT)

The APT is using DRS from PEO satellites to provide WX imagery for world customers. The first APT was pioneered on TIROS-VIII satellite launched in December 1963. Today, US TIROS PEO satellites continue to transmit images of the Earth by APT. These have been joined by Russian Meteor and the Chinese Feng Yun spacecraft, providing similar transmissions. Because of that, an international transmission standard has been agreed upon so a GES capable of receiving data from the US PEO can receive images from satellites of other countries as well.

On the latest TIROS-N (ATN) series PEO satellites the APT images are produced by a special instrument called the Advanced Very High Resolution Radiometer (AVHRR). This instrument is designed to detect five channels of energy reflection from the Earth ranging from the visible spectrum, the near-infrared and infrared spectra. The analog APT signal is derived from the original five channel digital data and multiplexed so that only two of the original channels appear in the APT format. These two images are selected from GCS, and during daylight passes they usually consist of the visual channel and one of the infrared channels. At night, two infrared images are usually found in the APT. The APT imagery consists of two pictures, side by side, representing the same view of the Earth in two different special bands. The Russian Meteor is transmitting daylight visible pictures only.

The APT signal is transmitted continuously from the PEO satellites to the GES. The result in image strip as long as the data transmission is received at the GES and as wide as the scanning instrument is designed to operate at a particular altitude. Radio reception of the APT signals is limited to line-of-sight from GES and can only be received when the PEO is above the horizon, namely when is in view. This is determined by both the altitude of the satellite and its particular path during the orbit across the GES reception range. At present, the US, Russian, and Chinese PEO operate at altitude between 810 and 1200 km. Thus, at these altitudes the maximum time of signal reception during an overhead pass is about 16 minutes. During this time a GES can receive a picture strip equivalent to about 5800 km along the satellite path under the best reception conditions.

In order to obtain APT video data using direct reception, accurate information concerning locations, movements and times that the satellite can be received must be available. This is necessary because signal reception is possible only while the satellites are above that GES horizon. Although all PEO satellites have basic orbital characteristics in common, so each spacecraft is unique in its orbital parameters and needs to be tracked individually. The data necessary to locate and track the WX satellites is generally not difficult to obtain. At any rate, the generation of future orbits of a given satellite can be easily calculated and, if a directional antenna is used, determining the azimuth and elevation of the satellite as it passes over the GES is not difficult after the basic orbital pattern are understood.

The last generation of US ATN satellites represents the current PEO spacecraft available for receiving DRS data. The basic operational concept of this series is only to maintain two satellites in a polar orbit at all time. One will maintain an orbit so that it will pass over the GES, traveling from North to South during the morning, having a southbound Equator crossing at about 0830 local solar time. While, the next satellite will pass from South to North during the afternoon, having a northbound Equator crossing at about 1430 local solar time. In fact, each of these satellites will also pass over the GES circa 12 hours later traveling in the opposite direction. During Winter 1997/98 of North Hemisphere, the US is operating NOAA-12 as a morning spacecraft and NOAA-14 as the afternoon spacecraft.

With a two satellite constellation such as the NOOA-12 and NOAA-14, one of which will pass over the observing/receiving station about every six hours. Most stations can receive two consecutive passes (about 100 min apart) from each satellite, day and night. Being able to receive imagery about every six hours is more than adequate to track the development and movement of large weather system within an 800 to 1000 km radius of the station.

Technological advances in microelectronics and computer software applications over the past two decades have made it rather simple to assemble and use a basic DRS GES suitable for APT. However, a basic DRS station consists the same components except is additional a method to predict when the PEO satellite has to be in view of the GES. This is because PEO satellites are not stationary like GEO over the adequate positions on the Earth.

The APT direct transmission service from PEO satellites is more suitable for smaller ships. The antenna is small and omni-directional therefore does not need to be aimed at the satellite direction. Several different types of antennas may be used for APT reception of data. One can be directional and requires tracking of the moving satellite, and the second type is omnidirectional. It has the advantage of being less expensive and not requiring tracking system, but will give a slightly reduced reception range. A turnstile reflector type of omnidirectional antenna is one of the simplest and least expensive antennas to use for APT. The next quadrifilar helix (QFH) antenna is a special type of omnidirectional antenna that provides a much better radiation pattern compared to previous, and does not suffer from the loss of signal strength exhibited in simple turnstile antenna.

The APT signals from PEO satellites are transmitted on radio frequency between 137 and 138 MHz FM. Two US TIROS PEO satellites transmit APT on 137.5 and 137.62 MHz, and Russian Meteor has been using 137.85 MHz. Beginning in early 2003, the European Organization for the Exploration of Meteorological Satellites (EUMETSAT) will launch their first PEO meteorological satellite Metop-1. It will curry a suite of advanced sensors and DRS transmission system. Otherwise, of special note to the DRS user community is that the current APT service will be replaced by Low Rate Picture Transmission (LRPT) by Metop-1 spacecraft. The LRTP service will be digital rather than analog; requiring same adequate modification to present installed receiving stations.

1.7.8.3. Applied Weather Technology (AWT)

The AWT focus is on providing high quality maritime global weather routing data service through Inmarsat-A, B and M SES equipment. This includes providing the shipmaster with initial routing recommendation, continuous enroute weather forecast advisories (WX) and post-voyage analyses. The AWT is PC aided weather routing system and is committed to research and development in worldwide weather information system. Thus, a number of fully operational software packages are under continuous development in order to provide up-to-date technology solution to maritime and other clients:

1) Voyage Simulation Engine – Is PC-based computer software for dead reckoning vessel position commensurate with prospective WX and adjusted by true position reported by vessel. An integrated database engine enables instant access to ship's actual and potential performance enroute using MSS.

2) Bonvoyage System (BVS) – Is actually network compatible shore-based company fleet management and monitoring system or can be onboard ship graphical marine WX data briefing system. Color Enhanced WX maps on PC screen overlaying with vessels route information not only provides up-to-the minute fleet status at sea or in port, but also allows visual recognition of enroute WX conditions. On the other hand, the master of vessel is routinely facing real time WX, not only bearing the paramount burden of safety of his ship, cargo and crew, but also struggling for the most economical management of the voyages. For example, the forecast of an imminent storm system ahead of vessel track is a very difficult task. However, it is impossible for the traditional text-based WX routing to explain fully the detailed shape of the storm system, dangerous wave generating area, detail grid information, and to find out the best solution for safe and economical ship's routes. To explore the BVS is necessary to supply a modern PC configuration running Windows, minimum 8 Mb RAM and 10 Mb free disk space, Super VGA video card with at least 1 Mb RAM, and popular asynchronous 9600 baud modem or faster recommended. The BVS can be easily tailored for pleasure boat and yacht operators.

The AWT system is actively conducting research and development in area of ocean wave modeling, oceanographic data compression, optimum ship routing, high speed shipboard data transmission, and so on. These future developments, when available, will become an integral part of the overall weather information service.

1.7.8.4. Global Meteorological Technologies (GMT)

The GMT system has been in operation for approximately 8 years as a manufacturer and service provider of software supplying WX data. Such data is provided to ships or land based operations through the Internet, Inmarsat, HF/VHF radio or PSTRN/PSDN networks. This system utilizes the backup network used by NOAA and Environment Canada for their WX service communication system.

The GMT WeatherWise software product is a more economical and superior alternative to Weatherfax as a means of receiving WX charts. Thus, charts are as near to real time WX conditions as can be technically be achieved, and are available within the hour of their measurement. Charts are presented in full colour using NOAA and WMO standard formats and codes, and are available 24 hours per day. Users have to supply corresponding PC configuration and peripherals. At all events, the faster is PC; the sooner analyses can be calculated. Otherwise, all action of the mentioned software can be activated by a mouse, trackball, joystick or touch screen equipped PC. A keyboard is not required to use GMT WeatherWise, other than to reconfigure active map and WX databases.

Data WeatherWise services are broadcast via Inmarsat-C or VHF radio and are available on GMT bulletin board via HF radio and Internet network. However, any GMDSS type approved Inmarsat-C SES can be used to receive Inmarsat FleetNET broadcast. A Hayes compatible 2400 baud, or higher, modem is required to log onto GMT service via Internet. Thus, the GMT WeatherWise is available in six modular regions: The North Pacific, North Atlantic, Eurasia, Far East, Good Hope and Horn.

1.7.8.5. Noble Denton Weather Services (NDWS)

The NDWS system is private initiative to provide special marine WX forecasting services to the international offshore and marine industry. Global meteorological data is gathered and processed from NDWS centre in London within 24 hours to enable the production of site or route specific forecasts for any location worldwide. The WX forecasts are updated once or twice daily by Fax, Tlx or E-mail. Weather routing for ocean towages can also be provided by Noble Denton in-house Master Mariners.

1.7.8.6. Global Sea State Information via Internet (GSSII)

The ERS satellites operated by the ESA measure significant sea wave heights and period, 10 m wind speeds and direction benefits the satellites 24 hours. These data are continuously received by UK Company Satellite Observing Systems where they are routinely processed and corrected before necessary calibration factors are applied. Results in the form of image maps and text summaries are generally available within 2 to 3 hours of data acquisition for direct delivery over the Internet or retrieval on a daily subscription service via the World Wide Web (WWW). However, a Sea State Alarm system also operates to give immediate WX warning by E-mail messages of regional conditions in any part of the world in excess of 10 m significant wave height or 30 m/sec wind speed or 6 m minimum swell.

1.7.8.7. Aeronautical Weather Applications

The most popular WX service for aircraft is provided by SITA and ARINC meteorological systems for aircraft in flight or in the airport as follows:
1. SITA Aircom Weather System – This WX system offer several alternatives via Radio, Inmarsat Aero terminals, Internet, On-site or Hosted solutions such as: Graphical WX charts created with country boundaries, airports and various other aeronautical features, thus providing maximum information within one viewable image; Graphical representation of WX data, mapped over airline routes, facilitates alternative route selection based on wind patterns and other critical WX information; Each instance of wind within user's environment can be programmed to receive a unique set of charts; Aircom surface WX with all meteorological parameters and NOTAM information important for safety flight; and Different WX charts adapted to pilot needs like Surface WX, Visibility, Satellite Imaginary, Radar Imaginary and Lighting. The Aircom WX charts service includes, but is not limited, the following types of weather information: significant weather, upper air weather (wind speed and direction), temperatures, icing, turbulence and precipitation.
2. ARINC Value Added Service (VAS) – Except other corporate and safety contribution the VAS system provides similar media as SITA to transfer all meteorological parameters and information to civil aircraft including Meteorological Aviation Reports (METAR) and Terminal Area Forecast (TAF). The ARINC network organize as well as Meteorological Data Collection and Reporting System (MDCRS), which collects information, organizes and disseminates real-time automated position and WX reports from participating airlines and forwards them to the National Weather Service (NWS) for input to their forecast models. Weather products include radar precipitation images, lighting, temperatures, icing, turbulence and accurate forecast of wind aloft are used to define areas of severe WX and contribute to flight planning efficiencies and aviation safety. Otherwise, this service is similar to GOES meteorological satellite system for WX data collection from aircraft, ships or other mobile or stationary platforms and transfer them via satellite to NOAA/NESDIS stations, as is shown in **Figure 1.11.** On the other hand, Terminal Weather Information for Pilots (TWIP) provides valuable situational awareness of weather conditions within 15 Nm of the airport. On the other hand, the TWIP service collects information from airport sensors and transmits severe WX warnings such as wind shear microbursts, gust fronts, heavy-to-moderate precipitation to aircraft and ground operations computers.

2

SPACE SEGMENT

This chapter describes orbital mechanics and their significance with regard to satellite use for mobile communications. Namely, here are introduced the fundamental laws governing satellite orbits and the principal parameters that describe the motion of the Earth's artificial satellites. The types of satellite orbits are also classified, presented and compared from the MSC system viewpoint in terms of coverage and link performances.

During the last two decades, commercial MSC networks have utilized GEO extensively to the point where orbital portions have become crowded; coordination between satellites is becoming constrained and could never solve the problem of polar coverage. On the other hand, Non-GEO MSC solutions have recently grown in importance because of their orbit characteristics and coverage capabilities in high latitudes and polar regions.

2.1. Platforms and Orbital Mechanics

The platform is an artificial object located in orbit around the Earth at a minimum altitude of about 20 km in the stratosphere and a maximum distance of about 36,000 km. The artificial platforms can have a different shape and designation but usually they have the form of aircraft, airship or spacecraft. In addition, there are special space stations and space ships, which are serving on more distant locations from the Earth's surface for scientific exploration and research and for cosmic expeditions.

Orbital mechanics is a specific discipline describing planetary and satellite motion in the Solar system, which can solve the problems of calculating and determining the position, speed, path, perturbation and other orbital parameters of planets and satellites. In fact, a space platform is defined as an unattended object revolving about a larger one. Although it was used to denote a planet's Moon, since 1957 it also means a man-made object put into orbit around a large body (planet), when the former USSR launched its first spacecraft Sputnik-1. Accordingly, man-made satellites are sometimes called artificial satellites.

Orbital mechanics support a communication satellites project in the phases of orbital design and operations. The orbital design is based on a generic survey of orbits and at an early stage in the MSC project is tasked to identify the most suitable orbit for the objective MSC service. The orbital operation is based on rather short-term knowledge of the orbital motion of the satellite and starts with TT&C maintenances after the satellite is located in orbit. In effect, only a few types of satellite orbits are well-suited for MSC and navigation systems.

2.1.1. Space Environment

The satellite service begins when a spacecraft is located as a space platform in the desired orbital position in a space environment around the Earth. This space environment is a very specific part of the Universe, where many factors and determined elements affect the planet and satellite motions. The Earth is surrounded by a thick layer of many different gasses known as the atmosphere, whose density decreases as the altitude increases. Hence, there is no air and the atmosphere disappears at about 180 km above the Earth, where the Cosmos begins. The endless environment in space is not very friendly and is extremely destructive, mainly because there is no atmosphere, the cosmic radiation is very powerful, the vacuum creates very high pressure on spacecraft or other bodies and there is the negative influence of very low temperatures.

The Earth's gravity keeps everything on its surface. All the heavenly bodies such as the Sun, Moon, planets and stars have gravity and reciprocal reactions. Any object flying in the atmosphere continues to travel until it meets forces due to the Earth's gravity or until it has enough speed to surpass gravity and to hover in the stratosphere. However, to send an object into space, it first has to overcome gravity and then travel at least at a particular minimum speed to stay in space. In this case, an object traveling at about 5 miles/sec can circle around the Earth and become an artificial spacecraft.

An enormous amount of energy is necessary to put a satellite into orbit and this is realized by using a powerful rocket. Rockets or launchers are defined as an apparatus consisting of a case containing a propellant (fuel) and reagents, by the combustion of which it is projected into the atmosphere or space. As the payload is carried on the top, the rocket is usually separated and drops each stage after burn-out. Finally, a rocket brings a payload up to the required velocity and leaves it in orbit. Sometimes, a rocket is also known as a booster, as a rocket starts with a low velocity and attains some required height, where air drag decreases and it attains a higher velocity.

2.1.2. Laws of Satellite Motion

A satellite is an artificial object located by rocket in space orbit following the same laws in its motion as the planets rotating around the Sun. In this sense, three so important laws for planetary motion were derived by Johannes Kepler, as follows:

1. First Law – The orbit of each planet follows an elliptical path in space with the Sun in one focus. Motion lies in the plane around the Sun (1602).

2. Second Law – The line from the Sun to planet or radius vector (r) sweeps out equal areas in equal intervals of time. This is the Law of Areas (1605).

3. Third Law – The square of the planet's orbital period around the Sun (T) is proportional to the cube of the semi-major axis (a = distance from the Sun) of the ellipse for all planets in the Solar system (1618).

Kepler's laws only describe the planetary motion if the mass of central body insofar as it is considered to be concentrated in its centre and when its orbits are not affected by other systems. However, these conditions are not completely fulfilled in the case of Earth motion and its artificial satellites. Namely, the Earth does not have an ideal spherical shape and the different layers of mass are not equally concentrated inside of the Earth's body. Because of this, the satellite motions are not ideally synchronized and stable, the motions are namely slower or faster at particular orbital sectors, which presents certain exceptions to the rule of Kepler's Laws. Furthermore, in distinction from natural satellites, whose orbits are almost elliptical, the artificial satellites can also have circular orbits, for which the basic relation can be obtained by the equalizing the centrifugal and centripetal Earth forces.

Thus, Kepler's Laws were based on observational records and only described the planetary motion without attempting an additional theoretical or mathematical explanation of why the motion takes place in that manner. In 1687, Sir Isaac Newton published his breakthrough work "Principia Mathematica" with own syntheses, known as the Three Laws of Motion:

1. Law I – Every body continues in its state of rest or uniform motion in a straight line, unless it is compelled to change that state by forces impressed on it.

2. Law II – The change of momentum per unit time of a body is proportional to the force impressed on it and is in the same direction as that force.

3. Law III – To every action there is always an equal and opposite reaction.

On the basis of Law II, Newton also formulated the Law of Universal Gravitation, which states that any two bodies attract one another with a force proportional to the products of their masses and inversely proportional to the square of the distance between them. This law may be expressed mathematically for a circular orbit with the relations:

$$F = m \, (2\pi/t)^2 \, (R + h) = G \, [M{\cdot}m/(R + h)^2] \qquad (2.1.)$$

where parameter m = mass of the satellite body; t = time of satellite orbit; R = equatorial radius of the Earth (6.37816×10^6 m); h = altitude of satellite above the Earth's surface; G = Universal gravitational constant (6.67×10^{-11} N m^2/kg^{-2}); M = Mass of the Earth body (5.976032×10^{24} kg) and finally, F = force of mass (m) due to mass (M).

2.1.2.1 Parameters of Elliptical Orbit

The satellite in circular orbit undergoes its revolution at a fixed altitude and with fixed velocity, while a satellite in an elliptical orbit can drastically vary its altitude and velocity during one revolution. The elliptical orbit is also subject to Kepler's Three Laws of satellite motion. Thus, the characteristics of elliptical orbit can be determined from elements of the ellipse of the satellite plane with the perigee (Π) and apogee (A) and its position in relation to the Earth, see **Figure 2.1. (A).** The parameters of elliptical orbit are presented as follows:

$$e = c/a = \sqrt{[1 - (b/a)^2]} \quad \text{or} \quad e = (\sqrt{a^2 - b^2}/a) \qquad p = a \, (1 - e^2) \quad \text{or} \quad p = b^2/a \qquad (2.2.)$$

$$c = \sqrt{(a^2 - b^2)} \qquad\qquad a = p/1 - e^2 \qquad\qquad b = a \sqrt{(1 - e^2)}$$

where e = eccentricity, which determines the type of conical section; a = large semi-major axis of elliptical orbit; b = small semi-major axis of elliptical orbit; c = axis between centre of the Earth and centre of ellipse and p = focal parameter. The equation of ellipse derived from polar coordinates can be presented with the resulting trajectory equation as follows:

$$r = p/1 + e \cos \Theta \quad [m] \qquad (2.3.)$$

where r = distance of the satellites from the centre of the Earth (r = R+h) or radius of path; Θ = true anomaly and E = eccentric anomaly. In this case, the position of the satellite will be determined by the angle called "the true anomaly", which can be counted positively in the direction of movement of the satellite from $0°$ to $360°$, between the direction of the perigee and the direction of the satellite (S). The position of the satellite can also be defined by eccentric anomaly, which is the argument of the image in the mapping, which transforms the elliptical trajectory into its principal circle, an angle counted positively in the direction of movement of the satellite from 0 to $360°$, between the direction of the perigee and the direction of the satellite. The relations for both mentioned anomalies are given by the following equations:

Figure 2.1. Elliptical and Circular Satellite Orbits

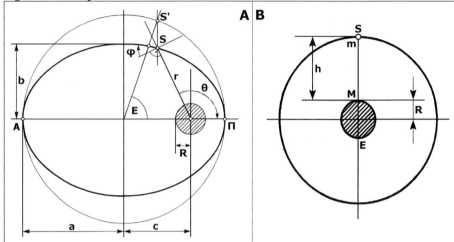

Courtesy of Book: "Telekomunikacije satelitima" by R. Galić

$$\cos \Theta = \cos E - e/1 - e \cos E \qquad \cos E = \cos \Theta + e/1 + e \cos \Theta \qquad (2.4.)$$

The total mechanical energy of a satellite in elliptical orbit is constant; although there is an interchange between the potential and the kinetic energies. As a result, a satellite slows down when it moves up and gains speed as it loses height. Thus, considering the termed gravitation parameter $\mu = GM$ (Kepler's Constant $\mu = 3.99 \times 10^5$ km^3/sec^2), the velocity of a satellite in an elliptical orbit may be obtained from the following relation:

$$v = \sqrt{[GM (2/r) - (1/a)} = \sqrt{\mu (2/r) - (1/a)]} \qquad (2.5.)$$

Applying Kepler's Third Law the sidereal time of one revolution of the satellite in elliptical orbit is as follows:

$$t = 2\pi \sqrt{(a^3/GM)} = 2\pi \sqrt{(a^3/\mu)} \qquad (2.6.)$$

$$t = 3.147099647 \sqrt{(26,628.16 \cdot 10^3)^3} \cdot 10^{-7} = 43,243.64 \quad [s]$$

Therefore, the last equation is the calculated period of sidereal day for the elliptical orbit of Russian-based satellite Molnya with apogee = 40,000 km, perigee = 500 km, revolution time = 719 min and a = 0.5 (40,000 + 500 + 2 x 6,378.16) = 26,628.15 km

2.1.2.2. Parameters of Circular Orbit

The circular orbit is a special case of elliptical orbit, which is formed from the relations a = b = r and e = 0, see **Figure 2.1. (B).** According to Kepler's Third Law, the solar time (τ) in relation with the right ascension of an ascending node angle (Ω); the sidereal time (t) with the consideration that $\mu = GM$ and satellite altitude (h), for a satellite in circular orbit will have the following relations:

$$\tau = t/(1 - \Omega t / 2\pi) \tag{2.7.}$$

$$t = 2\pi \sqrt{(r^3/\mu)} = 3.147099647 \sqrt{(r^3 \cdot 10^{-7})} \quad [s] \tag{2.8.}$$

$$h = [\sqrt[3]{(\mu t^2/4\pi^2)}] - R = 2.1613562 \cdot 10^4 (\sqrt[3]{t^2}) - 6.37816 \cdot 10^6 \quad [m] \tag{2.9.}$$

The time is measured with reference to the Sun by solar and sidereal day. Thus, a solar day is defined as the time between the successive passages of the Sun over a local meridian. In fact, a solar day is a little bit longer than a sidereal day, because the Earth revolves by more than $360°$ for successive passages of the Sun over a point $0.986°$ further. On the other hand, a sidereal day is the time required for the Earth to rotate one circle of $360°$ around its axis: $t_E = 23$ h 56 min 4.09 sec. Therefore, a geostationary satellite must have an orbital period of one sidereal day in order to appear stationary to an observer on Earth. During rotation the duration of sidereal day $t = 85,164,091$ (s) and is considered in such a way for synchronous orbit that $h = 35,786.04 \times 10^3$ (m). The speed is conversely proportional to the radius of the path (R+h) and for the satellite in circular orbit it can be calculated from the following relation:

$$v = \sqrt{(MG/R + h)} = \sqrt{(\mu/r)} = 1.996502 \cdot 10^{-7}/\sqrt{r} = 631.65 \sqrt{r} \quad [m/s] \tag{2.10.}$$

From equation (2.8.) and using the duration of sidereal day (t_E) gives the relation for the radius of synchronous or geostationary orbits:

$$r = \sqrt[3]{[(\mu t) / 2\pi)^2]} \tag{2.11.}$$

The satellite trajectory can have any angle of orbital planes in relation to the equatorial plane: in the range from PEO up to GEO plane. Namely, if the satellite is rotating in the same direction of Earth's motion, where (t_E) is the period of the Earth's orbit, the apparent orbiting time (t_a) is calculated by the following relation:

$$t_a = t_E \cdot t/t_E - t \tag{2.12.}$$

This means, inasmuch as $t = t_E$ the satellite is geostationary ($t_a = \infty$ or $\tau=0$). In **Table 2.1.** several values for times different than synchronous orbital time are presented.

Table 2.1. The Values of Times Different than the Synchronous Time of Orbit.

Parameter	Values of time					Unit
t	86,164.00	43,082.05	21,541.23	10,770.61	6,052.00	s
h	35,786.00	20,183.62	10,354.71	4,162.89	800.00	km
(R+h)	42,164.00	26,561.78	16,732.87	10,541.05	7,178.00	km
v	3,075.00	3,873.83	4,880.72	5,584.12	7,450.00	km/s^{-1}

According to **Table 2.1.** and equation (2.9.) it is evident that a satellite does not depend so much on its mass but decreases with higher altitude. In addition, satellites in circular orbits with altitudes of a 1,700, 10,400 and 36,000 km, will have t/ τ values 2/2,18, 6/8 and 24/zero, respectively. In this case, it is evident that only a satellite constellation at altitudes of about 36,000 km can be synchronous or geostationary.

Figure 2.2. Geometric Projection of Satellite Orbits

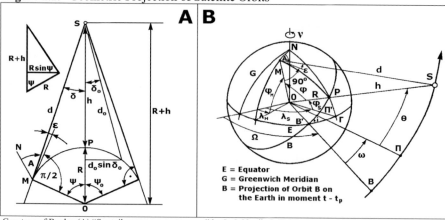

Courtesy of Books: (A) "Sputnikovaya svyaz na more" by L. I. Novik, I. D. Morozov and V. I. Solobev and
(B) "Mezhdunarodnaya sputnikovaya sistema morskoy svyazi – Inmarsat" by V. A. Zhilin

2.1.3. Horizon and Geographic Satellite Coordinates

The geographical and horizon coordinates are very important to find out many satellite
parameters and equations for better understanding the problems of orbital plane, satellite
distance, visibility of the satellite, coverage areas, etc. The coverage areas of a satellite are
illustrated in **Figure 2.2. (A)** with the following geometrical parameters: actual altitude (h),
radius of Earth (R), angle of elevation (ε), angle of azimuth (A), distance between satellite
and the Earth's surface (d) and central angle (Ψ) or sub-satellite angle, which is similar to
the angle of antenna radiation (δ).

The geographical and horizon coordinates of a satellite are presented in **Figure 2.2. (B)**
with the following, not yet mentioned, main parameters: angular speed of the Earth's
rotation (v), argument of the perigee (ω), moment of satellite pass across any point on the
orbit (t_o), which can be perigee (Π), projection of the perigee point on the Earth's surface
(Π'), spherical triangle (B'ГP), satellite (S), the point of the observer or mobile (M),
latitudes of observer and satellite (φ_M and φ_S), longitudes of observer and satellite (λ_M and
λ_S), inclination angle (i) of the orbital plane measured between the equatorial and orbital
plane and the right ascension of an ascending node angle in the moment of t_o (Ω_o).

Otherwise, the right ascension of an ascending node angle (Ω) is the angle in the equatorial
plane measured counter clockwise from the direction of the vernal equinox to that of the
ascending node, while the argument of the perigee (ω) is the angle between the direction of
the ascending node and the direction of the perigee.

2.1.3.1. Satellite Distance and Coverage Area

The area coverage or angle of view for each type of satellite depends on orbital parameters,
its position in relation to the LES and geographic coordinates. This relation is very simple
in the case where the sub-satellite point is in the centre of coverage, while all other samples
are more complicated. Thus, the angle of a GEO satellite inside its range has the following
regular reciprocal relation:

$$\delta + \epsilon + \Psi = 90° \tag{2.13.}$$

The circular sector radius can be determined by the following relation:

$$R_s = R \sin \Psi \tag{2.14.}$$

When the altitude of orbit h is the distance between satellite and sub-satellite point (SP), the relation for the altitude of the circular sector can be written as:

$$h_s = R (1 - \cos \Psi) \tag{2.15.}$$

From a satellite communications point of view, there are three key parameters associated with an orbiting satellite: **(1)** Coverage area or the portion of the Earth's surface that can receive the satellite's transmissions with an elevation angle larger than a prescribed minimum angle; **(2)** The slant range (actual line-of-sight distance from a fixed point on the Earth to the satellite) and **(3)** The length of time a satellite is visible with a prescribed elevation angle. Elevation angle is an important parameter, since communications can be significantly impaired if the satellite has to be viewed at a low elevation angle, that is, an angle too close to the horizon line. In this case, a satellite close to synchronous orbit covers about 40% of the Earth's surface. Thus, from the diagram in **Figure 2.2. (A)** a covered area expressed with central angle (2δ or 2Ψ) or with arc (MP≈RΨ) as a part of Earth's surface can be derived with the following relation:

$$C = \pi (R_s^2 + h_s^2) = 2\pi R^2 (1 - \cos \Psi) \tag{2.16.}$$

Since the Earth's total surface area is $4\pi R^2$, it is easy to rewrite C as a fraction of the Earth's total surface:

$$C/4\pi R^2 = 0{,}5 (1 - \cos \Psi) \tag{2.17.}$$

The slant range between a point on Earth and a satellite at altitude (h) and elevation angle can be defined in this way:

$$z = [(R \sin \epsilon)^2 + 2Rh + h^2]^{\frac{1}{2}} - R \sin \epsilon \tag{2.18.}$$

This determines the direct propagation length between LES, (h) and (ε) and will also find the total propagation power loss from LES to satellite. In addition, (z) establishes the propagation time (time delay) over the path, which will take an electromagnetic field as:

$$t_d = (3.33) z \quad [\mu sec] \tag{2.19.}$$

To propagate over a path of length (z) km, it takes about 100 msec to transmit to GEO. If the location of the satellite is uncertain ± 40 km, a time delay of about ± 133 μsec is always present in the Earth-to-satellite propagation path. When the satellite is in orbit at altitude (h), it will pass over a point on Earth with an elevation angle (ε) for a time period:

$$t_p = (2\Psi/360) (t/1 \pm (t/t_E) \tag{2.20.}$$

The quotations for right ascension of the ascending node angle (Ω) and argument of the perigee (ω) are as follows:

$$\Omega = 9{,}95 \ (R/r)^{3.5} \cos i \quad \text{or} \quad \Omega = \Omega_0 + v \ (t - t_0) \tag{2.21.}$$

$$\omega = 4{,}97 \ (R/a)^{3.5} \ [5 \cos^2 i - 1/(1 - e^2)^2]$$

The limit of the coverage area is defined by the elevation angle from LES above the horizon with angle of view $\varepsilon = 0°$. In this case, the satellite is visible and its maximal central angle for GEO will be as follows:

$$\Psi = \text{arc cos } (R \cos \varepsilon/r) - \varepsilon \quad \text{or} \tag{2.22.}$$

$$\Psi = \pi/2 - \text{arc sin } (R/r) = \text{arc cos } (R/r) - \varepsilon = \text{arc cos } k - \varepsilon$$

$$\Psi = \text{arc cos } 6{,}376.16/42{,}164.20 = \text{arc cos } 0.15126956 = 81° \ 17' \ 58.18"$$

$$C_{max} = 255.61 \cdot 10^6 \ (1 - 0.15126956) = 216.94 \cdot 10^6 \ (km^2)$$

Therefore, all MES and LES with a position above $\Psi = 81°$ will be not covered by GEO satellites. Since the Earth's square area is 510,100,933.5 km^2 and the extent of the equator is 40,076.6 km, only with three GEO mutually moved apart in the orbit by 120° it is possible to cover a great area of the Earth's surface, see **Figure 2.3. (A).** The zero angles of elevation have to be avoided, even to get maximum coverage, because this increases the noise temperature of the receiving antenna. Owing to this problem, an equation for the central angle with minimum angle of view between 5° and 30° will be calculated with the following relation:

$$\Psi_s = \text{arc cos } (k \cos \varepsilon) - \varepsilon \tag{2.23.}$$

The arch length or the maximum distant point in the area of coverage can be determined in the following way:

$$l = 2\pi R \ (2\Psi/360 = 222.64\Psi \ \ [km] \tag{2.24.}$$

The real altitude of satellite over sub-satellite point is as follows:

$$h = r - R = 42{,}162 - 6{,}378 = 35{,}784 \ \ [km] \tag{2.25.}$$

The view angle under which a GEO satellite can see LES/MES is called the "sub-satellite angle". More distant points in the coverage area for GEO satellites are limited around $\varphi = 70°$ of North and South geographical latitudes and around $\lambda = 70°$ of East and West geographical longitudes, viewed from the sub-satellite's point. Theoretically, all Earth stations around these positions are able to see satellites by a minimum angle of elevation of $\varepsilon = 5°$. Such access is very easy to calculate, using simple trigonometry relations:

$$\delta_{\varepsilon=0} = \text{arc sin } k \approx 9° \tag{2.26.}$$

Figure 2.3. GEO Satellite Configuration and Look Angle Parameters

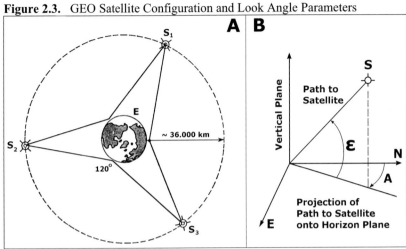

Courtesy of Books: (A) "Telekomunikacije satelitima" by R. Galić and
(B) "Satellite Communications" by T. Pratt and Ch. W. Bostian

At any rate, the angle (Ψ) is in correlation with angle (δ), which can determine the aperture radiation beam. For example, the aperture radiation beam of satellite antenna for global coverage has a radiation beam of $2\delta=17.3°$. According to **Figure 2.2. (A)** it will be easy to find out relations for GEO satellites as follows:

$$\text{tg } \delta = k \sin \Psi/1 - k \cos \Psi = 0.15126956 \sin \Psi/1 - \cos \Psi/1 - 0.15126956 \cos \Psi) \qquad (2.27.)$$

$$\delta_s = 90° - \Psi_s = 8° \, 42' \, 1.82"$$

Differently to say, the width of the beam aperture ($2\delta_s$) is providing the maximum possible coverage for synchronous circular orbit. The distance of LES and MES with regard to the satellite can be calculated using **Figure 2.2 (A).** and equations (2.13.) and (2.22.) by:

$$d = R \sin \Psi/\sin \delta = r \sin /\cos \varepsilon \qquad (2.28.)$$

The parameter (d) is quite important for transmitter power regulation of LES, which can be calculated by the following equation:

$$d = \sqrt{[(R + r)^2 - 2R \, r \cos \Psi]} \quad \text{or} \qquad (2.29.)$$

$$d = h \sqrt{[1 + 2 \, (1/k) \, (R/h)^2 \, (1 - \cos \varphi \cos \Delta\lambda)]} \quad \text{or}$$

$$d = r \, [1 - (R \cos \varepsilon/r)^2]^{1/2} - R \sin \varepsilon$$

Accordingly, when the position of any MES is near the equator in sub-satellite point (P) or right under the GEO satellite, then its distance is equal to the satellite altitude and takes out value for d=H of 35,786 km. Thus, every MES will have a further position from (P) when the central angle exceeds $\Psi = 81°$, when $d_{max}=41,643$ km.

Figure 2.4.   Elevation and Azimuth Angle Maps

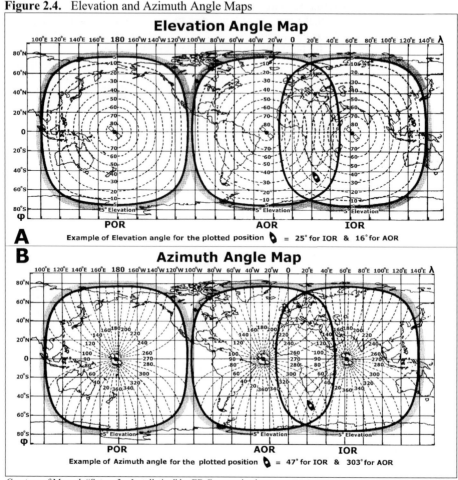

Courtesy of Manual: "Saturn 3 – Installation" by EB Communications

2.1.3.2. Satellite Look Angles (Elevation and Azimuth)

The horizon coordinates are considered to determine satellite position in correlation with an Earth observer, LES and MES terminals. These specific horizon coordinates are angles of satellite elevation and azimuth, shown in **Figure 2.2. (A and B)** and **Figure 2.3. (B)**.
The satellite elevation (ε) is the angle composed upward from the horizon to the vertical satellite direction on the vertical plane at the observer point. From point (M) shown in **Figure 2.2. (A)** the look angle of ε value can be calculated by the following relation:

$$tg \ \varepsilon = \cos \Psi - k/\sin \Psi \tag{2.30.}$$

In **Figure 2.4. (A)** is illustrated the Mercator chart of the 1st Generation Inmarsat space segment, using three ocean coverage areas with projection of elevation angles and with one example of a plotted position of a hypothetical ship (may also be aircraft or any mobile). Thus, it can be concluded that SES or any type of MES at designated position ($\varepsilon=25°$ for IOR and $\varepsilon=16°$ for AOR) has the possibility to use either GEO satellites over IOR or AOR to communicate with any LES inside the coverage areas of both satellites.

The satellite azimuth (A) is the angle measured eastward from the geographical North line to the projection of the satellite path on the horizontal plane at the observer point. This angle varies between 0 and $360°$ as a function of the relative positions of the satellite and the point considered. The azimuth value of the satellite and sub-satellite point looking from the point (M) or the hypothetical position of MES can be calculated as follows:

$$\text{tg A'} = \text{tg } \Delta\lambda_M - k/\sin \Psi \qquad\qquad (2.31.)$$

Otherwise, the azimuth value, looking from sub-satellite point (P), can be calculated as:

$$\text{tg A} = \sin \Delta\lambda/\text{tg } \varphi \quad \text{or} \quad \sin A = \cos \varphi \sin \Delta\lambda \text{ cosec } \Psi \qquad\qquad (2.32.)$$

In **Figure 2.4. (B)** is illustrated the Mercator chart of 1st Generation Inmarsat 3-satellite or ocean coverage areas with projection of azimuth angles, with one example for the plotted position of a hypothetical ship ($\varepsilon=47°$ for IOR and $\varepsilon=303°$ for AOR). Any mobile inside of both satellites' coverage can establish a radio link to the subscribers on shore via any LES.

However, parameter (A') is the angle between the meridian plane of point (M) and the plane of a big circle crossing this point and sub-satellite point (P), while the parameter (A) is the angle between a big circle and the meridian plane of point (P). Thus, the elevation and azimuth are respectively vertical or horizontal look angles, or angles of view, in which range the satellite can be seen.

In **Figure 2.5. (A)** is presented a correlation of the look angle for three basic parameters (δ, Ψ, d) in relation to the altitude of the satellite. Inasmuch as the altitude of the satellite is increasing as the values of central angle (Ψ), distance between satellite and the Earth's surface (d) and duration of communication (t_c) or time length of signals are increasing, while the value of sub-satellite angle (δ) is indirectly proportional. An important increase of look angle and duration of communication can be realized by increasing the altitude to 30 or 35,000 km, while an increase in look angle is unimportant for altitudes of more than 50,000 km. The duration of communication is affected by the direction's displacement from the centre of look angle, which will have maximum value in the case when the direction is passing across the zenith of the LES. The single angle of the satellite in circular orbit depends on the t/2 value, which in area of satellite look angle, can be found in the duration of the time and is determined as:

$$t_c = \Psi t/\pi \qquad\qquad (2.33.)$$

Practical determination of the geometric parameters of a satellite is possible by using many kinds of plans, graphs and tables. It is possible to use tables for positions of SES (φ, λ), by the aid of which longitudinal differences can be determined between MES and satellite for four feasible ship's positions: N/W, S/W, N/E and S/E in relation to GEO.

One of the most important practical pieces of information about a communications satellite is whether it can be seen from a particular location on the Earth's surface. In **Figure 2.5. (B)** a graphic design is shown which can approximately determine limited zones of satellite

Figure 2.5.   Look Angle Parameters and Graphic of Geometric Coordinates for GEO

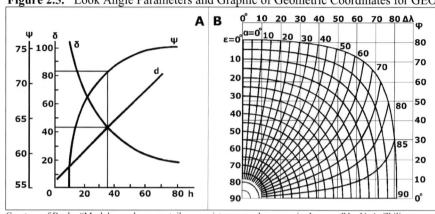

Courtesy of Book: "Mezhdunarodnaya sputnikovaya sistema morskoy svyazi – Inmarsat" by V. A. Zhilin

visibility from the Earth (MES) by using elevation and azimuth angles under the condition that $\delta = 0$. This graphic contains two groups of crossing curves, which are used to compare (φ) and ($\Delta\lambda$,) coordinates of mobile positions. Thus, the first group of parallel concentric curves shows the geometric positions where elevation has the constant value ($\varepsilon=0$), while the second group of fan-shaped curves starting from the centre shows the geometric positions where the difference in azimuth has the constant value ($a = 0$). This diagram can be used in accordance with **Figure 2.2. (B)** in the following order:

1. First, it is necessary to note the longitude values of satellite (λ_S) and mobile (λ_M) and the latitude of the mobile (φ_M), then calculate the difference in longitude ($\Delta\lambda$) and plot the point into the graphic with both coordinates (φ_M & $\Delta\lambda$).

2. The value of elevation angle (ε) can then be determined by a plotted point from the group of parallel concentric curves.

3. The difference value of azimuth (a) can be determined by a plotted point from the group of fan-shaped curves starting from the centre.

4. Finally, depending on the mobile position, the value of azimuth (A) can be determined on the basis of the relations presented in **Table 2.2.**

Table 2.2.   The Form for Calculation of Azimuth Values

The GEO direction in relation to MES	Calculating of Azimuth Angles
Course of MES towards S & W	A = a
Course of MES towards N & W	A = 180° – a
Course of MES towards N & E	A = 180° + a
Course of MES towards S & E	A = 360° – a

Inasmuch as the position of SES is of significant or greater height above sea level (if the bridge or ship's antenna is in a very high position) or according to the flight altitude of AES, then the elevation angle will be compensated by the following parameter:

$$x = \arccos\left(1 - H/R\right) \qquad\qquad (2.34.)$$

where H = height above sea level of observer or MES. Let us say, if the position of LES is a height of H = 1,000 m above sea level, the value of $x \approx 1°$. This example can be used for the determination of AES compensation parameters, depending on actual aircraft altitude. In such a way, the estimated value of elevation angle has to be subtracted for the value of the compensation parameter (x).

2.1.3.3. Satellite Track and Geometry (Longitude and Latitude)

The satellite track on the Earth's surface and the presentation of a satellite's position in correlation to the MES results from a spherical coordinate system, whose centre is the middle of Earth, is illustrated in **Figure 2.2. (B)**. In this way, the satellite position in any time can be decided by the geographic coordinates, sub-satellite point and range of radius. Thus, the sub-satellite point is a determined position on the Earth's surface; above it is the satellite at its zenith.

The longitude and latitude are geographic coordinates of the sub-satellite point, which can be calculated from the spherical triangle (B'ГР), using the following relations:

$$\sin \varphi = \sin (\Theta + \omega) \sin i \qquad\qquad (2.35.)$$

$$\operatorname{tg} (\lambda_S - \Omega) = \operatorname{tg} (\Theta + \omega) \cos i$$

With the presented equation in previous relation it is possible to calculate the satellite path or trajectory of sub-satellite points on the Earth's surface. The GEO track breaks out at the point of coordinates $\varphi = 0$ and $\lambda = \text{const.}$

Furthermore, considering geographic latitude (φ_M) and longitude (λ_M) of the point (M) on the Earth's surface presented in **Figure 2.2. (B)**, what can be the position of the MES, taking into consideration the arc (MP) of the angle illustrated in **Figure 2.2. (A)**, the central angle can be calculated by the following relations:

$$\cos \Psi = \cos \varphi_S \cos \Delta\lambda \cos \varphi_M + \sin \varphi_S \sin \varphi_M \quad\text{or} \qquad (2.36.)$$

$$\cos \Psi = \cos \text{arc } MP = \cos \varphi_M \cos \Delta\lambda$$

The transition calculations from geographic to spherical coordinates and vice versa can be computed with the following equations:

$$\cos \Psi = \cos \varphi \cos \Delta\lambda \quad\text{and}\quad \operatorname{tg} A = \sin \Delta\lambda/\operatorname{tg} \varphi, \quad\text{respectively} \qquad (2.37.)$$

$$\sin \varphi = \sin \Psi \cos A \quad\text{and}\quad \operatorname{tg} \Delta\lambda = \operatorname{tg} \Psi \sin A$$

These relations are useful for any point or area of coverage on the Earth's surface, then for a centre of the area if it exists, as well as for spot-beam and global area coverage for MSC systems. The optimum number of GEO satellites for global coverage can be determined by:

$$n = 180°/\Psi \qquad\qquad (2.38.)$$

For instance, if $\delta = 0$ and $\Psi = 81°$ it will be necessary to put into orbit only 3 GEO, and to get a global coverage from 70° N to 70° S geographic latitude. Hence, in a similar way the number of satellites can be calculated for other types of satellite orbits.

The trajectory of radio waves on a link between an MES and satellite at distance (d) and the velocity of light ($c = 3 \times 10^8$ m/s) requires a propagation time equal to:

$$T = d/c \quad (s) \tag{2.39.}$$

The phenomenon of apparent change in frequency of signal waves at the receiver when the signal source moves with respect to the receivers (Earth), was explained and quantified by Johann Doppler (1803–53). Namely, the frequency of the satellite transmission received on the ground increases as the satellite is approaching the ground observer and reduces as the satellite is moving away. This change in frequency is called Doppler effect or shift, which occurs on both the uplink and the downlink. This effect is quite pronounced for LEO and compensating for it requires frequency tracking in a narrowband receiver, while its effect are negligible for GEO satellites. In effect, the Doppler shift at a transmitting frequency (f) and radial velocity (v_r) between the observer and the transmitter can be calculated by the following relation:

$$\Delta f_D = f \, v_r / c \quad \text{where} \quad v_r = dR/dt \tag{2.40.}$$

For an elliptical orbit, assuming that $R = r$, the radial velocity is given by:

$$v_r = dr/dt = (dr/\Theta)(d\Theta/dt) \tag{2.41.}$$

The sign of the Doppler shift is positive when the satellite is approaching the observer and vice versa. Doppler effect can also be used to estimate the position of an observer provided that the orbital parameters of the satellite are precisely known. This is very important for development of Doppler satellite tracking and determination systems.

2.2. Spacecraft Launching and Station-Keeping Techniques

The launch of the satellite and controlling support services are a very critical point in the creation of space-based communication technology and the most expensive phase of the total system cost. At the same time, the need to make a satellite body capable of surviving the stresses of the launch stages is a major element in their design phase. Satellites are also designed to be compatible with more then one model of launch vehicle and launching type. In a more determined sense, there are multi-stage expendable and, manned or unmanned, reusable launchers. Owing to location and type of site there are land-based and sea-based launch systems. Additional rocket motors, such as perigee and apogee kick propulsion systems, may also be required.

The process of launching a communications satellite is based mostly on launching into an equatorial circular orbit, in particular the GEO but broadly similar processes or phases are used for all types of orbits. Otherwise, the processes involved in the launching technique depend on the type of satellite launcher, the geographical position of the launching site and constraints associated with the payload. In order to successfully put the satellite into the transfer and drift orbit, the launcher must operate with great precision with regard to the magnitude and orientation of the velocity vector. On the other hand, launching operations necessitate either TT&T facilities at the launching base or at the stations distributed along the trajectory.

Figure 2.6. Satellite Installation in Circular and Synchronous Orbit

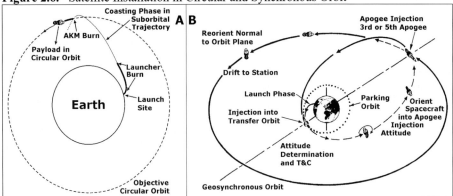

Courtesy of Book: "Commercial Satellite Communications" by S.C. Pascall and D.J. Withers

2.2.1. Satellite Installation and Launching Operations

Satellites are usually designed to be compatible with more than one prototype of launchers. Launching, putting and controlling satellites into orbit is very expensive operation, so the expenses of launcher and support services can exceed the cost of the satellites themselves.
The basic principle of any launch vehicle is that the rocket is propelled by reaction to the momentum of hot gas ejected through exhaust nozzles. Thus, for a spacecraft to achieve synchronous orbit, it must be accelerated to a velocity of 3,070 m/s in a zero-inclination orbit and raised a distance of 42,242 km from the centre of the Earth. Most rocket engines use the oxygen in the atmosphere to burn their fuel but solid or liquid propellant for a launcher in space must comprise both a fuel and an oxygen agent. There are two techniques for launching a satellite, namely by direct ascent and by Hohmann transfer ellipse.

2.2.1.1. Direct Ascent Launching

A satellite may be launched into a circular orbit by using the direct ascent method, shown in **Figure 2.6. (A).** The thrust of the launch vehicle is used to place the satellite in a trajectory, the turning point of which is marginally above the altitude of the desired orbit. The initial sequence of the ascent trajectory is the boost phase, which is powered by the various stages of the launch vehicle. This is followed by a coasting phase along the ballistic trajectory, the spacecraft at this point consisting of the last launcher stage and the satellite. As the velocity required to sustain an orbit will not have been attained at this point, the spacecraft falls back from the highest point of the ballistic trajectory.
When the satellite and final stage have fallen to the desired injection altitude, having in the meantime converted some of their potential energy into kinetic energy, the final stage of the launcher, called the Apogee Kick Motor (AKM) is activated to provide the necessary velocity increase for injection into the chosen circular orbit. In effect, the AKM is often incorporated into the satellite itself, where other thrusters are also installed for adjusting the orbit or the altitude of the satellite throughout its operating lifetime in space. The typical launch vehicles for direct ascent satellite launching are US-based Titan IV, Russian-based Proton and Ukrainian-based Zenit.

2.2.1.2. Indirect Ascent Launching

A satellite may be launched into an elliptical or synchronous orbit by using the successive or indirect ascent sequences, known as the Hohmann transfer ellipse method, illustrated in **Figure 2.6. (B)**. The Hohmann transfer ellipse method enables a satellite to be placed in an orbit at the desired altitude using the trajectory that requires the least energy. At the first sequence the launch vehicle propels the satellite into a low parking orbit by the direct ascent method. The satellite is then injected into an elliptical transfer orbit, the apogee of which is the altitude of the desired circular synchronous orbit. At the apogee, additional thrust is applied by an AKM to provide the velocity increment necessary for the attainment of the required synchronous orbit. In practice it is usual for the direct ascent method to be used to inject a satellite into a LEO and for the Hohmann transfer ellipse to be used for higher types of orbits.

2.2.2. Satellite Launchers and Launching Systems

Two major types of launch vehicles can be used to put a satellite into LEO, HEO and GEO constellation: Expendable and Reusable Vehicles. There are also two principal locations or site-based types of launching centres: Land-based and Sea-based launch systems.

2.2.2.1. Expendable Launching Vehicles

The great majority of communication satellites have been launched by expendable vehicles and this is likely to continue to be the case for many years to come. There are two types of these vehicles: expendable three-stage vehicles and expendable direct-injection vehicles.
1) Expendable Three-Stage Vehicles – Typical series of three-stage vehicles are Delta and Atlas (USA), Ariane (Europe), Long March (China) and H-II (Japan). In addition, a new generation of launchers have already been developed with two-stages such as Delta III and Ariane 5. Both stages are propellant systems using cryogenic liquid fuel, while the first stage is assisted by nine strap-on solid-fuel motors.
The first and second stages of three-stage expandable launch vehicles are usually designed to lift it clear of the Earth's atmosphere, to accelerate horizontally to a velocity of about 8,000 m/s and enters a parking orbit at a height of about 200 km. The plane of the parking orbit will be inclined to the equator at an angle not less than the latitude of the launch site. The most efficient way of getting from the parking orbit to a circular equatorial orbit is to convert the parking orbit into an elliptical orbit in the same plane, with the perigee at the height of the parking orbit and the apogee at about 36,000 km and then to convert the transfer ellipse to the GEO. Thus, the third stage is fired as the satellite crosses the equator, which ensures that the apogee of the Geostationary Transfer Orbit (GTO) is in the equatorial plane. When the satellite is placed in the GTO, the third stage has completed its mission and is jettisoned. The final phase of the Hohmann transfer three-stage launch sequence is carried out by means of AKM built into the satellite. The propulsion of this motor is required to provide at the apogee of the GTO a velocity increment of such a magnitude and in such a direction as to reduce the orbit to zero and make the orbit circular. Once the satellite is in the GEO trajectory, the attitude is corrected, the antennas and solar panels are deployed and the satellite is drifted to the correct longitude (apogee position) for operation.

2) Expendable Direct-Injection Vehicles – Typical models of direct-injection launchers are the USA-based Titan IV and the Russian-based Proton, illustrated **in Figure 2.7. (A)** and **(B)**, respectively and also Zenit (Ukraine). Otherwise, these types of vehicles do not need an AKM because direct-injection launchers have a fourth stage, which converts directly from GTO to GEO constellation. The Proton rocket is one of the most capable and reliable heavy lift launch vehicles in operation today. Proton D-1 and D-1-E launcher variants have three and four stages, respectively. At lift-off the total weight of Proton is about 688 tons and this vehicle has the capability of placing a maximum of 4,500 and 2,600 kg into GTO and GEO, respectively.

2.2.2.2. Reusable Launch Vehicles

Reusable launch vehicles have already been developed in the USA (Space Shuttle) and former USSR (Energia/Buran), which have as their aim the development of vehicles that could journey into space and return, all or much of their structure being reusable and thus, the satellite launching will cost less. Moreover, in using these launchers there will be less burnt-out upper stages than with expendable vehicles. What remains in space, the small pieces in transfer orbits for many years and much small debris, remains in LEO for a long time, adding to the growing space junk hazard for operational satellites and future space operations.

There are other projects for development of similar vehicles such as a small manned reusable space shuttle called Hermes (Europe) and Hope (Japan). In the UK an unmanned space plane Hotol is proposed, while in Germany and the USA two similar vehicles are projected: TAV (TransAtmospheric Vehicle) and Sanger Space plane, respectively. Thus, in development of these small vehicles it is important to realize whether any of them could carry sufficient weight and be able to put communication satellites into the desired orbits.

1. Space Shuttle – The US-based NASA developed a fleet of manned reusable vehicles of Space Transportation System (STS) called Space Shuttle, which are capable of lifting a satellite of up to 29.5 tons into a parking orbit, inclined at 28.5°, with an altitude of up to 431 km, shown in **Figure 2.7. (C)**. A Shuttle has three main elements: **(1)** the orbiter for carrying the satellite and crew; **(2)** a very large external tank containing propellant for the main engine of the orbiter and **(3)** two solid-propellant boosters. The reusable Space Shuttle plane is 37.2 m long, the fuselage is 4.5 m in diameter, the wingspan is 23.8 m and the mass is about 84.8 tons. This STS is designed to accommodate in total 7 crewmembers and passengers on board plane.

The system came into service in 1981 and made over twenty successful operational flights until January 1986, when the Shuttle Challenger was destroyed by a fault in the solid-propellant booster and all the crew were killed in a tragic accident. Following this disaster, NASA redesigned the booster but decided to use STS only for regular launches programme of government and scientific vehicles.

2) Energia/Buran Space plane – The launcher Energia is the most powerful operational reusable vehicle in the world, capable of carrying about 100 tons into space, whose four first-stage booster units are recoverable for reuse. In particular, it can launch the Buran space plane, enabling it to acquire a LEO and to land with the aid of its own rocket engine, shown in **Figure 2.7. (D)**. The main purpose for which those very heavy lift vehicles were developed was to ferry personnel and supplies for the Russian space station Mir, and also to retrieve or repair satellites already in orbit. The Energia vehicle can also carry into space a side-mounted canister containing an upper stage and a payload compartment suitable, for example, for a large heavy spacecraft or group of communication satellites to be placed in orbit. Thus, Energia flew for the first time on 15 May 1987, carrying a spacecraft mock-up and later on 18 November 1988 carrying an unmanned version of Buran space plane. The reusable Buran space plane is 36.3 m long, the fuselage is 5.6 m in diameter, the wingspan is 24 m and the mass is about 100 tons. It can be flown in automatic configuration or under the control of a pilot to place satellites in LEO or to retrieve them and come back to base for the next use. Up to ten people, crew and passengers, can be accommodated and it can carry in the cargo bay up to 30 tons into an orbit of 200 km altitude and $51.6°$ inclination. In fact, this plane enables large satellites to be put into orbit and construction of space stations to be considered for both for telecommunication purposes and for scientific missions.

The Energia Launch Vehicle was also the successor to the N-1 Moon Rockets, except that Buran was also used to launch Polyus from Baikonur Cosmodrome in Kazakhstan (former Soviet Union). Energia was 60 m high and 18 m in diameter, consisting in a central core and four strap-on boosters, while the core was 58.1 m high and 7.7 m diameter. It used 4 RD-0120 rocket engines. The propellants were liquid hydrogen and oxygen. The strap-on boosters were then 38.3 m high and 3.9 m in diameter, with a single four-chamber RD-170 kerosene/liquid oxygen rocket engine.

In 1992, the Russian Space Agency decided to terminate the Energia/Buran Program due to Russia's economic difficulties after disintegration of former Soviet Union. At that stage, the second Orbiter had been assembled and assembly of the third Orbiter with improved performance was nearing completion. Although the Energia project has been abandoned, it may return to service if a market is found, or adequate partners. Consideration is being given in Russia to the development of a more compact winged space plane designed to ferry personnel and their luggage into space. This compact shuttlecraft could be placed on top of a Proton launcher.

Figure 2.7.   Expendable and Reusable Launch Vehicles

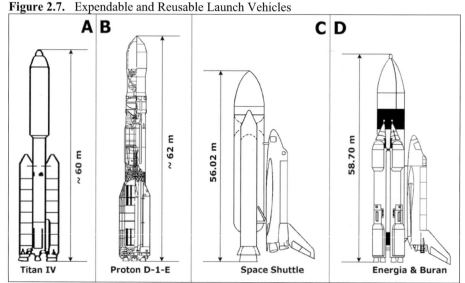

2.2.2.3. Land-Based Launching Systems

Most satellite launches have taken place from the following launch facilities:

1. US-Based Launch Centres – The USA launches satellites from two main locations, in Florida Cape Canaveral, suitable for direct equatorial orbit and the Vandenberg Air Force Base in California, suitable for polar orbit missions.

2. Russian Launch Centres – Russian satellites are launched from two main launch centres named Baikonur and Northern Cosmodrome. Baikonur lies north of Tyuratam in Kazakhstan, with all the launching support infrastructure for launching Proton and Energia heavy launchers. The Northern Cosmodrome is located near Plesetsk, south of the town Archangelsk, suitable for launching satellites for all purposes in high inclination orbits. This Cosmodrome is the world's busiest launch site.

3. European Launch Centres – The main European launch Cosmodrome is the Guiana Space Centre in French Guiana, using Ariane vehicles. The position of this Cosmodrome enables the best advantage to be taken of the Earth's rotation for direct equatorial orbit.

4. Chinese Launch Centres – The principal launch sites in China are Jiuquan and Xi Chang, for launching Long March vehicles. In the meantime, the Xi Chang launch centre has also become most used for launches into the GEO for the international market.

5. Japanese Launch Centres - The Japan's Tanegashima Space centre is situated in the prefecture of Kagashima. The facilities include the Takesaki Range for small rockets and the Osaki Range was used for the launch of H-I vehicles until the termination of program in 1992. After renovation the Osaki Range will be used as the launching for next generation of J-I Japanese vehicles. The new Yoshinobu launch complex has been constructed next to the Osaki centre to satisfy the requirements of the new H-II launcher.

2.2.2.4. Sea-Based Launch Systems

The Sea Launch Multinational Organization was developed in March 1996 to overcome the cost of land-based launch infrastructure duplication around the world. The newly formed Sea Launch system is owned by the Sea Launch Partnership Limited in collaboration with international partners such as US Boeing Commercial Space Company, Russian RSC Energia, Ukrainian KB Yuzhnoye/PO Yuzhmash, Shipping Anglo-Norwegian Kvaerner Group and Sea Launch Company, LLC. The Sea Launch Company, partner locations and operating centres, has US-based headquarters in Long Beach, California and is manned by selected representatives of each of the partner companies.

The Sea Launch Partners have the following responsibilities and tasks:

1. Boeing responsibilities include designing and manufacturing the payload fairing and adapter, developing and operating the Home Port facility in Long Beach, integrating the spacecraft with the payload unit and the Sea Launch system, performing mission analysis and analytical integration, leading operations, securing launch licensing documents and providing range services.

2. RSC Energia is responsible for developing and qualifying the Block DM-SL design modifications, manufacturing the Block DM-SL upper stage, developing and operating the automated ground support infrastructure and equipment, integrating the Block DM-SL with Zenit-2S and launch support equipment, planning and designing the CIS portion of launch operations, developing flight design documentation for the flight of the upper stage and performing launch operations and range services.

Figure 2.8.   Sea Launch Modules

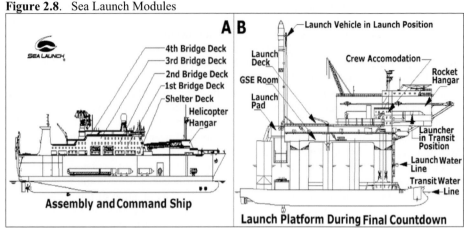

Courtesy of Manual: "User's Guide" by Sea Launch Company

3. KB Yuzhnoye/PO Yuzhmash are responsible for developing and qualifying Zenit-2S vehicle design modifications, integrating the launch vehicle flight hardware, developing flight design documentation for launch with respect to the first two stages, supporting Zenit processing and launch operations. Several significant configuration modifications have been made to allow the basic Zenit design to meet Sea Launch's unique requirements.

4. The Anglo-Norwegian Kvaerner Group is responsible for designing and modifying the Assembly and Command Ship (ACS), designing and modifying the Launch Platform (LP) and integrating the marine elements. Furthermore, Barber Moss Marine Management is responsible for marine operations and maintenance of both vessels.

The partner team of contractors has developed an innovative approach to establishing Sea Launch as a reliable, cost-effective and flexible commercial launch system. Each partner is also a supplier to the venture, capitalizing on the strengths of these industry leaders. The System consists in two main modules: Assembly (Command and Control Ship) and Launch Platform, both illustrated in **Figure 2.8. (A)** and **(B)**, respectively. However, transit for the ACS and the LP from Home Port in Long Beach to the launch site on the equator takes 10 to 12 days, based on a speed of 10.1 knots.

The Sea Launch Home Port complex is located in Long Beach, California. The Home Port site provides the facilities, equipment, supplies, personnel and other procedures necessary to receive, transport, process, test and integrate the spacecraft and its associated support equipment with the Sea Launch system. The Home Port also serves as the marine base of operations for both of the Sea Launch vessels. The personnel providing the day-to-day support and service during pre-launch processing and launch conduct to Sea Launch and its customers are located at the Home Port. The ACS performs four important functions for Sea Launch operations: **(1)** It serves as the facility for assembly, processing and checkout of the launch vehicle; **(2)** It houses the Launch Control Centre (LCC), which monitors and controls all operations at the launch site; **(3)** It acts as the base for tracking the initial ascent of the launch vehicle and **(4)** It provides accommodation for the marine and launch crews during transit to and from the launch site.

Therefore, the ACS is designed and constructed specifically to suit the unique requirements of Sea Launch. The ship's overall dimensions are nearly 200 m in length, 32 m in beam and a displacement of 34,000 tons. Major features of the ACS include: a rocket assembly compartment; the LCC with viewing room; helicopter capability; spacecraft contractor and customer work areas and spacecraft contractor and customer accommodation. The rocket assembly compartment, which is located on the main deck of the ACS, hosts the final assembly and processing of the launch vehicle. This activity is conducted while the vessels are at the Home Port and typically in parallel with spacecraft processing. The bow of the main deck is dedicated to processing and fuelling the Block DM-SL of the Zenit launch vehicle. After the completion of spacecraft processing and encapsulation the encapsulated payload is transferred into the rocket assembly compartment, where it is integrated with the Zenit-2S and Block DM. The launchers and the satellite are assembled horizontally in the ACS before sailing from the port of Long Beach to the designated launch site. A launcher with a payload will then be transferred in the horizontal position to the launch pad on LP and raised to a vertical position for fueling and launching.

During the launch sequence, the crew of the LP will be transferred to the ACS, which will initiate and control the launch from a position about 3 miles away from the LP pad. The LP is an extremely stable sea platform from which to conduct the launch, control and other operations. The LP rides catamaran-style on a pair of large pontoons and is self-propelled by a four-screw propulsion system (two in each lower hull, aft), which is powered by four direct-current double armature-type motors, each of which are rated at 3,000 hp. The LP in navigation has normal draft at sea water level but once at the launch location, the pontoons are submerged to a depth of 22.5 m to achieve a very stable launch position, level to within approximately 1°. The ballast tanks are located in the pontoons and in the lower part of the columns. Six ballast pumps, three in each pontoon, serve them. The LP has an overall length of approximately 133 m at the pontoons and the launch deck is 78 by 66.8 m. The Zenit-3SL launcher is a two-stage liquid propellant launch vehicle solution capable of transporting a spacecraft to a variety of orbits. The original two-stage Zenit was designed by KB Yuzhnoye quickly to reconstitute former Soviet military satellite constellations. The design emphasizes robustness, ease of operation and fast reaction times. The result is a highly automated launch capability using a minimum complement of launch personnel. The launcher as an integrated part of the Sea Launch system is designed to place spacecraft into a variety of orbits and is capable of putting 5,250 kg of payload into GEO.

The Sea Launch mission provides a number of technical support systems that are available for the customer's use in support of the launch process, including most importantly the following:

1. Communications – Internal communications systems are distributed between the ACS and LP. This system includes CCTV, telephones, intercom, video teleconferencing, public address and vessel-to-vessel radiocommunications, known as the Line-of-Sight (LOS) direct system.

This system links with the external communication system and provides a worldwide network that interconnects the various segments of the Sea Launch program. The external communication system includes Intelsat and two ground stations. The LES are located in Brewster, Washington and Eik, Norway and provide the primary distribution Gateways to the other communication nodes. Customers can connect to the Sea Launch communication network through the convenient Brewster site. The Intelsat system ties in with the ACS and launch platform PABX systems to provide telephone connectivity. Additionally, critical telephone, Fax, Tlx or data capability can be ensured by the Inmarsat system through SES Standard-B.

2. Tracking and Data Relay Satellite System (TDRSS) – The Sea Launch system uses a unique dual telemetry stream with the TDRSS. Telemetry is simultaneously received from the Zenit stages, the Block DM upper stage and the payload unit during certain portions of the flight. The Block DM upper stage and payload unit data are combined but the Zenit data is sent to a separate TDRSS receiver. Zenit data is received shortly after lift-off at approximately 9 sec and continues until Zenit Stage 2/Block DM separation, at around 9 minutes. These data are routed from the NASA White Sands LES to the Sea Launch Brewster LES and to the ACS. Otherwise, the data are also recorded at White Sands and Brewster for later playback to the KB Yuzhnoye design centre.

When the payload fairing separates, the payload unit transmitter shifts from sending high-rate payload accommodation data by LOS to sending combined payload unit/Block DM by TDRSS. The combined data is again routed from White Sands to Brewster, where it is separated into Block DM and payload unit data and then sent on to the ACS. The data are received on board the ship through the Intelsat communications terminal and are routed to Room 15 for upper-stage data and Room 94 for PLU data. Simultaneously, Brewster routes Block DM data to the Energia Moscow control centre station. However, the TDRSS coverage continues until after playback of the recorded Block DM data.

3. Telemetry System – Sea Launch uses LOS telemetry systems for the initial flight phase, as well as the TDRSS for later phases. The LOS system, which includes the Proton antenna and the S-band system, is located on the ACS. Other telemetry assets include Russian ground tracking stations and the Energia Moscow control centre. The following subsections apply to launch vehicle and payload unit telemetry reception and routing.

4. Weather (WX) Data System and Forecast – The ACS has a self-contained WX station, which includes a motion-stabilized C-band Doppler radar, surface wind instruments, wave radar, upper-atmospheric balloon release station, ambient condition sensors and access to satellite imagery and information from an on-site buoy.

2.3. Types of Orbits for Mobile and Other Satellite Systems

An orbit is the circular or elliptical path that the satellite traverses through space. This path appears in the chosen orbital plane in the same or different angle to the equatorial plane. All communication satellites always remain near the Earth and keep going around the same orbit, directed by centrifugal and centripetal forces. Each orbit has certain advantages in terms of launching (getting satellite into position), station keeping (keeping the satellite in place), roaming (providing adequate coverage) and maintaining necessary quality of communication services, such as continuous availability, reliability, power requirements, time delay, propagation loss and network stability.

There is a large range of satellite orbits but not all of them are useful for fixed and mobile satellite communication systems. In general, the one most commonly used orbit for satellite communications is GEO constellations, after which HEO and latterly GIO, PEO, LEO and MEO, shown in **Figure 2.9. (A).**

Otherwise, it is essential to consider that satellites can serve all communication, navigation, meteorological and observation systems for which they cannot have an attribute such as fixed or mobile satellites and the only common difference is which type of payload or transponder they carry on board. For example, its name can be satellite specified for fixed communications but in effect it can carry major transponders for fixed communications and others for mobile or other purposes and vice versa.

Figure 2.9. Type of Satellite Orbits and Tracks

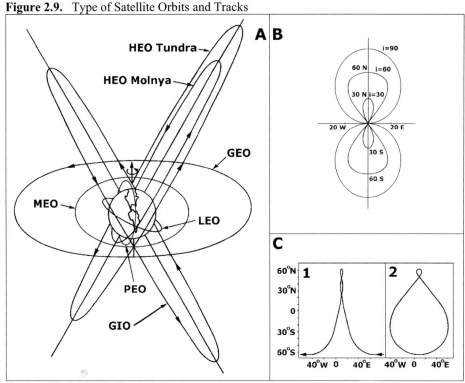

Courtesy of Book: "Satellite Communication Systems" by B.G. Evans

After many years of research and experiments spent on finding the global standardization for spatial communications, satellites remained the only means of providing near global coverage, even in those parts which other communications systems are not able to reach. There is always doubt about the best orbital constellation that can realize an appropriate global coverage and a reliable communications solution. Unfortunately, there is no perfect system today; all systems have some advantages or disadvantages. The best conclusion is to abridge the story and to say briefly that today the GEO system is the best solution and has only congestion as a more serious problem. The extensive use of GEO is showing that it provides something good. Inmarsat is the biggest GEO operator whose service and revenue confirm this point of view. The advantages of Inmarsat MSC solutions can be realized if someone uses them such, as operators on board mobiles and finds out how powerful they are. Most of other regional GEO worldwide networks, such as ACeS, Optus and Thuraya are also more successful than other Non-GEO constellations.

Especially Big LEO and ICO systems or hybrid constellations such as Ellipso have had several years of serious economical and concept difficulties. It is sufficient to see **Table 2.3.** to understand that the major reasons for LEO problems are enormous satellite cost, complex network and short satellite visibility and lifetime. The LEO/PEO constellations are the same or similar and because of differences in inclination angle of orbital plane and type of coverage they will be considered separately.

Table 2.3. The Properties of Four Major Orbits

Orbital Properties	LEO	MEO	HEO	GEO
Development Period	Long	Short	Medium	Long
Launch & Satellite Cost	Maximum	Maximum	Medium	Medium
Satellite Life (Years)	3–7	10–15	2–4	10–15
Congestion	Low	Low	Low	High
Radiation Damage	Zero	Small	Big	Small
Orbital Period	<100 min	8-12 hours	½ Sidereal Day	1 Sidereal Day
Inclination	90°	45°	63.4°	Zero
Coverage	Global	Global	Near Global	Near Global
Altitude Range (km^{-3})	0.5–1.5	8–20	40/A – 1/P	40 (i=0)
Satellite Visibility	Short	Medium	Medium	Continuous
Handover	Very Much	Medium	No	No
Elevation Variations	Rapid	Slow	Zero	Zero
Eccentricity	0 to High	High	High	Zero
Handheld Terminal	Possible	Possible	Possible	Possible
Network Complexity	Complex	Medium	Simple	Simple
Tx Power/Antenna Gain	Low	Low	Low/High	Low/High
	Short	Medium	Large	Large
Propagation Delay	Low	Medium	High	High
Propagation Loss	High	Medium	Low	Zero

The track of the satellite varies from 0 to 360°, see **Figure 2.9. (B).** The track of the GEO satellite is at a point in the centre of the coordinate system; two tracks are apparent movements of the GIO satellite with respect to the ascending node of both 30° and 60° inclination angles and the last is the track of the PEO satellite with an inclined orbit plane to the equator of 90°. The tracks of HEO Molniya (part of the track) and Tundra (complete track) orbits are shown in **Figure 2.9. (C-1/C-2)**, respectively. These two tracks pass over the African Continent and almost all of Europe.

This is very important for MSC systems that the orbit used can provide satellite view during 24 hours with less handovers and network difficulties. However, for other types of broadcasting a communication satellite must be visible from the region concerned during the periods when it is desired to provide a communication service, which can vary from a few hours to 24 hours a day. When the service is not continuous, it is desirable that the intervals during which the service is available repeat each day at the same time.

2.3.1. Low Earth Orbits (LEO)

The LEO systems are either elliptical or more usually circular satellite orbits between 500 and 2,000 km above the surface of the Earth and bellow the Inner Van Allen Belt. The orbit period at these altitudes varies between ninety minutes and two hours. The radius of the footprint of a communications satellite in LEO varies from 3,000 to 4,000 km. Therefore, the maximum time during which a satellite in LEO orbit is above the local horizon for an observer on the Earth is up to 20 minutes. In this case, the traffic to a LEO satellite has to be handed over much more frequently than all other types of orbit. Namely, when a satellite which is serving a particular user, moves below the local horizon, it needs to be able to quickly handover the service to a succeeding one in the same or adjacent orbit. Due to the relatively large movement of a satellite in LEO constellation with respect to an observer on the Earth, satellite systems using this type of orbit need to be able to cope with large Doppler shifts. In fact, satellites in LEO are not affected at all by radiation damage but are affected by atmospheric drag, which causes the orbit to gradually deteriorate. Satellites in LEO and MEO constellation are subject to orbital perturbation. For very LEO satellites the aerodynamic drag is likely to be significant and in general, some of the other perturbations, such as precession of the argument of the perigee, resolve to zero in the orbit is circular or polar. On the other hand, a perturbation is unlikely to have a serious effect on the operation of a multi-satellite constellation since it will usually affect all satellites of the configuration in equal measure.

The major advantages of LEO are as follows:

a) The LEO system may become important in the field of MSC using handheld terminals with global roaming and to be exceedingly useful in areas not covered by cellular systems. The LEO constellations cover almost the entire Earth's surface and some of them provide polar coverage and show promise in the fields of mobile data and Internet and FSS networks for broadband data transmission and communications.

b) High Doppler shift allows the LEO system to be used for satellite positioning, tracking and determination.

c) The relatively small distance between LES and LEO results in much lower power and smaller user terminals. Furthermore, the one-way speed-of-light propagation delay of at least 0,25 sec using GEO is obviated with LEO, which effect can be annoying in two-way voice transmission. For example, for two-way voice via a satellite at an altitude of about 1,000 km, the delay is only 13 ms in total for uplink and downlink.

d) Satellite path diversity eliminates signal interruption due to path obstruction. In **Figure 2.10.** handover from satellite A to satellite B is demonstrated and path diversity between satellite B and C. This figure illustrates the LEO MSS space and ground architecture with utilization of handheld personal terminals (PES). On the other hand, the Satellite Access Node (SAN) is the LES providing a link between PES terminals through satellites and ground telecommunications infrastructures.

The disadvantages of LEO are as follows:

a) The orbit period at about 1,000 km altitude is in the order of approximately 100 min and the visibility at a point on the Earth is only some 10 min, requiring 40 to 80 satellites in six to seven planes for global coverage. Thus, in reality a GMSC system using this type of orbit requires a large number of satellites, in a number of different inclined orbits, which increases the total cost of the network.

b) Frequently handover is necessary for uninterrupted communications. Satellite visibility for MES could be improved by using more satellites. The optimum number of satellites of about 48 inclined in the constellation in a carefully optimized pattern of orbit planes will provide continuous visibility of one or other of the satellites at any location on Earth.

c) During times of the year that the orbital plane is in the direction of the Sun, a satellite in LEO is eclipsed for almost one-third of the orbit period. Consequently, there is a significant demand on battery power, with up to 5,000 charge/discharge cycles per year, which with existing NiCd types of batteries, reduces satellite lifetimes to 3–7 years.

d) The launch cost is low, with direct injection into the orbit of several satellites but the total cost is very high, with a minimum of 40 satellites being produced.

The first generation of LEO satellites was used for military communications because single GEO could be an easy target for an opponent. The large number of LEO satellites will reduce enormously the risk of vulnerability if someone wishes to destroy only one satellite.

Figure 2.10. LEO MSS Diversity and Handover

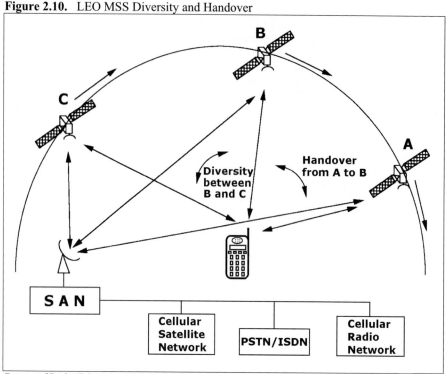

Courtesy of Book: "Telecommunications Transmission Systems" by Huurdeman

Because this orbit configuration is in the initial phase of its exploitation, it is still free of congestion problems. There are two types of LEO constellations known as Little LEO, useful for messaging and satellite tracking systems and Big LEO, suitable for voice, video and data communications.

2.3.1.1. Little LEO

The Little LEO mobile satellite systems are a category of LEO solutions that utilize birds of small size and low mass for low-bit-rate transmission under 1 Kb/s. Thus, the Little LEOs are constellations of very small Non-GEO satellites, which operate in LEO orbits providing mainly mobile data messaging and tracking services for vehicles and ships and other FSS and broadcasting services. The FCC has allocated a frequency band of 137–138 MHz for the downlink and 148–149.9 MHz for the uplink to Little LEO systems, which is a heavily utilized spectrum for private and government services worldwide, such as Orbcomm, Falsat, Leo One, VITASat, Starnet and other systems. The mass of satellites in these solutions range from 40 kg in Orbcomm to 150 kg in the Starnet system. Otherwise, these systems prefer a spectrum below 1 GHz, because it enables the use of cost-effective equipment. In this way, nonvoice two-way messaging and positioning with low cost transceiver, which would be equipped with an alphanumerical display, are the major characteristics of these systems.

2.3.1.2. Big LEO

The Big LEO is a larger Non-GEO satellite system, which operates in LEO constellations and provides mainly mobile telephony, Fax, data and RDSS services. Compared to the Little LEO systems, satellites in Big LEO systems are expected to be bigger in body and to have more power and bandwidth to provide a different service to their subscribers. This system will use the underutilized spectrum available in the L-band, because of the commercial failure of proposed RDSS service. Currently, the frequency spectrum of 1610–1626.5 MHz for uplinks and 2483.5–2500 MHz for downlinks are assigned to these MSC systems. It is interesting to note that although the names of these systems include LEO, their frequencies are the ones usually utilized in MEO and GEO satellite systems. For this reason a new ICO system is systematized in the category of Big LEO constellations together with real Big LEO systems, such as Iridium and Globalstar, which are located at a lower altitude than ICO, at about 700–1,500 km from the Earth's surface. Thus, all of the new proposed Big LEO systems would offer global handheld telephone service by means of satellites on lower altitudes moving very fast, instead of fixed GEO relays. The bigger size of the satellites enables them to carry a transponder on board with more complex data processing facilities than the simple store-and-forward feature of the Little LEO satellite configuration. Hence, an important fact is that these systems are networking with cellular and spreading their roaming and billing capabilities in real global coverage.

2.3.2. Circular Orbits

The GEO satellite constellation has great advantages for MSS communication applications where polar coverage is not required but there are solutions for providing polar roaming. Satellite orbits in 63.4° inclined high-apogee HEO have some advantages from GEO also providing polar coverage. In fact, the most popular circular equatorial orbit with zero inclination is the GEO satellite constellation. The period of rotation is equal to that of the Earth and has the same direction. However, both of these orbits exhibit high line-of-sight (LOS) loss and long transmission times and delays. Using new technology, these problems can be solved, or as an alternative to these orbits there are LEO and MEO constellations with their good and bad characteristics. The choice of orbit depends on the nature of the MSC mission, the acceptable interference in an adequate environment and the performance of the launchers.

2.3.2.1. Medium Earth Orbit (MEO)

The MEO satellite constellations, known also as Intermediate Circular Orbits (ICO), are circular orbits located at an altitude of around 10,000 to 20,0000 km between the Van Allen Belts. A LEO constellation for MSS global coverage requires around 10 satellites in two or three orbital planes, each plane inclining 45° to the equator. Their orbit period measures about 6 to 8 hours, providing slightly over 1 hour local visibility above the horizon for an observer on the Earth and handover from one to the next satellite is every 6 hours minimum. The MEO satellites are operated in a similar way to Big LEO systems providing global coverage. However, compared to a LEO system a MEO constellation can only be in circular orbit; Doppler effect and handover is less frequent; propagation delay is about 70 ms and free space loss is greater; satellites are affected by radiation damages from the Inner

Van Allen Belt only during the launching period; fewer eclipse cycles means that battery lifetime will be more than 7 years; cosmic radiation is lower, with subsequently longer life expectancy for the complete MEO configuration; higher average elevation angle from users to satellite minimizes probability of LOS blockage and higher RF output power required for both indoor and handheld terminals.

An example of MEO satellite constellation is ICO system in currently developing cycles as a former Inmarsat-P system (10 operational + 2 spare satellites in 2 inclined planes of 45° and at 10,355 km altitude) and the abandoned Odyssey system (12 + 3 satellites in 3 inclined planes at 10,355 km). Accordingly, a Non-GEO satellite system known as the ICO Global Communications network, which operates in MEO constellation, will provide MSC mobile telephony, Fax and data services, including Internet access and was scheduled to be operational in 2004.

There is in exploitation a special model of MEO constellation known in practice as Highly Inclined Orbit. This particular orbit is of interest because it has been chosen for existing and proposed GSNN systems such as Navstar (GPS), Navsat, GLONASS and the newly developed Galileo. In all, complete implementation of this orbit configuration would have 24 satellites in 3 orbital planes equidistant from each other, at an altitude of 20,000 km and at an inclination of 55°. In comparison with existing GSNN the new Galileo system will have 30 satellites in high MEO of about 28,000 km and at a similar inclination of 56°. At this point, its interest to polar MSC would be the eventual prospect of satellite sharing with navigation services, in a similar fashion to a high PEO with minimum of 3 satellites in the same orbital plane.

2.3.2.2. Geostationary Earth Orbit (GEO)

A GEO has a circular orbit in the equatorial plane, with an orbital period equal to the rotation of the Earth of 1 sidereal day, which is achieved with an orbital radius of 66,107 (Equatorial) Earth Radii, or an orbital height of 35,786 km. Otherwise, a satellite in a GEO will appear fixed above the surface of the Earth, and remain in a stationary position relative to the Earth itself. Theoretically, this orbit is with zero inclination and track as a point but in practice, the orbit has small non-zero values for inclination and eccentricity, causing the satellite to trace out a small figure eight in the sky.

The footprint or service area of a GEO satellite covers almost 1/3 of the Earth's surface or 120° in longitude direction and up to 75°–78° latitude North and South of the Equator but cannot cover the polar regions. In this way, near-global coverage can be achieved with a minimum of three satellites in orbit moved apart by 120°, although the best solution is to employ four GEO satellites for better overlapping. This type of orbit is essentially used for commercial communication services for both FSS and MSS with the following advantages:

a) The satellite remains stationary with respect to one point on the Earth's surface and so the LES antennas can be beamed exactly towards the focus of the GEO satellite without periodical tracking. Only mobile high-gain antennas need auto-tracking systems, while low-gain omnidirectional antennas are free of tracking systems.

b) The new Inmarsat GEO space constellation consisting in four satellites can cover all three-ocean regions with four overlapping longitudes, except for the polar regions beyond latitudes of 75° North and South. Otherwise, the polar regions can be covered for maritime and aeronautical MSS applications with current HF Radio systems or in combination with PEO or HEO satellite constellations.

c) The Doppler shift, affecting synchronous digital systems caused by satellites to drift in orbit (affected by the gravitation of the Moon and to a lesser extent of the Sun) is small for all LES and MES within satellite coverage.

The disadvantages of GEO compared with LEO and MEO operation are as follows:

a) The long signal delay is due to the large distance of about 35,800 km if the satellite is in zenith for MES and about 41,000 km at the minimum elevation angle of about 5°. For the EM waves traveling at the speed of light this causes a round-trip signal delay of 240 to 270 ms and full duplex delay of 480 to 540 ms. Thus, the voice used via satellite can experience some disturbance but echo cancellation devices developed in the 1980s can reduce the problem. Besides, for data transmission equipment, especially when using error-correction protocols that require retransmission of blocks with detected errors, complex circuitry with special high-capacity buffer devices is required to overcome delay problems. In addition, practical experience has shown that given good control of the echo, a telephone connection which includes one hop in each direction via a GEO satellite is acceptable to public users.

b) The required higher RF output power and the use of directional antennas aggravate GEO operation slightly for use with handheld terminals, although it is not critical, because some GEO operators provide this service, such as Thuraya and ACeS.

c) The launch procedure to put a satellite in GEO is expensive but the total cost of 4 satellites is less than the cost of a minimum of 12 or 40 for MEO and LEO, respectively.

As stated earlier, the major disadvantage of a GEO satellite in a voice transmission is the round-trip delay between satellite and LES of approximately 2.5 sec, which can be successfully solved with current and newly advertised echo cancellation circuits. Because of the enormous use of the GEO constellation for many space applications, some parts of the GEO are becoming congested, owing to only one radius and latitude. This orbit is geostationary and so its track is one point called the sub-satellite point and obviously, handover and Doppler effect does not apply to GEO. A GEO satellite is at essentially fixed latitude and longitude, so even a narrow-beam Earth antenna can remain fixed. Satellites in GEO can use high and recently low-gain antennas, which helps to overcome the great distances in achieving the required EIRP at ground level. On the other hand, using satellite spot beam antennas, coverage can be confined to smaller spot areas, bigger power and higher speed of transmission, such as new generation of Inmarsat-3 spacecraft.

Furthermore, a variety of perturbing forces causes the GEO satellite to drift out of its path and assigned position towards so-called inclined orbit (GIO). By far the most important perturbations are the lunar and at a lesser degree the solar gravitational forces, which cause the satellite to drift in latitude or North–South direction. However, the longitudinal drift in East–West direction is caused by fluctuations in the gravitational forces from the Earth, due to its nonspherical shape and by fluctuations in solar radiation pressure. Thus, to counteract these perturbations the satellite needs station-keeping devices. The GEO satellites pass through both Van Allen Belts only on launch, so their effect is insignificant. After reaching the end of operational life a satellite has to be removed from its orbital slots into a graveyard orbit some 200 km above the GEO plane.

Thus, the GEO satellite constellation seems likely to continue to dominate in the satellite communications world, especially in MSS, providing near global coverage with low and high-power transmission. Especially attractive is the reliable and economical use of the Inmarsat standard-C low-power transceiver and low-gain omnidirectional antenna for maritime, land vehicles and aeronautical two-way data/messaging and telex and one-way E-mail service.

The major existing GEO mobile systems in the world are Inmarsat and GEOSAT of the Cospas-Sarsat system as global solutions and ACeS, AMSC, MSAT, Artemis, Emsat, Optus, N-Star, Solidaridad and Thuraya as regional networks. Some of these systems, such as ACeS and Thuraya, also provide a service for handheld and mini indoor terminals, which makes it obvious that some authors made the mistake of assuming that for GEO it is very difficult to provide a handheld service and that Inmarsat mini-M is the smallest terminal for GEO, as is mentioned on 13 page of "Low Earth Orbital Satellites for Personal Communication Networks" written by A Jamalipour.

Early in 1995, Pasifik Satelit Nusantara of Indonesia along with Philippine Long Distance Telephone and Jasmine International Public Co Ltd of Thailand came together and formed a joint venture for MSS today known as Asia Cellular System (ACeS). The ACeS handheld dual mode (GSM/ACeS) terminal is manufactured by Ericsson.

Thuraya is Private Joint Stock Company registered on 26 April 1997 in UAE under federal Law No. 8 of 1984 as a Regional GEO Mobile Satellite Communication System Operator providing voice, low bit rate data and facsimile services. The two prototypes of Thuraya handheld terminals are being manufactured by Hughes Network Systems (USA) and ASCOM (Switzerland).

2.3.2.3. Geosynchronous Inclined Orbit (GIO)

This system would consist in four satellites at six-hour intervals around the Earth orbit at an inclination of 45° to the equatorial plane. The satellites provide polar coverage for six hours either side of their most northerly and southerly movement. Special LES with full tracking antennas are needed, therefore this system in general must be considered complex and expensive for a polar communication system.

Otherwise, a GIO satellite has a period of orbit equal to or very little different from a sidereal day (23h 56 min and 4.1 sec), which is time for one complete revolution of the Earth. The satellite movement speed has only very little difference from the angular velocity of the Earth, so this movement also has constant angular velocity. Otherwise, the projection of this movement on the equatorial plane is not at a constant velocity. There is an apparent movement of the satellite with respect to the reference meridian on the surface of the Earth and that of the satellite on passing through the nodes. The orbit may be inclined at any angle, which produces a repeating ground track. In **Figure 2.9. (B)** are presented tracks of 30° and 60° inclined orbits. The coupled N–S and E–W motion of GIO satellites is shown as a figure eight pattern, while the patterns could also be distorted circles. Depending on the inclination angle, the GIO satellite shows points on the equator at various longitudes.

A satellite may operate in this orbit for several reasons. First, it is often desirable to save the inclination control fuel required for GEO circle. Sometimes there is no need to control inclination because tracking LES antennas are required for other reasons, while mobiles such as ships and aircraft require tracking antennas. Some GEO satellites may last beyond their planned lifetime if run low on fuel and cease inclination control. In effect, the GIO constellation with non-zero inclination can be chosen because of easy launching and placing of the satellite into orbit. This satellite must move with an angular velocity equal to the Earth and be in a prograde orbit, that is, revolving eastward in the same direction as the Earth rotates. Otherwise, the only requirements for a GIO constellation are the right period and direction of rotation.

2.3.3. Highly Elliptical Orbits (HEO)

Using inclined HEO configuration, both polar areas can be effectively covered with four satellites; two in each polar orbit. The elevation angle to the HEO satellites remains high for most of the 12-hour period of visibility, which is especially required for continuous Euro-Asian regional coverage providing land MSC service. At this point, blocking of the beam due to occlusion of the satellite by buildings, mountains, hills and trees is minimized. Besides, multiple trajectories caused by successive reflection of various obstacles are also reduced in comparison with systems operating with low elevation angles, like GEO.

The apogee altitude combines polar coverages with nearly synchronous advantages. Thus, minimum two special LES in both northern and southern polar regions are required to serve MES terminals. The LES tracking can be reached by a fairly directive fixed antenna while the satellite is in its slow apogee sector, the HEO space constellation is namely designed to cover the area under the apogee. Tracking of the satellite is facilitated on account of the small apparent movement and the long visibility duration. Otherwise, it is even possible to use antennas whose 3 dB bandwidth is a few tens of degrees, with fixed pointing towards the zenith, which permits the complexity and cost of the terminal to be reduced while retaining a high gain. A satellite in HEO constellation near the apogee can also use a high gain antenna to overcome the great distances in achieving the required EIRP values. The noise captured by the LES antenna, from the ground or due to interference from other radio systems and atmosphere, is also minimized due to the high elevation angles. At any rate, these advantages have led the former USSR to use these orbits for a long time in order to provide coverage of high latitude territories for mobile systems.

The HEO satellite two-way voice transmission has a similar delay as a GEO at the apogee of about 0.25 sec. Therefore, free space loss and propagation delay for HEO is comparable to that of the GEO constellation. Compared with GEO, the launch and satellite cost of the HEO constellation is reasonably low; this constellation is free of congestion because of only a few current and projected new HEO systems and provides high elevation angles for LES, which reduces atmospheric losses. Due to the relatively large movement of a satellite in HEO with respect to an observer on Earth, satellite systems using this model of orbit need to be able to cope with large Doppler shifts, 14 kHz for Molnya and 6 kHz for Tundra orbits in L-band 1.6 GHz. However, as the former USSR's experience has shown, satellites in this orbit tend to have rather a short lifetime due to the repetitive crossing of both Van Allen Belts. The rest of the disadvantages are the necessity of constant satellite tracking at the MES, compensation of signal loss variation, long eclipse periods and complex control system of MES and spacecraft.

The HEO satellite typically has a perigee at about 500 km above the Earth's surface and an apogee as high as 50,000 km. The orbits are inclined at 63.4° in order to provide services to locations at high northern latitudes. The particular inclination value is selected in order to avoid rotation of the apses, i.e., the intersection of a line from the Earth's centre to the apogee and the Earth's surface will always occur at latitude of 63.4°N. Orbit period varies from eight to 24 hours. Owing to the high eccentricity of the orbit, a satellite will spend about two thirds of the orbital period near apogee and during that time it appears to be almost stationary for an observer on Earth (this is referred to as apogee dwell). After this period, a switchover needs to occur to another satellite in the same orbit in order to avoid loss of communications. There have to be at least three HEO satellites in orbit, with traffic being handed over from one to the next every eight hours at a minimum.

When there is an orbit in HEO plane of non-zero inclination, the satellite passes over the region situated on each side of the equator and will possibly cover the polar regions if the inclination of the orbit is close to 90°. By orienting the apsidal line, namely the line between perigee and apogee, in the vicinity of the perpendicular to the line of nodes (when ω is close to 90° or 270°), the HEO satellite at the apogee systematically returns above the regions of a given hemisphere. In this way, it is possible to establish satellite links with LES or MES located at high latitudes. Although the satellite remains for several hours in the vicinity of the apogee, it does move with respect to the Earth and after a time dependent on the position of the MES, the satellite disappears over the horizon as seen from the mobiles. However, to establish permanent links it is necessary to provide several suitably phased satellites in similar orbits, which are spaced around the Earth (with different right ascensions of the ascending node and regularly distributed between 0 and 2π) in such a way that the satellite moving away from the apogee is replaced (handover) by another satellite in the same area of the sky as seen from the MES. However, the problems of satellite acquisition and tracking by the MES are simplified. Finally, there only remains the problem of handover and switching the links from one satellite to other, so the RF link frequencies of the various satellites can be different in order to avoid interference.

Examples of HEO systems are Molnya, Tundra, Loopus, Borealis of the Ellipso system and Archimedes. The ESA proposed Archimedes system employs a so-called "M-HEO" 8-hour orbit. This produces three apogees spaced at 120°. In effect, each apogee corresponds to a service area, which could cover a major population centre, for example the whole European continent, the Far East and North America.

2.3.3.1. Molnya Orbit

The first prototype HEO Molnya satellite was launched in 1964 and to date more than 150 have been deployed, primarily produced by the Applied Mechanics NPO in Krasnoyarsk, former USSR. The HEO Molnya satellites weigh approximately 1.6 metric tons at launch and stand 4.4 m tall, with a base diameter of 1.4 m. Electrical energy is provided by 6 windmill-type solar panels, producing up to 1 kW of power. A liquid propellant attitude control and orbital correction configuration maintains satellite stability and performs orbital maneuvers, although the latter usage is rarely needed. Sun and Earth sensors are used to determine proper spacecraft attitude and antenna pointing. The first Molniya 3 spacecraft appeared in 1974, primarily to support civil communications (domestic and international), with a slightly enhanced electrical power system and a communications payload of three 6/4 GHz transponders with power outputs of 40/80 W.

The second stratum of the Russian spacebased communications system consists of 16 HEO Molniya-class spacecraft in highly inclined 63° semi-synchronous orbit planes. With initial perigees between 450 and 600 km fixed deep in the Southern Hemisphere and apogees near 40,000 km in the Northern Hemisphere. In fact, Molniya satellites are synchronized with the Earth's rotation, making two complete revolutions each day with orbital period of 718 minutes. The laws of orbital mechanics dictate that the spacecraft orbital velocity is greatly reduced near apogee, allowing broad visibility of the Northern Hemisphere for periods up to eight hours at a time. Thus, by carefully spacing 3 or 4 Molniya spacecraft, continuous communications can be maintained. This type of orbit was pioneered by the USSR and is particularly suited to high latitude regions, which are difficult or impossible to service with GEO satellites.

The 16 operational Molniya satellites are divided into two types and four distinct groups. Namely, eight Molniya 1 satellites were divided into two constellations of four vehicles each. Both constellations consist of four orbital planes spaced 90° apart, but the ascending node of one constellation is shifted 90° degrees from the other, i.e., the Eastern Hemisphere ascending nodes are approximately 65° and 155°E, respectively. Although the system was designed to support the Russian Orbita TV network, a principal function was to service government and military communications traffic via a single 40 W 1.0/0.8 GHz satellite transponder.

The hypothetical Russian Molniya network can employ minimum 3 HEO satellites in three 12-hour orbits separated by 120° around the Earth, with apogee distance at 39,354 km and perigee at 1,000 km. This orbit takes the name from the communication system installed by the former USSR, whose territories are situated in the Northern Hemisphere at high latitudes. The orbital period (t) is equal to (t_E /2), or about 12 hours. The characteristics of an example Molnya orbit are given in **Table 2.4.**

Table 2.4. Molnya and Tundra Orbit Parameters

Characteristics	Molnya Orbit	Tundra Orbit
Orbital Period (t)	12 h	24 h
Sidereal Period	11 h 58 min 2 s (half day)	23 h 56 m 4 s (full day)
Semi-major Axis (a)	26,556 km	42,164 km
Inclination (i)	63.4°	63.4°
Eccentricity (e)	0.6 to 0.75	0.25 to 0.4
Perigee Altitude (h_p)	$a(1 - e) - R$	$a(1 - e) - R$
(e.g.: e=0.71)	1,250 km	25,231 km
Apogee Altitude (h_a)	$a(1 + e) - R$	$a(1 + e) - R$
(e.g.: e=0.71)	39,105 km	46,340 km

The only one-track cycles of a total of two satellite tracks on the surface of the Earth is shown in **Figure 2.9. (C-1)** for a perigee argument equal to 270°. The shape of this track is cycles of one orbit only near Greenwich Meridian, so the centre, of the next identical track is around 180° westward. Therefore, the satellite at apogee passes successively on each orbit above two points separated by 180° in longitude. The apogee is situated above regions of 63° latitude (the altitude of the vertex is equal to the value of the inclination and the apogee coincides with the vertex of the track when the argument of the perigee is equal to 270°). The large ellipticity of the orbit results in a transit time for the period of the orbit situated in the Northern Hemisphere greater than that in the Southern Hemisphere. The value of inclination, which makes the drift of the argument of the perigee equal to zero, is 63.45°. A value different from this leads to a drift, which is non-zero but remains small for value of inclination, which does not deviate too greatly from the nominal value. By way of example, for an inclination i = 65°, that is variation of 1.55°, the drift of argument of the perigee has a value of around 6.5° per annum.

It is evident that the Molnya HEO orbit has the advantage of high-elevation-angle coverage of the Northern Hemisphere, because of a need to completely cover a great part of the Russian territory. Three satellites in this orbit and phasing are chosen so that at least one satellite is available at any time over the horizon. Thus, with three satellites, each satellite is used (or handover is) 8 hours per day, while with four satellites handover is every 6 hours. The LES must use tracking antenna systems, so a terminal with only one antenna will have an outage during handover (switching) from one satellite to another.

The disadvantages of the Molnya orbit include the need for multiple satellites (which the system does not need), the poor, virtually useless coverage of the Southern Hemisphere and the need for tracking antennas at each LES. Since the distance from terminal to satellite is continually changing, the received power and frequency vary (Doppler effect). The former may require automatic uplink power control and scheduling is needed to allow LES to switch satellites simultaneously. As the satellite altitude varies, the beam also coverage changes, so the satellite carries a tracking antenna that must be kept continuously pointed at operating LES.

2.3.3.2. Tundra Orbit

The Russian Tundra HEO system employs 2 satellites in two 24-hour orbits separated by 180° around the Earth, with apogee distance at 53,622 km and perigee at 17,951 km, which provides visibility duration of more than 12 hours with high elevation angles. The Tundra orbit can be useful for regional coverage for both FSS and MSS applications. Similar to the Molniya orbit, this orbit is particularly useful for LMSS where the masking effects caused by surrounding obstacles and multiple path are pronounced at low elevation angles, (> 30°). The period (t) of the orbit is equal to t_E, which is around 24 hours. The characteristics of an example orbit of this type are given in **Table 2.4.** This orbit has only one track on the Earth's surface, as shown in **Figure 2.9. (C-2),** for a perigee argument equal to 270°, inclination i = 63.4° and eccentricity e = 0.35. The latter parameter can have three values of eccentricity e = 15, e = 25 and e = 45.

According to the value of orbital eccentricity, the loop above the Northern Hemisphere is accentuated to a greater or lesser extent. For eccentricity equal to zero, the track has a form of a figure 8, with loops of the same size and symmetrical with respect to the equator. When the eccentricity increases, the upper loop decreases, while the lower loop increases and the crossover point of the track is displaced towards the North. This loop disappears for a value of eccentricity of the order of e = 0.37 and the lower loop becomes its maximum size. The transit time of the loop represents a substantial part of the orbital period and varies with the eccentricity. The position of the loop can be displaced towards the East or West, with respect to the point of maximum latitude, by changing the value of argument of the perigee (ω) and the eccentricity.

2.3.3.3. Loopus Orbit

The proposed Loopus system, which employs 3 satellites in three 8-hour orbits separated by 120° around the Earth, has an apogee distance at 39,117 km and perigee at 1,238 km. This orbit has similar advantages and disadvantages as for the Molnya orbit. One of the problems encountered by the LES is that of repointing the antenna during the handover (changeover) from one satellite to another. With orbits whose track contains a loop, it is possible to use only the loop as the useful part of the track in the trajectory. Handover between two satellites is performed at the crossover point of the track. At this instant the two satellites are seen from the LES in exactly the same direction and it is not necessary to repoint the antenna. To achieve continuous coverage of the region situated under the loop, the transmit time of the loop must be a sub-multiple of the orbit period and the number of satellites. Hence, the coverage can be extended to one part of the hemisphere by increasing the number of satellites in orbit regularly spaced about the globe.

2.3.4. Polar Earth Orbits (PEO)

The PEO constellation is today a synonym for providing coverage of both polar regions for different types of meteorological observation and satellite determination services. Namely, a satellite in this orbit travels its course over the geographical North and South Poles and will effectively follow a line of longitude. Certainly, this orbit may be virtually circular or elliptical depending upon requirements of the program and is inclined at about $90°$ to the equatorial plane, covering both poles. The orbit is fixed in space while the Earth rotates underneath and consequently, the satellite, over a number of orbits determined by its specific orbit line, will pass over any given point on the Earth's surface. Therefore, a single satellite in a PEO provides in principle coverage to the entire globe, although there are long periods during which the satellite is out of view of a particular ground station. Accessibility can of course be improved by deploying more than one satellite in different orbital planes. If, for instance, two such satellite orbits are spaced at $90°$ to each other, the time between satellites passes over any given point will be halved.

The PEO system is rarely used for communication purposes because the satellite is in view of a specific point on the Earth's surface for only a short period of time. Any complex steerable antenna systems would also need to follow the satellite as it passes overhead. At any rate, this satellite orbit may well be acceptable for a processing store-and-forward type of communications system and for satellite determination and navigation.

There are four primary requirements for PEO systems as follows:

1. To provide total global satellite visibility for worldwide LEOSAR Cospas-Sarsat distress and safety satellite beacons EPIRB, PLB and ELT applications;

2. To provide global continuous coverage for current or newly developed and forthcoming satellite navigation systems;

3. To provide at L-band or any convenient spectrum the communication requirements of ships and aircraft in the polar regions not covered by the Inmarsat system; and

4. To provide global coverage for meteorological and synoptic observation stations.

The Inmarsat team has studied two broad ranges of orbit altitude of PEO for both distress and communication purposes, first, low altitudes up to 1,400 km and second, high altitudes above 11,000 km. In reality, these two orbit ranges are separated by the Inner Van Allen radiation belt. In the regions of the radiation belt the radiation level increases roughly exponentially with height at around 1,000 km, reaching a peak at about 5,000 km altitude. Therefore, a critical requirement to reduce high-energy proton damage to the solar cell arrays of the satellite system constrains the PEO to low and high altitudes. As is evident, another Outer Van Allen Belt has no negative influence on these two PEO constellations because it lies far a way between MEO and GEO satellite planes.

These two specific systems studied by Inmarsat are Cospas-Sarsat Low PEO at 1,000 km altitude and High PEO at 12,000 km altitude, similar to that studied by ERNO, named SERES (Search and Rescue Satellite) system. Thus, it is considered that these two systems demonstrate clearly the solutions tradeoff and constraints on a joint PEO distress, SAR and communication mission. Other possible orbits for polar coverage can be an inclined HEO Molnya constellation of four satellites; GIO $45°$ inclined orbit of four satellites and $55°$ inclined circular MEO at 20,000 km altitude for GPS and GLONASS satellite navigation systems. In the meantime, the Cospas-Sarsat system has developed a special GEOSAR system using three GEO satellites for global distress communications satellite beacons in combination with already developed LEOSAR systems using four PEO satellites.

Figure 2.11. Type of Satellite Orbits and Tracks

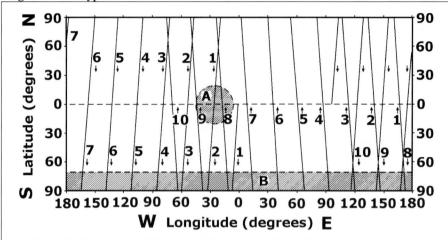

Courtesy of Book: "Commercial Satellite Communication" by S.C. Pascall and D.J. Withers

For both the Low and High PEO systems the number of operational satellites required to provide adequate Earth coverage needs to be minimized in order to achieve minimum system costs. An IMO and ICAO requirement for the GMDSS/Cospas-Sarsat mission is that there should be no time delay in distress alerting anywhere in the globe.

2.3.4.1. Low PEO

The Low PEO satellite constellation similar to the LEO mostly employs both polar and near-polar orbits for communications and navigation utilities. Thus, a particular example of a system that uses this type of orbit is the Cospas-Sarsat SAR system for maritime, land and aeronautical applications. This system uses 8 satellites in 4 near-polar orbits: four US-based Sarsat satellite constellations at 860 km orbits, inclined at 99°, which makes them sun-synchronous and four Cospas satellite configurations at 1,000 km orbits, inclined at 82°. However, this orbit was also suitable for the first satellite navigation systems Transit and Cicada, developed by the USA and the former USSR, respectively.

Otherwise, with a limited number of low altitude PEO satellites it is impossible to provide continuous coverage to polar region, because the view of individual spacecraft is relatively small and their transit time is short. However, because the time for a single orbit is low, less then two hours and a different section of the polar region is covered at each orbit due to Earth rotation, this drawback is somewhat offset. For a given number of satellites, preferably about eight, it is possible to optimize the constellation that maximizes total system coverage, to improve handover and minimize waiting time between transits.

In **Figure 2.11.** is illustrated the Earth track of ten successive orbits of satellite in Low PEO with an altitude of 1,000 km. The MES in shaded area A (4,200 km in diameter) would see the satellite, in the absence of environmental screening, at an angle of elevation not less than 10°, while the satellite was passing through the equatorial plane. The coverage area has the same size and shape wherever the satellite is in the orbit but its apparent size and shape would change with latitude, being distorted by the map projection used in the figure.

The South Pole coverage area at a single pass of the satellite is shown by shaded area B. The same figure shows that a single PEO satellite in a polar orbit will have a brief sighting of every part of the Earth's surface every day. There will be 2 or 3 of these glimpses per day near the equator, the number increasing as the poles approach. The period of visibility as seen from the MES range from about 10 min, the satellite passing overhead, down to a few seconds when the satellite appears briefly above the horizon. If the orbital plane of the satellite is given an angle of inclination differing from $90°$ of the PEO, a similar Earth track is obtained but the geographical distribution of the satellite visibility changes. One LEO satellite with an orbital inclination of $50°$ would have better visibility between $60°$ N and $60°$ S latitude than a PEO satellite but it would have no visibility at all of the polar regions.

The Low PEO configuration is attractive for mobile distress communications for two reasons. First, the transmission path loss is relatively low, allowing reliable communication with a low powered satellite beacon and PEO spacecraft. An altitude of about 1,000 km is the upper limit for good reception of signals at 243/406 MHz sent from emergency distress beacons. Secondly, the Doppler shift is high, approximately 30 kHz at 1.6 GHz, allowing accurate location of the distress transmitter. On the other hand, there are several significant disadvantages. However, as mentioned earlier, PEO coverage is not continuous unless there is simultaneous communication between a distress buoy and a ground terminal because of the small footprint of each individual satellite. Accordingly, storage and retransmission of distress messages on-board processing would be necessarily adding to the distress alert delay time and also to satellite mass and complexity.

The short visibility period during a transit and the uneconomic need for large numbers of satellites for continuous coverage makes a Low PEO unattractive for communications considerations. If this orbit configured well as an economic solution for distress coverage in polar regions to be used for communications purposes, users would have to operate with the following restrictions: **(1)** Only burst mode, non-simultaneous data communication would be possible; **(2)** Transmission time and/or bit rate would be limited by satellite message storage capability; **(3)** Replies to the message would require an interrogation or polling system from the MES expecting a reply; **(4)** Depending on the PEO constellation and MES position, a reply could take some hours.

However, many of these PEO communication limitations would be removed if a system of inter-satellite links, possibly in addition to inter-GEO infrastructure, were used to provide a near-continuous, simultaneous two-way communication system. The complexity and likely cost of such system would almost certainly not be justified by the expected low level of polar communication traffic. Thus, in considering the possible integration of PEO and GEO for communication purposes, it is necessary to determine the additional requirements and constraints arising from polar operation. In this context, for reliable communications the number of additional LES required for operation to PEO is a significant element of the overall system. For example, a constellation of eight Low PEO would require about six LES worldwide for polar coverage assuming message storing and forwarding techniques, where a High PEO would require a minimum of two LES located in North and South polar latitudes for continuous polar coverage with simultaneous two-way communications. In addition, it would be necessary to obtain reliable terrestrial links between the LES of each system, as well as inter-satellite links between the PEO and GEO satellites.

In any case, by using the store and transmit method, a Low PEO system could effectively be served for the relay of mobile distress, safety and urgency messages, for maritime, land and aeronautical applications via satellite beacons to receive-only terminals on shore.

2.3.4.2. High PEO

The High PEO constellation would consist of three satellites separated by 120° in the same circular orbit of 12,000 km altitude, geometrically similar to the GEO and as orbit similar to MEO configuration. This orbit provides continuous coverage to all polar regions above 59° latitude. Thus, six satellites (in two orbital planes of three satellites each) would provide continuous and real global coverage if that were required, which GEO cannot obtain.

By comparison with Low PEO systems transmission path losses are higher at an altitude of 12,000 km but not to the extent that a distress beacon need be especially high powered to transmit successfully to a high PEO satellite. Reception of the Cospas-Sarsat existing two very low-powered distress frequencies will be interfered, but not impossible. The Doppler shift is lower (about 10 kHz at 1.6 GHz), not allowing very accurate area location of the distress transmitters. Single high latitude LES in both Arctic and Antarctic polar regions allows reception with no delay of all distress messages transmitted from above 59° latitude. Furthermore, using these two LES positioned at high latitude with continuous visibility of at least one of the three satellites and collocated or linked with an Inmarsat LES, can offer a full range of near continuous communication services to the polar regions.

2.3.5. Hybrid Satellite Orbits (HSO)

The Hybrid satellite constellation can be configured by several types of combinations between existing orbital solutions today. Namely, any of these combinations can provide better global coverage for both hemispheres, including both polar regions. In this context will be introduced shortly five hybrid constellation systems, which are currently using or developing MSC and navigation systems as follows:

1) Combination of GEO and HEO Constellations – The development of a MSS which would provide reliable communications with MES, such as vessels, road vehicles, trains and aircraft, rural areas and remote terminals. This MSC system, called Marathon, includes five GEO Arcos-type satellites and four Mayak-type satellites in a HEO, as well as a ground segment that is composed of base stations and terminals installed at fixed or mobile users premises. Therefore, the combination of GEO and Non-GEO satellite constellations makes it possible to render GMSC services, including those at high latitudes and in the both polar areas, this is especially important for Russia, with its vast northern Eurasian territories and to provide the most reliable satellite communication between the territories of the western and Eastern Hemispheres. This hybrid constellation can be useful as well as for the Alaska, Greenland and northern territory of Canada.

2) Combination of GEO and PEO Constellations – This current combination of orbits has been developed by the efforts of the Cospas-Sarsat organization, with the assistance of IMO, Inmarsat and other international and regional contributors. At the other words, the Cospas-Sarsat space segment is a combination of three GEO operational satellites of the subsystem called GEOSAR and four PEO operational satellites of the subsystem called LEOSAR, with spare spacecraft for all participants. The GEOSAR employs one satellite type of INSAT-2A and two GOES type GOES-E and GOES-W, while the LEOSAR configuration has two satellites supplied by Cospas and two by Sarsat. Otherwise, the GEOSAR project in the future has to include the European MSC and two Russian Luch-M spacecraft. This system is responsible for providing distress alert and to help SAR forces on-scene determinations for maritime, land and aeronautical applications.

Figure 2.12.   Spacecraft Sub-System

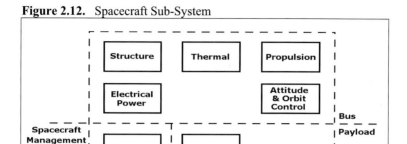

Courtesy of Book: "Satellite Communications Systems" by M. Richharia

3) Combination of GEO and LEO Constellations – Celestri is the Motorola trademark name for a proposed GEO and LEO satellite hybrid communication network. The network will combine 9 GEO and 63 LEO satellites in 7 planes with Earth-based control equipment and will provide interfaces to existing telecommunication infrastructures, the Internet and corporate and personal networks. The system will offer a 64 Kb/s voice circuit from anywhere in the world. The architecture is not limited to fixed sized channels but permits dynamic bandwidth assignment based on application demand. Business users will benefit by Celestri's ability to offer remote access to LAN infrastructures.

4) Combination of MEO and HEO Constellations – The new proposed MSS Ellipso is developing in combination with an initial complement of seven Concordia satellites deployed in a circular equatorial MEO at an altitude of 8,050 km and ten Borealis satellites in two HEO planes inclined at $116.6°$. They have apogees of 7,605 km and perigees of 633 km and a three-hour orbital period. This combination of two constellations would provide coverage of the entire Northern Hemisphere including North Pole areas and part of the Southern Hemisphere up to $50°$ latitude South. Thus, the HEO satellites can spend a greater proportion of their orbital periods over the northern latitudes and, together with the MEO constellation; the Ellipso hybrid system will provide voice, data and Fax communication and navigation RDSS services to areas with large landmasses, enormous populations with a large density of users and potentially widespread markets. This system is also planned to cooperate with the terrestrial PSTN and other services.

5) Combination of MEO and LEO Constellations – The Kompomash consortium for space systems in Russia is preparing the Gostelesat satellite system for MSS, using 24 satellites in MEO and 91 in LEO constellation. The launches of these two constellations are planned to commence in 2001 to 2004. However, this MSC project is provided for future global mobile communications with possibility to cover both pole regions.

2.4. Spacecraft Sub-Systems

A communications satellite essentially consists in two major functional units: payload and bus. The primary function of the payload is to provide communication between LES and MES, while the bus provides all the necessary electrical and mechanical support to the payload and all satellite missions illustrated in **Figure 2.12.**

Figure 2.13. Configuration of Spacecraft Transponders

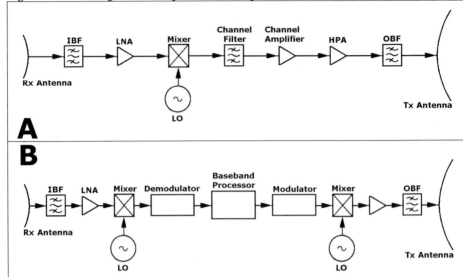

Courtesy of Handbook: "Mobile-Satellite Service" by ITU

2.4.1. Satellite Payload

The payload is made up of a repeater and antenna system. The repeater performs the required processing of the received signal and the antenna system is used to receive signals from LES and to transmit signals to MES in the coverage area and vice versa.

2.4.1.1. Satellite Repeaters

The function of a repeater is to receive the uplink RF signals from either ground segment service or feeder links, to convert these signals to the appropriate downlink frequency and power and to retransmit them towards the service or feeder links ground segment. Two types of repeater architecture are possible in on-board utilization of communication satellites: Transparent or Bent-pipe, and Regenerative transponders.

1. Transparent Transponder – The basic function of the satellite transponder is to isolate individual carriers or groups of carriers of signals and to boost their power level before they are retransmitted to the ground stations. The carrier frequencies are also altered as the carriers pass through the satellite. Satellite repeaters that process the carrier in this way are typically referred to as transparent or bent-pipe transponders, shown in **Figure 2.13. (A)**. Only the basic RF characteristics of the carrier (amplitude and frequency) are altered by the satellite. However, the detailed signal carrier format, such as the modulation characteristics and the spectral shape, remains completely unchanged. In such a way, transmission via a transparent satellite transponder is often likened to a bent-pipe because the satellite simply channels the information back to the ground stations. Therefore, a bent-pipe is a commonly used satellite link when the satellite transponder simply converts the uplink RF into a downlink RF, with its power amplification.

Initially, the received uplink signals from LES or MES by Rx antenna are filtered in an Input Bandpass Filter (IBF) prior to amplification in a Low Noise Amplifier (LNA). The output of the LNA is then fed into a Local Oscillator (LO), which performs the required frequency shift from uplink to downlink RF and the bandpass Channel Filter after the Mixer removes unwanted image frequencies resulting from the down conversion, prior to undergoing two amplification stages of signals in the Channel and High Power Amplifier (HPA). Finally, the output signal of the HPA is then filtered in the Output Bandpass Filter (OBF) prior to transmission through Tx antenna. The IFB is a bandpass filter which blocks out all other RF used in satellite communications. After that, the receiver converts the incoming signal to a lower frequency, using an LO which is controlled to provide a very stable frequency source. This is needed to reduce all noises to facilitate processing of the incoming signal and to enable the downlink frequencies to be established. The channel filter isolates the various communications channels contained in the waveband allowed through by the input filter. Filtering often leads to large power losses, creating a need for extra amplification, usually followed by a main amplifier. In order to attain the required gain of HPA this segment may employ either a Solid-State Power Amplifier (SSPA) or a Traveling Wave Tube Amplifier (TWTA). In a more complex design, in order to achieve higher RF power, it may be possible to combine the output of several amplifiers. To do this the incoming signal must be divided in such a way so as to provide separate identical input to each amplifier, see 6 TWTA presented in **Figure 2.14. (A).** A power combiner then recombines the RF signals from the amplifiers to produce a single RF output. The output filter removes all unwanted signals from the transmitted downlink returning to the Earth stations. High reliability throughout the lifetime of the satellite is achieved by duplicating critical units in the receiver, such as TWTA, etc.

2. Regenerative Transponder – Other satellite system designs go through a more complex process to manipulate the carrier's formats, by using on-board processing architecture. This payload architecture offers advantages over the transparent alternative, including improved transmission quality and the prospect of compact and inexpensive MES and handheld user terminals. A typical on-board processing system will implement some or all of the functions that are performed by the ground-based transmitter and/or receiver in a transparent satellite system. Therefore, these functions may include recovery of the original information on board the satellite and the processing of this information into a different carrier format for transmission to the ground stations. In fact, any satellite transponder that recreates the signals carrier in this way is usually referred to as a regenerative transponder, illustrated in **Figure 2.13. (B).** This type of satellite transponder provides demodulation and modulation capacity completely on board the satellite.

The received uplink signal goes along the down-converter segment prior to coming into the on-board demodulator, where it is demodulated and processed in the base band processor. This technology provides flexible functions, such as switching and routings. The downlink signal generated by an on-board modulator passes along the up-converter segment and is transmitted via the antenna. For this type of system link design can be separately conducted for the uplink and downlink because link degradation factors are decoupled between the uplink and downlink by the on-board demodulator and modulator, supported by the base band processor. A regenerative transponder with base band processing permits reformatting of data without limitation to MES Rx, while the bent-pipe system requires a satellite link design for the entire link, involving both uplink and downlink, but the forward link burst rate is limited by the MES G/T and demodulation performance.

Figure 2.14.   Spacecraft C/L and L/C-band Transponders

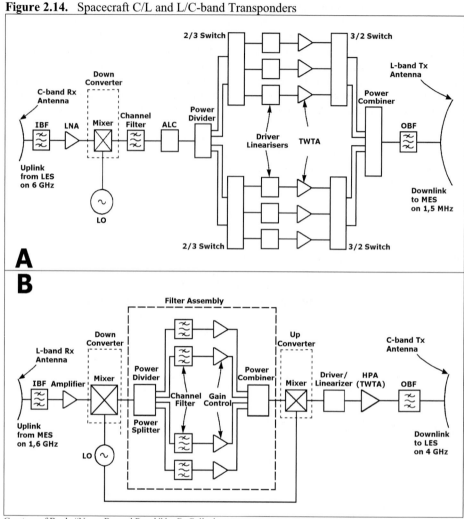

Courtesy of Book: "Never Beyond Reach" by B. Gallagher

Advanced MSS include intersatellite links to establish a direct connection between satellite transponders and in this way to enlarge system coverage and help the reception of signals from other GEO satellite coverage not visible for particular LES and MES terminals. Moreover, an intersatellite link can help to solve the problems associated with some TTN infrastructures or to reduce landline charges. In the same way, intersatellite links can also provide a connection from a satellite to neighboring satellites in a constellation of Non-GEO space segment and so, they are beneficial in reducing the number of satellite hops when Earth coverage of each satellite is limited. Intersatellite links are usually implemented with regenerative satellite transponder systems, mainly due to the flexible connection to intersatellite links.

2.4.1.2. Satellite Transponders for MSC

A transponder for MSC systems cannot be a Mobile Satellite Transponder, which has been determined by some authors; because a transponder cannot be de facto mobile but serves MSS or even FSS. It is, however, an electronic segment made up of repeaters (receiver and transmitter) on board the satellite. The Inmarsat-2 payload consists in two transponders: the C/L-band, shown in **Figure 2.14. (A)** and the L/C-band, shown in **Figure 2.14. (B).**

a) Inmarsat-2 C/L-band Transponder – This transponder receives uplink signals in the C-band of 6.4 GHz from LES and retransmits downlink signals in the L-band of 1.5 GHz to MES, after frequency conversion and signal amplification by a HPA. The signals received by a C-band antenna are fed via IBF and LNA to a down-converter section. A signal channel is followed by an Automatic Level Control (ALC) device, which limits the level of the signal to the amplifier. The HPA consists in six TWTA and their associated power supplies. In front of each TWTA is a driver/linearizer, predisposed to compensate the nonlinear RF properties of the TWTA. The signal driver supplies an equal drive signal to each of the four TWTA that are active at any given time and the other two can be activated for backup if the operating TWTA malfunctions. For this reason, the signal driver is preceded by an amplitude equalizer. However, the active TWTA are selected by 2/3 and 3/2 switches and their output powers are combined by a power combiner. The total power is fed to an L-band transmission antenna via OBF.

b) Inmarsat-2 L/C-band Transponder – This transponder receives uplink signals in the L-band of 1.6 GHz from MES and retransmits downlink signals in the C-band of 3.6 GHz to LES, after frequency conversion and signal amplification by the HPA. The signals received by an L-band antenna are fed to a down-converter via IBF and LNA. At the down -converter, signals are converted into 60 MHz IF by LO. A filter assembly then provides the required characteristics divided into four channels. Following up-conversion the signal passes to an ALC unit and the power for four channels is combined and signals are up-converted from 60 MHz to 3.8 GHz by activated TWTA. The amplified signal in HPA then goes through bandpass and harmonic filters in OBF before being distributed among the 7 cup-dipole elements of the C-band transmit antenna for radiation to the Earth's surface.

2.4.1.3. Satellite Antenna System for MSC

The antenna array system of Inmarsat-2 satellite for MSC is shown in **Figure 2.15 (A).** The satellite antenna system mounted on the spacecraft structure, similar to the transponders, is composed of two main integrated elements: the C/L-band and the L/C-band antenna.

a) Inmarsat-2 C/L-band Arrays – This uplink is actually the feeder link, which operates in the 6 GHz RF range. The signals sent by LES are detected by a C-band receiving array, comprising seven cup-dipole elements in the smallest circle. On the other hand, the L-band transmit antenna is the biggest segment of the whole system, consisting in 43 individual dipole elements, arranged in three rings around a single central element. Thus, this antenna is providing near-global coverage service downlink for MES in the 1.5 GHz RF spectrum.

b) Inmarsat-2 L/C-band Arrays – This arrays is actually the service uplink and operates in the 1.6 GHz RF range. The signals sent by MES in adjacent global coverage region are detected by L-band receiving array, comprising nine cup-dipole elements arranged in a circle. Finally, the C-band transmit antenna consists in seven cup-dipoles for radiation of the feeder downlink to LES in the 3.6 GHz RF spectrum.

Figure 2.15.   Spacecraft Antenna Systems

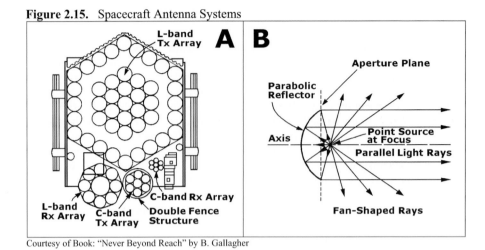

Courtesy of Book: "Never Beyond Reach" by B. Gallagher

2.4.1.4. Characteristics of Satellite Antennas

Both transmit antenna array systems are providing a global (wide) footprint on the Earth's surface. However, narrow circular beams from GEO or Non-GEO can be used to provide spot beam coverage. For instance, from GEO the Earth subtends an angle of 17.4°. Antenna beams 5.8° wide can reuse three frequency bands twice in providing Earth disc coverage. The directional properties of antenna arrays can be exploited to permit RF reuse in space communications, which is similar to several radio stations using the same RF being geographically far apart. Earth coverage by seven spot beams (six spots are set out around one spot in the centre) can be arranged by three pairs of beams: 1 and 4, 2 and 5 and 3 and 6, operating on frequencies f2, f3 and f4, respectively. Mutual interference within pairs is avoided by pointing one beam as far away from the other as possible. Coverage of the centre of the disc is provided by a single beam operating on frequency f1. The main advantage with this spot footprint that is specific Earth areas can be covered more accurately than with wide beams. Furthermore, a greater power density per unit area for a given input power can be achieved very well, when compared with that produced by a global circular beam, leading to the use of much smaller receiving MES antennas. The equation that determines received power (P_R) is proportional to the power transmitted (P_T) separated by a distance (R), with gain of transmit antenna (G_T) and effective area of receiving antenna (A_R) and inverse proportional with 4π and square of distance. The relations for P_R and G_T are presented as follows:

$$P_R = P_T\, G_T\, A_R/4\pi R^2 \tag{2.42.}$$

$$G_T = 4\pi\, A_T/\lambda^2$$

where G_T = effective area of transmit antenna and λ = wavelength. The product of P_T and G_T is gain, generally as an increase in signal power, known as an EIRP. Signal or carrier power received in a link is proportional to the gain of the transmit and receive antennas (G_R) presented as:

$$P_R = P_T G_T G_R \lambda^2/(4\pi R)^2 \quad \text{or} \quad P_R = P_T G_T G_R/L_P L_K \quad [W] \tag{2.43.}$$

The last relation can be derived with the density of noise power giving:

$$P_R/N = P_T G_T (G_R/T_R) (1/K L_P L_K) \tag{2.44.}$$

where L_P = coefficient of energy loss in free space, L_K = coefficient of EMW energy absorption in satellite channels, T_R = temperature noise of receiver, G_R/T_R is the figure of merit and K = Boltzmann's Constant (1.38×10^{-23} J/K or its alternatively value is -228.6 dBW/K/Hz).

At any rate, P_R has a minimum allowable value compared with system noise power (N), i.e., the carrier and noise (C/N) or signal and noise (S/N) ratio must exceed a certain value. This may be achieved by a trade-off between EIRP ($P_T G_T$) and received antenna gain (G_R). If the receive antenna on the satellite is very efficient, the demands on the LES/MES are minimized. Similarly, on the satellite-to-Earth link, the higher the gain of the satellite transmit antenna, the greater the EIRP for a given transmitter power. Satellites often have parabolic dish antennas, though there are also other types, such as phased arrays. The principal property of a parabolic reflector is its ability to turn light from a point source placed at its focus into a parallel beam, mostly as illustrated in **Figure 2.15 (B).** In practice the beam can never be truly parallel, because rays can also be fan-shaped, i.e., a car headlamp is a typical example. In a microwave antenna the light source is replaced by the antenna feed, which directs waves towards the reflector. The length of all paths from feed to aperture plane via the reflector is constant, irrespective of their angle of parabolic axis. The phase of the wave in the aperture plane is constant, resulting in maximum efficiency and gain. The gain of an aperture (G_a) and parabolic (G_p) type of antennas are:

$$G_a = \eta (4\pi A/\lambda^2) = 4\pi A_E/\lambda^2 \tag{2.45.}$$

$$G_p = \eta (\pi D)^2 /\lambda^2$$

where η = efficiency factor, A = projected aperture area of antenna, $A_E = \eta^A$ is the effective collecting area and D = parabolic antenna diameter. Thus, owing to correlation between frequency and wavelength, $f = c/\lambda$ is given the following relations:

$$G_p = \eta (\pi D f/c)^2 = 60,7 (D f)^2 \tag{2.46.}$$

where the second relation comes from considering that $\eta \approx 0.55$ of numerical value. If this value is presented in decibels the gain of antenna will be calculated as follows:

$$G_T = 10 \log G_p. \tag{2.47.}$$

For example, a parabolic antenna of 2 m in diameter has a gain of 36 dB for a frequency at 4 GHz and a gain of 38 dB for a frequency at 6 GHz. Parabolic antennas can have aperture planes that are circular, elliptical or rectangular in shape. Thus, antenna with circular shape and homogeneous illumination of aperture with a gain of -3 dB has about 47.5% of effective radiation, the rest of the power is lost. To find out the ideal characteristics it is necessary to determine the function diagram of radiation in the following way:

$$F (\delta_o) = s (\delta_o)/s (\delta_o=0) \tag{2.48.}$$

where parameter s (δ_o) = flow density of radiation in the hypothetical satellite angle (δ_o) and s $(\delta_o=0)$ = flow density in the middle of the coverage area. Looking the geometrical relations in **Figure 2.2. (A)** follows the relation:

$$F (\delta_o) = d_o/h = \cos \delta \sqrt{(k^2 - \sin^2 \delta_o)}/1 - k \tag{2.49.}$$

where, as mentioned, k = R/(R + h) = sin δ and if $\delta_o = \delta$, the relation is defined as:

$$F (\delta) = k \cos \delta \tag{2.50.}$$

For GEO satellite the value of ΔL is given as a function of angle δ, which is the distance from the centre of the coverage area, where the function diagram of the radiation is as follows:

$$F (\delta) = \Delta L = 20 \log = 20 \log R/(R + h) \cos \delta = 10 \log R/(1 + 2R/h) \quad [dB] \tag{2.51.}$$

Therefore, in the case of GEO satellites the losses of antenna propagation are greater around the periphery than in the centre of the coverage area for about 1.32 dB. The free-space propagation loss (L_P) and the input level of received signals (L_K) are given by the equations:

$$L_P = (4\pi d/ \lambda)^2 \tag{2.52.}$$

$$P_R/S = P_T G_T/4\pi d^2 L_K$$

The free-space propagation loss is caused by geometrical attenuation during propagation from the transmitter to the receiver.

2.4.2. Satellite Bus

The satellite bus is usually called a platform and consists in several sections, shown in **Figure 2.12.** The function of the satellite platform is to support the payload operation reliably throughout the mission of primary construction section, such as Structure Platform (SP), Electric Power (EP), Thermal Control (TC), Attitude and Orbit Control (AOC), Telemetry, Tracking and Command (TT&C) and Propulsion Engine.

2.4.2.1. Structure Platform (SP)

The structure has to house and keep together all components of bus and communications modules, enable protection from the environment and facilitate connection of the satellite to the launcher. It comprises a skeleton on which the equipment modules are mounted and a panel, which covers and provides protection for sensitive parts during the operational phase from micrometers and helps to shield the equipment from extremes of heat, coldness, vacuum and weightlessness, including the relatively small dynamic forces produced by the station-keeping, attitude control engines and inertial momentum devices.

The spacecraft is protected during the launch phase with an enclosure, or nose cone. At the end, the nose cone is jettisoned, at which time the spacecraft must survive the inertial and thermal stress of an additional propulsion stage until it is inserted into orbit. In this sense, a spacecraft is virtually free of gravitational stress when in orbit, which allows the use of very large deployable arrays, which would collapse under their own weight on the Earth's surface without problems. Thus, large stresses are developed during launch as a result of massive acceleration and intense vibration, so the SP body must be sufficiently strong to withstand all external forces. On the other hand, all large structures such as antenna and solar arrays have to be folded and protected during a launch sequence and must have a deployable mechanism. The deployment of structures requires a special technique in the vacuum of space because of the lack of a damping medium, such as air.

Most satellites are either cylindrically-shaped and are stabilized by spinning the whole of main body, or box shaped and three-axis body stabilized. Spin stabilized structures have a cylindrical part, which rotates at a speed of 50–100 rpm and the despun stabilized part has mounted antennas always facing to the Earth. The spinning part of the cylinder is covered with solar cells and its spin axis is oriented perpendicularly to the Sun. Body stabilized structures rotate once for every rotation of the Earth, so that the side with mounted antenna will always face the Earth. This platform utilizes a deployed set of solar panels with solar cells mounted on one side of the panel surface relative to the Sun.

Materials in space are not subject to gravitational stress or atmospheric corrosion and the effects of the space environment are not all benign by any means. The high vacuum causes some materials to sublime or evaporate and some to weld together on contact. The latter behavior means that special attention has to be given to the materials used for bearings. The basic materials for the main frame are aluminum or magnesium alloys and special plastic or fibre materials and for other components carbon fibre, epoxy resins and carbon nanotube filaments are used.

2.4.2.2. Electric Power (EP)

The primary source of power for a communications satellite is the Sun. Hence, solar cells are used to convert energy received from the Sun into an electrical source. The principal components of the power supply system include: (1) The power electric generator, usually solar cell arrays located on the spinning body of a spin-stabilized satellite or on the paddles for a three-axes stabilized satellite; (2) Reliable electrical storage devices, such as batteries, for operating during periods of solar eclipses; (3) The electrical harness for conducting electricity to all of the devices demanding power; (4) The special converters and regulators delivering regulated voltage and currents to the devices on board the spacecraft and (5) The electrical control and protection section is associated with the remote monitoring TT&C satellite system.

The solar arrays are the motor during entire life of the satellite, providing sufficient power to all active components. Each cell delivers about 150 mA at a few hundred millivolts and an array of cells must be connected in series or parallel together to give the required voltage and current for operating the equipment until the end of its life and to recharge the batteries when the satellite moves out of an eclipse. Charge is applied via the main electric power bus or a small section of the solar cell. In the course of exploitation, batteries are sometimes reconditioned by intentionally discharging them to a low charge level and recharging again, which usually prolongs their life.

The operational status of batteries including recharge, in-service or reconditioning is remotely controlled by a special ground segment. The mass of a battery constitutes a significant portion of the total satellite mass. Therefore, a useful figure of merit to evaluate the performance of a battery is capacity in W/h per unit weight taken at the end of its life. Until recently virtually all satellites used Ni-Cd (Nickel-Cadmium) batteries because of their high reliability and long lifetime. These batteries provide a low specific energy of about 30 to 40 Wh/kg. The latest type of Ni-H (Nickel-Hydrogen) batteries can store at least 50% more energy per kilogram.

When a satellite passes through the Earth's shadow, the solar arrays stop producing power and the satellite structures use the energy from batteries. The GEO satellite undergoes around 84 eclipses in a year, with a maximum duration of 70 min. Thus, the eclipse occurs twice a year for 42 consecutive days each time. The percentage of eclipses' duration for GEO and HEO is much less than for lower satellite orbits. The LEO satellites can undergo several thousand eclipses in a year. For example, a LEO satellite in equatorial orbit at an altitude of 780 km can remain in the Earth's shadow for 35% of the orbital period. For a MEO under similar conditions, the maximum eclipse duration would be about 12.5% of the orbital period and a total duration of about 3 hours a day, with about 4 eclipses per day. Otherwise, the Sun can also be sometimes eclipsed by the Moon's shadow, which is less predictable.

2.4.2.3. Thermal Control (TC)

Thermal control of a communication satellite is very important factor during entire satellite lifetime, which is necessary to achieve normal temperature balance and proper performance of all subsystem. Thermal stress results from high temperature effects from the Sun and from low temperatures occurring during eclipse period. The obvious objective of the TC is to assure that the spacecraft structure and all equipment is maintained within temperatures that will provide successful operations. A satellite undergoes different thermal and other conditions during the launch and operational phase. The vacuum in space limits all heat transfer mechanisms to and from a spacecraft and its external environment to that of radiation. However, some main parts are usually in direct sunlight with a flux density of over 1 kW/m^2, while other parts are facing the shadow side at a temperature of about $-270°C$. In addition, an eclipse causes temperature variation from around $-180°$ to $+60°C$, when the ambient temperature falls well below $0°C$ and rises rapidly from the moment the satellite emerges from the eclipse. All these extremes have to be eliminated or moderated for normal satellite operations, especially because all electronic devices need optimum temperatures between $-5°$ and $+45°C$.

These problems can be solved by remote TC techniques, using both passive and active means of controlling and regulating the temperature inside spacecraft. The passive means are simple and reliable, using surface finishes, filters and insulation blankets. The active means are necessary to supplement the passive systems, which include louvers and blinds operated by bimetallic strips, heat pipes, thermal louvers and different electrical heaters. Heat pipes are used to transfer heat from internal hot spots or devices to remote radiator surfaces or must be transported to the outside surface where it can be dissipated. On the other hand, special electric heaters are used to maintain minimum component or structure temperatures during cold conditions. Accordingly, the TC subsystem ensures temperature regulation for optimum efficiency and satellite performance.

2.4.2.4. Attitude and Orbit Control (AOC)

The attitude and orbital control subsystem checks that a spacecraft is placed in its precise orbital position, and maintains, thereafter, the required attitude throughout its mission. Control is achieved by employing momentum wheels, which produce gyroscopic torques, combined with an auxiliary reaction control gas thruster system. Many various sensors are employed to detect attitude errors, including Sun's initial orientation purposes. The AOC system performs satellite orientation and accurate orbital positioning throughout its lifetime, because loss of attitude renders a spacecraft useless. There are in use two common AOS, such as attitude control and orbit or station keeping control systems. The objective of attitude control is to keep the antenna RF beam pointing at the intended areas on the Earth, which procedure involves as follows: **(1)** Measuring the attitude of the satellite by sensors; **(2)** Comparing the results of measurements with the required values; **(3)** Calculating the corrections to reduce eventual errors and **(4)** Introducing these corrections by operating the appropriate torque units.

a) Attitude Control – Currently, all types of attitude stabilization systems have relied on the conservation of angular momentum in a spinning element, which can be classified into the two categories already mentioned, such as spin-stabilized and three-axis stabilization. The satellite is rapidly spun around one of its principal axes of inertia. Thus, in the absence of any perturbing torque, the satellite attains an angular momentum in a fixed direction in an absolute frame of reference. For the GEO satellite, the spin (pitch direction) axis must be parallel to the axis of the Earth's rotation. The perturbation torques reduces the spin of the satellite and they affect the orientation of the spin axis. The second system of attitude control is a body-stabilized design in a three-axis stabilized satellite, whose body remains fixed in space. This solution is the simplest method of attitude control using a momentum wheel, which simultaneously acts as a gyroscope, in a combination of spin and drive stabilization. Certain perturbing torques can be resisted by changing its spin speed and the resulting angular momentum of the satellite.

b) Orbit or Station-Keeping Control – On-board propulsion requirements for both GEO and Non-GEO are important to keep a satellite in the correct orbital attitude and position. For this reason several types of propulsion systems are used, such as arc jet thrusters, ion and solar electrical propulsion, pulsed plasma thrusters, iridium-coated rhenium chambers for chemical propellants, etc. In order that the appropriate station-keeping corrections can be applied, it is essential that the orbit and position of a satellite are accurately determined. This may be done by making measurements of the angular direction and distance of the satellite from the Earth station, or a number of LES. When the orbit and position of the satellite have been determined, it is possible to calculate the velocity increments required to keep the N–S and E–W excursion of the satellite within the tolerated limits. The frequency with which N–S correction must be made depends on the maximum allowable value of the orbital inclination but the total increment required each year to cancel out the attraction of the Sun and Moon is 40 to 50 m/s. Otherwise, E–W station-keeping is usually achieved by allowing the satellite to drift towards the nearest point of equilibrium until it reaches the maximum tolerable error in longitude, then the process is repeated on the other side of the nominal longitude and finally, the satellite drifts back once more towards the point of equilibrium and the process is repeated. The frequency and magnitude of the velocity increments required depend on the angular distance between the satellite and the points of equilibrium and on the tolerable error, which is a maximum of about 2 m/s.

2.4.2.5. Telemetry, Tracking and Command (TT&C)

The telemetry, tracking, command and communication equipment enables data to be send continuously to the Earth stations, received from these stations and allows ground control stations to track the spacecraft and to monitor the health of the spacecraft and also to send commands to carry out various tasks like switching the transponders in and out of service, switching between redundant units, etc.

The TT&C system supports the function of spacecraft management for successful operation of payload and bus sub-systems. The main functions of a TT&C are as follows:

1) Telemetry Sub-System – The function of telemetry is to monitor various spacecraft parameters and performances such as voltage, current, temperature, output from attitude sensors, reaction wheel speed, pressure of propulsion tanks and equipment status and to transmit the monitored data to the SCC on Earth. The telemetered data are analyzed at the SCC and used for routine operational and failure diagnosis purposes, to provide data about the amount of fuel remaining, to support determination of orbital parameters, etc.

2) Tracking Sub-System – The function of tracking is to provide necessary sources to Earth stations for the tracking and determination of orbital parameters. In such a way, to maintain a satellite in its assigned orbital slot and provide look angle information to LES in the network, it is necessary to estimate the orbital parameters regularly. These parameters can be obtained by tracking the communications satellite from the ground and measuring its angular position and range. Most SCC employ angular and range or range-rate tracking to control satellite orbits.

3) Command Sub-System – This sub-system receives commands transmitted from the ground SCC, verifies reception and executes commands to perform various functions of the satellite during its operational mission, such as: Satellite transponder and beacon switching, Antenna pointing control, Switch matrix reconfiguration, Controlling direction and speed of solar arrays drive, Battery reconditioning, Thruster firing and Switching heaters of the various systems.

2.4.2.6. Propulsion Engine (PE)

The functions of the propulsion motors are to generate the thrust required for the attitude and orbital control of errors caused by solar and lunar gravity and other influences, or possibly the adequate assistance of the satellite into its final orbit. Hence, these errors are normally corrected at set intervals in response to commands from SCC. The necessary impulse is provided by thrusters, which operate by ejecting hot or cold gas under pressure. The thrust requirements for orbital control are provided by mono or bi-propellant fuels. The attitude control thrusters are positioned away from the centre of the mass to achieve the maximum thrust, the thrust being applied perpendicular to the direction of a spacecraft's centre of mass. The orbit control thrusters are mounted so that the thrust vector passes through the centre of mass. The relocation of a satellite from transfer orbit into GEO may be performed by apogee boost motor. In some satellites this is achieved by a solid or liquid fuel engine. Moreover, the choice between these two motors has a significant effect on the internal arrangements of the satellite.

3

TRANSMISSION TECHNIQUES

Early satellite communication systems, primarily fixed and later mobile networks used the analog transmission technique. Although modern MSC networks are still using the analog technique, the rapid development of high-speed digital mobile equipment is fostering a trend toward completely digital MSC. This chapter includes techniques and technology that enable signals to be sent from one user to another and vice versa via satellites and LES.

There are several methods of modulation, multiplexing and multiple access techniques used in MSC and their reverse processes, with some overlapping. Modulation is the process by which the baseband signal in EM form can be impressed upon a carrier, so Phase (PM) and Frequency (FM) modulations are used heavily in satellite communications because of their positive ability to deal with nonlinear distortion, noise and interference. The amplitude of the PM and FM carrier is held constant, so there is no apparent change in the power level. Most nonlinear distortion is the result of amplitude variations on the carrier, while PM and FM is able to perform better in this environment than Amplitude Modulation (AM), so because of that problem there is a major limitation in directly using AM for MSC.

Multiplexing and multiple access require sharing the resources of the satellite. Facilities are shared based on spectrum assignment (frequencies), by time sharing (time domain) and by spatial separation (antenna beam and polarization). Theoretically, any method can be used for the transmission of analog and digital signals. In practice, frequency is easier used with analog signals, whereas time division is easier to use with digital signals. Multiplexing consists in combining the signals from several users into a single signal, which then forms the signal used for modulation of the carrier. After demodulation, the individual signals are separated by an inverse operation called demultiplexing. All types of coding, decoding and error corrections are presented. Finally, up-to-date technologies of Internet and Broadband with conclusive explanations of the major protocols used in MSS are discussed.

3.1. Baseband Signals

The information transmitted over an RF communications link consist in signals conveyed from one user to another via telephone, video telephone, via facsimile, telex, data, packets, radio and PC terminals and via television cameras and receivers. Such signals are called baseband signals, which consist in basic electrical impulses carried by the radio path of a satellite or via other communication networks. If the baseband signal is analog, the voltage which represents it can take any value within a given range but if the signal is digital, the voltage takes discrete values within a given range.

The type of baseband signals are determined by the communication requirements of the final users and the nature of the network. The main types of baseband used in MSS are voice (Tel), Fax and Tlx signals on Tel channels, then data, video and image signals. The baseband signals of mobile broadcast satellite services consist in direct TV (television) and sound transmissions, while in a broadband wireless or satellite service baseband the signal is IT Internet and multimedia, such as data, video retrieval and image transfer.

Each of these radio signals has to be arranged in a form suitable for transmission over some physical layer (air or wire) and such a technique is called baseband signal processing. The modified baseband signal is then superimposed onto a higher frequency carrier wave, when the signal modulates the carrier to a value suitable for propagation over the many different transmission links, such as radio, satellite, etc. The analog process at the transmit end of the link is called modulation and the circuit that performs modulation is the modulator, while at the receive end of the link, it is called demodulation and the circuit that recovers the baseband information is the demodulator. Digital systems use a modulator for transmission and demodulator for reception of the RF carriers within one unit, known as a modem.

The process of modulation/demodulation applies to radio, satellite and terrestrial links using either analog or digital transmission methods. Where GMSC are concerned there could also be a terrestrial RF link involved between service subscribers' terminals on shore and LES for maritime, land and aeronautical service. The RF terrestrial link may be radio and relay microwave link or landline (wire/optical fibre) or a combination of all three mediums. In both government and E-commerce, the need for communications to be secure is of vital importance. Thus, by attaching an encryption device to the transmitter, MSC customers can be confident that their baseband voice, data or video signals are authentic, not tampered with and secure from eavesdroppers almost anywhere in the world.

3.1.1. Voice Signals

The first commercial voice service via satellites was established by Intelsat at a time when underwater cables could not keep pace with the rapid growth of telephone traffic. In spite of data traffic growth in recent years, voice traffic has become dominant and will remain in the future integrated with videophones. In general, signals from most telephone sets are analog but these signals can be transformed into digital mode and multiplexed at a local telephone exchange for transmission on trunk routes.

Based on extensive studies of voice signals, it is now well known that most of the speech energy during normal conversation lies between 0 and 3400 Hz and the range of voice frequency that can be registered by the human ear is up to about 20 kHz. Therefore, a bandwidth of 3 to 4 kHz is allocated to each voice channel. The CCITT recommends the range of telephony signals as 300–3400 Hz. The maximum energy of a signal representing speech is in the range of 800 Hz and about 99% of the energy is situated below 3000 Hz. The signal power of an average talker implied to a zero relative level point is given by:

$$P_m = P_a + 0.115\sigma^2 + 10\log\tau \quad [dBm_o] \tag{3.1.}$$

where $P_a = -12,9$ dBm$_o$ represents the average power of the speech signal; $\sigma = 5.8$ dB is the standard deviation of the normal distribution of the active speech power; $\tau = 0.25$ is the active factor of a talker. Thus, in total the value of $P_m = -15$ dBm$_o$ (Article G.223 of CCITT Recommendation). The quality of a received analog voice signal has been specified by the CCITT to give a worst-case baseband signal-to-noise ratio for voice (Tel) signals, for transmission over a long distance, as 50 dB and with the maximum allowable noise in the baseband of 10,000 picowatts. Speech is characterized by having a large, dynamic range of up to 50 dB to accommodate the volume difference between a whisper and a shout.

Speakers also tend to pause often and on average, a speaker will talk for only 40% of the period available; the remainder of the time the link is idle.

In recent years, considerable effort has been spent in digitizing radio voice and telephone systems. Some of these techniques are applicable to transfer any type of analog into digital signal and vice versa, using various analog-to-digital conversion techniques. The main need has arisen from the advent of MSS and cellular systems, which both operate on a RF bands that have great demand, hence considerable effort is being spent on improving the spectral efficiency of transmission as the spectrum becomes scarcer. The use of spectrally efficient voice coding coupled with an effective combination of modulation and multiple access techniques, provides a viable transmission solution, so Inmarsat-M mobile standard enables voice communications using voice codes operating at about 6 Kb/sec. The quality of the reconstituted speech for digital transmission at the receive end, among other factors, will depend on the number of bits transmitted per second and the number of bits received in error. This is bit error rate (BER), which value is considered to be about 10^{-4} (1 bit error per 10,000 bits) to provide good speech quality. The BER can be used as a design threshold, although some systems commonly have values superior to 10^{-5} or even more.

Today, many modern transmission techniques are in use for the transfer of analog and digital interactive voice signals between fixed and mobile transmitters and receivers and for direct transmitting of voice/Fax signals to the FES and MES. Despite rapid technological advances and the diverse range of data capabilities available, voice is very important for quick, instant and easy communication, meeting the needs of different markets worldwide. Therefore, Inmarsat users can rely on clear, digital voice radiocommunications up to full broadcast quality. Voice GMSC terminals on board ships, in vehicles or inside aircraft cockpits enable direct dealing with telephone subscribers on land via satellites, LES and TTN. The voice service of Globalstar GMPSC infrastructure is based on a Code Exciter Linear Prediction (CELP) variable rate vocoder. The voice quality has to meet the value provided by IS-96, which is the terrestrial CDMA standard. The superior voice quality can be offered at low data rates in large part due to the adaptable vocoder. In marginal areas, where the user terminal cannot generate sufficient power to close the link, the peak data rate is reduced to 4.8 or 2.4 Kb/s. Hence, this will provide intelligible voice service in areas that otherwise could not be served. The vocoder will incorporate echo cancellation, which can be disabled if this function is provided by the satellite network. On the other hand, the forthcoming new ICO Global Communications has a technical project that can also enable voice roaming inside buildings or tunnels, which will eliminate the blocking effect of objects, which is the current situation in personal satellite communications, which is a very serious problem of GSM or other cellular systems.

The radiobroadcasting program occupies a spectrum from 40 Hz to 15 kHz, enabling a high quality of voice signal. The test signal is a pure sinusoid at a frequency of 1 kHz. When the broadcast radio program is transmitted by communication satellite in digital form the performance objective is stipulated in terms of error probability. These errors have the effect of generating audible "clicks". To limit their frequency to about 1 per hour, a binary error rate in the order of 10^{-9} is required by CCIR Report 648. The present broadcasting satellite system operates at 12 GHz and is designed for community fixed stations with large antennas, while Direct Broadcasting Satellite (DBS) is serving the individual reception of signals by fixed or semi-fixed terminals with small antennas called VSAT. The present system is broadcasting both radio and TV programs only for fixed terminals. Although some developed countries have started to use Direct Audio Broadcasting (DAB) satellite systems in the L, S and Ka-band for MSS. At any rate, these models of broadcasting systems are suitable for mobile utilization on ships, vehicles and aircraft.

3.1.2. Data Signals

Data transmission is composed using digital signals consisting in a series of bits. Actually, bits are bi-state pulses with low state called "0" (–V) and high state "1" (+V). Information is superimposed on a digital stream by arranging groups of bits called words, which can be used for the transmission of analog signals from a telephone or alphanumeric characters from a PC keypad. Therefore, facsimile, telex, data and PC E-mail networking terminals are used for the transfer of data through the medium of different transmission techniques and applications that are expanding rapidly, including GMSC systems.

Data signals used by Inmarsat MSS can be broadly classified into three ranges:

1) Narrowband data of about 300 b/s can be transmitted via MES TDMA 24 Kb/s Tlx channel with possibly 16 data bursts directed to LES and PSTN via a modem.

2) Full duplex voice and 9.6 Kb/s data signals can be transmitted in the SCPC data channel with a rate of 24 Kb/s in O-QPSK modulation. In fact, the data channel enables packet data transfer using the CCITT X.25 recommendation for interface Data Terminal Equipment (DTE) and Data-Circuit Terminating Equipment (DCTE) operating in packet mode to the PSDN by dedicated circuits. This channel also supports CCITT Group-3 Fax, which is also available in the SCPC voice channel; using a 2.4 b/s data rate and APC voice codes.

3) Wideband one-way High Speed Data (HSD) transmission can be supported with a rate of 56 or 64 Kb/s via voice channels on a dedicated frequency with a special type of V.32 (9600 b/se) modem. This service can be used for the transfer of PC data, high quality digital audio and compressed video. The Duplex HSD will offer two-way 64 Kb/s in both, namely, in ship-to-shore and shore-to-ship directions. The two-way HSD can be also used for digital multiplexing data channels and video conferencing between ships and shore.

There are several known techniques in transmitting digital data, facsimile, E-mail and PC networking data applications, whose utility is also expanding rapidly for fixed and mobile systems. The new Global Area Network (GAN) portfolio of the HSD service developed by Inmarsat offers a cost-effective extension to corporate LAN and WAN even in the world's most rugged regions. Moreover, mobile packet data allows users to pay according to the amount of data they send rather than how long they spend online, suited to Internet web-based applications such as Intranet access and E-commerce via satellite.

Meanwhile, the quality of data service provided by GMPSC systems enables transmission speeds up to 7.2 Kb/s. The Bit Error Rate contributed for instance by the Globalstar system is less than 1×10^{-6}. Hence, a higher terminal rate of about 9.6 Kb/s can be processed if the equipment incorporates an elastic buffer to accommodate the required flow control.

3.1.3. Video Signals

Prior to the advent of satellite communications it was very impracticable to relay video TV programs across oceans because of the limited bandwidth of submarine coaxial cables. Today optical fibre cables have enough channels but cannot, like satellite systems, transmit DBS TV video programs to many widely separated fixed or VSAT stations and especially to mobile terminals. Direct broadcast by satellite has been in use about 15 years, because the first such satellite for Western Europe was launched in November 1987. Moreover, the demand for this type of satellite service continues to gain interest for the distribution of TV programs direct to homes (households) or other fixed stations and later, to provide a service for mobile stations.

There are various inexpensive ways in which it is possible to add data transmissions to the DBS and to provide cheap information services and E-mail for both domestic and business purposes, to onshore or offshore TV subscribers.

More then a decade ago, Inmarsat started to provide still pictures mode from the oil fields to any office using a wide range of analog and digital portable devices for transmitting, receiving and processing high resolution color photographs in less then five minutes. In addition, Inmarsat developed the HSD Store-and-Forward video system. Namely, with the aid of advanced video codecs it is possible to digitize and compress video material and then transmit it at 56/64 Kb/s to a receiver, which decompresses and buffers the data in almost full motion video. The store-and-forward technique also ensures that the received material is error-free, since the data transmission is achieved by using error detection and repeat transmission mode. There is a new possibility today for Inmarsat MES to also use ISDN interfaces for the transfer of intensive data interactive applications such as Video Conferencing (VC) and Digital Image Transfer (DIT). Therefore, utility of standard ISDN interfaces will enable one to easily connect mobile fleet offices with corporate applications. Video baseband allows Inmarsat users to relay video from almost anywhere on the globe. This provides an effective flexible solution for media companies and the growing number of corporation who also use the VC system to facilitate "real time" discussion on a regular basis. Users benefit from ISDN speeds and easy-to-use portable equipment, which delivers immediate live action, training or medical and equipment diagnostic support, as required.

3.2. Analog Transmission

Analog transmission is characterized by processing performed on the baseband signal before and after modulation in order to improve the quality of the link. The carrier can support only one or few channels for the transmission of baseband signals. In the case of a carrier transmitted from the station representing only a single user channel, this is Single Channel Per Carrier (SCPC) transmission. On the other hand, if the carrier represents a number of multiplexed users, it is designated Multiple Channel Per Carrier (MCPC) and use several channels is Frequency Division Multiplexing (FDM) transfer.

A signal needs to be impressed on an RF carrier for transmission through the satellite and for this purpose uses a process known as modulation. The objective of any communications is to transmit the modulated carrier to Rx as reliably as possible, so that the demodulated signals can be satisfactorily recovered. In analog transmission systems, the information waveform in the form of voice, data or video signal are modulated directly from the source onto the carrier at the modulator of Tx, by using methods of Amplitude (AM), Frequency (FM) and Phase Modulation (PM). Frequency and phase modulation are used most widely for direct analog transmission in satellite communications, while amplitude modulation is a process used indirectly in the satellite link.

Modulation may also be used at very low frequencies, like the more common and various forms of Phase Shift Keying (PSK). In addition, modulation is often done on a carrier with an RF of about 70 MHz lower than the transmission RF. This RF is then up-converted to the transponder frequency on 6/4 GHz for amplification and retransmission. Previous types of satellite do not change the received modulation before retransmission. Satellites are now being designed to allow only one modulation method to be employed in the uplink and another for the downlink; each link can be optimized. The space link between two LES can generally be accomplished by the combination of modulation and multiplexing techniques.

3.2.1. Baseband Processing

The purpose of baseband processing is to improve the quality of the space (satellite) link using different methods whose cost is less than that arising from modification of one of the parameters involved in the link budget. The principal methods of baseband processing for telephone transmission are speech activation, pre- and de-emphasis and companding and for TV transmission pre- and de-emphasis only.

3.2.1.1. Speech Activation

The principle of speech activation is to establish the space link only when the subscriber is actually speaking. As the activity factor is $\tau = 0.25$, its application to a multicarrier SCPC system should permit a reduction of the power required on the satellite by about 6 dB.
In practice the reduction is only in the order of 4 dB, allowing guard times for activation and deactivation of the carrier and the sensitivity of S/N spikes. Therefore, the activation threshold parameter has to be from –30 to –40 dBm$_0$, the carrier activation time can be 6 to 10 ms and deactivation time can be 150 to 200 ms.

3.2.1.2. Pre- and De-Emphasis

The noise at the output of the demodulator of a FM transmission has a parabolic spectral density, when the high frequency components of the signal are more affected by noise than the low frequencies. **Figure 3.1.** shows the principle of the pre- and de-emphasis effects. The pre-emphasis filter prior to modulation of the signal increases the amplitude of the high frequency components. The gain of the cross-over frequency in the pre-emphasis filter is 0 dB. After demodulation the de-emphasis filter is performing inverse effects, namely, reducing the amplitude of the high frequency signal and noise power. In such a way, the signal is restored without spectral distortion and the S/N is improved. The improvement can be 4 to 5 dB for telephony and around 13 dB for TV signals.

Figure 3.1.   The principle of Pre- and De-emphasis

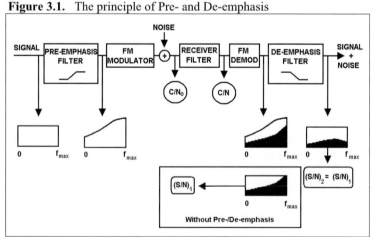

Courtesy of Book: Satellite Communications Systems by G. Maral & other

Figure 3.2.   Characteristics of typical Compandor and Effect of Companding

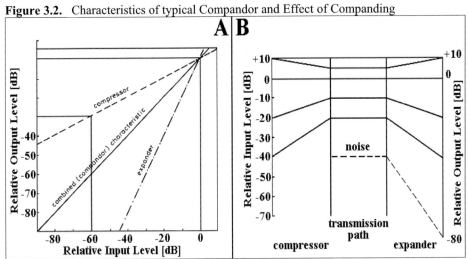

Courtesy of Book: "An introduction to Satellite Communications" by D.I. Dalgleish

3.2.1.3.   Companding Process

Companding is a process of compression and expansion for reducing the effect of noise on speech channels in accordance with the specification of CCITT Recommendations G.162 and article G.166. In effect, a compandor comprises a compressor and an expander and an improvement in the S/N ratio at the output of the demodulator is obtained by reducing the dynamic range of the signal before modulation (compression) and performing the inverse operation after demodulation (expansion) in order to restore the original speech signal to its correct relative level. These circuits perform the task of modifying the speech signal in analog voice channels, thus, when the gain of the compressor and expander are controlled by the speech power at a syllabic rate, the technique is referred to as syllabic companding.
The characteristics of a typical compressor and expander are show in **Figure 3.2. (A)**. The compressor varies in such a way that an input level of x dB results in an output level of x/2 dB. For instance, input levels to the compressor of +10 by 0 (the reference level) and –60 dB result in output levels from the compressor of +5 in relation to the 0 and –30 dB, respectively. The signal characteristic of the expander is inverse to that of the compressor. It is not practicable to use compressors and expanders with a ratio of greater than 2:1. The effect of companding on the relative levels of a signal and the noise added during transmission, i.e., after the compression and before the expander, is shown in **Figure 3.2 (B)**. The compressor raises the power of weak signals relative to the noise. In this way, for an input signal at a level of –40 dB the effect of the compressor is to raise the S/N power ratio from 0 to 20 dB. Otherwise, in the absence of companding, a signal would be at the same level as the noise. The expander restores the original level of the speech signal and it reduces the level of the noise. For example, if the noise level at the receiver input between syllables is –40 dB with respect to the zero reference level, the corresponding noise level after expansion is reduced from –40 to –80 dB. The improvement in S/N ratio is the result of attenuation and is subjective, since it is associated with the absence of perceived noise during silences in the conversation. It is considered to be in the order of 15 dB.

Figure 3.3. Modulation Options

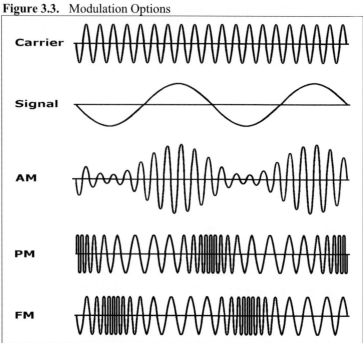

Courtesy of Book: "Guide to Telecommunications Transmission Systems" by A. A. Huurdeman

3.2.2. Analog Modulation and Multiplexing

Source waveforms of analog signals are directly modulated onto RF or IF carriers at the transmitter, using any form of three types of modulation: AM, FM or PM.

Analog modulation is a process in which some characteristics of a HF carrier are varied in accordance with the baseband signal, which is the creation, transmission and reception of EM fields. In reality, EM fields and waves can be used to communicate all kinds of information from place to place. Therefore, an analog modulated signal can be represented as the following sinusoidal wave:

$$c(t) = A \cos[2\pi f t + \varphi] \tag{3.2.}$$

where A = amplitude, $2\pi f$ = form of ω_c = angular frequency of the carrier, f = frequency and φ = carrier phase. Modulation can be achieved by altering the amplitude, frequency or phase of the wave in accordance with the information signals. Consequently, AM, FM and PM analog carriers are simply forms of modulated carriers in which either the amplitude, frequency or phase is modulated by the information waveform. Thus, by changing the amplitude, frequency or phase results in AM, FM or PM, respectively, see **Figure 3.3.** In fact, since EM fields are vector fields, it is also possible to modulate the polarization of a wave to communicate information. However, polarization modulation is only normally used for special purposes, as the polarization state of a wave can easily be "scrambled" during free space transmission by effects such as unwanted reflections from buildings.

The AM system is a type of linear modulation in which the baseband signal is linearly related to the modulated signal, while FM and PM are kinds of angle modulation in which the baseband signal is angularly related to the modulated signal. However, angle modulated FM and PM signals require more carrier bandwidth than AM but achieve a higher demodulated C/N ratio for the same carrier C/N value. The reverse process of recovering the information signal from the modulated carrier at the demodulator of the receiver is known as demodulation.

Multichannel operation of FM for analog high-capacity satellite telephone transmission is normally accomplished by a system called Frequency Division Multiplexing (FDM). In such a system different channels are separated from one another by being assigned different subscribers, which are then combined to fill the total bandwidth of the transponder.

3.2.2.1. Amplitude Modulation (AM)

Amplitude modulation is a type of linear modulation and is not used as a transmission modulating process in the satellite link. However, it can be used to modulate individual voice channels before combining them using FDM technique.

In AM, the amplitude of the carrier frequency is modified by the amplitude of a modulating signal, namely a carrier has to be amplitude modulated when the amplitude of the carrier varies in accordance with the signals. The side bands are displaced at either side of the carrier frequency by the magnitude of the modulation frequency and the amplitude depends on the modulation amplitude. The fundamental equation of AM modulation signals is:

$$c(t) = A[1 + \Delta_a \, m_{AM}(t)] \cos(\omega_c t + \varphi) \qquad\qquad (3.3.)$$

where $m_{AM}(t)$ = AM signal represents the information waveform to be transmitted and Δa = AM index (coefficient giving the degree of modulation of angular frequency of the carrier). The AM method is employed usually in AM radio broadcasting and radiocommunication transmissions. In these systems the intensity or amplitude of the carrier wave varies in accordance with the modulating signal. When the carrier is thus, modulated, a fraction of the power is converted to side band extending above and below the carrier frequency by an amount equal to the highest modulating frequency. If the modulated carrier is rectified and the carrier frequency filtered out, the modulating signal can be recovered. This form of AM is not a very efficient way to send information because the power required is quite large and because the carrier, which contains no information, is sent along with the information. In AM, the information is carried only in the side bands and therefore power in the carrier remains unutilized. In a Double Side Band Suppressed Carrier (DSB-SC), modulation of the carrier is suppressed and only side bands are used. The amplitude of this wave does not follow the signal amplitude and consequently, the inherent simplicity of using envelope detection is lost. Therefore, this modulation is not used in satellite communications.

In a variant of amplitude modulation, called Single Side Band (SSB), the modulated signal contains only one side band and no carrier. Hence, the information can be demodulated only if the carrier is used as a reference. This is normally accomplished by generating a wave in the receiver (Rx) at the carrier frequency. The SSB is used for long-distance HF radiotelephony and telegraphy over land, such as in the maritime HF radio frequency bands and submarine cables. In the other words, the SSB scheme is called an SSB Suppressed Carrier (SSB-SC) when the carrier is suppressed.

In fact, the most common application of SSB modulation in satellite communications is to multiplex voice carriers into a composite baseband signal. Thus, the required C/N ratio and the occupied bandwidths are two aspects considered in assessing the suitability of SSB for satellite transmission. The occupied RF bandwidth of an SSB transmission is the same as the baseband bandwidth, which is 4 to 5 kHz for a single telephone channel transmission. Typical bandwidth of –30 kHz is necessary for FM and 20 kHz for the O-QPSK, so both schemes are widely used in satellite communications.

However, a typical satellite communication link can economically provide C/N ratios in the order of 10 to 12 dB, making SSB transmission inefficient from power considerations. That is to say, this disadvantage can be offset to a large extent by the use of compandors, which offer an S/N ratio advantage of about 15 to 20 dB and in such a manner, SSB transmission appears attractive. This scheme is called Amplitude Companded SSB (ACSSB). Bandwidth efficiency is essential for mobile satellite service and for this reason ACSSB has been considered favorably for such applications.

3.2.2.2. Frequency Modulation (FM)

In frequency modulation (FM), the amplitude of the carrier frequency is modified by the frequency deviation of a modulating signal. The side bands are displaced at either side of the carrier frequency with an infinite number of separate bands, most resolving near zero. The amplitude depends on the modulation frequency change and the frequency depends on the rate of frequency change of the modulation. Thus, the fundamental equations of FM modulation are given below:

$$c(t) = A \cos[\omega_c t \, 2\pi\Delta_f \int m_{FM}(t)dt + \varphi] \tag{3.4.}$$

where $m_{FM}(t)$ = FM signal representing the information waveform to be transmitted and Δf = FM index (coefficient giving the degree of modulation angular frequency of the carrier).

Using FM the frequency of the carrier wave is varied in such a way that the change in frequency at any instant is proportional to another signal that varies with time. The FM band has become the choice of music listeners and for the audio portion of TV broadcasting because of its low-noise and wide-bandwidth qualities; however, the FM schemes are also extensively used in satellite communications. Examples of FM applications are multiplexed telephony, SCPC systems and TV broadcasting. Thus, FM systems are well suited for those cases where the baseband signal is in analog form. An example is FM with companding used in Rx and Tx voice channels of the analog Inmarsat-A system of the SES transceiver. Furthermore, this scheme also offers advantages for transmission of digital data in applications where simple receivers are essential, such as the Inmarsat paging system. At any rate, an important requirement of a paging system is the need for simple, low cost and rugged receivers.

3.2.2.3. Phase Modulation (PM)

Phase modulation (PM) like frequency modulation is a form of angle modulation so-called because the angle of the sine wave carrier is changed by the modulating wave. The two methods are very similar in the sense that any attempt to shift the frequency or phase is accomplished by a change in the other. The PM relation can be expressed with:

$$c(t) = A \cos[\omega_c t + \Delta_p \, m_{PM}(t) + \varphi] \tag{3.5.}$$

where $m_{PM}(t)$ = PM signal representing the information waveform to be transmitted and Δp = PM index (coefficient giving the degree of modulation angular frequency of the carrier).

3.2.3. Frequency Division Multiplexing (FDM)

The process of combining baseband signals and sharing the communication channels is known as multiplexing, while the reverse process of extracting individual baseband signals is called demultiplexing. The CCITT proposed all multiplexing standards including FDM, which is applicable only to telephony baseband signals. Whereas, the digital Time Division Multiplexing (TDM) standard is applicable to all types of baseband signals.
Each type of MES usually transmits and receives many kinds of signal transmissions to and from spacecraft. Multiplexing enables the division of channels or the combination of two or more input signals into a single output for transmission. In effect, the most common and known analog multiplexing method is FDM, used in satellite communication transmissions. The simplest approach to multiplexing is to assign a specific part of the available frequency or bandwidth spectrum to each signal. If two signals initially have the same spectrum, the frequency of one or both is shifted, so in such a way they will not overlap. Therefore, the FDM is a multiplexing solution where signals occupy the channel at the same time but on different frequencies.

3.3. Digital Transmission

Digital transmission relates to the link for which the MES terminal is designed to produce the digital signals by PC or modem and to send them through a transceiver. Moreover, it is possible to transmit analog signals (voice or broadcast) in digital form. Although this choice implies an increased baseband, it permits signals from diverse origins to be transmitted on the same satellite channels and the satellite link to be incorporated in ISDN, which implies the use of TDM. The digitization of analog signals implies the stages of sampling, quantization and source coding. The simple digital transmission chain includes a transmitting and receiving segment. However, the first unit of the transmitting segment is TDM with input signals from digital (direct) and analog (via encoder) sources and after multiplexed signals pass devices such as: Data Encryption, Channel Encoding, Scrambling and Digital Modulation, where the digital signal is transmitted through up or down satellite links. On the other hand, in the receiving segment of reverse mode incoming signal goes through devices such as: Demodulator, Descrambling, Channel Decoding, Data Decryption and TDM Demultiplexing in its direction to different users.
Digital signals use the same principle to modulate a carrier as analog signals. The AM, FM and PM schemes are all applicable to digital modulation; digital equivalents are ASK, FSK and PSK and special hybrid modulation solutions have also been developed to optimize digital modulation, such as QAM. The pioneering generation of MSS of Marisat and the first Inmarsat-A SES used analog FM for the voice circuits and digital PSK for the telex and signaling traffic. However, the current generation of Inmarsat standards for all mobile applications uses a combination of analog and digital transmission techniques and/or full digital modulation and multiplexing. In addition, newly developed MSS Non-GEO systems such as Iridium, Globalstar and others usually employ digital solutions.

3.3.1. Delta Modulation (DM)

Analog signals, such as speech and video signals, generally have a considerable amount of redundancy, namely, there is a significant correlation between successive samples. Thus, when these correlated samples are coded as in the PCM system, the resulting digital stream contains redundant information. The redundancy in these analog signals makes it possible to predict a sample value from the preceding sample values and to transmit the difference between the actual sample value and the predicted sample value estimated from the past samples. This result in a technique called difference encoding. One of the simplest forms of this is DM, which provides a staircase approximation of the sampled version of the analog input signal. This type of modulation is a way of digitizing a voice waveform, transmitting the digits and reconstructing the original analog waveform that avoids the quantizer and the A/D and D/A converters employed in PCM. In Linear Delta Modulation (LDM) a circuit of DM determines the difference between an incoming waveform m(t) and estimated waveform e(t), where a difference of signal error voltage is as follows:

$$\Delta m(t) = m(t) - e(t) \tag{3.6.}$$

The quantizer output is a positive constant when $\Delta m(t)$ is positive and vice versa. Namely, the difference between the input and the approximation is quantized into two levels, $+\Delta$ and $-\Delta$, corresponding to a positive and a negative difference, respectively. Moreover, at any sampling instant the approximation can be increased or decreased by Δ value, depending on whether it is below or above the analog input signal. Furthermore, a digital output of 1 or 0 can be generated according to whether the difference is $+\Delta$ or $-\Delta$. These pulses go to a conventional PSK digital modulator for transmission. Increasing the step size (Δ) will result in poor resolution and increasing the sampling rate will lead to a higher digital bit rate.

In LDM the step size is fixed at a value that provides performance near the peak, while better final performance may be achieved through a scheme of Adaptive Delta Modulation (ADM) in which the value of step size is varied during the modulation process. On the other hand, is difficult to make comparisons between the performance of PCM and DM because the latter is continually improving. Thus, most commercial satellite links to date use PCM for digital transmission of speech, while DM has been used for military satellite applications.

3.3.2. Coded Modulation (CM)

The CM is a combination of modulation and error correction codes without degrading the power of bandwidth efficiency. Thus, using forward error correction, such as block and convolutional codes, the bit error performance is improved by expanding the required bandwidth. Obtaining the power efficiency requires twice the bandwidth of the original uncoded signal because of the increase in the symbol rate of modulation and complex implementation. This can be done by increasing the number of phases in PSK modulation without expanding signal bandwidth. In this case, the 8-PSK signals with 2/3 rate of convolutional code have the same bandwidth as uncoded 4-PSK. However, the bit error performance degrades by about 4 dB due to an increase in phase but will be referable if the coding gain becomes more than 4 dB. There are two practical CM in use for MSS: the Trellis Coded Modulation (TCM) and the Block Coded Modulation (BCM).

1. Trellis Coded Modulation (TCM) – The TCM uses the combination of convolutional coding and expanded signal sets of 8-PSK to transmit two information bits per symbol. Thus, the modulation signals in TCM are assigned to each one- or two-satellite trellis branch, although binary code symbols are assigned in the convolutional codes. Here is a very important definition of measuring the distance between modulation signals assigned to each trellis branch. The modulation signal assignment in TCM can be designed either by the Euclidean or Hamming distances. The performance of TCM can be improved by increasing the number of states or by modifying the signal constellation. At this point, one solution is provided by multidimensional signals and another is performed by multiple coded modulations, known as Multiple TCM (MTCM). A TCM 8-DPSK modem with a rate of 2/3 and 16 states for 4800 b/s has been implemented in the NASA MSAT-X experimental program for LMSC.

2. Block Coded Modulation (BCM) – Instead of convolutional coding this scheme uses short binary block codes, which could be simpler and faster to decode. Similarly to TCM, this type of modulation can improve performance by using Multiple BCM (MBCM). Thus, MBCM with two symbols per branch has a coding gain of 3 dB relative to conventional BCM. The two symbols per branch MBCM has a performance for coding gains of 1.1 dB and 2.2 dB at BER = 10^{-3} relative to the BCM 8-PSK and the uncoded QPSK, respectively.

3.3.3. Pulse Code Modulation (PCM)

The digitization starts with the conversion of analog voice signals into a digital format. An analog can be converted into a digital signal of equal quality if the analog signal is sampled at a rate that corresponds to at least twice the signal's maximum frequency. A technique for converting an analog signal to a digital form is PCM, which requires three operations:

1) Sampling – This operation converts the continuous analog signal into a set of periodic pulses, the amplitudes of which represent the instantaneous amplitudes of the analog signal at the sampling instant. Thus, the process of sampling involves reading of input signals at discrete points in time. Hence, the sampled signals consist in electrical pulses, which vary in accordance with the amplitude of the input signal. In accordance with the Nyquist sampling rate, an analog signal of bandwidth B Hz must be sampled at a rate of at least 1/2B to preserve its wave shape when reconstructed.

2) Quantizing – This technique is the process of representing the continuous amplitude of the samples by a finite set of levels. If V quantizer levels are employed to represent the amplitude range it take the $\log_2 V$ bit to code each sample. In voice transmission 256 quantized levels are employed, hence each sample is coded using $\log_2 256 = 8$ bits and thus, the digital bit rate is 8,000 x 8 = 64,000 b/s. Thus, the process of quantization introduces distortion into the signal, making the received voice signals raspy and hoarse. This type of distortion is known as quantization noise, which is only present during speech. When a large number of quantization steps, each of ΔS volts, are used to quantize a signal having an rms signal level S_{rms}, the signal-to-quantization noise ratio is given by:

$$S_{qn} = (S_{rms})^2/(\Delta S)^2/12 \qquad\qquad (3.7.)$$

A large number of bits are necessary to provide an acceptable signal-to-quantization noise ratio throughout the dynamic amplitude range. Some analysis of speech signals shows that smaller amplitude levels have a much higher probability of occurrence than high levels.

3) Coding – This solution protects message signals from impairment by adding redundancy to the message signal.

Another important approach in digital coding of analog signals is Differential PCM. This is basically a modification of DM where the difference between the analog input signals and their approximation at the sampling instant is quantized into V levels and the output of the encoder is coded into $\log_2$ V bits. In such a way, it combines the simplicity of DM and the multilevel quantizing feature of PCM and in many applications can provide good reproduction of analog signals comparable to PCM, with a considerable reduction in the digital bit rate.

3.3.4. Quadrature Amplitude Modulation (QAM)

Each higher PSK modulation requires a better S/N ratio performance, which is difficult to achieve without special schemes. Instead of 16 or higher PSK modulation QAM is used, which is a combination of amplitude and phase modulation. In effect, modulation can be achieved in a similar manner to that of QPSK, by which the in-phase and quadrature carrier components are independently amplitude modulated by the incoming data streams. The incoming signals are detected at the receiver using matched filters. In terms of bandwidth, it is a highly efficient method for transmitting data flow. However, the sensitivity of the QAM method to variation in amplitude, limits its applicability to satellite communication systems in practice, where non-linear payload characteristics may distort the waveform, resulting in the reception of erroneous messages. **Figure 3.6. (d)** shows a 16-QAM signal space diagram, which shows a 16-state grid that permits both carrier amplitude and phase change. This scheme is not yet considered favorable for satellite communication systems.

3.3.5. Time Division Multiplexing (TDM)

Satellite links normally relay many signals from many MES but to avoid interfering with each other it is necessary for some kind of separation or division. This separation is known as multiplexing and its common forms are FDM (already explained) and TDM. The TDM is easier to implement with digital modulation and to form hybrid solutions applicable to all type of baseband signals.

The TDM is a time multiplexing solution where a group of various transmission signals on the same frequency at different times take turns using a channel. In this way, a group of pulses from a number of channels may be interleaved to form a single high rate bit stream of multiplexed assembly directly modulated onto the RF carrier. Since digital signals are precisely timed and consist in short pulse groups with relatively long intervals between them, TDM is the only natural way to combine digital signals for transmission. This system has the advantage that less equipment is required than is needed to modulate each channel onto a separate carrier and that the transmission efficiency of a satellite is usually better when it is carrying a few transmissions on many channels and vice versa.

In addition, accurate timing is essential to the correct operation of digital systems. For that reason the TDM system uses a synchronous clock, which controls the timing of all slave clocks and plesiochronous independent clocks with very good accuracy. Namely, if the transmission is without errors and breaks, in the synchronous system it would be only necessary to provide single markers at the beginning of transmission where the decoder could identify all streams of bits.

Figure 3.4. Comparison of: A) ASK; B) FSK and C) PSK

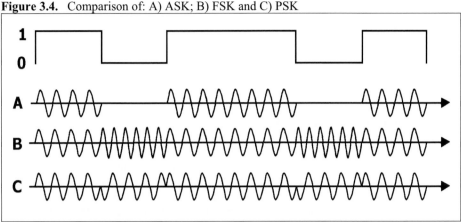

Courtesy of Book: "Mobile Satellite Communication Networks" by R.E. Sheriff and Y.F. Hu

However, in any practical communication system, regular markers must be provided in the bit stream so that the decoder can extract groups of digitally representative samples, identify the channels of a TDM assembly and resynchronize the system after errors, breaks of transmission or drift of the clock. Therefore, when plesiochronous time multiplexes are combined to form higher order of TDM, it is necessary, in order to preserve synchronism over the long term, to allow for the occasional addition of dummy or padding bits.

3.3.6. Types of Digital Shift Keying

Digital signals can be used to modulate the amplitude, frequency and phase and therefore the solutions of shift keying available for digital modulation are Amplitude Shift Keying (ASK), Frequency Shift Keying (FSK) and Minimum Shift Keying (MSK) as applications of FSK and Phase Shift Keying (PSK).

In terms of performance, ASK and FSK, both illustrated in **Figure 3.4. (A)** and **(B)**, respectively, require twice as much power to attain the same bit error rate performance as PSK, shown in **Figure 3.4. (C)**. At the top of the same figure a stream of a digital signal with 1 and 0 binary state is presented. Consequently, the vast majority of MSS employ a method of phase modulation known as PSK.

3.3.6.1. Amplitude Shift Keying (ASK)

This scheme can be accomplished simply by the on-off gating of a continuous carrier. The simplest ASK technique is to represent one binary level (binary 1) by a single signal of fixed amplitude and the other level (binary 0) by switching off the signal. The absence of the signal for one of the binary levels has the disadvantage that if fault conditions exist it could be misinterpreted as received data. Waveform for ASK, using different amplitude signals for the logic levels, are an alternative method to prevent this disadvantage. As with speech telephony circuits, the upper side band and carrier may be suppressed to reduce the bandwidth requirement and concentrate the available power on the signal containing the information.

3.3.6.2. Frequency Shift Keying (FSK)

This solution may be used whereby the carrier frequency has one value for a 1 bit and another for a 0 bit. The main difficulty in the use of this FM technique is that the gap between the frequencies used must be increased as the modulation rate increases. More exactly, for a restricted channel bandwidth, especially using in-band supervisory signaling, there is a limit to the maximum bit rate that is possible with this technique.

3.3.6.3. Minimum Shift Keying (MSK)

The MSK is a binary form of Continuous Phase Frequency Shift Keying (CPFSK), where the frequency deviation (Δf) from the carrier is set at half the reciprocal data rate of 1/2T. The MSK scheme may also be viewed as a special form of Offset-QPSK, consisting in two sinusoidal envelope carriers, employing modulation at half the bit rate. It is for this reason the MSK demodulator is usually a coherent quadrature detector, similar to that for QPSK. In this case, the error rate performance is the same as that of BPSK and QPSK. Similarly, differentially encoded data has the same error performance as D-PSK. This solution can also be received as an FSK signal using coherent or non-coherent methods, however, this will degrade the performance of the link. At any rate, the side lobes of MSK are usually suppressed, using Gaussian filters and the modulation method scheme adopted by GSM cellular systems.

3.3.6.4. Phase Shift Keying (PSK)

The PSK scheme is a technique using a multistate signaling stream in which the rate of data transmission can be increased without having to increase the bandwidth. In this shift keying system the phase of the carrier changes in accordance with the baseband digital stream or information content. Hence, a general form of a PSK scheme is given by the following expression:

$$s_m(t) = A \cos (\omega t + \phi) \tag{3.8.}$$

$$\phi = (2m + 1) \, \pi/M$$

where A = amplitude; ω = frequency angle; ϕ = phase angle varied in accordance with the information signal; m = integer in the range from 0 to (M − 1) and M = number of states. Depending on how many bits can be combined in a group of information as a symbol, there are a number of combination possibilities for PSK digital carriers.

3.3.7. Combinations of PSK Digital Carriers

There are several types of hybrid solutions used for the combination of PSK Digital Carriers. The PSK family is most popular for satellite communications and especially for MSC. A real-valued band bass signal s(t), which has a common form for these types of PSK methods, is expressed as follows:

$$s(t) = A(t) \cos [2\pi f_c t + \phi(t)] = A(t) \cos [2\omega_c t + \phi(t)] \quad [V] \tag{3.9.}$$

where A (t) = amplitude; f_c = carrier frequency of signal s(t); ɸ (t) = phase angle and ω_c = frequency angle varied in accordance with the signal s(t). Including information waveform m(t) the previous relation can be determined as follows:

$$s(t) = Re\ [m(t)\ e^{j2\pi fct} \tag{3.10.}$$

In linear modulation methods such as pulse, PAM and PSK can be expressed by:

$$m(t) = A(t)\ exp\ [j\phi(t)] = \sum_{n=0}^{\infty} a_n\, p(t)\ (t - nT) \tag{3.11.}$$

where Re = real part of the complex in the next bracket; a_n = information carrying symbols; p(t) = signal pulse and T = time interval of symbol.

3.3.7.1. Binary PSK (BPSK)

The simplest form of PSK is Binary PSK (BPSK), where the digital information modulates a sinusoidal carrier. For a general case of M-ary PSK (number of states) in BPSK M=2, so the baseband bit rate and the symbol rates are the same. Binary data is expressed by a_n = exp (jϕ_n) for ϕ_n = 0 or π and the phase changes every data bit of duration period T_b. When p(t) is a rectangular pulse over symbol duration T = T_b, the BPSK signal is expressed by:

$$s(t) = Aa_n \cos 2\pi f_c t\ =\ A\cos (2\omega_c t + \phi_n)\quad for\ nT \leq t < (n+1)\ T \tag{3.12.}$$

In **Figure 3.5. (A)** a diagram is illustrated of how, theoretically, the phase of the carrier changes instantaneously by 180° when the baseband signal switches from 0 to 1, while **Figure 3.6. (a)** represents the two states of the carrier by two vectors with a phase difference of 180°. In this sense, where ɸ = 0° and ɸ =180° when the baseband signal is 0 and 1, respectively and where A = +V and A = –V when the baseband signal is 0 and 1, respectively.
Except for standard-A, the TDM/BPSK scheme is used by almost all Inmarsat standards for process such as: Forwarding Signaling/Assignment Channels, Return Request Channels use Aloha BPSK in Inmarsat standard-A, C and Aero, while Slotted Aloha BPSK is dedicated for Inmarsat standard-M. On the other hand, Tlx message channel modulation in the forward direction uses TDM BPSK for Inmarsat standard-A and B and in the return link serves only for Inmarsat standard-A. In the similar manner, Inmarsat standard-C uses the BPSK (1/2-FEC) scheme in both directions.

3.3.7.2. Quadrature PSK (QPSK)

For M-ary, PSK is selected from M signals like: exp [j2π (m – 1)/M], where m = 1, 2,..., M and the resulting signal s(t) is written as follows:

$$s(t) = A\cos [2\pi f_c t + 2\pi/M\ (m-1)]\quad for\ m = 1, 2,..., M \tag{3.13.}$$

A slightly more complex form PSK is QPSK or 4-PSK, for which ϕ_n is a set of 0, π/2, π and 3/2π. Then, the signal s(t) is given by:

$s(t) = A/\sqrt{2}\ (a^I_n \cos (2\pi f_c t + \pi/4) + A/\sqrt{2}\ (a^Q_n \cos (2\pi f_c t + \pi/4)$ $\hspace{2cm}$ (3.14.)

where a^I_n and a^Q_n are the ± 1 value data, which are converted from input data sequence a_n into the in-phase channel (I channel) and quadrature channel (Q channel), respectively. In such a way, the relation (a^I_n, a^Q_n) is $(1, 1)$ for $\phi_n = 0$; $(1, -1)$ for $\phi_n = \pi/2$; $(-1, -1)$ for $\phi_n = \pi$ and $(-1, 1)$ for $\phi_n = 3\pi/2$.

In **Figure 3.5. (B)** the shape of QPSK modulated signals is presented and a QPSK scheme in the I-Q plane is shown in **Figure 3.6. (b)**.

Only two binary digits are needed to describe four possible states as follows: the ϕ values of 45°, 135°, 225° and 315° correspond to 00, 01, 11 and 10, respectively. In fact, each state of the signal carries two bits of information. A combination of two bits or more, which corresponds to a discrete state of a signal, is called a symbol, in the case of QPSK; the symbol rate is half the bit rate. Namely, a bit rate of 120 Mb/s corresponds, with QPSK, to a symbol rate of 60 Mb/s or Mbauds.

The octal phase modulation known as 8-phase PSK is a constant amplitude scheme with a higher bandwidth efficiency of 3 b/s/Hz, shown in **Figure 3.6. (c)**. In fact, the demands of high bit-rate applications are related to images, TV and HDTV transmission.

3.3.7.3. Offset QPSK (O-QPSK)

The O-QPSK scheme delays the quadrature bit stream by T sec relative to the in-phase bit stream to restrict the phase transition to phase changes of 0 or $\pi/2$ every T sec. Using a^I_n and a^Q_n, the equivalent low-pass and resulting signals are expressed by:

$$u(t) = \sum_{n=0}^{\infty} a^I_n\, p(t)\, (t - 2nT) - ja^Q_n\, p(t)\, (t - 2nT - T)\, e^{j\pi/4} \hspace{2cm} (3.15.)$$

$$s(t) = [\sum_n a^I_n\, p(t)\, (t - 2nT)] \cos (2\pi f_c t + \pi/4) + [\sum_n a^Q_n\, p(t)\, (t - 2nT - T)] \sin (2\pi f_c t + \pi/4)$$

Figure 3.5. Hybrid PSK Modulations: A) BPSK; B) QPSK and C) O-QPSK

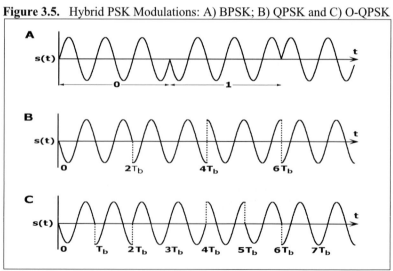

Courtesy of Book: "Mobile Satellite Communications" by S. Ohmori and other

The data transmission occurs in the conventional QPSK at the same time in both I and Q channels. This scheme has larger phase changes than O-QPSK, which has phase changes of at most $\pm\pi/2$ data transmission and large envelope fluctuations do not occur as they do with Π-phase changes in QPSK. **Figure 3.5. (C)** shows the O-QPSK modulated signal's shape. The Aloha O-QPSK (1/2-FEC) serves in return request channels by a transmission speed of 24 Kb/s for Inmarsat-B and 600 b/s for Aero standards. The Inmarsat-B and Aero standards use 16 Kb/s or 9.6 Kb/s, respectively of the APC O-QPSK scheme in voice channel coding/modulation, while Inmarsat standard-M uses 4.8 Kb/s IMBE in coding O-QPSK. The Inmarsat-B in Tlx return link uses the TDMA O-QPSK (1/2-FEC) modulation scheme.

3.3.7.4. Differential PSK (DPSK)

Channel conditions in MSC are more severe than additive white Gaussian noise channels due to multipath fading, shadowing and Doppler effects. These problems can be solved by using differentially coherent detection of encoded signals. The received signal phase is generally an 180° ambiguity sign, which cannot be resolved unless some known reference signal is transmitted and a comparison made. In the worst case, if such a situation is left unresolved, the received signal could end up being the complement of the transmitted signal.

When the phase condition is constant for 2T seconds, the DPSK demodulator can obtain the optimum a posteriori probability. The DPSK scheme can also be used to remove sign ambiguity at the receiver. Differentially encoding data prior to modulation occurs when a binary 1 is used to indicate that the current message bit and prior code bit are of the same polarity and 0 to represent the condition when the two pulses are of opposite polarity. The equivalent low-pass signal in the interval of m is given by:

Figure 3.6.   Representation of PSK Signals in I-Q Plane

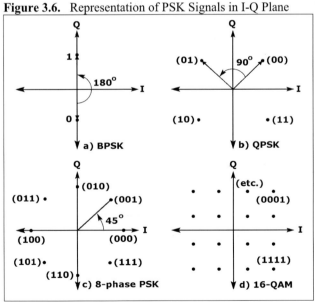

Courtesy of Book: "Mobile Satellite Communications" by M. Richharia

$$v_m = \alpha e^{j(\phi m - \phi)} u_m(t) + n_m(t) \qquad (3.16.)$$

where α = loss factor; ϕ_m = phase difference in m interval; $u_m(t)$ = signal in m interval and $n_m(t)$ = Gaussian noise in m interval. The noncoherent detection technique is useful when the carrier phase is difficult to estimate at the receiver. In MSC channels, the transmitting oscillator that generates the carrier cannot be completely stabilized and an MSC link has a low S/N ratio. Besides, for coherent detection, the bandwidth of the carrier-tracking loop at the Rx must be decreased in the proper manner to increase the S/N and obtain a good phase reference. Since propagation characteristics in MSC such as multipath fading and the Doppler effect cause rapid phase variation, the carrier phase of the Rx signal does not remain fixed long enough to be estimated. The problem can be solved with noncoherent detection, which is equivalent low-pass signal in relation to random phase value (ϕ_n) and Gaussian noise is given by:

$$v_m = \alpha e^{-j\phi n} u_n(t) + n(t) \qquad (3.17.)$$

Therefore, noncoherent detection results in the degradation of the bit error performance with respect to coherent detection. In this case, the bit error rate probability (P_b) for binary DPSK is determined with the following relation:

$$P_b = \tfrac{1}{2} \exp(-\gamma_b) \qquad (3.18.)$$

where $\gamma_b = E_b/N_0$ is the value of the S/N density ratio per bit signal. In this sense, when considering the use of DPSK schemes, a trade-off between simplified receiver complexities against reduced performance characteristics in the presence of noise, particularly when employing higher order modulation techniques, needs to be made.

3.3.7.5. π/4-QPSK

Recently, π/4-QPSK, or π/4 shift QPSK, has become very popular for MSC as well as for cellular systems because it has a compact spectrum with small spectrum restoration due to nonlinear amplification and can perform differential detection. Thus, the phase point of this scheme always shifts its phase over successive time intervals by $\pm\pi$/4 or $\pm 3\pi$/4. Therefore, the spectrum of this scheme is the same as that of a QPSK that undergoes an instantaneous $\pm\pi$/24 or $\pm\pi$ phase transition. In such a way, the π/4-QPSK signal can reduce the envelope fluctuation due to band-limited filtering or nonlinear amplification more than can QPSK. In a more general sense, this is because differential detection can be used since π/4-QPSK is not an offset scheme.

For MSC channels, a strong line-of-sight signal can be expected even in the Rician fading channel, so coherent demodulation is desirable for improved power efficiency. The bit error performance of coherently detected π/4-QPSK is the same as that of QPSK. On the other hand, differential detection is also desirable for simple hardware implementation. The bit error probability for differential coherer detected π/4-QPSK is given by the following relation:

$$P_b = e^{-2\gamma b} \sum_{k=0}^{\infty} (\sqrt{2} - 1)^k I_k (\sqrt{2}\gamma_b) - 1/2\, I_0 (\sqrt{2}\gamma_b)\, e^{-2\gamma b} \qquad (3.19.)$$

where $\gamma_b = E_b/N_0$ and I_k = modified Bessel function of the first kind in k interval order. Thus, the differential detection is about 2 to 3 dB inferior to the coherent detection in AWGN and fading channels and noncoherent detection is also applicable. Because of these advantages, $\pi/4$-QPSK has been chosen as the standard modulation technique for several systems of cellular and satellite-based mobile communications. Some experimental results show that in fully saturated amplifier systems, $\pi/4$-QPSK still has significant spectral restoration. In effect, to reduce this restoration, $\pi/4$-controlled transition QPSK (CTPSK) uses both sinusoidal shaping pulses and timing offsets of the phase transition between I and Q channels.

3.3.8. Digital Voice Coding

The MSC transmission applications need the utilization of digital voice (speech) coding technique which require high-quality speech coding at bit rates as low as possible and that they be hard under severe propagation conditions, such as multipath fading and shadowing. Digital voice coding techniques can be classified in groups of waveform and parametric coding. The waveform coding group is achieved by matching as closely as possible the waveform of the original and the reconstructed signals, while the parametric coding group represents the speech signal using a model for speech production, such as time-varying linear predictive filters and transmission parameters of the filter only. At the decoder, the speech signal is reconstructed by the inverse prediction filters, using received parameters. Several hybrid-coding methods are proposed as a combination of waveform and parametric coding. In the hybrid coding, both the filter parameters and the quantized residual sample, which is the signal that remains after filtering the speech signal, are transmitted.

3.3.8.1. Adaptive Predictive Coding (APC)

The APC digital technique is based on an initial estimation and prediction of the input speech (voice) waveform samples, plus a residual (error) signal derived by comparing the estimated samples with the actual speech input. This initial estimation and residual signal are quantized and transmitted to the receiver, where an inverse quantized residual signal is fed into a synthesized filter in order to reconstruct the speech signal. Coding efficiency is achieved by removing the waveform redundancy in the input signals. Furthermore, a quantizer operating only on the residual signal requires fewer bits per sample. The effect of quantization is to distort the reconstructed output speech signals. Maximum Likelihood Quantization (MLQ) techniques are used to minimize quantization noise. On the other hand, an adaptive quantization scheme with a noise-shaping filter is used to provide the subjective performance. A combination of short-term and long-term predictors reduces the variance of the residual signal and the subsequent quantization error.

The Inmarsat digital MSS utilize the APC/O-QPSK speech (voice) coding technique, which major parameters are systematized within the framework of the APC system for both standard-B and Aero of 16 Kb/s and 9.6 Kb/s, respectively. The typical APC digital codec configuration consists in transmitting and receiving sides, both including two main components. The Tx side is an APC coder containing an inverse filter with predictor of negative predisposition and quantizer (Q), which produces output signals while the Rx side is an APC decoder containing a synthesis filter with an inverse quantizer (Q^{-1}), which takes input signals and then forwards them in the predictor with positive predisposition.

3.3.8.2. IMBE Coding

After extensive testing in trials during which seven candidates' voice codecs (coder and decoder) were evaluated, a codec manufactured by Digital Voice Systems Incorporated (DVSI) was selected: the Improved Multi Band Excitation (IMBE) model of voice coding algorithm. Namely, it was judged to provide: reasonable speech (voice) quality and speaker recognition; excellent intelligibility; acceptable tolerance to burst and random errors up to 4% error rate; good performance in the presence of background acoustic and landline noise; relative immunity to the effect of non linear microphones and relatively straightforward implementation.

In use, the IMBE codec speech is sampled and digitized, and then fast Fourier transform is performed to determine a smoothed frequency spectrum (narrow bandwidth). Thus, where the detected energy is above a predetermined threshold, the harmonic is declared voiced and where the energy is declared unvoiced, then the amplitudes are measured in both cases. The smoothed spectral envelope, the voice/unvoiced status of each harmonic and the amplitudes of voice/unvoiced harmonics are transmitted to the speech synthesizer, located at the receiver. Accordingly, at the receiver the pitch period and voiced harmonics are reconstructed in the speech synthesizer, using sinusoidal oscillators whose amplitude are controlled by voiced amplitude information from the transmitter. Besides, the unvoiced harmonics are simulated by band-limited noise from a noise source in the receiver, the amplitude of the noise at each harmonic being controlled by the unvoiced amplitude information transmitted over the communication channel. The sampled smoothed spectrum information is multiplied with the synthesized harmonic spectrum and an inverse fast Fourier transform performed to construct a digitized and synthesized voice signal.

At any rate, by employing speech analyses and coding at the transmitter, together with decoding and speech syntheses in the receiver, a large reduction in bandwidth and hence the transmitted bit-rate, can be achieved. The HF speech components no longer have to be sent over the communication channel but can be regenerated in the synthesizer, which is the only mathematical model that includes frequency sources and modulators, thereby reducing occupied channel bandwidth. More exactly, to reproduce the speaker's voice, it is only necessary to send the receiver a coded set of instructions, which instruct the speech synthesizer how to:

1) Simulate the distinguishing characteristics of the speaker's voice, namely, to set the pitch frequency and selected harmonics and
2) Modulate these to reproduce the speaker's message.

Both types of information change only slowly during speech and can be transmitted in a narrow bandwidth. Inmarsat-M uses digital voice coding at 6.4 Kb/s speech codec rates, including error correction coding to protect the most critical speech syntheses parameters. The coding rate has been optimized to preserve voice (speech) quality in a typical mobile environment where high background acoustic noise levels, multipath distortion (maritime) and blockage of the line-of-sight path to the satellite (land) are often encountered.

3.3.8.3. Adaptive Differential PCM (ADPCM)

The Pulse Code Modulation (PCM) is the simplest method of waveform coding, and the voice signal s(t) quantized to one of many (2^k) amplitude levels, however, where (k) is the number of binary digits only to represent each sample.

When the sampling rate is chosen to be several times the Nyquist rate, the adjacent samples become highly correlated and the signal does not change rapidly from sample to sample. At this point, using this correlation between adjacent speech samples, Differential PCM (DPCM) quantizes the differences between the input sample and the predicted value, which is estimated by a linear predictor. Since such differences are smaller than the sampled amplitude themselves, fewer bits are required to represent the Tel signal.

The DPCM scheme with fixed predictors can provide from 4 to 11 dB improvement of gain over direct quantized PCM. The ADPCM method uses both adaptive quantization and adaptive prediction to reduce coding errors. In this sense, the typical configuration of an ADPCM encoder is composed of sample, quantizer and predictor, providing feed-forward quantization where the quantizer step size is proportional to the variance of the input of the quantizer. This type of ADPCM provides about 10 to 11 dB improvement of S/N ratio over PCM. Several adaptation algorithms have been proposed for this scheme. Since ADPCM uses a scalar quantization, it is difficult to reduce the bit rate to less than 8 Kb/s.

3.3.8.4. Linear Predictive Coding (LPC)

The LPC method synthesizes the speech using a linear predictive filter, which is excited by appropriate signals such as an impulse sequence for voiced speech or random noise for unvoiced speech. The LPC codec transmits only the parameters of the linear predictive filter and the index of its selected excitation signals. The basic configuration of LPC codec is composed of excitation generator, long-term predictor, short-term predictor and weighting filter. The input speech signals come into the weighting filter and are assumed to be produced by an all-pole filter and returned in the minimization procedure.

The short-term predictor removes the redundancy in the speech signal by the predicted value, using the past (p) sample. However, some periodicity, which is related to the pitch period of the original signal in the 50 to 400 Hz pitch frequency, still remains. This residual signal is removed by the long-term predictor (pitch predictor) and turns into a noise-link signal. The pitch predictor is not essential for medium bit rate LPC codecs, although it can improve their performance but it is very essential for low bit rate codecs such as CELP, which use the excitation signal modeled by a random Gaussian noise process.

3.3.8.5. Multipulse Excited LPC (MELPC)

The configuration of the MELPC encoder is the same as CELP, the only difference being the excitation generator. In the MELPC mode, the excitation generator produces a sequence of pulses located at nonuniformly spaced intervals with different amplitudes. Both the amplitudes and positions of these pulses are determined using a closed-loop analysis-by-synthesis method. Thus, the synthesized signal is reconstructed using a sequence of pulses produced by the excitation generator and the long-term and short-term predictor.

While MELPC assumes that both amplitudes and positions of excitation pulses are initially unknown, the regular pulse-excited LPC assumes that the pulses are regularly spaced but the amplitudes are unknown. Using a suitable error criterion, the error between the original and the synthesized signal is minimized. Two types of codecs need similar bit rates for the same speech quality because MELPC needs less excitation pulses, due to the optimization of the pulse position but it needs the transmission of pulse positions as well as their amplitudes.

3.3.8.6. Code Excited Linear Prediction (CELP)

The LPC and MELPC can produce good voice quality at bit rates as low as 9.6 Kb/s but they cannot maintain their quality below this rate, because they have to expend a large number of bits for encoding the excitation pulse. Besides, as excitation signals, the CELP voice coder uses the collection of code vectors, which are previously produced using vector quantization techniques based on Gaussian processes and are stored as a large codebook. Using each code vector and the predictors, synthesized speech is produced and then the most suitable code vector that produces the lowest error between the original and the reconstructed signal is selected. The index that is assigned to the code vector, the voice gain and the values of the parameters for short- and long-term predictors are transmitted.

3.4. Channel Coding and Decoding

Voice, video, data and telex information are transmitted in digital form through a channel that can cause degradation of these transmission signals. The noise, interference, fading and other obstacle factors experienced during transmission could increase the probability of bit error at the receiver. Differently to say, the data signal may be encoded in such a way as to reduce the likelihood of bit error. Anyway, the coding process uses redundant bits, which contain no information to assist in the detection and correction of errors. The subject of coding emerged following the fundamental concepts of information theory laid down by Shannon in 1948, which is the relationship between communication channel and the rate at which information can be transmitted over it. Basically, the theorems laying down the fundamental limits on the amount of information flow through a channel are given.

3.4.1. Channel Processing

Channel processing is composed of special activities, which can improve the transmission techniques throughout satellite channels in connection with gain, errors, noise, interference, concentration and authenticity.

3.4.1.1. Digital Speech Concentration and Channel Multiplication

The system for digital speech concentration (interpolation) uses the activity factor of telephone channels in order to reduce the number of satellite channels required to transmit a given number of terrestrial channels. The Digital Speech Interpolation (DSI) technique is based on the fact that in a normal telephone conversation each participant monopolizes the circuit for only around half the time. As the silence between syllables, words and phrases increases so does the unoccupied time. Hence, on average, the activity time of a circuit is from 35% to 40% of the connection time. By making use of the actual activity of the channels, several users can be permitted to share the same telephone circuit. Certain numbers of terrestrial satellite channels require only half the satellite channels and the gain is about 2. By adding a low rate encoder to the digital speech concentrator, the gain can be further increased. For example, with encoding at 32 Kb/s a gain increases by a factor of 2 can be obtained in voice channels used alternately for speech or data transmission. The theoretical DSI gain is defined by the ratio between the actual number of speakers (input trunks) and the number of transmission channels (bearers) required to service them.

On the other hand, the function of the Digital Circuit Multiplication (DCM) equipment is to concentrate a number of input digital lines (trunks) onto a smaller number of digital output channels (bearers), thereby achieving a higher digital efficiency of the link or channels. This technique is qualified by the circuit multiplication gain, which is defined as the ratio of the input channels number over the number of DCM output channels. It is used in digital circuit multiplication equipment of the Intelsat/Eutelsat system.

3.4.1.2. Channel Encoding

The two fundamental problems related to reliable transmission of information via channels were identified by C.E. Shannon as follows:
1) The use of minimal numbers of bits to represent the information given by a source in accordance with a fidelity criterion. In reality, this issue is usually identified as a problem of inefficient MSC, to which the source coding provides most practical solutions.
2) The recovery as exactly as possible of the information after its transmission through a communication channel in the presence of noise and other interference. This is a problem of unreliable MSC, to which channel (error) coding is the basic solution.
Shannon proved that by proper encoding these two objectives can always be achieved, provided that the transmission rate (R_b) verifies the fundamental expression $H<R_b<C$, where H = source entropy and C = channel capacity. The Bit Error Rate (BER) of a digital system may be improved either by increasing E_b/N_0 or by detecting and correcting some of the errors in the received data. For the Additive White Gaussian Noise (AWGN) channel, the Shannon Hartley law states that capacity of a channel is given by the following relation:

$$C = B \log (1 + C/N) \quad [b/s] \tag{3.20.}$$

where B = channel bandwidth in Hz and S/N = signal-to-noise ratio at the receiver. Thus, the channel capacity is the measure rate of the maximum information quantity that two parties can communicate without error via a probabilistically modeled channel. Namely, this chain of channels is composed of information data on input rate (R_b), channel encoder with redundancy data (r) and encoded data symbols on output rate (R_c). A reverse channel model contains input encoded data symbols, channel decoder and output information data symbols. According to Shannon, if information is provided at rate R, which is less than the capacity of the channel, then a means of coding can be applied such that the probability of error of the received signal is arbitrarily small. If this rate is greater than the channel capacity, then it will not be possible to improve the link quality by means of coding techniques. Indeed, its application could have a detrimental effect on the link. Rearranging the above equation in terms of energy-per-bit and information rate, where the information rate is equal to the channel capacity, results in the following:

$$C/B = \log (1 + E_b C/N_0 B) \tag{3.21.}$$

where E_b = information bit rate; E_b = related to the carrier power and the information bit rate and C/N_0 = carrier-to-noise density ratio. Moreover, the above expression can be utilized to derive the Shannon limit, the minimum value of Eb/N_0 below which there can be no error-free transmission of information. As C/B tends to zero, this can be shown to be equal to −1.59 dB ($1/\log_2 e$). The code rate and input rate can be defined as:

$$c = n/n + r \quad \text{and} \quad R_c = R_b/n \quad [b/s] \tag{3.22.}$$

The capacity of the channel is independent of the coding/modulation scheme used. Hence Shannon's channel coding theorem exactly stated that, for a given carrier-to-noise ratio, the error probability could be made as small as desirable, provided that the information rate (R_b) is less than the capacity (C) and a suitable coding is used. In any case, in MSC systems, channel coding is especially interesting because of the severe power, bandwidth and propagation limitations. Moreover, the considerable progresses in multiple access modulation schemes, resource assignment algorithms, signal processing techniques and advanced error control coding provide the most efficient means to realize highly reliable information transmission.

3.4.1.3. Digital Compression

Digital transmission in general uses compression techniques for data and video signals. The effective data transfer via the Inmarsat MSC system can be significantly increased by using data compression software. Essential results were provided on PC by the PKZIP/PKUNZIP program developed by US-based PKWARE, which in a fraction of a second gives a 2 to 3 times reduction in size of ASCII files and 1.5 times for many types of binary files. The ARJ compression software from Robert K. Jung is slower but more effective than PKZIP. It can also be recommended for the compression of data files containing graphic information. Thus, the real-time data compression incorporated into the most advanced modems can also increase the effective data rate of ASCII files transmission but for transmission of already compressed files with information it is better not to use the compression in the modem.

Their use in compressed video systems, where a TV Receive Only (TVRO) can also receive many channels of video from one transponder (about 6 to 8), has become very widespread, first in the USA and then in Europe. The compression system that has now become standard refers to a Moving Picture Expert Group (MPEG), formed under the auspices of the ISO and the International Electrotechnical Commission (IEC). In such a system a number of digitized videos are combined into a single bit stream in a source coder. That bit stream is then sent to a channel coder for FEC and then to a QPSK modulator, an up-converter, amplifier and an antenna for up linking to the satellite transponder. Since only one signal is present in the transponder at any on time, there is no need for back-off and full transponder power is used. At the reverse side, the compressed video downlink comprises a line in the chain of an antenna, tuner including down-converter and QPSK demodulator, then FEC detector, demultiplexer and MPEG decoder. The MPEG is determined to provide standard compression that allows video and accompanying audio signals to be compressed in channel width. The packetizer function is to enter a suitable code in the bit stream for the individual digitized TV program so that it can be separated in the receiving chain, allowing the enabled user to select the desired program. Thus, the BER is determined from the (E_b/N_0) obtained for a combination of whatever transponder EIRP, FEC coding, transmission symbol rate and receiver system are used. If a plot of that FEC system is not available, then the Viterbi mode FEC coding performance could be used for a good estimate of results. This type of compression has effects like: MPEG-2 compression results in the removal of most audio and video redundancy; the FES utilization scheme resulting in a rapid BER increase and the resultant (E_b/N_0) should be high enough to achieve a BER of 10^{-6} for a TVRO.

3.4.1.4. Voice Encryption

Encryption is used when it is wished to prevent exploitation or tampering with transmitted messages or voice conversation by unauthorized users, in the form of algorithmic operation in real time, bit-by-bit, on the binary stream. Thus, the set of parameters, which defines the transformation, is called a key. Its use is often associated with military communications but commercial satellite systems are increasingly induced by customers to propose encrypted links, particularly for administrative and government sectors. In fact, due to the extended coverage of satellite networks and the easy access to them by small MES, eavesdropping and message falsification are potentially within the reach of a large number of agents of modest means.

The encryption transmission chain is composed of an encryption unit with plain text input, satellite channel with intruder and key distribution for retransmission of cipher (encrypted) text and de-encryption with a unit for production of plain output text. The encryption and de-encryption units operate with a key provided by the key generation unit. Acquisition of a common key implies a secure method of key distribution. This key is entered into the encryption unit through a key injector about the size of a matchbox. Without this key, a potential eavesdropper attempting to listen in on the conversation would hear nothing but a noise made up of digital signals. The technique used for most voice encryption consists in speech compressing and digitizing, using a very complicated coding process. In such a way, the voice signal is sliced into small bits, which are processed by an algorithm into bits of voice with a very complex structure. At the other end of the process, using the same key pattern, the voice is reproduced as it was before the encryption.

A typical example of voice encryption for MSC is Satsec A1 for secure voice transmission, of Inmarsat standard-A MES. This unit is housed in a modern smart telephone, which uses a very sophisticated Swiss encryption technique. The Satsec A1 features include full digital voice and facsimile encryption and LPC voice compression, using full duplex operations CCITT V.22bis and V.27 Modem with a rate of 2400 b/s, giving business users a security level comparable to that used by government agencies. In case of transmission only in half duplex mode, the unit automatically falls back to the built-in VOX-controlled quasi-duplex operational mode. Hence, this device should not be confused with a voice scrambler, even a digital one, which is no longer a competitive alternative to high level encryption. The unit digitizer uses linear predictive coding and has, unlike self-synchronizing stream ciphers, no bit error multiplication. In a more general sense, under the same circumstances, this device may even have an enhanced effect on the channel is quality of transmission.

The aspects of encryption are confidential to avoid exploitation of the voice/message by unauthorized persons and to provide authentic protection against any modification of the message by an intruder. This system uses the following technique:

1. On-line encryption (Stream Cipher) - Each bit of the original binary message stream (plain text) is combined using a simple operation, for example modulo-2 addition, with each bit of a binary stream (keystream) generated by a key device. Otherwise, the latter could be a pseudorandom sequence generator whose structure is defined by the key.

2. Encryption by Block (Block Cipher) - The transmission of the original binary stream message into an encrypted stream is performed simply block-by-block, according to the logic defined by the key.

Besides, encryption is commonly used in direct TV broadcasting to avoid illegal reception and military applications to minimize the probability of message interception.

3.4.2. Coding

As is known, satellite communication systems are generally limited by the available power
and bandwidth. Thus, it is of interest if the signal power can be reduced while maintaining
the same grade of service (BER). As mentioned, this can be achieved by adding extra or
redundant bits to the information content by using a channel coder. Otherwise, excepting
several main classes of channel coder, the three most widely used in MSC are block, cyclic
and convolutional encoders.

3.4.2.1. Block Codes

Binary linear block codes are expressed in the (n, k) form, where (k) is the information bits
number that is converted into (n) code word bits. There are (n, k) party bits in each encoded
block, where the difference between (n) and (k) bits are added by the coder as a number of
redundancy bits (r). In the other words, a coded block comprising (n) bits consists in (k)
information and (r) redundant bits expressed as follows:

$$n = k + r \qquad\qquad (3.23.)$$

Such a code is designated as a (n, k) code, where the code rate or code efficiency is given
by the ratio of (k/n). Mapping between message sequences and code words can be achieved
using look-up tables; although as the size of the code block increases such an approach
becomes impractical. This is not such a problem as linear code words can be generated
using some form of linear transformation of the message sequence. Thus, a code sequence
(c) comprising of the row vector elements (c_1, c_2,..., c_n) is generated from a message
sequence (m), comprising the row vector elements (m_1, m_2,..., m_k) by a linear operation:

$$c = m \, G \qquad\qquad (3.24.)$$

where G = generator matrix. Thus, in general, all (c) code bits are generated from linear
combinations of the (k) message bits.
A special category known as a systematic code occurs when the first (k) digits of the code
are the same as the first (k) message bits, namely if input message bits appear as part of the
output code bits. The remaining n-k code bits are then generated from the (k) message bits
using a form of linear combination, and they are termed the party data bits.
The generator matrix for a linear block code is one of the bases of the vector space of valid
codewords. The generator matrix defines the length of each codeword (n), the number of
information bits (k) in each codeword and the type of redundancy that is added; the code is
completely defined by its generator matrix. The generator matrix is a (k · n) matrix that is
the row space of V_k. Thus, one possible generator matrix for a typical (7, 4) linear block
code has to be presented in four rows as blocks: G = 1101000/0110100/1110010/1010001.
Thus, the distance between two coded words (for example, first 2 and second 2 digits) in a
block is defined as the number of bits in which the words differ and is called the Hamming
distance (d_h). The Hamming distance has the capability to detect all coded words having
errors (e_d), where $e_d < (d_h -1)$; to detect and correct (e_{dc}) bits, where $e_{dc} = (d_h -1)/2$ and to
correct t and detect (e) errors, where the Hamming distance as a minimum space between
two coded blocks is given by:

$$d_h = t + e + 1 \tag{3.25.}$$

In the detection process, two coded words separated by (d_h) are most likely to be mistaken for each other. The extended Golay code offers superior performance to Hamming codes but at a cost of increased receiver complexity. In practice, code words are conveniently generated using a series of simple shift registers and modulo-2 adders.

3.4.2.2. Cyclic Codes

These code methods are a subclass of linear codes, where a code word is generated simply by performing a cyclic shift of its predecessor. In other words, each bit in a code sequence generation is shifted by one place to the right and the end bit is fed back to the start of the sequence, hence the term cyclic. Both the linear Hamming and extended Golay codes have equivalent cyclic code generators. Thus, non-systematic cyclic codes are generated using a unique generator polynomial g(p) and message polynomials in the forms as follows:

$$g(p) = p^{n-k} + g_{n-k}p^{k-1} + \dots g_1p + 1 \tag{3.26.}$$

$$m(p) = m_{k-1}p^{k-1} + m_{k-2}p^{k-2} + \dots + m_1p + m_0 \tag{3.27.}$$

where the generator polynomial is a factor of p^{n+1} and the value ($m_{k-1}\dots m_0$). When this is multiplied by the generator polynomial, it results in the generation of a code word by:

$$c(p) = (m_{k-1}p^{k-1} + m_{k-2}p^{k-2} + \dots + m_1p + m_0)\, g(p) \tag{3.28.}$$

Thus, an alternative to this approach is to generate systematic cyclic codes, which can be generated in three steps, involving the use of feedback shift register:
a) The message polynomial is multiplied by p^{n-k}, which is equivalent to shifting the message sequence by (n − k) bits. This is necessary to make space for the insertion of the party bits.
b) The product of step 1, $p^{n-k}m(p)$ is divided by the generator polynomial, g(p).
c) The remainder from step 2 is the party bit sequence, which is then added to the message sequence prior to transmissions.
The cyclic codes scheme has two methods used in MSC: Bose-Chadhury-Hocquenghem (BCH) and Reed-Solomon (RS).
1. BCH Codes – The BCH codes are the most powerful of all cyclic codes with a large range of block length, code rates, alphabets and error correction capability. These codes have been found to be superior in performance to all other codes of similar block length and code rate. Most commonly used BCH codes have a code word block length as $n = 2^m$ − 1, where (m = 3, 4 …). For instance, the Inmarsat standard-A uses 57 bits plus 6 party bits encoded with BCH (63, 57 code in TDM channels and for the return request channel burst employs Aloha BPSK (BCH) 4800 b/s.
2. RS Codes – The RS codes are a subset of the BCH codes specially suited for correcting the effect of the burst errors. The latter consideration is particularly important in the context of the MSC channels and hence, RS codes are usually incorporated into the system design. This set of codes has the largest possible code minimum distance of any linear code with the same encoder input and output block length. Thus, the RS codes are specified using the

convention RS (n, k), where n = number of code symbols word length per block; k = data symbols encoded and the difference between (n) and (k) is the number of parity symbols added to the data. The code minimum distance is given by:

$$d_{min} = n - k + 1 \qquad\qquad (3.29.)$$

The code is capable of correcting errors such as: $e = 1/2\ (d_{min} - 1)$ and $e = (n - k)/2$, or to use an alphabet of 2^m symbols with: $n = 2^m - 1$ and $k = 2^m -1 - 2e$, where $m = 2,3 \ldots$ and so on. The advantage of RS codes is the reduction in the number of words (n) symbols, which are code words, producing a possibly large value of minimum distance (d_{min}).

3.4.2.3. Convolutional Codes

The second family of commonly used codes is known as convolution codes. Unlike block codes, which operate on each block independently, these codes retain several previous bits in memory, which are all used in the coding process. They are generated by a typed-shift register and two or more modulo-2 adders connected to particular stage of the register. The number of bits stored in the shift register is termed the constraint length (K). Bits within the register are shifted by (k) input bits. Each new input generates (n) output bits, which are obtained by sampling the outputs of the modulo-2 adders. The ratio of (k) to (n) is known as the code rate. These codes are usually classified according to the following convention: (n, k, K), for example (2, 1, 7), refers to a half-rate encoder of constraint length 7. It is important to know what sequence of output code bits will be generated for a particular input stream. There are several techniques available to assist with this question, the most popular being connection pictorial, state diagram, tree diagram and trellis diagram.

However, to illustrate how these methods are applied, the simple example of half-rate (1/2) encoder will be considered with constraint length k = 3. The system has two modulo-2 adders, so that the code rate is 1/2. The input bit (m) placed into the first of the shift register causes the bits in the register to be moved one place to the right. The output switch samples the output of each modulo-2 adder, one after the other, to form a bit pair for the bit just entered. The connections from the register to the adders could be one, two or three interfaces for either adder. The choice depends on the requirement to produce a code with good distance properties. A similar encoder used by the Inmarsat standard-A is a half-rate convolutional encoder. Therefore, in terms of connections to the modulo-2, adders can be defined using generator polynomials in the encoder configuration. The polynomial for the generating arm (n) of the encoder $g_n(p)$ and the generator polynomials representing encoder $g_1(p)$ can have the following relations:

$$g_n(p) = g_0(p) + g_1 p^1 + \ldots g_n p^n \qquad\qquad (3.30.)$$

$$g_1(p) = 1 + p + p^2 = 1 + p^2 \qquad\qquad (3.31.)$$

where the value of g_1 takes on the value of 0 or 1 and a 1 is used to indicate that there is a connection between a particular element of the shift register and the modulo-2 adder. Thus, to provide a simple representation of the encoder, generator polynomials are used to predict the output coded message sequences for a given input sequences. For instance, the input sequence 10110 can be represented by the polynomial relation:

$$m(p) = 1 + p^2 + p^3 \qquad (3.32.)$$

Combining this with the respective generator polynomials and using the rules of module-2 arithmetic results in the following:

$$m(p)g_1(p) = (1 + p^2 + p^3)(1 + p + p^2) = 1 + p + p^5 \qquad (3.33.)$$

$$m(p)g_2(p) = (1 + p^2 + p^3)(1 + p^2) = 1 + p^3 + p^4 + p^5 \qquad (3.34.)$$

The output code sequence c(p) obtains by interleaving the above two products as follows:

$$c(p) = [1,1]p^0 + [1,0]p^1 + [0,0]p^2 + [0,1]p^3 + [0,1]p^4 + [1,1]p^5 \qquad (3.35.)$$

Here, the number between brackets represents the output code sequence.

The Inmarsat analog standard-A uses a HSD channel encoding configuration for the information data stream at 56 Kb/s. The scrambling sequence on the input data stream shall be provided by the scrambler before the convolutional encoder described in CCITT Recommendation V.35 scheme. The data stream then passes differential encoder state stage 1 followed by 1/2 (half) convolutional encoding with constant length k = 7. The half (1/2) rate convolutional encoder can provide two data streams to the QPSK modulator using two generator polynomials rates as follows: $G_1 = 1 + x^2 + x^3 + x^5 + x^6$ and $G_2 = 1 + x + x^2 + x^3 + x^6$. The encoder provides two parallel data streams to the modulator: I and Q, while (Q) should lag (I) by 90° in the modulator.

The Inmarsat digital standard-B for transmission and out-of-band signaling channels uses digital modulation and FEC in order to efficiently utilize satellite power and bandwidth. The basic modulation and coding techniques are filtered by 60% roll-off O-QPSK and 40% roll-off BPSK, both with convolutional coding at either rate: 1/2 or 3/4 FEC and 8-level soft decision Viterbi decoding (constraint length = 7). Hence, punctured coding is used to derive 3/4 and 1/2 rates. All BPSK channels are differentially encoded outside the FEC. The Inmarsat standard-M for all transmissions, with the exception of those fields carrying digitally coded voice, employs FEC with convolutional encoding of constraint length k = 7 and 8-level soft decision Viterbi decoding.

There are two generator polynomials rates: G_1 (133 octal) and G_2 (171 octal). The transmitted bit is nominated by 1 and deleted by 0. However, the first bit in each transmission frame is the output from the G_1 polynomial and all bits are transmitted at the rate of 1/2 code. The output data from the bit selector are punctured coded data of 3/4 rate.

3.4.2.4. Concatenated Codes

These codes were originally developed for deep space communications and occur when two separate coding techniques are combined to form a large code. The inner decoder is used to correct most of the errors introduced by the channel, the output of which is then fed into the outer decoder, which further reduces the bit error rate to the target level. That is to say, a typical concatenated coding scheme would employ half-rate convolutional encoding of constraint length 7 (2, 1, 7) – Viterbi decoding as the inner scheme and RS (255, 223) block encoding and decoding as the outer scheme. Interleaving between the inner and outer coders can be used to further improve the performance.

3.4.2.5. Turbo Codes

Turbo Codes are a new class of error correction codes that were introduced in 1993 by a group of researchers from France, along with a practical decoding algorithm. The major importance of these codes is that they enable reliable transmission with power efficiencies close to the theoretical limit predicted by Claude Shannon. Since their introduction, turbo codes have been proposed for low-power applications such as deep space and satellite communications, as well as for interference limited applications such as third generation cellular/personal communication services. Due to the use of a pseudo-random interleaver, turbo codes appear randomly to the channel, yet possess enough structure so that decoding can be physically realized.

Thus, developed for deep space and satellite communication applications, turbo codes offer a performance significantly better than concatenated codes. They are generated using two or more recursive systematic convolutional code generators concatenated in parallel. Here, the term recursive implies that some of the output bits of the convolutional encoder are fed back and applied to the input bit sequence. The first encoder takes the information bits as input. The key to the turbo code generation is the presence of a permuter, which performs a function similar to an interleaver; with the only difference that here the output sequence is pseudo-random. The permuter takes a block of information bits, which should be large to increase performance, for example more than 1000 bits and produces a random, delayed sequence of output bits, which is then fed into the second encoder. In such a manner, the outputs of the two encoders are partly bits transmitted along with the original information bits. Hence, in order to reduce the number of transmitted bits, the party bits are punctured prior to transmission.

From various simulation results it is recognized that turbo codes are capable of achieving an arbitrarily low BER of 10-5 at an Eb/No ratio of just 0.7 dB. For instance, in order to achieve this level of performance, large block sizes of 65,532 data bits are required. Because of this prohibitively enlarged block size, an original turbo code is not well suited for real time Tel communication systems such as IS-95 CDMA cellular standard. For that reason, the work on this problem has focused on the design of short block length codes, compatible with IS-95 standard.

3.4.3. Decoding

The complete transmission loop requires any type of encoder followed by modulation and transmitter via transmission channel to receiver, namely to demodulator and decoder. In such a manner, decoding is the reverse method of coding and every type of decoding on the transmit side needs the same convenient decoding method on the receive side.

3.4.3.1. Block Decoding

The simplest means of decoding block codes is by a method of correlation whereby the decoder makes a comparison between the received code word and all permissible code words, selecting that word that gives the nearest match. Decoding of such codes will also depend on whether error detection or error correction is required. Decoders of block codes generally cannot use soft decision outputs from the demodulator, unlike the decoders for convolutional codes

3.4.3.2. Convolutional Decoding

The effect of the transmission channel on the signal and the probability of detection of a 1 or 0 in the presence of Gaussian noise are important factors during detection. In such a manner, an output from the demodulator can be configured to give a correct decision regarding whether the incoming signal is 1 or 0. The process of decoding then depends on the two state inputs it receives.

An alternative demodulator configuration allows quantization of the predicted level which gives the decoder more necessary information regarding the probable state of the demodulator output. For example, if 3 bit ($2^3 = 8$ levels) quantization occurs then 0 0 0 would suggest a firm valuation of the level received as a 0. On the other hand, an 0 0 1 scheme suggests the 0 is received close to the threshold and this valuation as a 0 is made with less certainly. The reason for quantization is to provide the convolutional decoder with more information in order to correctly recover the transmitted information with better error performance probability.

3.4.3.3. Turbo Decoding

The turbo decoder operates by performing an interactive decoding algorithm, resulting in the partial transfer of an a priori likelihood estimate of the decoded bit sequence between the constituent decoders. Initially, the received information bits, which may be in some error due to the influence of the channel, are used to perform a priori likelihood estimates by the respective decoders. In a more precise sense, the decoders employ a Maximum a Posteriori (MAP) algorithm to perform converge on the likely sequence of data transfer, after which the interaction between decoders ceases and the output sequence is obtained from one of the decoders.

An interleaver can be placed between the output of Decoder 1 and the input of Decoder 2, to provide an additional weighted decision input into Decoder 2; similarly, a de-interleaver is placed at the output of Decoder 2, to provide feedback to Decoder 1. The decoding time is proportional to the number of interactions between decoders.

3.4.3.4. Sequential Decoding

A sequential decoder may be used for convolutional decoding and it operates in a similar manner to the Viterbi decoder. On receipt of the incoming code word sequence this decoder will penetrate into the tree according to a decision made regarding the best path to follow. For that reason, using a trial and error technique, the decoder will progress as long as the chosen path appears correct, otherwise it will backtrack to try a different route. At this point, either soft decision or hard decision decoding is possible with the sequential decoder, although soft decision would considerably increase the computational time and storage space required.

A major advantage of sequential decoding is that the number of states examined is independent of constraint length, allowing the use of large constraint lengths and low error probability. A disadvantage is the need to store input sequences while the decoder searches for its preferred route through the tree. If the average decodes rate falls below that of the average symbol arrival rate, there is a danger that the decoder cannot cope, causing a loss of input information.

3.4.3.5. Viterbi Decoding

Viterbi maximum likelihood decoding of convolutional codes provides the best possible results in the presence of random errors. Thus, in an attempt to match the output sequence received by the decoder, Viterbi's algorithm models the possible state transition through a trellis identical to that used by the encoder. Accordingly, the Viterbi decoding algorithm is a maximum likelihood path algorithm that takes advantage of the remaining path structure of convolutional codes. This method works by modeling the possible state transitions of the encoder and finding the output sequence that matches most closely to that received by the decoder. Its task is to realize that not all paths through the encoder states can contribute to the final decoded output and that many paths can be rejected after each frame is received, which keeps the problem to manageable proportions.

If the encoder remembers (v) bits, then there are 2^v possible memory states to be modeled by the decoder. Hence, this term dominates expressions for speed, complexity and cost of the decoder and currently imposes an upper limit of 8 to 10 on constraint length. By path maximum likelihood decoding means that of all the possible paths through the trellis, a Viterbi decoder chooses the convenient path, most likely in the probabilistic sense to have been transmitted. Viterbi decoders easily make use of either hard or soft decision making. This decoding can incorporate soft decisions very simply, which will almost double the error correction power of the code and this can provide an additional gain of up to 3 dB. Otherwise, the procedure for choosing the best Viterbi scheme is to maximize constraint length within the limits of cost and speed, to find a nonsymmetrical code with the best value of d_∞ and to use soft decisions. The maritime Inmarsat standard-B and multipurpose M utilize an 8-level soft decision Viterbi decoding in their channels (constraint length = 7).

3.4.4. Error Correction

There are several methods (such as ADPCM) that reduce the number of redundant bits in speech, audio and visual signals in order to make more economic use of bandwidth. First of all, it is necessary to consider methods that require the deliberate addition of redundant bits to messages. The added bits are very carefully chosen and error correction systems make it possible to achieve large savings in the power required to realize low BER. At the receiver the additional bits are used to detect any errors introduced by channels. To achieve this technique in MSC are employed: FEC, ARQ and Pseudo-noise and Interleaving. In such a way, it is also possible to combine FEC and ARQ in an integration form known as a Hybrid Error Correction (HEC) transmitting scheme. At this point, however, the HEC method is used to reduce BER and the number of retranslated blocks. Such an arrangement could also be used to provide feedback information to the transmitter regarding slow variations, such as a fading.

3.4.4.1. Forward Error Correction (FEC)

The FEC is a technique where errors are detected and corrected at the receiver. Thus, this scheme requires only a one-way transmission link, since the message contains parity bits used for detection and correction of errors. In such a way, it is working only on receiving Tlx mode in radio and satellite one-way transmissions. The basic FEC technique used in MSC can be classified into two major (already explained) categories such as Convolutional and Block codes. The FEC coding as a result of convolutional coding is used in Inmarsat standards for some voice, telex and signaling channels. For example, Inmarsat standard-B uses convolutional encoder of constraint length 7- and 8-level soft decision Viterbi decoder. The coding rate is either 3/4 or 1/2, while for voice channel the code rate 3/4 is used and is derived by puncturing the rate 1/2 with k = 7 convolutional code.

On the other hand, the association of both basic coding techniques results in an even more powerful FES scheme known as the concatenated coding system. This powerful FEC scheme has been introduced in recent years, for a considerable increase of the service quality without appreciable expansion of bandwidth. While the inner code, with Viterbi decoding, can correct a large part of the random errors and very short error bursts, the residual errors at the outputs of the Viterbi decoder tend to be grouped in bursts. Thus, using a properly chosen interleaving that cuts the error bursts into shorter ones, a high rate Reed-Solomon code can be used as the outer code in order to correct most of these dispersed errors bursts to achieve a very low bit error rate. Thus, the introduction of concatenated coding and trellis-coded modulation into MSC is the most remarkable event in the domain. **Table 3.1.** shows a list of FEC techniques along with their performance.

Table 3.1. Performances of FEC Techniques

Code	Decoding Technique	Gain (BER=10^{-5})	Gain (BER=10^{-8})	Bit Rate	Complexity
Convolutional	Threshold	1.5 – 3.0	2.5 – 4.0	Very High	Low
Convolutional	Viterbi (Soft Decision)	4.0 – 5.0	5.0 – 6.5	High	High
Convolutional	Sequential (Hard Decision)	4.0 – 5.0	6.0 – 7.0	High	Low
Convolutional	Sequential (Soft Decision)	6.0 – 7.0	8.0 – 9.0	Medium	Low
Concatenated (Convol./RS)	Viterbi Inner and Algebraic Outer	6.5 – 7.5	8.5 – 9.5	High	Medium
Concatenated (Short Block/ RS)	Soft Inner and Algebraic Outer	4.5 – 5.5	6.5 – 7.5	Medium	High
Short Block Linear	Soft Decision	5.0 – 6.0	6.5 – 7.5	Medium	High
Block (BSH/RS)	Algebraic (Hard Decision)	3.0 – 4.0	4.5 – 5.5	High	Medium

An FEC scheme can improve the quality of a digital transmission link by the following two aspects: **a)** A bit error rate reduction, closely related to the service quality criterion and **b)** a saving in the E_b/N_0 or C/N_0 to be considered in the link budget. The E_b/N_0 or C/N_0 saving is often called the coding gain, expressed in dB as a difference at certain BER values, of the coded system and the reference noncoded one. In the comparison between different transmission schemes, E_b/N_0 is usually used because it is independent of the coding scheme, where the gain is given as follows:

$$G = (E_b/N_0) \text{ ref} - (E_b/N_0) \text{ cod} \quad [\text{dB}] \tag{3.36.}$$

The merit of a coding system can also be appreciated in terms of the savings in C/N_0 and C/N, then, considering information rate (R_b) and information transmission bandwidth, the equations for carrier-to-noise density ratio (C/N_0) and the carrier-to-noise ratio (C/N) are:

$$(C/N_0) = (E_b/N_0) + \log R_b \quad \text{and} \quad (C/N) = (E_b/N_0) + \log R_b - 10 \log W \quad [\text{dB}] \tag{3.37.}$$

A coding gain in E_b/N_0 means in general a gain in C/N but the coding in C/N depends on the bandwidth expansion with respect to the reference system. It is however possible to have a coding gain without bandwidth expansions using Trellis Coded Modulation (TCM).

3.4.4.2. Automatic Request Repeat (ARQ)

The ARQ is a technique with which a high degree of data integrity is required and latency is not a significant factor. In reality, the ARQ scheme, based on error detection coding and a retransmission protocol, is well adapted to the situation where a two-way channel is available.

Typical examples of such systems can be encountered in a computer data network using satellite links. However, it is worthy of notice that the ARQ and improved ARQ, as well as HEC techniques, are widely used in modern digital communications and storage systems.

The ARQ method requires a two-way link, since a receiver, detecting an error, does not attempt to correct it but simply requests the transmitter to retransmit the message. Thus, the ARQ scheme basically works with the following modus:

1) Stop and Wait ARQ – After each message block is sent via satellite link, the transmitter waits for acknowledgement. If the message block received is in error, the transmitter will retransmit that block but if this message is correctly received, the next message block is transmitted. A half-duplex link is required, transmission on the link is possible in both directions but not at the same time.

2) Continuous ARQ with Repeat – The transmitter sends and the receiver acknowledges message blocks continuously. Hence, any message block not correctly received causes the transmitter to return to the block in question (incorrect received block) and recommence continuous retransmission from there. A full-duplex link is necessary for transmission in both directions simultaneously.

3) Continuous ARQ with Selective Repeat – In this ARQ arrangement only the block received in error is retransmitted and the transmitter continues from where it left off at the last block, instead of repeating all the subsequent even correctly received messages. In such a manner, however, full-duplex link is also necessary for transmission in both directions simultaneously.

A major advantage of ARQ compared with FEC is that decoding equipment for error corrections can be simpler and the redundancy in the total message stream is less. The ARQ efficiency is good for low error ratios but for high ratios requiring retransmission of a large number of message blocks, the system becomes inefficient. A disadvantage of ARQ is the variability of the delays experienced from end-to-end of the link and so, the possible requirement for large data stores of incoming data blocks.

The Inmarsat standard-C uses packets of data and each one transmitted contains a 16-bit checksum field. After that, the receiver completes an expected checksum for each packet and compares this with the actual packet received in order to verify that the packet has been correctly received. The ARQ method is used if the packet received is in error.

3.4.4.3. Pseudo Noise (PN)

The PN generator will produce a set of cyclic codes with good distance properties. Thus, the name of the sequence is given, because the sequence, although deterministic, appears to have the properties of sampled white noise. Furthermore, a PN sequence is easily generated using shift registers, and has a correlation function that is high packet for zero delay and approximates to zero for other delays. The PN sequence, being deterministic, is usually for synchronization purposes between a transmitter and receiver.

Some Inmarsat standards use a scrambler circuit before FEC encoding and a descrambler at the receive end following FEC decoding. For Inmarsat standard-B and M, for instance, the scrambler/descrambler circuits are PN generators using 15 stages.

The scrambler/descrambler circuits are clocked at the rate of one shift per information bit. The first bit into the scrambler at the beginning of a frame is modulo-2, added with the output of the scrambler shift generator, corresponding to the initial state-scrambling vector. The initial state of the shift register is located at the beginning of a burst and a frame.

Considering the Inmarsat standard-M, the initial state of the scrambler shift registers at the LES (for SCPC channel operating in voice mode) and is sent to the LES by the MES at the start of a call as part of a call set-up sequence. The MES chooses any initial state (except all zeros) on a random basis for each call and signals this scrambling vector message (8D in hexadecimal form or 10001101) for implementation at the LES with the Least Significant Bit (LSB) in shift register No 1 and the Most Significant Bit (MSB) in shift register No 15 of the scrambler. The MES simultaneously sets the descrambler shift register with the same scrambling vector. Otherwise, for MES-to-LES channels, a fixed initial state default value of 6,959 in hexadecimal form or 110100101011001, is used in MES scramblers and LES descramblers.

3.4.4.4. Interleaving

As is well known, the MSC transmission channel introduces errors of a bursting nature. Hence, in the short term, the errors introduced by the channel cannot be considered to be statistically independent or memory less, the criterion upon which most coders (block and convolutional) optimally operate. In order to mimic a statistically independent channel, a technique known as interleaving is incorporated into the transmitter chain after the output of the encoder and from the interleaver the input signal passes via the modulator. In reverse mode the output signal goes through the demodulator, deinterleaver and channel decoder. This circle presents the interleaver/deinterleaver segment within the transmission/reception chain in the satellite link. Hence, the role of the interleaver is to re-order the transmission sequence of the bits that make up the code words in some predetermined fashion, such that the effect of an error burst is minimized.

Interleaving can be performed for both block and convolutional codes. Block interleaving is achieved by firstly storing the output code words of the encoder into a two-dimensional array. Consider the case of an $(m \cdot n)$ array, where (m) is the number of code words to be interleaved and (n) is the number of code word bits. Thus, each row of the array comprises a generated code word. Once the array is full, the contents are then output to the transmitter but in this case, data is read out on a column-by-column basis.

Generally speaking, the transmission of each symbol of a particular code word will be non-sequential. Namely, the input signal goes via the input sequence into the interleaver block and after processing the output sequence would correspond to the chain starting with C_{11}, C_{21}, C_{31}, C_{41}, C_{51}, C_{61}, C_{71}, C_{12}-C_{72}, … until C_{18}-C_{78}. At this point, the effect of any error bursts will have been dispersed in time throughout the transmitted code words. Convolutional interleavers work along similar lines, achieving performance characteristics similar to block interleaving.

At the receiver, the inverse of the interleaving function is performed by a deinterleaver and the original code words are reconstituted prior to feeding into the encoder. Namely, a burst of error affecting the transmitted bits indicated by the chain coming from interleaver block would be dispersed among the code words at the receiver.

3.5. Multiple Access Technique

In satellite communication systems, as a rule, many users are active at the same time. The problem of simultaneous communications between many single or multipoint satellite users, however, can be solved by using Multiple Access (MA) technique. Since the resources of the systems such as the transmitting power and the bandwidth are limited, it is advisable to use the channels with complete charge and to create a different MA to the channel. This generates a problem of summation and separation of signals in the transmission and reception parts, respectively. Deciding this problem consists in the development of orthogonal channels of transmission in order to divide signals from various users unambiguously on the reception part. There are five the following principal forms of MA techniques:

1) Frequency Division Multiple Access (FDMA) is a scheme where each concerned LES or MES is assigned its own different working carrier radio frequency inside the spacecraft transponder bandwidth.

2) Time Division Multiple Access (TDMA) is a scheme where all concerned Earth stations use the same carrier frequency and bandwidth with time sharing, non-overlapping intervals.

3) Code Division Multiple Access (CDMA) is a scheme where all concerned Earth stations simultaneously share the same bandwidth and recognize the signals by various processes, such as code identification. Actually, they share the resources of both frequency and time using a set of mutually orthogonal codes, such as a Pseudorandom Noise (PN) sequence.

4) Space Division Multiple Access (SDMA) is a scheme where all concerned Earth stations can use the same frequency at the same time within a separate space available for each link.

5) Random (Packet) Division Multiple Access (RDMA) is a scheme where a large number of satellite users share asynchronously the same transponder by randomly transmitting short burst or packet divisions.

Currently, these methods of multiple access are widely in use with many advantages and disadvantages, together with their combination of hybrid schemes or with other types of modulations. Hence, multiple access technique assignment strategy can be classified into three methods as follows: **(1)** Preassignment or fixed assignment; **(2)** Demand Assignment (DA) and **(3)** Random Access (RA); the bits that make up the code words in some predetermined fashion, such that the effect of an error burst is minimized.

In the preassignment method channel plans are previously determined for chairing the system resources, regardless of traffic fluctuations. Therefore, this scheme is suitable for communication links with a large amount of steady traffic. However, since most mobile users in MSC do not communicate continuously, the preassignment method is wasteful of the satellite resources. In Demand Assignment Multiple Access (DAMA) satellite channels are dynamically assigned to users according to the traffic requirements. Due to high efficiency and system flexibility, DAMA schemes are suited to MSC systems. In RA a large number of mobile users use the satellite resources in bursts, with long inactive intervals. In effect, to increase the system throughout, several mobile Aloha methods have been proposed.

Figure 3.7. Multiple Access Techniques

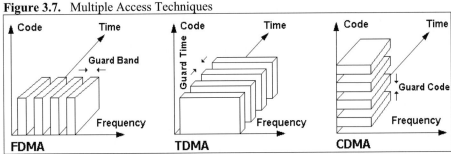

Therefore, the MA techniques permit more than two Earth stations to use the same satellite network for interchanging information. Several transponders in the satellite payload share the frequency bands in use and each transponder will act independently of the others to filter out its own allocated frequency and further process that signal for transmission. Thus, this feature allows any LES located in the corresponding coverage area to receive carriers originating from several MES and vice versa and carriers transmitted by one MES can be received by any LES. This enables a transmitting Earth station to group several signals into a single, multi-destination carrier. Access to a transponder may be limited to single carrier or many carriers may exist simultaneously. The baseband information to be transmitted is impressed on the carrier by the single process of multi-channel modulation.

3.5.1. Frequency Division Multiple Access (FDMA)

The most common and first employed MA scheme for satellite communication systems is FDMA concept shown in **Figure 3.7. (FDMA),** where transmitting signals occupy non-overlapping frequency bands with guard bands between signals to avoid interchannel interference. The bandwidth of a repeater channel is therefore divided into many sub-bands each assigned to the carrier transmitted by an Earth station. The MES transmit continuously and the channel transmits several carriers simultaneously at a series of different frequency bands. Because of interchannel interference, it is necessary to provide guard intervals between each band occupied by a carrier to allow for the imperfections of oscillators and filters. The downlink Rx selects the required carrier in accordance with the appropriate frequency. When the satellite transponder is operating close to its saturation, nonlinear amplification produces intermodulation (IM) products, which may cause interference in the signals of other users. In order to reduce IM, it is necessary to operate the transponder by reducing the total input power according to input back off and that the IF amplifier provides adequate filtering.

Therefore, FDMA allocates a single satellite channel to one user at once. In fact, if the transmission path deteriorates, the controller switches the system to another channel. Although technically simple to implement, FDMA is wasteful of bandwidth because the voice channel is assigned to a single conversation, whether or not somebody is speaking. Moreover, it cannot handle alternate forms of data, only voice transmissions. This system's advantages are that it is a simple technique using equipment proven over decades to be reliable and it will remain very commonly in use because of its simplicity and flexibility.

It does have some disadvantages however:

1) A FDMA method is the relatively inflexible system and if there are changes in the required capacity, then the frequency plan has to change and thus, involve many LES.

2) Multiple carriers cause IM in both the MES HPA and in the transponder HPA. Reducing IM requires back off of the HPA power, so it cannot be exploited at full capacity.

3) As the number of carriers increase, the IM products between carriers also increase and more HPA back off is needed to optimize the system. The throughput decreases relatively rapidly with the number of transmission carriers, therefore for 25 carriers it is about 40% less than with 1 carrier.

4) The FM system can suffer from what is known as a capture effect, where if two received signals are very close in frequency but of different strengths, the stronger one tends to suppress the weaker one. For this reason the carrier power has to be controlled carefully.

Therefore, with the FDMA technique, the signals from the various users are amplified by the satellite transponder in a given allocated bandwidth at the same time but at different frequencies. Depending on the multiplexing and modulation techniques employed, several transmission hybrid schemes can be considered and in general may be divided into two categories, based on the traffic demands of Earth stations on MCPC and SCPC.

3.5.1.1. Multiple Channels Per Carrier (MCPC)

The main elements of the MCPC are multiplexer, modulator and transmitter using a satellite uplink, when LES multiplexes baseband data is received from a terrestrial network and destined for various MES. Moreover, the multiplexed data are modulated and transmitted to the allocated frequency segment, when the bandwidth of the transponder is shared among several MES, each with different traffic requirements. The transponder bandwidth is divided into several fixed segments, with several time frequency divisions allocated to these MES terminals. Namely, between each band segment is a guard band, which reduces the bandwidth utilization efficiency and the loss is directly related to the number of accessing MES in the network as shown in **Figure 3.7. (FDMA)**. Depending on the number of receiving MES, a total number of carriers will pass through the satellite transponder.

On the other hand, the signals received from different MES extract the carrier containing traffic addressed to LES by using an appropriate RF filter, demodulator, baseband filter and demultiplexer. The output of the demodulator consists in multiplexed telephone channels for a few MES together with the channels addressed to them. A baseband filter is used to filter out the desired baseband frequency segment and finally, a demultiplexer retrieves individual telephone channels and feeds them into the terrestrial network for onward transmission.

Each baseband filter of LES receive stations in this scheme corresponds to a specific one in the LES transmitting station. However, any change in channel capacity requires the return of this filter, which is difficult to implement. Thus, many schemes may be categorized according to the type of baseband signal.

3.5.1.2. Single Channel Per Carrier (SCPC)

For certain applications, such as the provision of MSC service to remote areas or individual MES, traffic requirements are low. In reality, assigning multiple channels to each MES is wasteful of bandwidth because most channels remain unutilized for a significant part of the day. For this type of application the SCPC type of FDMA is used. In the SCPC system each carrier is modulated by only one voice or by low to medium bit rate data channel. Some old analog systems use Companded FM but most new systems are digital PSK modulated.

In the SCPC scheme, each carrier transmits a single carrier. The assignment of transponder channels to each MES may be fixed Pre-Assigned Multiple Access (PAMA) or variable Demand-Assigned Multiple Access (DAMA), the channel slots of the transponder are assigned to different MES according to their instantaneous needs. In the case of PAMA, a few SCPC channels, about 5 to 10, are permanently assigned to each MES. In case of DAMA, a pool of frequency is shared by many MES terminals. When necessary, each MES requests a channel from frequency management of the Network Control Station (NCS), which may always attempt to choose the best available channel or a lower quality channel until an unoccupied channel has been found. The allocation is then announced on a signaling channel known as a broadcast channel. The announcement is received by the calling and called MES, which then tune to the allocated channel. The communication takes place on the allocated channel and the end of call is announced by a signaling message, following which the NCS returns the channel to the common pool.

In addition, the SCPS solution requires an Automatic Frequency Control (AFC) pilot to maintain the spectrum centering on a channel-by-channel basis. This is usually achieved by transmitting a pilot tone in the centre of the transponder bandwidth. It is transmitted by designated reference LES and all the MES use this reference to correct their transmission frequency. A receiving station uses the pilot tone to produce a local AFC system which is able to control the frequency of the individual carriers by controlling the frequency of the LO. Hence, drift in MSC translation frequency and frequency variations caused by the Doppler effect and the carriers retain their designated frequencies relative to each other. This feature is essential, because if uncorrected, the sum of the total frequency error can cause carrier overlapping, as carrier bandwidths are small. Thus, a stable receive frequency permits the LES demodulator design to be simplified. Centrally controlled networks, such as Inmarsat MSS A, B, C, M and other networks are simple to manage missions, because they provide a higher usage of channels and can use simple demand-assignment equipment. The SCPS scheme is cost-effective for networks consisting in a significant number of Earth stations, each needing to be equipped with a small number of channels.

3.5.2. Forms of FDMA Operations

There are several hybrid schemes of multiplexed FDMA in combination with SCPS, PSK, TDM and TDMA techniques.

3.5.2.1. SCPC/FM/FDMA

The baseband signals from the network or users each modulate a carrier directly, in either analog or digital form according to the nature of SCPC signal in question. Therefore, each carrier accesses the satellite on its particular frequency at the same time as other carriers on the different frequencies from the same or other station terminals. Information routing is thus, performed according to the principle of one carrier per link.

The Inmarsat-A standard use SCPS, utilizing analog transmission with FM for telephone channels. Thus, in calculating the channel capacity of the SCPC/FM system it is necessary to ensure that the noise level does not exceed specified defined values. Therefore, the CCIR

Recommendations for an analog channel state that the noise power at a point of zero, the relative level should not exceed 10,000 WOP with a 50 dB test tone, namely the noise ratio. In this way, it is assumed that the minimum required carrier-to-noise ratio per channel is at least 10 dB.

3.5.2.2. SCPC/PSK/FDMA

In this arrangement, each voice or data channel is modulated onto its own RF carrier. The only multiplexing occurs in the transponder bandwidth, where frequency division produces individual channels within the bandwidth. Various types of this multiplex scheme are used in channels of the Inmarsat standard-B system. In this case, the satellite transponder carrier frequencies may be PAMA or DAMA. For PAMA carriers the RF is assigned to a channel unit and the PSK modem requires a fixed-frequency LO input. For DAMA, the channels may be connected according to the availability of particular carrier frequencies within the transponder RF bandwidth. For this arrangement, the SCPC channel frequency required is produced by a frequency synthesizer.

The forward link assigned by TDM in shore-to-ship direction uses the SCPC/DA/FDMA solution for Inmarsat standard-B voice/data transmission. This standard in the return link for channel request employs Aloha O-QPSK and for low speed data/telex uses the TDMA scheme in ship-to-shore direction. The Inmarsat-Aero in forward ground-to-aircraft direction uses packet mode TDM for network broadcasting, signaling and data and the circuit mode of SCPS/DA/FDMA with distribution channel management for service communication links. Thus, the request for channel assignment, signaling and data in the return aircraft-to-ground direction the Slotted Aloha BPSK (1/2 – FES) of 600 b/s is employed and consequently, the TDMA scheme is reserved for data messages.

3.5.2.3. TDM/FDMA

This arrangement allows the use of TDM groups to be assembled at the satellite in FDMA, while the PSK is used as a modulation process at the Earth station. Systems such as this are compatible with FDM/FDMA carriers sharing the same transponders and the terminal requirements are simple and easily incorporated. The Inmarsat standard-B system for telex low speed data uses this scheme in the shore-to-ship direction only and in the ship-to-shore direction uses TDMA/FDMA. The CES TDM and SES TDMA carrier frequencies are pre-allocated by Inmarsat. Each CES is allocated at least one forward CES TDM carrier frequency and a return SES TDMA frequency. Thus, additional allocations can be made depending on the traffic requirements.

The channel unit associated with the CES TDM channel for transmission consists in a multiplexer, different encoder, frame transmit synchronizer and modulator. So at the SES, the receive path of the channel has the corresponding functions to the transmitted end. The CES TDM channels use BPSK with differential coding, which is used for phase ambiguity resolution at the receive end.

3.5.2.4. TDMA/FDMA

As previously stated, however, the TDMA signals could occupy the complete transponder bandwidth. In fact, a better variation of this is where the TDMA signals are transmitted as a sub-band of transponder bandwidth, the remainder of which being available for example for SCPC/FDMA signals. Thus, the use of a narrowband TDMA arrangement is well suited for a system requiring only a few channels and has the all advantages of satellite digital transmission but can suffer from intermodulation with the adjacent FDMA satellite channels. Accordingly, the practical example of this scheme is the Tlx service of the Inmarsat-B system in ship-to-shore direction, which, depending on the transmission traffic, offers a flexible allocation of capacity for satellite communication and signaling slots.

3.5.3. Time Division Multiple Access (TDMA)

The TDMA application is a digital MA technique that permits individual Earth station transmissions to be received by the satellite in separate, non-overlapping time slots, called bursts, which contain buffered information. The satellite receives these bursts sequentially, without overlapping interference and is then able to retransmit them to the MES terminal. Synchronization is necessary and is achieved using a reference station from which burst position and timing information can be used as a reference by all other stations. Each MES must determine the satellite system time and range so that the transmitted signal bursts, typically QPSK modulated, are timed to arrive at the satellite in the proper time slots. The offset QPSK modulation is used by Inmarsat-B SES. So as to ensure the timing of the bursts from multiple MES, TDMA systems use a frame structure arrangement to support telex in the ship-to-shore direction. Therefore, a reference burst is transmitted periodically by a reference station to indicate the start of each frame to control the transmission timing of all data bursts. A second reference burst may also follow the first in order to provide a means of redundancy. In the proper manner, to improve the imperfect timing of TDMA bursts, several synchronization methods of random access, open-loop and closed-loop have been proposed.

In **Figure 3.7. (TDMA)** a concept of TDMA is illustrated, where each mobile terminal transmits a data burst with a guard time to avoid overlaps. Since only one TDMA burst occupies the full bandwidth of the satellite transponder at a time, input back off, which is needed to reduce IM interference in FDMA, is not necessary in TDMA. At any instant in time, the transponder receives and amplifies only a single carrier. Thus, there can be no IM, which permits the satellite amplifier to be operated in full HPA saturation and the transmitter carrier power need not be controlled. Because all MES transmit and receive at the same frequency, tuning is simplified. This results in a significant increase in channel capacity. Another advantage over FDMA is its flexibility and time-slot assignments are easier to adjust than frequency channel assignments. The transmission rate of TDMA bursts is about 4,800 b/s, while the frame length is about 1.74 seconds and the optimal guard time is approximately 40 msec, using the open-loop burst synchronization method.

There are some disadvantages because TDMA is more complex than FDMA:

1) Two reference stations are needed and complex computer procedures, for automated synchronizations between MES terminals.

2) Peak power and bandwidth of individual MES terminals need to be larger than with FDMA, owing to high burst bit rate.

Therefore, in the TDMA scheme, the transmission signals from various users are amplified at different times but at the same nominal frequency, being spread by the modulation in a given bandwidth. Depending on the multiplexing techniques employed, two transmission hybrid schemes can be introduced for use in MSS.

3.5.3.1. TDM/TDMA

The Inmarsat analog standard-A uses the TDM/TDMA arrangement for telex transmission. Each SES has at least one TDM carrier and each of the carriers has 20 telex channels of 50 bauds and a signaling channel. Moreover, there is also a common TDM carrier continuously transmitted on the selected idle listening frequency by the NCS for out-of-band signaling. The SES remains tuned to the common TDM carrier to receive signaling messages when the ship is idle or engaged in a telephone call. When an SES is involved in a telex forward call it is tuned to the TDM/TDMA frequency pair associated with the corresponding CES to send messages in shore-to-ship direction. Telex transmissions in the return ship-to-shore direction form a TDMA assembly at the satellite transponder. Each frame of the return TDMA telex carrier has 22 time slots, while each of these slots is paired with a slot on the TDM carrier. The allocation of a pair of time slots to complete the link is received by the SES on receipt of a request for a telex call. Otherwise, the Inmarsat-A uses for forward signaling a telex mode, while all other MSS Inmarsat standards for forward signaling and assignment channels use the TDM BPSK scheme.

The new generation Inmarsat digital standard-B (inheritor of standard-A) uses the same modulation TDM/TDMA technique but instead of Aloha BPSK (BCH) at a data rate of 4800 b/s for the return request channel used by Inmarsat-A, new standard-B is using Aloha O-QPSK (1/2 – FEC) at a data rate of 24 Kb/s. This MA technique is also useful for the Inmarsat standard-C terminal for maritime, land and aeronautical applications. In this case, the forward signaling and sending of messages in ground-to-mobile direction use a fixed assigned TDM carrier. The return signaling channel uses hybrid, slotted Aloha BPSK (1/2 FEC) with a provision for receiving some capacity and the return message channels in the mobile-to-ground direction are modulated by the TDMA system at a data rate of 600 b/s.

3.5.3.2. FDMA/TDMA

The Iridium system employs a hybrid FDMA/TDMA access scheme, which is achieved by dividing the available 10.5 MHz bandwidth into 150 channels introduced into the FDMA components. Each channel accommodates a TDMA frame comprising eight time slots, four for transmission and four for reception. Each slot lasts about 11.25 msec, during which time data are transmitted in a 50 Kb/s burst. Each frame lasts 90 msec and a satellite is able to support 840 channels. Thus, a user is allocated a channel occupied for a short period of time, during which transmissions occur.

3.5.4. Code Division Multiple Access (CDMA)

The CDMA solution is based on the use the modulation technique also known as Spread Spectrum Multiple Access (SSMA), which means that it spreads the information contained in a particular signal of interest over a much greater bandwidth than the original signal.

In this MA scheme the resources of both frequency bandwidth and time are shared by all users employing orthogonal codes, shown in **Figure 3.7. (CDMA).** Therefore, the CDMA is achieved by a PN (Pseudo-Noise) sequence generated by irreducible polynomials, which is the most popular CDMA method. In this way, a SSMA method using low-rate error correcting codes, including orthogonal codes with Hadamard or waveform transformation has also been proposed.

Concerning the specific encoding process, each user is actually assigned a signature sequence, with its own characteristic code, chosen from a set of codes assigned individually to the various users of the system. This code is mixed, as a supplementary modulation, with the useful information signal. On reception, from all the signals that are received, a given user is able to select and recognize, by its own code, the signal, which is intended for it, and then to extract useful information. The other received signal can be intended for other users but they can also originate from unwanted emissions, which gives CDMA a certain anti-jamming capability. For this operation, where it is necessary to identify one CDMA transmission signal among several others sharing the same band at the same time, correlation techniques are generally employed.

From a commercial and military perspective this MA is still new and has significant advantages. Interference from adjacent satellite systems including jammers is better solved than with other systems. This scheme is simple to operate as it requires no synchronization of the transmitter and is more suited for a military MES. Small mobile antennas can be very useful in these applications, without the interference caused by wide antenna bandwidths. Using multibeam satellites, frequency reuse with CDMA is very effective and allows good flexibility in the management of traffic and the orbit/spectrum resources. The Power Flux Density (PFD) of the CDMA signal received in the service area is automatically limited, with no need for any other dispersal processes. It also provides a low probability of intercept of the users and some kind of privacy, due to individual characteristic codes. The main disadvantage of CDMA by satellite is that the bandwidth required for the space segment of the spread carrier is very large, compared to that of a single unspread carrier, so the throughput is somewhat lower than with other systems.

Therefore, in the CDMA scheme, the signals from various users operate simultaneously, at the same nominal frequency but are spread in the given allocated bandwidth by a special encoding process. Depending on the multiplexing techniques employed the bandwidth may extend to the entire capacity of the transponder but is often restricted to its own part, so CDMA can possibly be combined in the hybrid scheme with FDMA and/or TDMA. The SSMA technique can be classified into two methods: Direct Sequence (DS) and Frequency Hopping (FH). A combined system of DC and FH is called a hybrid CDMA system and the processing gain can be improved without increases of chip rate. The hybrid system has been used in the military Joint Tactical Information Distribution System (JTIDS) and OmniTRACS, which is Ku-band MSS developed by the Qualcomm Company. In a more precise sense, the CDMA technique was developed by experts of the Qualcomm Company in 1987. At present, the CDMA system advantages are practically effective in new satellite systems, such as Globalstar, also developed by Qualcomm, which is devoted to MSS handheld terminals and Skybridge, involved in FSS. This type of MA is therefore attractive for handheld and portable MSS equipment with a wide antenna pattern.

Antennas with large beam widths can otherwise create or be subject to interference with adjacent satellites. In any case, this MA technique is very attractive for commercial, military and even TT&C communications because some Russian satellites use CDMA for command and telemetry purposes.

The Synchronous-CDMA (S-CDMA) scheme proves efficiently to eliminate interference arising from other users sharing the same carrier and the same spot beam. Interference from other spot beams that overlap the coverage of the intended spot is still considerable. This process to ensure orthogonality between all links requires signalling to adjust transmission in time and frequency domains for every user independently.

3.5.4.1. Direct Sequence (DS) CDMA

This dominant DS-CDMA technique is also called Pseudo-Noise (PN) modulation, where the modulated signal is multiplied by a PN code generator, which generates a pseudo-random binary sequence of length (N) at a chip rate (R_c), much larger than information bit rate (R_b), with a relation as follows:

$$R_c = N \cdot R_b \tag{3.38.}$$

This sequence is combined with the information signal cut into small chip rates (R_c), thus, speeding the combined signal in a much larger bandwidth ($W \sim R_c$), namely, the resulting signal has wider frequency bandwidth than the original modulated signal. The transmitting signal can be expressed in the following way:

$$s(t) = m(t)\, p(t)\, \cos(2\pi f_c t) = m(t)\, p(t)\, \cos \omega_c t \tag{3.39.}$$

where $m(t)$ = binary message to be transmitted and $p(t)$ = spreading NP binary sequence. At the receiver the signal is coherently demodulated by multiplying the received signal by a replica of the carrier. Neglecting thermal noise, the receiving signal at the input of the detector of Low-Pass Filter (LPF) is given by:

$$r(t) = m(t)\, p(t)\, \cos \omega_c t\, (2 \cos \omega_c t) = m(t)\, p(t) + m(t)\, p(t)\, \cos 2\omega_c t \tag{3.40.}$$

The detector LLF eliminates the HF components and retains only the LW components, such as $u(t) = m(t)\, p(t)$. This component is then multiplied by the local code [$p(t)$] in phase with the received code, where the product $p(t)^2 = 1$. At the output of the multiplier this gives:

$$x(t) = m(t)\, p(t)\, p(t) = m(t)\, p(t)^2 = m(t) \quad [V] \tag{3.41.}$$

The signal is then integrated over one bit period to filter the noise. The transmitted message is recovered at the integrator output, so in fact, only the same PN code can achieve the despreading of the received signal bandwidth. In this process, the interference or jamming spectrum is spread by the PN codes, while other user's signals, spared by different PN codes, are not despread. Interference or jamming power density in the bandwidth of the received signal decreases from their original power. Otherwise, the most widely accepted measure of interference rejection is the processing gain (G_p), which is given by the ratio R_c/R_b and value of $G_p = 20 - 60$ dB. The input and output signal-to-noise ratios are related as follows:

$$(S/N)_{Output} = G_p\, (S/N)_{Input} \tag{3.42.}$$

In the forward link, the LES (Hub Station) transmits the spread spectrum signals, which are spread with synchronized PN sequence to different MSC users. Since orthogonal codes can be used, the mutual interference in the network is negligible and the channel capacity is close to that of FDMA. Conversely, in the return link, the signals transmitted from different

MES users are not synchronized and they are not orthogonal. The first case is referred to as synchronous and the second case as asynchronous SSMA.

However, the nonorthogonality causes interference due to the transmission of other MES in the satellite network and as the number of simultaneously accessing users increases, the communication quality gradually degrades in a process called Graceful Degradation.

3.5.4.2. Frequency Hopping (FH) CDMA

The FH-CDMA system works similarly to the DS system, since a correlation process of de-hopping is also performed at the receiver. The difference is that here the pseudo-random sequence is used to control a frequency synthesizer, which results in the transmission of each information bit rate in the form of (N) multiple pulses at different frequencies in an extended bandwidth. The transmitted and received signals have the following forms:

$$s(t) = m(t) \cos \omega_c(t) \, t \tag{3.43.}$$

$$r(t) = m(t) \cos \omega_c(t) \, t \cdot 2 \cos \omega_c(t) \, t = m(t) + m(t) \cos 2\omega_c(t) \, t$$

Thus, at the receiver the carrier is multiplied by an unmodulated carrier generated under the same conditions as at the transmitter. The second term in the receiver is eliminated by the LPF of the demodulator. The relation of processing gain for FH is:

$$G_p = W/\Delta f \tag{3.44.}$$

where W = frequency bandwidth and Δf = bandwidth of the original modulated signal. At this point, coherent demodulation is difficult to implement in FH receivers because it is a problem to maintain phase relation between the frequency steps. Due to the relatively slow operation of the frequency synthesizer, DS schemes permit higher code rates than FH radio systems.

3.5.5. Space Division Multiple Access (SDMA)

The significant factor in the performance of MA in a satellite communications system is interference caused by different factors and other users. In the other words, the most usual types of interference are co-channel and adjacent channel interference. The co-channel interference can be caused by transmissions from non-adjacent cells or spot beams using the same set of frequencies, where there is minimal physical separation from neighboring cells using the same frequencies, while the adjacent channel interference is caused by RF leakage on the subscriber's channel from a neighboring cell using an adjacent frequency. This can occur when the user's signal is much weaker than that of the adjacent channel user. Signal to Interference Ratio (SIR) is an important indicator of call quality; it is a measure of the ratio between the mobile phone signal (the carrier signal) and an interfering signal. A higher SIR ratio means increasing overall system capacity.

Taking into account that within the systems of satellite communications, every user has their own unique spatial position, this fact may be used for the separation of channels in space and as a consequence, to increase he SIR ratio by using SDMA. In effect, this method is physically making the separation of paths available for each satellite link.

Terrestrial telecommunication networks can use separate cables or radio links but on a single satellite, independent transmission paths are required. Thus, this MA control radiates energy into space and transmission can be on the same frequency: such as TDMA or CDMA and on different frequencies, such as FDMA.

In using SDMA, either FDMA or TDMA are needed to allow LES to roam in the same satellite beam or for polarization to enter the repeater. Thus, the frequency reuse technique of same frequency is effectively a form of SDMA scheme, which depends upon achieving adequate beam-to-beam and polarization isolation. Using this system reverse line means that interference may be a problem and the capacity of the battery is limited.

On the other hand, a single satellite may achieve spatial separation by using beams with horizontal and vertical polarization or left-hand and right-hand circular polarization. This could allow two beams to cover the same Earth surface area, being separated by the polarization. Thus, the satellite could also have multiple beams using separate antennas or using a single antenna with multiple feeds. For multiple satellites, spatial separation can be achieved with orbital longitude or latitude and for intersatellite links, by using different planes. Except for frequency reuse, this system provides on-board switching techniques, which, in turn, enhance channel capacity. Additionally, the use of narrow beams from the satellite allows the Earth station to operate with smaller antennas and so produce a higher power density per unit area for a given transmitter power. Therefore, through the careful use of polarization, beams (SDMA) or orthogonal (CDMA), the same spectrum may be reused several times, with limited interference among users.

The more detailed benefits of an SDMA system include the following:

1) The number of cells required to cover a given area can be substantially reduced.

2) Interference from other systems and from users in other cells is significantly reduced.

3) The destructive effects of multipath signals, copies of the desired signal that have arrived at the antenna after bouncing from objects between the signal source and the antenna can often be mitigated.

4) Channel reuse patterns of the systems can be significantly tighter because the average interference resulting from co-channel signals in other cells is markedly reduced.

5) Separate spatial channels can be created in each cell on the same conventional channel. In other words, intra-cell reuse of conventional channels is possible.

6) The SDMA station radiates much less total power than a conventional station. One result is a reduction in network-wide RF pollution. Another is a reduction in power amplifier size.

7) The direction of each spatial channel is known and can be used to accurately establish the position of the signal source.

8) The SDMA technique is compatible with almost any modulation method, bandwidth, or frequency band including GSM, PHP, DECT, IS-54, IS-95 and other formats. The SDMA solution can be implemented with a broad range of array geometry and antenna types.

Another perspective of the realization of SDMA systems is the application of smart antenna arrays with different levels of intelligence consisting in the antenna array and digital processor. Since the frequency of transmission for satellite communications is high enough (mostly 6 or 14 GHz), that the dimensions of an array placed in orbit is commensurable with the dimensions of the parabolic antenna, is a necessary condition to put such systems into orbit. Thus, the SDMA scheme mostly responds to the demands of LEO and MEO constellations, when the signals of users achieve the satellite antenna under different angles ($\pm22°$ for the MEO). In this instance, ground level may be split into the number of zones of service coverage determined by switched multiple beam pattern lobes in different satellite

Figure 3.8. The Beam Patterns and Adaptive Antenna Applications for SDMA

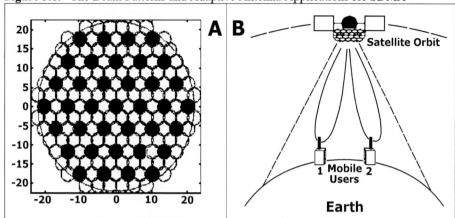

Courtesy of Paper: "Smart Antenna Application for Satellite Communications with SDMA" by V. Zaharov & other

directions, or by adaptive antenna separations, which is shown in **Figure 3.8. (A).** There are two different beam-forming approaches in SDMA for satellite communications: **(1)** The multiple spot beam antennas are the fundamental way of applying SDMA in large satellite systems including MSS and **(2)** Adaptive array antennas dynamically adapt to the number of users.

3.5.5.1. Switched Spot Beam Antenna

Switched Multi-Beam Antennas are designed to track each subscriber of a given cell with an individual beam pattern as the target subscriber moves within the cell (spot). Therefore, it is possible to use array antennas and to create a group of overlapping beams that together result in omnidirectional coverage. This is the simplest technique comprising only a basic switching function between separate directive antennas or predefined beams of an array. Beam-switching algorithms and RF signal-processing software are incorporated into smart antenna designs. For each call, software algorithms determine the beams that maintain the highest quality signal and the system continuously updates beam selection, ensuring that customers get optimal quality for the duration of their call. One might design overlapping beam patterns pointing in slightly different directions, similar to the ones shown in **Figure 3.8. (A).** Every so often, the system scans the outputs of each beam and selects the beam with the largest output power. The black cells reuse the frequencies currently assigned to the mobile terminals, so they are potential sources of interference. In fact, the use of a narrow beam reduces the number of interfering sources seen at the base station. Namely, as the mobile moves, the smart antenna system continuously monitors the signal quality to determine when a particular beam should be selected.

Switched-beam antennas are normally used only for the reception of signals, since there can be ambiguity in the system's perception of the location of the received signal. In fact, these antennas give the best performance, usually in terms of received power but they also suppress interference arriving from directions away from the active antenna beam's centre, because of the higher directivity, compared to a conventional antenna, some gain is achieved. In high-interference areas, switched-beam antennas are further limited since their pattern is fixed and they lack the ability to adaptively reject interference. Such an antenna will be easier to implement in existing cell structures than the more sophisticated adaptive arrays but it gives only limited improvement.

3.5.5.2. Adaptive Array Antenna Systems

Adaptive Array Antenna Systems select one beam pattern for each user out of a number of preset fixed beam patterns, depending on the location of the subscribers. At all events, these systems continually monitor their coverage areas, attempting to adapt to their changing radio environment, which consists in (often mobile) users and interferers. Thus, in the simplest scenario, that of a single user and no interferers, the system adapts to the user's motion by providing an effective antenna system pattern that follows the mobile user, always providing maximum gain in the user's direction. The principle of SDMA with adaptive antenna system application is quite different from the beam-forming approaches described in **Figure 3.8. (B)**.

The events processed in SDMA adaptive array antenna systems are as follows:

1) A "Snapshot", or sample, is taken of the transmission signals coming from all of the antenna elements, converted into digital form and stored in memory.

2) The SDMA digital processor analyzes the sample to estimate the radio environment at this point, identifying users and interferers and their locations.

3) The processor calculates the combining strategy for the antenna signals that optimally recovers the user's signals. With this strategy, each user's signal is received with as much gain as possible and with the other users/interferers signals rejected as much as possible.

4) An analogous calculation is done to allow spatially selective transmission from the array. Each user's signal is now effectively delivered through a separate spatial channel.

5) The system now has the ability to both transmit and receive information on each of the spatial channels, making them two-way channels.

As a result, the SDMA adaptive array antenna system can create a number of two-way spatial channels on a single conventional channel, be it frequency, time, or code. Of course, each of these spatial channels enjoys the full gain and interference rejection capabilities of the antenna array. In theory, an antenna array with (n) elements can support (n) spatial channels per conventional channel. In practice, the number is somewhat less because the received multipath signals, which can be combined to direct received signals, takes place. In addition, by using special algorithms and space diversity techniques, the radiation pattern can be adapted to receive multipath signals, which can be combined. Hence, these techniques will maximize the SIR or Signal to Interference and Noise Ratio (SINR).

3.5.5.3. SDMA/FDMA

This modulation arrangement uses filters and fixed links within the satellite transceiver to route an incoming uplink frequency to a particular downlink transmission antenna. A basic arrangement of fixed links may be set up using a switch that is selected only occasionally. Thus, an alternative solution allows the filter to be switched using a switch matrix, which is controlled by a command link. Because of the term SS (Switching Satellite) this scheme would be classified as SDMA/SS/FDMA. The satellite switches are changed only rarely, only when it is desired to reconfigure the satellite, to take account of possible traffic changes. The main disadvantage of this solution is the need for filters, which increase the mass of the payload.

3.5.5.4. SDMA/TDMA

This solution is similar to the one previously explained in that a switch system allows a TDMA receiver to reconfigure the satellite. Under normal conditions, a link between beam pairs is maintained and operated under TDMA conditions. The utilization of time slots may be arranged on an organized or contention basis. Switching is achieved by using the RF signal. Thus, on board processing is likely to be used in the future, allowing switching to take place by the utilization of baseband signals. The signal could be restored in quality and even stored to allow transmission in a new time slot in the outgoing TDMA frame. This scheme is providing up and downlinks for the later Intelsat VI spacecraft, known as SDMA/SS/TDMA.

3.5.5.5. SDMA/CDMA

This arrangement allows access to a common frequency band and may be used to provide the MA to the satellite, when each stream is decoded on the satellite in order to obtain the destination addresses. Thus, on-board circuitry must be capable of determining different destination addresses, which may arrive simultaneously, while also denying invalid users access to the downlink. However, on-board processors allow the CDMA bit stream to be retimed, regenerated and stored on the satellite. Because of this possibility the downlink CDMA configurations need not be the same as for uplink and the Earth link may thus, be optimized.

3.5.6. Random Division Multiple Access (RDMA)

For data transmission, a bit stream may be sent continuously over an established channel without the need to provide addresses or unique words if the channel is not charred. In fact, where charring is implemented, data are sent in bursts, which thus, requires unique words or synchronization signals to enable time-sharing with other users, to be affected in the division of channels. Each burst may consist in one or more packets comprising data from one or more sources that have been assembled over time, processed and made ready for transmission. However, this type of multiplex scheme is also known as Packet MA. Packet access can be used in special RDMA solutions, such as Aloha, where retransmission of blocked packets may be required.

Random access can be achieved to the satellite link by contention and for that reason is called a contention access scheme. This type of access is well-suited to satellite networks containing a large number of stations, such as MES, where each station is required to transmit short randomly-generated messages with long dead times between messages. The principle of RDMA is to permit the transmission of messages almost without restriction, in the form of limited duration bursts, which occupy all the bandwidth of the transmission channel. Therefore, in other words, this is MA with time division and random transmission and an attribute for the synonym Random Division Multiple Access is quite assessable.

A user transmits a message irrespective of the fact that there may be other users equally in connection. The probability of collisions between bursts at the satellite is accepted, causing the data to be blocked from receipt by the Earth station. In case of collision, the destination Earth station receiver will be confronted with interference noise, which can compromise message identification and retransmission after a random delay period. The retransmissions can occur as many times as probably are carried out, using random time delays. Therefore, such a scheme implies that the transmitter vies for satellite resources on a per-demand basis. In this case, it will provide that no other transmitter is attempting to access the same resources during the transmission burst period, when an error-free transmission can occur. The types of random protocols are distinguished by the means provided to overcome this disadvantage. The performance of these protocols is measured in terms of the throughput and the mean transmission delay. Throughput is the ratio of the volume of traffic delivered at the destination to the maximum capacity of the transmission channels. The transmission time, i.e., delay is a random variable. Its mean value indicates the mean time between the generation of a message and its correct reception by the destination station.

3.5.6.1. Aloha

The most widely-used contention access scheme is Aloha and its associated derivatives. This solution was developed in the late 1960s by the University of Hawaii and allows usage of small and inexpensive Earth stations (including MES) to communicate with a minimum of protocols and no network supervision. This is the simplest mode of operation, which time shares a single RF divided among multiple users and consists in stations randomly accessing a particular resource that is used to transmit packets. When an Aloha station has something to transmit, it immediately sends a burst of data pulses and can detect whether its transmission has been correctly received at the satellite by either monitoring the retransmission from the satellite or by receiving an acknowledgement message from the receiving party. Should a collision with another transmitting station occur, resulting in the incorrect reception of a packet at the satellite, the transmitting station waits for a random period of time, prior to retransmitting the packet. Otherwise, a remote station (MES) uses Aloha to get a hub station (LES) terminal's attention. Namely, the MES terminal sends a brief burst requesting a frequency or time slot assignment for the main transmission. Thus, once the assignment for MES is made, there is no further need for the Aloha channel, which becomes available for other stations to use. After that, the main transmissions are then made on the assigned channels. At the end, the Aloha channel might be used again to drop the main channel assignments after the transmission is completed. The advantages of Aloha are the lack of any centralized control, giving simple, low-cost stations and the ability to transmit at any time, without having to consider other users.

In the case where the user population is homogenous, so that the packet duration and message generation rate are constant, it can be shown that the traffic carried S (packet correctly interpreted by the receiver), as a function of total traffic G (original and retransmitted message) is given by the relation:

$$S = G \exp(-2G) \quad [\text{packet/time slot}] \tag{3.45.}$$

where (S = transmission throughput) and (G) are expressed as a number of packets per time slot equal to the common packet duration. The Aloha protocol cannot exceed a throughput of 18% and the mean transmission time increases very rapidly as the traffic increases due to an increasing number of collisions and packet retransmissions.

The Aloha mode is relatively inefficient with a maximum throughput of only 18.4% (1/2). However, this has to be counter-weight against the gains in simple network complexity, since no-coordination or complex timing properties are required at the transmitting MES.

3.5.6.2. Slotted Aloha

This form of Aloha or S-Aloha, where the time domain is divided into slots equivalent to a single packet burst time; there will be no overlap, as is the case with ordinary Aloha. The transmissions from different stations are now synchronized in such a way that packets are located at the satellite in time slots defined by the network clocks and equal to the common packet duration. Hence, there cannot be partial collisions; every collision arises from complete superposition of packets. In effect, the timescale of collision is thus, reduced to the duration of a packet, whereas with the Aloha protocol, this timescale is equal to the duration of both packets. This situation divides the probability of collision by two and the throughput becomes:

$$S = G \exp(-G) \quad \text{[packet/time slot]} \tag{3.46.}$$

This protocol enables collisions between new messages and retransmission to be avoided and increases the throughput of S-Aloha in the order of 50–60% by introducing a frame structure, which permits the numbering of time slots. Each packet incorporates additional information indicating the slot number reserved for retransmission in case of collision. For the same value of utilization as basic Aloha, the time delay and probability of packet loss are both improved. The major disadvantages of S-Aloha are that more complex equipment in the Earth station is necessary, because of the timing requirement and because there are fixed time slots, customers with a small transmission requirement are wasting capacity by not using the time slot to its full availability.

3.5.6.3. Slot Reservation Aloha

This solution of an extension for the slotted-Aloha scheme allows time slots to be reserved for transmission by an Earth station. In general terms this mode of operation is termed a Packed Reserved Multiple Access (PRMA). Slot reservation basically takes two forms:

1) Implicit – When a station acquires a slot and successfully transmits, the slot is reserved for that station for as long as it takes the station to complete its transmission. The network controller then informs all stations on the network that the slot is available for contention once more. There is only the problem that a station with much data to transmit could block the system to other users.

2) Explicit – Every user station may send a request for the reservation of a time slot prior to transmission of data. A record of all time slot occupation and reservation requests is kept. Actually, a free time slot could be allocated on a priority basis. Some kind of control for the reservation of slots is necessary and this could be accomplished by a single or all stations being informed of slot occupancy and reservation requests.

3.6. Mobile Broadband and Internet Protocols

Under the third and fourth-generation IMT-2000 services, high-speed and large-capacity multimedia data and Internet will be delivered through the satellite and radio networks and a wide variety of network services such as global roaming will be provided at the same time. In such a way many solutions and protocols will be employed in terrestrial radio and telecommunication systems, including ISDN, IP, TCP, IP/ATM, and UBR/ABR broadband

and Internet solutions and protocols. At all events, with its new advanced service menu, the IMT-2000 network is required to provide not only mobility management for individual terminals but also various service control functions. With capabilities such as Quality of Service (QoS), in which the desired communications quality is individually established for each communication session, the IP/ATM switching system provides for a sophisticated and economic network configuration.

Usage of different satellite fixed and mobile solutions has increased dramatically in past years. Along with that trend there has been an increasing demand for higher transmission rates. Hence, to cope with this demand, new technologies have emerged onto the radio and telecommunications stage. The IP/ATM over satellite is one fine example, which is creating waves in the global telecommunication world today and is ideal for a wide range of satellite applications, including data transfer, voice, imaging, full motion video and multimedias. Its characteristics include scalability, which make it ideal for the traffic demands of broadband multimedia expected on data networks of the future. Just as the Internet revolutionized worldwide communications, ATM brings new meaning to high-speed satellite networking.

The development of the Internet technologies is the concern of the Internet Engineering Task Force (IETF), which publishes its recommendations on the website under a series of Request for Comments (RFC) documents, with a specific identification number for each RFC. These documents are available for downloading from the IETF website in the form of information FYI RFC documents or those that specify Internet Standards (STD RFC).

3.6.1. Mobile Internet Protocol (IP)

The drive towards the establishment of an information society will bring together the two most successful solutions of all of the technological advances of the latter quarter of the 20th Century. This is the integration of Internet and mobile communications to deliver Mobile IP to all mobile customers. In this respect, fourth generation (4G) mobile networks will be based upon an IP-environment.

Significant effort around the world is now underway towards standardizing such a mobile environment through such organizations as the Mobile Wireless Internet Forum (MWIF), IETF, 3GPP and 3GPP2. Recently, two different approaches to the network architecture by 3GPP and 3GPP2 have been proposed. First, the 3GPP solution for the W-CDMA radio interface, is based on an evolution of the GMPSC network, with enhancements to the call control functionalities obtained though the introduction of a new network element, Called State Control Function (CSCF), to allow the provision of Voice over IP (VoIP) services. Second, the 3GPP2 solution for the CDMA2000 radio interface system has adopted a forthcoming packet network architecture incorporating Mobile IP functionalities in support of packet data mobility. Eventually, it is hoped that a single, harmonized solution will emerge from the two distinct approaches.

In the future mobile network, therefore, where satellites will operate alongside terrestrial mobile networks, various categories of service will co-exist, where "best effort", which is presently available over the Internet, will operate with guaranteed QoS classes. Moreover, the move towards an all IP network should facilitate the networking between satellite and terrestrial mobile networks, since the problem of providing mobile link connectivity across the different networks reduces to the level of attaining the appropriate route to direct the packets of information to the appropriate terminal.

Figure 3.9. WAAS CNS over IP

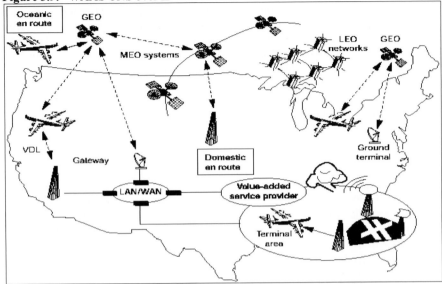

Courtesy of Paper: "Application of Mobile-IP to Space and Aeronautical Networks" by NASA

In the Mobile IP environment, the mobile terminal has a temporary IP address, known as the Care-of-Address (CoA), which is associated with a correspondent node of a particular access network. The CoA is in addition to the IP address that is permanently assigned to the mobile terminal, which is its home address, which is stored at its home sub-network. Thus, the mobile node's CoA is made known to the home agent. Packets addressed to the mobile node are thus, tunneled by the home agent to the correspondent node, identified by the CoA. Namely, when a mobile node moves from one access network to another, a new CoA will be assigned by the visiting network operator corresponding to the new correspondent node. The mobile node then makes the home agent aware of the change in the CoA, through the transmission of a binding update message.

Mobile IP and the mobile router will play a major role in NASA's aeronautics programs, including AATT, WINCOMM and SATS. Each program requires continuous connectivity, while dynamically traversing various sub networks. **Figure 3.9.** illustrates what a potential aeronautical network could entail in 2005. Although only the USA is shown, connectivity will however be global. In addition, the future aeronautical networks will use whatever wireless networking is available at reasonable cost. The LEO, MEO, HEO or GEO satellite constellation may be used, as will VDL, future Stratospheric Communication Platforms and perhaps next-generation cellular, if it survives the competition previously mentioned. Unique, cost-effective satellite techniques will also be deployed; examples would be using GEO direct broadband and broadcasting, known as DVB-RCS. Digital TV satellites with data embedded in the Moving Picture Expert Group 2 (MPEG-2), transport streams to transmit data to the aircraft. The return path would be an inexpensive, low-bandwidth, duplex channel. Under the AATT program NASA is working in alliance with the FAA to enable an increase in capacity, flexibility and efficiency, while maintaining flight safety of aircraft operations within the USA and the global airspace system. Similar to a WAAS, the goals are to increase terminals throughput by 40% and to increase en route throughput by 20%. Current operations in the National Air System (NAS) are severely constrained by an antiquated ATM, supported by a decades-old CNS infrastructure, using primarily analog radio voice, radar surveillance and ground-based navigational aids. Extensive ATC system delays and gridlock have resulted. A digital airspace WAAS/IP infrastructure is needed to meet the bandwidth, latency, security and integrity requirements of the future free-flight ATM system. The mobile router is an enabling technology for such a system.

3.6.1.1. IP Security Protocol (IPSec)

The IPSec protocol is a framework of open standards developed by the IETF, which can provide security for the transmission of sensitive information over unprotected networks such as the Internet. In this sense, the IPSec protocol acts at the network layer, protecting and authenticating IP packets between participating IPSec devices, such as Cisco or other IP routers. With IPSec, data can be transmitted across a public network without fear of observation, modification, or spoofing. This mode enables fixed applications such as Virtual Private Networks (VPN), including Intranets, Extranets and remote user access. Mobile users will be able to establish a secure connection back to their office, similarly to the encryption method.

For example, the mobile user can establish an IPSec known as a "tunnel" with a corporate firewall, requesting authentication services in order to gain access to the corporate network; all of the traffic between the user and the firewall will then be authenticated. The user can then establish an additional IPSec tunnel requesting data privacy services with an internal router or end system. Therefore, the IPSec method is a framework of open standards that provides data confidentiality, data integrity and data authentication between participating peers. IPSec provides these security services at the IP layer; it uses Internet Key Exchange (IKE) to handle the negotiation of protocols and algorithms based on local policy and to generate the encryption and authentication keys to be used by IPSec. In fact, IPSec can be used to protect one or more data flows between a pair of hosts, between a pair of security Gateways, or between a security Gateway and a host.

3.6.1.2. Mobile Transmissions over IP

The IP data link can be used for transmissions such as regularly made long-distance phone calls, videoconferencing connections or sending Fax messages. In such a way, IP telephony is known as a Voice over IP (VoIP), whilst similar to this, Videophone can also be established over IP (VPoIP) and Fax over IP (FoIP). Actually, these services are the transmission of telephone, video or Fax calls over a data network like one of the many terrestrial networks that make up the Internet or will be possible through Mobile Satellite Internet (MSI) as future MSC standards. On the other hand, a VoIP application meets the challenges of combining legacy voice networks and packet networks by allowing both voice and signaling information to be transported over the packet network. A FoIP application enables the networking of standard Fax machines with packet networks. It accomplishes this by extracting the Fax image from an analog signal and carrying it as digital data over the packet network. The VPoIP is a similar application, which can enable the networking of standard video cameras, videoconferencing or videophones with mobile packet network or via mobile ISDN links.

However, the TDM over IP (TDMoIP) is a transport technology that extends T1, E1, T3, E3, serial data or analog voice circuits transparently across IP or Ethernet networks. When used for voice mode, these TDM over IP circuits are transparent to signaling and provide superior voice quality with much lower latency than VoIP. The TDM scheme over IP supports all PBX features and all modem and fax rates. In other words, TDMoIP is not limited to voice and can extend to circuits carrying any protocol over IP including Frame Relay, ATM, ISDN, SS7, SNA, HDLC, Asynchronic, Synchronic, X.25, as well as H.320 and H.324 video over IP (VPoIP). The TDM scheme over IP was recently developed by the RAD Company to provide a simple, inexpensive migration strategy to IP-based networks.

3.6.1.3. Mobile IP version 6 (MIPv6)

The IP was introduced in the Arpanet in the mid 1970s; the version in common use today is IP version 4 (IPv4). Although several protocol suites (including Open System Interconnection) have been proposed over the years to replace IPv4, none have succeeded because of IPv4's large and continually growing, installed base. Nevertheless, IPv4 was never intended for the current Internet, either in terms of the number of hosts, types of applications, or security concerns. In the early 1990s, IETF recognized that the only way to cope with these changes was to design a new version of IP as a successor to IPv4. Thus, the IETF formed the next IP generation (IPng) Working Group to define this transitional protocol to ensure long-term compatibility between the current and new IP versions and support for current and emerging IP-based applications. The result of this effort was IP version 6 (IPv6), described in RFC 1883–1886; these four RFC were officially entered into the Internet Standard Track in December 1995. IPv6 is designed as an evolution from IPv4 rather than as a radical change. Useful features of IPv4 were carried over in IPv6 and less useful features were dropped.

Mobile IPv6 is an IP-layer mobility protocol for the IPv6 Internet, being standardized by the IETF. Thus, the idea is that when mobility, like any other functionality, is implemented in the network layer, it needs to be implemented only once and will then be transparently available for all higher layer protocols. It remains to be seen how well this promise is fulfilled in practice. There are, however, some applications such as mobile VPN access, for which Mobile IP is clearly a good solution.

The mobile network architecture is configured in such a way that the first half of an IPv6 address indicates the subnet to which the address belongs and it is used for routing IP packets across the Internet. Thus, when a mobile Internet host, known as a Mobile Node (MN) in the Mobile IPv6 terminology, moves to a different place in the network topology, its subnet and IP address necessarily change. This creates two kinds of problems: existing connections (e.g., TCP connections and IPSec security associations) between the mobile and other hosts known as a Correspondent Nodes (CN) become invalid and the mobile is no longer reachable at its old address for new connections. Moreover, the former problem is important in stateful protocols and has little effect on stateless protocols, such as HTTP. The latter problem typically concerns servers and not client computers.

Mobile IPv6 protocol has two basic goals: all transport-layer and higher layer connections and security associations between the mobile and its correspondents should survive the address change and the mobile host should be reachable as long as it is connected to the Internet somewhere in the world. Mobile IP makes some quite strong assumptions about the environments in which it is used. First, that all mobile hosts have a home network and a

Home Address (HoA) on that network. This is a reflection from a time when mobility was an exception: few Internet nodes would be mobile and even they would for most of the time remains stationary at home. In any case, Mobile IP solves the reachability problem by ensuring that the mobile is always able to receive packets sent to its home address. The IP address of a stationary IP node normally serves two purposes: it is both an identifier for the node and an address that is used for routing messages to the node. Mobile IP preserves this dual use of home addresses, which are an identifier for the mobile, as well as an address to which correspondents can send packets. The mobile's current location, called Care-of Address (CoA), on the other hand, is a pure address and serves no identification purposes. Any IPv6 address can be or become mobile and there is no way of distinguishing a mobile and stationary host by just looking at its address. This is because the Mobile IP protocol was originally designed to be transparent to the mobile's correspondents and the correspondent did not need to know that the mobile, in fact, was a mobile.

An additional technical advantage in the adoption of the MIPv6 IP solution as a final stage of standardization should be determined to provide the opportunity to perform seamless communications handover between satellite and TTN. Recent trials involving handover between TTN using MIPv6 have demonstrated the feasibility of such an approach. At any rate, researchers are now addressing the needs of integrated space/terrestrial mobile communication networks, based upon packet-oriented service delivery.

3.6.2. Transmission Control Protocol (TCP)

The transmission of Internet packets in Tx-Rx direction is primarily achieved using the TCP solution. The TCP mode is a connection-oriented transport protocol that sends data as an unstructured stream of bytes and provides the functionality to ensure that the transmission rate of data over the network is appropriate for the capabilities of the Rx device, as well as the devices that are used to route the data from the Tx to the Rx. Thus, by using sequence numbers and acknowledgment messages, TCP can provide a sending node with delivery information about packets transmitted to a destination node. Where data has been lost in transit from source to destination, TCP can retransmit the data until either a timeout condition is reached or until successful delivery has been achieved or can also recognize duplicate messages and will discard them appropriately. If the sending PC is transmitting too fast for the receiving computer, TCP can employ flow control mechanisms to slow data transfer. The TCP method can also communicate delivery information to the upper-layer protocols and the applications it supports and is responsible for ensuring that the network's resources are divided in an equitable manner among all users of the network. Applications such as File Transfer Protocol (FTP) and HTTP (the language of the Web) rely on TCP to transport their data over the network as quickly as the network will allow.

3.6.2.1. TCP/IP over Satellite

As far a TCP/IP connection is concerned, a mobile satellite network should be viewed as any other network connection. Given the way that TCP operates, the transmission of IP packets over the MSC link poses several problems that need to be overcome if services are to be delivered efficiently. In such a manner, the major difficulty is due to the latency of the satellite link, which when combined with a burst error channel and the characteristics of the TCP protocol itself, can result in an inefficient means of transmission. This is because TCP operates using a conservative congestion control mechanism, whereby new data can be transmitted only when an ACK (acknowledgement) from a previous transmission has been received. Therefore, with this in mind, the need for a high quality link between the satellite and MES is re-emphasized, since packets in error are presently deemed to be due to congestion on the satellite network. Hence, in this case, TCP responds by reducing its transmission rate accordingly. It can be seen that TCP operates on the basis of "best effort", founded on the available resources of the network.

When starting transmission, TCP enters the network in a restrained manner, whereby the initial rate of transmission is carefully controlled to avoid overloading the network with traffic. The TCP method achieves this by employing a congestion control mechanism such as slow start, congestion avoidance, fast retransmit and recovery. Slow start is used, as its name suggests, at the start of transmission or after congestion of the network has been detected and the data transmission is reduced. On the other hand, congestion control is used to gradually increase the transmission rate once the initial data rate has been ramped-up using the slow start algorithms. In this sense, the fast transmit and fast recovery algorithms are used to speed up the recovery of the transmission rate after significant congestion in the communication network has been detected.

3.6.2.2. TCP Intertwined Algorithms

Modern implementations of TCP contain four intertwined algorithms that have never been fully documented as Internet control mechanism standards, such as: Slow Start, Congestion Avoidance, Fast Retransmit (FRet) and Fast Recovery (FRec).

1. Slow Start – The old TCP mode would start a connection with the sender injecting multiple segments into the network, up to the window size advertised by the receiver. While this works when the two hosts are on the same LAN, if there are routers and slower links between the sender and the receiver, problems can arise. Some intermediate router must queue the packets and it is possible for that router to run out of space. Hence, the algorithm to avoid this is called slow start. It operates by observing that the rate at which new packets should be injected into the network is the rate at which the acknowledgments are returned by the other end. Furthermore, slow start adds a window to the sender's TCP known as Congestion Window (CWnd), which is initialized to one segment when a new connection is established with a host on another network. Each time an ACK is received, the CWnd is increased by one segment. Another important parameter is the Receiver Advertised Window (RWnd), which is the maximum amount of data that can be buffered at the receiver. The maximum transmission rate is determined by the minimum of CWnd and RWnd, thus, ensuring that the Rx is not overloaded with data that it is not able to ACK. The sender can transmit up to the minimum of the CWnd and the RWnd. The CWnd is flow control imposed by the sender, while the RWnd is flow control imposed by the Rx. The former is based on the sender's assessment of perceived network congestion; the latter is related to the amount of available buffer space at the Rx for this connection. The sender starts by transmitting one segment and waiting for its ACK. When that ACK is received, the CWnd is incremented from one to two and two segments can be sent. When each of those two segments is acknowledged, the CWnd is increased to four. This provides an exponential growth, although it is not exactly exponential because the receiver may delay its ACK, typically sending one ACK for every two segments that it receives. At some point the capacity of the Internet may be reached and an intermediate routers will start discarding packets. This tells the sender that its congestion window has become too large. However, from the above description of CWnd, it can be seen that after transmitting the first segment, the data sender remains idle until an ACK message has been received. For a GEO satellite, this would take in the region of roughly 500–570 ms.

2. Congestion Avoidance – Congestion can occur when data arrives on a big pipe (a fast LAN) and gets sent out via a smaller pipe (a slower WAN). It can also occur when multiple input streams arrive at a router whose output capacity is less than the sum of the inputs. Therefore, congestion avoidance is a way to deal with lost packets of data. The assumption of the algorithm is that packet loss caused by damage is very small (much less than 1%); therefore the loss of a packet signals congestion somewhere in the network between the source and destination. There are two indications of packet loss: a timeout occurring and the receipt of duplicate ACK. The congestion avoidance and slow start are independent algorithms with different objectives. But when congestion occurs TCP must slow down its transmission rate of packets into the network and then invoke slow start to get things going again. Hence, in practice they are implemented together. They require that two variables be maintained for each connection: a CWnd and a Slow Start Threshold (SSThres) size. If CWnd is less than or equal to SSThres, TCP is in slow start, while TCP is performing congestion avoidance. Slow start continues until TCP is halfway to where it was when congestion occurred and then congestion avoidance takes over. Slow start has CWnd begin at one segment and be incremented by one segment every time an ACK is received.

3. Fast Retransmit – Before describing the change, realize that TCP may generate an immediate ACK (a duplicate ACK) when an out-of-order segment is received, with a note that one reason for doing so was for the experimental fast retransmit algorithm. Thus, this duplicate ACK should not be delayed, the purpose is to let the other end know that a segment was received out of order and to tell it what sequence number is expected. Since TCP mode does not know whether a duplicate ACK is caused by a lost segment or just a reordering of segments, it waits for a small number of duplicate ACK to be received. It is assumed that if there is just a reordering of the segments, there will be only one or two duplicate ACK before the reordered segment is processed, which will then generate a new ACK. If three or more duplicate ACK are received in a row, it is a strong indication that a segment has been lost. The TCP mode then performs a retransmission of what appears to be the missing segment, without waiting for a retransmission timer to expire.

4. Fast Recovery – After fast retransmit sends what appears to be the missing segment, the congestion avoidance but not slow start is performed. This is the fast recovery algorithm. It is an improvement that allows high throughput under moderate congestion, especially for large windows. The reason for not performing slow start in this case is that the receipt of the duplicate ACK tells TCP that more than just a packet has been lost. Since the receiver can only generate the duplicate ACK when another segment is received, that segment has left the network and is in the receiver's buffer. That is, there is still data flowing between the two ends and TCP does not want to reduce the flow abruptly by going into slow start.

3.6.3. Mobile Asynchronous Transfer Mode (ATM)

The ATM solution is a network switching technology used by Broadband Integrated Service Digital Network (BISDN). It uses a technique called cell switching, breaking all data into cells, or packets and transmits them from one location on the network to another, connected by switches. So, the small, constant cell size allows ATM equipment to transmit audio, video and computer data over the same network and assure that no single type of data hogs the line. The latest implementations of ATM support data transfer rates from 25 to 622 Mb/s, which compares to a maximum of 100 Mb/s for Ethernet. It is a protocol designed to handle isochronous (time critical) data such as video and telephony (audio), in addition to more conventional data communications between computers. These protocols are capable of providing a homogeneous network for all traffic types. The same protocols are used regardless of whether the application is to carry conventional telephony (VoIP), Fax over IP (FoIP), entertainment video over IP (VPoIP) or computer network traffic over LAN, MAN, WAN or satellite networks.

Small and constant packet size allows switching to be implemented in hardware, rather than have routing done in the software. This makes ATM switches sufficiently rapid that multiple isochronous data can be statistically multiplexed together. The significance of this is that ATM protocols provide bandwidth on demand. Network management software will allow small amounts of bandwidth to be set aside for simple transactions, such as E-mails, while allowing more bandwidth for resource intensive multimedia applications. This sort of technology means that a level of bandwidth can be guaranteed and that a connection is not delayed or interrupted by network traffic, which in ATM parlance is known as QoS.

The ATM transport protocol is universal since it can be used for all kinds of networks be it physical networks, (twisted pair, coax and fibre optics) or virtual networks, such as radio and satellite systems. Today's emphasis on multimedia presentations, videoconferencing, remote lecturing, etc., has made ATM more attractive to network administrators. Simply throwing more bandwidth onto the network is not the solution to the problem, since not all demands on the network are equal. The ATM's ability to provide the bandwidth and QoS guarantees makes it the obvious upgrade choice.

Other areas where ATM is used include Digital TV, high definition television (HDTV) and video vending. A public ATM network would offer some entirely new services to people in their homes. The most significant type of new services will be the possible video services. What will be so unique about these services is the opportunity for a viewer to interact more with the programs they receive. At this point, the use of the ATM solution would reduce costs considerably, making ATM a very attractive technology for mobile applications, such as oil companies, shipping and airlines.

Many designers of satellite systems are thinking about implementing the ATM protocol, which transmits data that have been placed in cells of a constant length (53 bytes). Thus, the ATM guarantees data transmission at a rate ranging between 2 Mb/s and 2.4 Gb/s. The protocol acts on the principle that a virtual channel should be set up between two points whenever such a need appears. This is what makes the ATM protocol different from the TCP/IP protocol, in which messages are transmitted in packet form, where each packet may reach the recipient via a different route. The ATM protocol enables data transmission through various media. Taking into account the header of the cell (cell-tax), which takes 5 bytes, the application of the ATM protocol may not appear to be so cost-effective when the rate of transmission is low and the capacity of the link (e.g., in two-way modem channels) becomes a basic limitation.

As noted, the ATM technology is expected to provide QoS-based networks that support voice, video and data applications. Initially, the ATM protocol was originally designed for fibre-based terrestrial networks that exhibit low latencies and error rates. With the increasing demand for electronic connectivity across the world, satellite networks play an indispensable role in the deployment of global networks. The Ka-band communication link using the GHz frequency satellites spectrum can reach user terminals across most of the populated world. Thus, the ATM-based satellite networks can effectively provide real time as well as non-real time communications services to remote areas.

There are four different types of ATM service.

1. Constant Bit Rate (CBR) – The CBR ATM service specifies a fixed bit rate so that data is sent in a steady stream. This is analogous to a leased line.

2. Variable Bit Rate (VBR) – This ATM service provides a specified throughput capacity but data is not sent evenly. This is a popular choice for voice and videoconferencing data.

3. Unspecified/Undefined Bit Rate (UBR) – It does not guarantee any throughput levels. This is used for applications, such as file transfer, that can tolerate delays.

4. Available Bit Rate (ABR) – It provides a guaranteed minimum capacity but allows data to be burst at higher capacities when the network is free.

However, there is opinion that ATM holds the answer to the Internet bandwidth problem but others are skeptical. Thus, ATM creates a fixed channel, or route, between two points whenever a data transfer begins. This differs from TCP/IP, in which messages are divided into packets and each packet can take a different route from source to destination. This difference makes it easier to track and bill data usage across an ATM.

3.6.3.1. IP/ATM over Satellite

As mentioned, the ATM terrestrial technology can be successfully used in satellite communication networks to enhance the overall performance of the network. This network as usual, can comprise corresponding satellite configuration and several ground stations communicating via satellite on the one hand and ATM Switch TTN infrastructure on the other, interconnected with the main component of the network, known as ATM Satellite Internetworking Unit (ASIU). The ASIU is responsible for management and control of system resources as well as the overall system administrative functions, like real-time bandwidth allocation, network access control, system timing and traffic control. This fixed satellite configuration can be very easily implemented in MSC with connecting MES terminals via satellite with LES and ASIU to the ATM Switching TTN.

The internal architecture of the ASIU in the satellite direction stage is to extract the ATM cells from the various transport methods. After the ATM cells are extracted, they are error corrected and then buffered according to their type. After that, the cells are encoded and transmitted through the modem to the satellite. Hence, looking at the various components in a bit more detail, the first stage will be the extraction of ATM cells from the transport protocols, using the following methods:

1. Plesiochronous Digital Hierarchy (PDH) – The PDH protocol was developed to carry digitized voice efficiently. At any rate, it mainly operates by multiplexing various rates of bit streams at the highest allowed clock speed. When necessary, it adds a stuffing bit which the demultiplexer can later remove and therefore, the stuffing procedure is inefficient. Also, rerouting signals after network failures and managing remote network elements are extremely difficult.

2. Physical Layer Convergence Protocol (PLCP) – On the other hand, PLCP mode works differently. The PLCP combines 12 ATM cells in one frame format with a header in front. Consequently, this method is not very good for satellite transmission as it is very sensitive to burst errors, which may result in the loss of a whole frame and loss of synchronization of the PLCP device.

3. Synchronous Optical Network (SONET) or Synchronous Digital Hierarchy (SDH)
– For the reasons already explained above PDH and PLCP are not suitable to be used with ATM and engineers have been turning to SDH as the protocol for ATM. The SONET, similar to the SDH protocol, uses pointer bytes to indicate the location of the first byte in the payload of the SDH frame. Furthermore, the SDH protocol incorporates a cell delineation mechanism for the acquisition and synchronization of the ATM cells on the receiver side of the network and has some considerable advantages over PDH/PLCP, the main being: flexibility of service, allowing the network operator to respond quickly to customers' requirements; improved quality and supervision, permitting the operator to increase the quality of the offered services; reduced operations costs, by using efficient network management technology which makes the remote control of the network possible mostly without on-site activities and higher rates of transfer are well defined and direct multiplexing is possible without an intermediate multiplexing stage. In fact, despite these advantages SDH has an important problem in that incorrect pointer detection may produce an incorrect payload extraction and an error block. As a result, all received blocks may be incorrect and severely errored. To overcome this problem, an efficient error mechanism, error detection and recovery must be used in the next stage of the ASIU architecture.

However, satellite systems have several inherent constraints. The resources of the satellite network, especially the satellite and the Earth station, are expensive and typically have low redundancy; these must be robust and be used efficiently. Thus, the large delays in GEO and delay variations in LEO systems affect both real-time and non-real time applications. In an acknowledgment and timeout-based congestion control mechanism (like TCP), performance is inherently related to the delay-bandwidth product of the connection. Moreover, TCP Round-Trip Time (RTT) measurements are sensitive to delay variations that may cause false timeouts and retransmissions. As a result, the congestion control issues for broadband satellite networks are somewhat different from those of low-latency terrestrial networks. Both interoperability issues as well as performance issues need to be addressed before a transport-layer protocol like the TCP model can satisfactorily work over long latency satellite ATM networks.

Satellite ATM networks can be used to provide broadband access to remote locations, as well as to serve as an alternative to fibre-based backbone networks. In either case, a single satellite is designed to support thousands of Earth terminals. The Earth terminals set up Virtual Channels (VC) through the on board satellite switches to transfer ATM cells among one another. Because of the limited capacity of a satellite switch, each MES has a limited number of VC it can use for TCP/IP data transport. In backbone networks, these MES are IP/ATM edge devices that terminate ATM connections and route IP traffic in and out of the ATM network. Namely, these high-capacity backbone routers must handle thousands of simultaneous IP flows. As a result, the routers must be able to aggregate multiple IP flows onto individual VC. Flow classification may be done by means of a QoS that can use IP source–destination address pairs, as well as transport-layer port numbers. Therefore, the QoS manager can further classify IP packets into flows based on the differentiated services code points in the TOS byte of the IP header.

In addition to flow and VC management, all MES terminals also provide a means for congestion control between the IP and ATM networks. The on-board ATM switches must perform traffic management at the cell and VC levels. Hence, TCP hosts implement various TCP flow and congestion control mechanisms for effective network bandwidth utilization. The enhancements that perform intelligent buffer management policies at the switches can be developed for UBR service to improve transport layer throughput and fairness. A policy for selective cell drop based on per-VC accounting can be used to improve fairness.

Providing a minimum Guaranteed Rate (GR) to UBR traffic has been discussed as possible candidate to improve TCP performance over UBR. The goal of providing GR is to protect the UBR service category from total bandwidth starvation and provide a continuous minimum bandwidth guarantee. It has been shown that in the presence of high loads of higher priority Constant Bit Rate (CBR), Variable Bit Rate (VBR) and Available Bit Rate (ABR) traffic, TCP congestion control mechanisms benefit from a minimum GR. Guaranteed Frame Rate (GFR) has recently been proposed in the ATM Forum as an enhancement to the UBR service category. GFR will provide a minimum rate guarantee to VC at the frame level. The GFR service also allows for the fair usage of any extra network bandwidth. GFR is likely to be used by applications that can neither specify the traffic parameters needed for a VBR VC, nor have the capability for ABR (for rate-based feedback control). Current internetworking applications fall into this category and are not designed to run over QoS-based networks. Routers separated by satellite ATM networks can use the GFR service to establish VC between one another. GFR can be implemented using per-VC queuing or buffer management.

3.6.3.2. UBR over Satellite

The UBR service class is intended for delay-tolerant or non-real time applications, that is, those which do not require tightly constrained delay and delay variation, such as traditional computer communications applications. Sources are expected to transmit non-continuous bursts of cells. Namely, the UBR service supports a high degree of statistical multiplexing among sources and includes no notion of a per-VC allocated bandwidth resource. Thus, transport of cells in UBR service is not necessarily guaranteed by mechanisms operating at the cell level. However, it is expected that resources will be provisioned for UBR service in such a way as to make it usable for some sets of applications. The UBR service may be considered as an interpretation of the common term "best effort service".

The UBR service is typically used for data transmission applications such as file transfer and E-mail. Neither an ATM-attached router nor an ATM switch provides traffic or QoS guarantees to a UBR virtual circuit. As a result, UBR VC can experience a large number of cell drops or a high cell transfer delay as cells move from the source to the destination device. In its simplest form, an ATM switch implements a tail drop policy for the UBR service category. Hence, if cells are dropped, the TCP source loses time waiting for the retransmission timeout. Even though TCP congestion mechanisms effectively recover from loss, the link efficiency can be very low, especially for large delay-bandwidth networks. In general, link efficiency typically increases with increasing buffer size. In fact, the performance of TCP over UBR can be improved by using buffer management policies. In addition, TCP performance is also affected by TCP congestion control mechanisms and TCP parameters such as segment size, timer granularity, receiver window size, slow start threshold and initial window size.

The TCP Reno implements the fast retransmit and recovery algorithms that enable the connection to quickly recover from isolated segment losses. However, fast retransmit and recovery cannot efficiently recover from multiple packet losses within the same window. A modification to Reno TCP is proposed, so that the sender can recover from multiple packet losses without having to timeout. Thus, the TCP solutions with Selective Acknowledgment

(SACK) are designed to efficiently recover from multiple segment losses. With SACK, the sender can recover from multiple dropped segments in about one round trip. The studies show that in low-delay networks, the effect of network-based buffer management policies is very important and can dominate the effect of SACK. The throughput improvement provided by SACK is very significant for long latency connections.

When the propagation delay is large, timeout results in the loss of a significant amount of time during slow start from a window of one segment. Reno TCP (with fast retransmit and recovery) results in the worst performance (for multiple packet losses) because timeout occurs at a much lower window than in Vanilla TCP. With SACK TCP, a timeout is avoided most of the time and recovery is complete within a small number of round trips. The NewReno modification to the fast retransmit and fast recovery algorithms (TCP Reno) has been therefore proposed to counteract multiple packet drops where the SACK option is not available. For lower delay satellite networks (LEO), both NewReno and SACK TCP provide high throughput but as the latency increases, SACK significantly outperforms NewReno, Reno and Vanilla.

3.6.3.3. ABR over Satellite

The ABR service category is another option to implement TCP/IP over ATM. This service category is specified by a Peak Cell Rate (PCR) and a Minimum Cell Rate (MCR), which is guaranteed by the network. ABR connections use a rate-based closed-loop end-to-end feedback control mechanism for congestion control. The network tries to maintain a low cell loss ratio by changing the Allowed Cell Rate (ACR) at which a source can send. Switches can also use the Virtual Source/Virtual Destination (VS/VD) feature to segment the ABR control loop into smaller loops. Studies have indicated that ABR with VS/VD can effectively reduce the buffer requirement for TCP over ATM, especially for long delay paths. In addition to network-based drop policies, end-to-end flow control and congestion control policies can be effective in improving TCP performance over UBR. The fast retransmit and recovery mechanism can be used in addition to slow start and congestion avoidance to quickly recover from isolated segment losses. Otherwise, the SACK option has been proposed to recover quickly from multiple segment losses.

The TCP performance over ABR service is important for the satellite IP network. Namely, a key ABR feature, VS/VD, highlights its relevance to long delay paths. Most of the consideration assumes that the switches implement a rate-based switch algorithm like ERICA+. Credit-based congestion control for satellite networks has also been suggested, in long-latency satellite configurations, therefore the feedback delay is the dominant factor in determining the maximum queue length. A feedback delay of 10 ms corresponds to about 3,670 cells (at OC-3) of queue for TCP over ERICA+, while a feedback delay of 550 ms corresponds to 201,850 cells. This indicates that satellite switches need to provide at least one feedback delay of buffering to avoid loss on these high delay paths. A point to consider is that these large queues should not be seen in downstream workgroup or WAN switches, because they will not provide as much buffering. Satellite switches can isolate downstream switches from such large queues by implementing the VS/VD option, while VS/VD can effectively isolate nodes in different VS/VD loops. As a result, the buffer requirements of a node are bound by the feedback delay-bandwidth product of the upstream VS/VD loop. At this point, VS/VD helps to reduce the buffer requirements of terrestrial switches connected to Gateway terminals. The feedback delay-bandwidth products of the satellite hop are about

160,000 cells and dominates the feedback delay-bandwidth product of the terrestrial hop (about 3,000 cells). Without VS/VD, terrestrial switch S, a bottleneck, must buffer cells up to the feedback delay-bandwidth product of the entire control loop (including the satellite hop). With a VS/VD loop between the satellite and the terrestrial switch, the queue accumulation due to the satellite feedback delay is confined to the satellite switch. The terrestrial switch only buffers cells that are accumulated due to the feedback delay of the terrestrial link to the satellite switch.

3.6.4. Digital Video Broadcasting-Return Channel over Satellite (DVB-RCS)

The convergence of MSC and Internet technique has opened many opportunities to deliver new multimedia service over hybrid satellite systems to MES. The interactive nature of the Internet has paved the way for new generation MSS to support interactivity. Apart from the convergence between mobile and Internet technologies, the other major technical driver is the convergence between mobile and fixed technologies. By supplementing broadcasting systems with a narrowband uplink, new interactive services can be facilitated in DVB and DAB solutions. This is foreseen for fixed network operation and could equally be adopted onto a mobile network such as the UMTS, thus demonstrating the concept of convergence of personal mobile communications, Internet and broadcasting technology.

The recent major development in satellite broadcasting technology was the standardization of the DVB-RCS, which allows the users within broadcast terrestrial network (DVB-T) to communicate directly with the broadcast satellite network (DVB-S) through an assigned return channel. The DVB-T cell can be comprised by UMTS/GPRS, ISDN, B-ISDN, and ATM broadband networks and the DVB-S cells may include rural, Consumer broadcasting SoHo/SME LAN, Corporate WAN and multicasting networks. This greatly simplifies the overall network architecture and associated network management procedures, in that now all kind of communication solutions takes place over the same access network. Presently, the DVB-RCS system has been specified for indoor use only, however, it can be envisaged that very soon, research and standardization efforts will be directed towards establishing a corresponding mobile standards. In this instance, will be enough to develop corresponding mobile satellite antenna, and this could open up significant new opportunities in the mobile satellite sector.

As a sample of new such solution will be introduced Nera's SatLink DVB-RCS system, see **Figure 3.10.** It enables via SatLink HUB with C, Ku or Ka-band antenna to interface the TTN (DVB-T) cell via corresponding satellite connections (C, Ku or Ka-band GEO) to the SatLink Router Terminals (DVB-S) cells for the following services:

1. Regenerate rural communications: VoIP, IP/TV, Internet access and interactive TV/radio two separate way broadcasting (Telephony/Broadband/Broadcast);

2. Broadband access: Asymmetric Digital Subscriber Line (ADSL) anywhere and anytime, Internet access/E-mail, Consumer, SoHo/SME LAN, Corporate WAN, Intranet/VPN, File Transport Protocol (FTP) and Hyper-Text Transfer Protocol (HTTP); The FTP scheme is service for moving and copying an electronic file of any type from one computer to another over the Internet. It can be used both for downloads and uploads;

3. Teleservice E-medicine and E-education: Videoconference, Image/Video/Audio transfer and Interactive distance learning; and

4. Multicasting: Web casting, Video streaming, Satellite newsgathering and Push/Pull data delivery.

Figure 3.10. Nera SatLink DVB-RCS HUB and ODU/IDU with Router Terminals

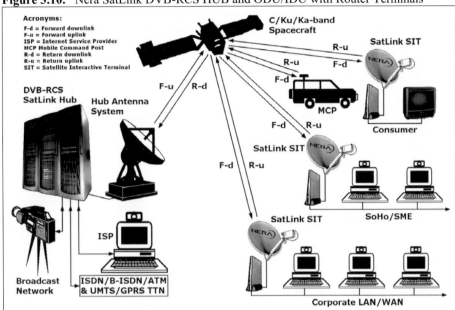

Courtesy of Pamphlet Paper: "Nera SatLink Two-way Satellite Broadband Solutions" by Nera

Nera SatLink Hub supports existing DVB-RCS compliant Forward Link System (FLS), but Nera also offers FLS solutions. The Nera SatLink Hub provides all the required interfaces and management functions necessary to set up a DVB-RCS service. The Hub operator can interface DVB-RCS terminals to a terrestrial network or service provider, and manage all the operational aspects of the system. The Hub station design optimized for robustness and stability in operations and can be delivered in a number of different configurations to suit customers precise applications. Its architecture is also designed to accommodate upgrades and expansions of DVB-RCS network. The DVB-RCS operators have a choice of C, Ku or Ka-band antenna and RF equipment. One ordinary DVB-RCS configuration can handless 1 to 10 forward link transmitters and from one to several hundred return link receivers.

Nera DVB-RCS Satellite Interactive Terminal (SIT) is composed by SatLink Outdoor Unit (ODU) and Nera SatLink DVB-RCS Router Satellite Terminal (RST). Standard ODU/IDU terminals interface ensures full compatibility with any DVB-RCS system or VSAT modem, ETSI/CE approval.

Nera SatLink DVB-RCS Router is desktop terminals and routers that provide two-way IP communications via satellite at Ku-Band frequencies. Nera SatLink Routers and Terminals for corporations, institutions, home offices and householders offer an open-interface for high-capacity broadband access that bypasses the "last mile" bottleneck associated with terrestrial infrastructure.

Therefore, DVB-RCS system offers broadband access to core IP networks using standard technologies such as DVB-S, DVB-RCS and IP interfacing DVB-T with user terminals. Thus, the broadband network delivers two-way IP connectivity both between user terminals in the satellite system and between user terminals and the terrestrial network. The broadcast

MPEG2/DVB-S service is available in one-way (unidirectional) from the SatLink HUB to the users terminals. With very low cost per bit, service prices become comparable to those offered by terrestrial networks and can be delivered where other technologies cannot reach. This network can provide solutions in vas variety of user segments, such as: Automotive, Mobile, Education and Learning, Financial services, Petroleum and convenience outlets, Restaurant and entertainment, Retail solutions, Remote areas, Teleworker, etc.

The main components of the Nera SatLink Hub are Network Control Centre, Forward Link Subsystem, Return Link Subsystem, Reference and Synchronization Subsystem.

1. Network Control Centre (NCC) – The NCC configuration assures traffic management, system supervision and protocol handling, and offers sophisticated bandwidth management of the return link through four different bandwidth allocation algorithms based on the DVB-RCS standard. The Network Management Subsystem is the interface between the Operator and the Hub, using a user friendly Web-interface. Call and management data is stored and made available towards other operational tools through standard interfaces, e.g. for billing purposes.

2. Forward Link Subsystem (FLS) – This link consists typically (but not necessarily) of an IP-encapsulator, providing the encapsulation of IP transfer of data packets into MPEG frames, an MPEG Multiplexer and a modulator according to the DVB-S specification. The SatLink Hub can be delivered with Nera providing the Forward Link System or integrated towards an existing Forward Link System.

3. Return Link Subsystem (RLS) – This link is the powerful bank of radio receivers, which collect the Turbo-coded MF-TDMA bursts transmitted by all the terminals in the network. Each one of Nera's RLS receivers can be individually configured to operate at any frequency, bit-rate or coding rate. In this way the SatLink hub offers unrivalled flexibility for the operator to maximize the use of air-interface resources. The modular design of the RLS enables the system to scale from small to very large networks.

4. Reference and Synchronization Subsystem (RSS) – This special equipment delivers the synchronization and timing information in the gateway for synchronization of the entire satellite network.

5. SatLink User Terminal (SUT) – This terminal consists of two main units, the Indoor Unit (IDU) and the Outdoor Unit (ODU).

a) SatLink IDU is offered in different models with scalable performance, ensuring that the customers get the optimum prices and performances for communications and applications. In effect, the IDU contains the DVB-S/DVB-RCS modem and the interface to the local network. It is equipped with a powerful command line interface and Web-based interface for remote management of terminals and safe software upgrade in order to keep the operating cost of the DVB-RCS network low. The IDUs can be supplied for either desktop use or for rack mounting and come in a scalable choice of performance with a range of data IP throughputs from 4 Mb/s to 15 Mb/s.

b) SatLink ODU consists of the RF transmitter, one or more RF receivers, along with an antenna. Nera SatLink terminals can be configured with a variety of ODUs dependent upon frequency band, bandwidth requirements and traffic pattern, satellite system, coverage and geographical locations. Typical antenna sizes arc 0.75, 0.9, 1.2, 1.8 and 2.4 metres in both C-band, Ku-band and Ka-band.

4

ANTENNA SYSTEMS FOR MOBILE SATELLITE APPLICATIONS

In the beginning of radio development, mobile communication systems were conceived for the transmission and receiving of telegraphy and telephony signals via antenna from ships, cars, trains and aircraft. The consideration of antenna transmission is inevitable, especially in GMSC systems, where their propagation characteristics are much affected by different and changeable local environments during movement of mobile and differ greatly from those observed in fixed satellite systems. To create antenna hardware for mobile GMSC systems, engineers have to consider all related factors in order to realize full mechanical and transmission potentials.

This chapter describes antenna characteristics, requirements and basic relations of antenna systems; considerable antenna classification for maritime, land, aeronautical and personal satellite applications for ships, vehicles (tracks, buses, cars and trains), aircraft (airplanes and helicopters) and integrated solutions with GSM antenna on the top of handheld phones; retrospective of all antenna configurations for GMSC, such as low-gain omnidirectional antennas, three principal divisions of medium-gain directional antennas and three types of high-gain directional aperture antennas and finally, all type of antennas are presented for particular MSC systems and antenna mounting and tracking systems.

4.1. Evolution of Antenna Systems for Mobile Radio communications

The Russian professor of physics Popov designed his first world's radio receiver in 1895 with antenna in the shape of wire mounted on a balloon in the air and transmitter with a lightning conductor as an antenna, including a metal filings coherer and a detector element with telegraph relay and a bell. Soon later, Marconi started commercially to deploy radio and antenna equipment on board different merchant ships and to establish his own company for the production of maritime radio and antenna equipment.

Since the initial use of mobile radio was for long distance wireless communications at LF and the first shipboard antenna were all made of haphazard lengths of wire strung as high as possible above the ship's topside, evidently the thinking was that the longer and higher the wire, the better the results should be. After those different kinds of wire and whip antennas were developed for MF/HF/VHF maritime and aeronautical applications.

4.1.1. Development of Antennas for MSC

The MSC systems introduced new complexities into the design of shipboard antennas. The direct line-of-sight between antenna and satellite requires the antenna to see from horizon to overhead (zenith – 90°) in elevation and 360° in azimuth angle, with total hemispherical coverage. This is fulfilled in the case of transceiver antenna through the use of tracking rotatable, high-gain antennas often installed in pairs on board ship to attain full coverage, irrespective of blockage in the form of the funnel, musts, stacks and other deck objects.

The ship platform itself imposes even more stringent requirements. Therefore, in spite of constant vibrating, pitching, rolling and yawing during bad weather conditions, the MSC antenna's narrow radiation beam must be pointed accurately from any position on the high seas. The situation regarding land or aero antenna is less complicated. As the 1970s dawned, optimism and enthusiasm about satellite communications was so great with ideas to virtually replace HF radio in the Navy with the new Fltsatcom military mobile system. For instance, since this time several types of UHF transceiving antennas developed for US naval shipboard service were used, such as the oldest crossed-dipole array, its improved version and the so-called wash-tube similar to an SBF antenna and one type of SHF parabolic dish antenna. However, as discussed in Chapter 1, the first real global MSC system was the Marisat system, which employed SES and L-band antenna systems similar to current Inmarsat-A and B terminals.

4.1.2. Classification and Types of Mobile Satellite Antennas (MSA)

The general classifications of MSA in connection with the service operator/provided are performed on the following four major types: Satellite Communications Antennas, Satellite Broadband (Multimedia) Antennas, Satellite Navigation Antennas and Military Satellite Antennas. However, the fundamental classification of MSA according to gain values and characteristics falls into three main groups: Low-Gain Omnidirectional Antennas; Medium-Gain Directional Antennas and High-Gain Directional Aperture Antennas.

4.2. Antennas Requirements and Technical Characteristics

This section describes important general requirements for mobile antenna solutions used in GMSC systems for maritime, land and aeronautical applications, including antennas for personal handheld terminals. At any rate, the mobile antenna has to be compact, flexible and lightweight and perform with good mechanical and electrical characteristics, especially for heavy mobiles such as ships and aircraft, owing to the special conditions of installation and the influence of changeable environmental conditions.

4.2.1. Mechanical Characteristics

Mobile antennas have to satisfy the requirements of mechanical characteristics in relation to construction strength and easy installation. Easy installation and appropriate physical shape are very important requirements in addition to compactness and lightweight. In the case of shipborne antennas, the installation requirements are not as severe compared to that of aircraft and cars because even in small ships there is a comfortable space to install an antenna set. Otherwise, the only problem is because all types of ships satellite antennas are sometimes under stress from vibration and sloping caused by strong winds, ship's rolling and pitch or is subject to corrosion by sea salt. Owing to these problems, a ship's antenna has to be well protected by plastic radome and properly mounted on a strong mast, specially designed for a certain size of antenna. However, in the case of road vehicles, especially small cars, low profile and lightweight equipment is required. The requirements are the same for aircraft, although more severe conditions are required to satisfy avionic standards. Namely, low air drag is one of the most important requirements for aircraft antennas. Vehicles and aircraft utilize smaller and more aerodynamic sizes of antenna.

4.2.2. Electrical Characteristics

Sometimes, the mechanical construction of antenna is perfect because of some functional or electrical characteristics; however, designers of antenna have to keep in mind that the compact design of antenna has two major disadvantages in electrical characteristics, such as low-gain and wide beam coverage. The gain is closely related to the beam width and a Low-Gain Antenna (LGA) should have a wide beam width. As the gain of antenna is theoretically determined by its physical dimensions, reducing the size of antenna means decreasing its gain. Because of low-gain and limited electric power supply, it is very difficult for mobile antennas to have enough receiving capability (G/T) and transmission power (EIRP). These disadvantages of MES can be compensated by a satellite that has a large antenna and HPA with enough electrical power. A powerful satellite with high G/T and EIRP performance should permit the fabrication of MES with compact and lightweight antennas. The next disadvantage is that a wide beam antenna is likely to transmit undesired signals to and receive them from an undesired direction, which will cause interference in and from other systems. The wide beam is also responsible for several fading effects, such as that from sea surface reflections in MMSC and AMSC and multipath fading in LMSC and so, a compact mobile antenna system is required to prevent fading and interference.

Accordingly, it is inevitable for mobile antennas not to have enough performance, such as gain, radiation power and receiving capability because of their small physical dimensions. Without consideration of this, the requirements of transmitting and receiving performance of mobile antennas mainly depend on the satellite transmission capability. The first and second generations of Inmarsat satellites have a global beam and the third generation has spot and global beams to provide global coverage. The regional or domestic MSS, such as AMSC, MSAT and Optus have spot beams. The spot gives higher satellite capacity than global beams although there are basically no big differences between requirements for mobile antennas in the global system, such as Inmarsat or the mentioned regional systems.

4.2.3. Basic Relations of Antennas

The basic relations of antenna systems are very important parameters to easily understand the mode of antenna functions in two-way (duplex) satellite transmission systems, such as MES transceiving antennas. Moreover, these characteristics of MES antenna systems are needed for link budget calculations and for good satellite up and downlink design, which can provide reliable and acceptable quality satellite communications. At this point, this implies that the signal transmitted via the MES Tx antenna must reach the Rx antenna of other MES or LES at a carrier level sufficiently above the unwanted signals generated by various unavoidable sources of noise and interference.

4.2.3.1. Frequency and Bandwidth

In almost all present and forthcoming MSS using GEO satellites, the L-band 1.6/1.5 GHz is used for a link between the satellite and MES. The required frequency bandwidth in L-band MSS is about 8% to cover transmission and receiving channels. In using a narrow-band antenna, such as an omnidirectional patch antenna, some efforts have to be made to widen the bandwidth. The S and L-band are allocated in WARC-92 for the Big LEO Iridium and Globalstar systems, which require frequency bandwidths of about 5%.

4.2.3.2. Gain and Directivity

The required antenna gain is determined by a link budget, which can be calculated by taking into consideration the required channel quality and the satellite capability. The channels are expressed as C/N_o and depend on the G/T and EIRP values of the satellite and MES. Thus, in the abandoned Inmarsat-P system and forthcoming ICO system, medium gains of 7 to 16 dBi are required for voice and HSD channels using a transmission speed of about 24 Kb/s. In the case of present systems such as AMSC or MSAT, Optus and similar systems using GEO constellation, a medium gain between 8 to 15 dBi is required for voice and HSD channels of 24 Kb/s.

On the other hand, in the case of the present Inmarsat-A and B MES, a comparative High-Gain Antenna (HGA) of minimum 24 dBi is required, due to the difference in satellite capabilities. Meanwhile, LGA of about 0 to 4 dBi are used in the Inmarsat-C and other similar omnidirectional systems to provide LSD of only about 600 to 1,200 b/s. The GPS system has adopted LGA because of the extremely low data rate of 50 b/s from the satellites. Because they have the same type of LGA system, it is possible to integrate Inmarsat-C MES with the GPS receiver.

There are no exact definitions to differentiate between characteristics of Low, Medium and High-gain antenna systems, except by the gain quantum, shape of the antenna and type of service. However, in the present and upcoming L-band GMSC applications, classification of L-Band MES antenna systems by their receiving and service capabilities is illustrated in **Table 4.1.**

Table 4.1.   Classification of L-Band Antenna Systems in GMSC

Type of Antenna	Gain Class	Typical Gain [dBi]	Typical G/T [dBK]	Typical Antenna (Dimension)	Typical GMSC Services
Omnidirectional	Low	0 – 4	–27 to –23	Quadrifilar Drooping-dipole Patch	LSD (Messages) Ship (Inmarsat-C) Vehicles & Aircraft
Semidirectional (Only in Azimuth)	Medium	4 – 8	–23 to –18	Array (2–4 elements) Helical, Patch SBF (0,4m Φ)	Voice/HSD Ship (Inmarsat-M) Vehicles
		8 – 16	–18 to –10	Phased Array (20 elements)	Aircraft (Inmarsat-Aero)
Directional	High	17 – 20	– 8 to –6	Dish (0,8m Φ)	Voice/HSD
		20 – 24	–4	Dish (1m Φ)	Ship (Inmarsat-A, B)

The ideal antenna gain can be defined with an isotropic (hypothetical) antenna, which has an isotropic radiation pattern without any losses and therefore radiates power in all directions in uniform intensities. Thus, if input power (P_{in}) is put into an isotropic antenna, the power flux-density per ideal unit area (P_{id}) at distance (r) from the antenna is given by the following relation:

$$P_{id} = P_{in}/4\pi r^2 \quad [W/m^2]$$
(4.1.)

However, if radiated power density is $P(\theta, \phi)/r^2$ in directions (θ = angle between the considered direction and the one in which maximum power is radiated, known as boresight;

and ϕ = phase) at distance (r) from the antenna under elevation, the gain of the antenna can be defined by the following equations:

$$G(\theta, \phi) = P(\theta, \phi)/r^2/P_{id} = P(\theta, \phi)/r^2/P_{in}/4\pi r^2 = 4\pi\, P(\theta, \phi)/P_{in} = P(\theta, \phi)/\, P_{in}/4\pi \quad [dBi] \quad (4.2.)$$

The above-defined gain is called an absolute gain or directive gain, which is determined only by the directivity (radiation pattern) of the antenna without taking account of any losses in the antenna system, such as impedance mismatch loss or spillover loss. Thus, if direction is not specified and the gain is not given a function of (θ, ϕ), it is assumed to be maximum gain. There is a general relationship between absolute gain and the physical dimensions of the antenna and this is given by the equation as follows:

$$G = 4\pi/\lambda^2\, \eta a \qquad\qquad\qquad (4.3.)$$

where η = aperture efficiency and a = physical aperture, which will denote the effective aperture of the antenna. According to the above relation it can be realized that compact antennas with small apertures must have low gain. If an antenna aperture is a dish a known diameter (d), can be written in normal and in decibel expression as follows:

$$G = (\pi d/\lambda)^2\, \eta = 10 \log \eta\, (\pi d/\lambda)^2 \quad [dBi] \qquad\qquad (4.4.)$$

Thus, it can be calculated that the gain in the Inmarsat shipborne antenna with a diameter of d = 1 m operated at 1.5 GHz is about 21 dBi.

The directivity of the antenna D(θ, ϕ) does not include dissipative losses and is defined as the ratio of P(θ, ϕ) to the power per unit solid angle from an isotropic antenna radiation, the same total antenna radiated power (P_r). The antenna directivity can be expressed by:

$$D(\theta, \phi) = P(\theta, \phi)/P_r/4\pi \qquad\qquad\qquad (4.5.)$$

The definition of antenna directivity does not take the efficiency of an antenna into account because ($P_r/4\pi$) is related to the actual power launched into space. The ratio of G(θ, ϕ) to D(θ, ϕ) is termed the radiation efficiency of the antenna.

4.2.3.3. Radiation Pattern, Beamwidth and Sidelobes

Radiation calculation is possible in principle if the EM field can be described quantitatively at all points of the antenna surface whose boundaries are those of the apertures. In this section, the radiation pattern from a circular aperture is considered as the aperture-type antenna has generally been used in MSC, especially in MMSC. This simple problem will give an insight into the characteristics of mobile antennas. For an antenna that generates a single focused beam, the principal parameter affecting the antenna radiation pattern E(θ, ϕ), after the aperture size (a), is the aperture illumination distribution $E_a(r, \psi)$, which is the amplitude of the far field radiation pattern E, at the point (θ, ϕ), being essentially the Fourier transform of the illumination distribution and is given by:

$$E(\theta, \phi) = 1/\pi a^2 \int_0^{2\pi} \int_0^a E_a(r, \psi)\, \exp\,[-jkr \sin \theta \cos (\phi - \psi)]\, r\,dr\,d\psi \qquad (4.6.)$$

Figure 4.1. Geometrical Parameters of Antenna Pattern and Gain Characteristics

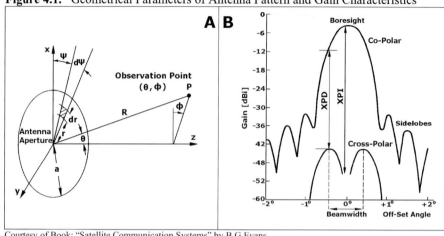

Courtesy of Book: "Satellite Communication Systems" by B.G Evans

An example considers an antenna employed in MSC, which utilizes a circular aperture, where for circularly symmetric aperture illumination distribution this relation reduces:

$$E(\theta) = 2/a^2 \int_0^a E_a(r) J_0 (kr \sin \theta \cos (\phi - \psi)] \, rdr \qquad (4.7.)$$

where a = d/2 denotes the radius of antenna aperture; J_0 = first kind and order zero of the Bassel function and k = 2 π/2 denotes the wave number. The other notations that denote distance and angles in coordinates are defined in the geometry illustrated in **Figure 4.1. (A)**. The antenna radiation pattern is three-dimensional in nature, so it usually has to be represented from the point of view of a single-axis plot.

The characteristics of the MES antenna radiation pattern affect interference levels directly. Any improvement in the pattern will therefore be fully reflected in the interference level and such improvement constitutes a very effective means of solving interference problems. To improve the pattern, one can either increase the antenna diameter or, with a constant diameter, use a specific technique for reducing the sidelobes. This method is therefore applicable when the MSC network is in the initial stages of development.

The antenna gain is normally calculated with reference to the boresight, i.e., the direction at which the maximum antenna gain occurs, in the case when θ, φ = 0°. Gain is usually expressed in (dBi), where component (i) refers to the fact that it is relative to the isotropic gain. In this instance, the matter of moment in a dual polarization frequency re-use satellite communication system is polarization discrimination between the co-polar and cross-polar signals, especially in the antenna main beam region, as illustrated in **Figure 4.1. (B)**. An important parameter that is used in an antenna's specification is the beam width evaluated by Half Power Beam width (HPBW) $2\theta_{HP}$, where θ_{HP} is the half-power angle when radiated power becomes half the maximum level (-3 dB). The HPBW (θ_0) is given by the following equation:

$$\theta_0 = 65 \, (\lambda/d) \qquad (4.8.)$$

Here it is possible to realize that the half-power bandwidth is inversely proportional to the operating frequency and the diameter of the antenna. For example, a 1 m receiver antenna operating in the C-band (4 GHz) has a 3 dB bandwidth of roughly 4.9°, while the same antenna operating in the Ku-band (11 GHz) has a 3 dB bandwidth of approximately 1.8°.

The antenna systems have co-polar and cross-polar gains, where the reception of unwanted, orthogonally polarized cross-polar signals will add as interference to the co-polar signal. The ability of an antenna to discriminate between a wanted polarized waveform and its unwanted orthogonal component is termed as its Cross-Polar Discrimination (XPD). When dual polarization is employed and the antenna's ability to differentiate between the wanted polarized waveform and the unwanted signal of the same polarization, introduced by the orthogonal polarized wave, it is termed as the Cross-Polar Isolation (XPI). In this context, an antenna typically would have an XPI > 30 dB.

The level of the antenna pattern's sidelobes is also important, as this tends to represent gain in an unwanted direction. For a transmitting gain this leads to the transmission of unwanted power, resulting in interference to other systems, or in the case of a receiving antenna, the reception of unwanted signals or noise. The sidelobe characteristic of MES is one of the main factors in determining the minimum spacing between satellites and therefore the orbit and spectrum utilization efficiency. The ITU-R S.465-5 recommendation gives a reference radiation diagram for use in coordination and interference assessment, which is defined by:

$$G = 32 - 25 \log \phi \quad [\text{dBi}] \quad \text{for } \phi_{min} \leq \phi < 48° \qquad\qquad (4.9.)$$

$$= 10 \text{ dBi} \qquad\qquad \text{for } 48° \leq \phi \leq 180°$$

where G = gain relative to an isotropic antenna; ϕ = off-axis angle referred to the main lobe axis and ϕ_{min} = 1° or 100 λ/d degrees, whichever is the greater. In this context, most of the effective power radiated by an antenna is contained in the so-called main lobes of the radiation pattern, while some residual power is radiated in the sidelobes. Sidelobes are an intrinsic property of antenna radiation and diffraction theory shows that they cannot be completely suppressed. However, sidelobes are also due partly to antenna defects which can be minimized by proper design. Conversely, due to the reciprocity theorem, the receive antenna gains and radiation patterns at the same frequency, are identical to the transmit antenna gains and radiation patterns. Unlikely, unwanted power can also be picked up by the antenna sidelobes during reception.

For large satellite antennas, with a diameter over 100 λ (wavelengths), a reference radiation pattern is recommended by the CCIR for interference to and from other satellite and terrestrial communication systems. At this point, the diameters of the vehicle antennas under discussion are, in many cases, below five wavelengths in the L-band. Further CCIR action is expected to define a reference radiation pattern for mobile antennas in MSC.

4.2.3.4. Polarization and Axial Ratio

The antenna and the EM field received or transmitted have polarization properties. Thus, the polarization of an EM wave describes the shape and orientation of the locus of the extremities of the field vectors as a function of time. A wave may be described as linearly, circularly or elliptically polarized. Linear polarization is such that the electric E-field is oriented at a constant angle as it is propagated that can be either vertically or horizontally.

If a plane wave is propagated along the (z) axis and electric field (E) is on the (x-z) or (y-z) planes, relations for linear vertical and horizontal polarization can be written as follows:

$$E_x = E_a \cdot e^{j(\omega t - kz + \phi a)} \quad \text{and} \quad E_y = E_b \cdot e^{j(\omega t - kz + \phi b)} \tag{4.10.}$$

where E_a; ω; k and ϕ_a denote the maximum amplitude of electric field, angular frequency $(2\pi f)$, wave number and initial phase, respectively, while E_b and ϕ_b are the maximum amplitude and the initial phase of the wave.

Circular polarization is the superposition of two orthogonal linear polarizations, such as vertical and horizontal, with a 90° $(\pi/2)$ phase difference. The tip of the resultant E-field vector may be imagined to rotate as it propagates in a helical path. There is a Left-Hand Circularly Polarized (LHCP) wave with anticlockwise rotation and a Right-Hand Circularly Polarized (RHCP) wave with clockwise rotation.

An elliptically polarized wave may be regarded as the result either of two linearly or two circularly polarized waves with opposite directions. This type of polarization is the case when the amplitudes and phase difference between the two waves are not equal $(\pi/2)$.

As discussed in a previous section, the signals fields can contain co- and cross-polar components. In this way, the cross-polarization of a source becomes of increasing interest to MSC antenna designers. In the case of Tx or Rx antennas with a linearly polarized field, the cross-polar component is the field at right angles to this co-polar component. Namely, if the co-polar component is vertical, then the cross-polar component is horizontal. Circular cross-polarization is that of the opposite hand to the desired principal or reference polarization. Impure circular polarization is in fact elliptical. The level of impurity is measured by the elliptical and known as the Axial Ratio (AR). The AR can be defined as the ratio of the major axis electric component to that of the minor axis by:

$$|AR| = |E_1/E_2| \quad (1 \leq |AR| \leq \infty) \tag{4.11.}$$

The signal for AR denotes the direction of rotation, however, an absolute value is usually used to evaluate circular polarized radiated waves and can be expressed in decibels by the following equation:

$$|AR| = 20 \cdot \log(|E_1/E_2|) \quad [dB] \quad \text{for} \quad (0 \leq |AR| \leq \infty) \tag{4.12.}$$

Accordingly, the AR is determined by the performance of the antenna, so the AR is one of the most important parameters of circular polarized antennas. It can easily be understood that the AR depends on direction with respect to the axis of the antenna. In general, the AR is best (smallest) in the boresight direction and is progressively worse further away from the boresight.

Circular polarized waves are used in order to eliminate the need for polarization tracking. RHCP has been used in the Inmarsat transmission system. Otherwise, in the case of aperture-type antennas, such as the parabolic reflector antenna, which is commonly used as a shipborne antenna in the current Inmarsat-A and B terminal, an axial ratio of below 1. 5 dB in the boresight direction is so easy to achieve that polarization mismatch loss is almost negligible. However, in the case of phased array antennas, a degradation of the axis ratio caused by beam scanning must be taken into account.

4.2.3.5. Figure of Merit (G/T) and EIRP

Although gain is an essential factor in considering antennas, the figure of merit ratio of a gain-to-noise temperature (G/T) is more commonly specified from the standpoint of MSS and satellite communications in general. The figure of merit for the receiving station is defined as the ratio between the gain of the antenna in the direction of the receiving signal and the receiving system noise temperature, the gain-to-noise temperature ratio (G/T) is generally given for the maximum gain derived from gain formula (4.4.) as follows:

$$G_{max} = P_{max}/P_0/4\pi = 10 \log G \quad [dB] \tag{4.13.}$$

The G_{max} is often called the antenna gain expressed in dB, where the total radiated power in all directions can be determined by the following integration:

$$P_0 = \int_0^{2\pi} \int_0^{\pi} P(\theta, \phi) \sin \theta \, d\theta \, d\phi \tag{4.14.}$$

The G/T value is expressed in decibels per Kelvin $[dB(K^{-1})]$ by the following relation:

$$(G/T) = 10 \log G - \log T_{SA} = 10 \log G - \log T_S \quad [dB(K^{-1})] \tag{4.15.}$$

The Earth station G/T typical values range from 35 $dB(K^{-1})$, for instance an LES receive antenna with a 15 to 18 m diameter has some 15.5 $dB(K^{-1})$. The G/T is a very important parameter of an Earth station, so the methods used for its measurement and the contribution to the noise temperature are the subject of the ITU-R S.733 Recommendation.

The noise temperature measured at the terminals of antenna pointed to the sky depends upon frequency of operation, elevation angle and the antenna sidelobe structure. In more formal terms, the noise temperature will be derived from a complete solid angle integration of the noise power received from all noise sources (terrestrial and galactic) and determined for clear weather conditions by the following integral:

$$T_A = 1/4\pi \int_\Omega P(\theta, \phi) T(\theta, \phi) \, d\theta \, d\phi \tag{4.16.}$$

Thus, to produce a low noise antenna, its sidelobes must be minimized, especially in the direction of the Earth's surface, where T = noise temperature.

System total noise temperature of the system (T_{SR}) at an input port of receiver LNA or at the antenna output (T_A), taking account of losses caused by tracking, feed lines and a radome is defined by:

$$T_{SR} = T_R + T_a (1 - 1/a) + T_A/a \quad \text{or} \quad T_{SA} = T_A + T_a (a - 1) + a \, T_A \tag{4.17.}$$

where T_R = noise temperature of the receiver (LNA) with a typical value of about 80 K to 100 K in the L-band; T_a = temperature of the environment of about 300 K; L_f = total loss of feed lines and components such as diplexer, cables and phase shifters if a phased-array antenna is used; a = attenuation expressed as a power ratio ($a \geq 1$ or in decibels $a_{dB} = 10 \log a$); T_A = antenna noise temperature that comes from such effects as the ionosphere and the Earth, which value of about 200 K depends on factors such as frequency and bandwidth and T_{SA} = antenna with a noise temperature. In such a manner, the noise of the antenna temperature must be kept as low as possible by proper design solution in order to obtain a high figure of merit (G/T).

With reference to the previously expressed formula ($P_{id} = P_{in}/4\pi r^2$) of transmitting antenna power density on the spherical surface if it has a transmitting gain (G_T) and where (P_{in}) is equal to the transmitted power (P_T), the power density (P_D) can be written as:

$$P_D = G_T \cdot P_T/4\pi r^2 \ [W/m^2] \tag{4.18.}$$

where ($G_T \cdot P_T$) related values are considered to be the radiation power transmitted by an ideal omnidirectional antenna. Therefore, this term is considered as an Effective (or also Equivalent) Isotropically Radiated Power (EIRP), which can be expressed in antilogarithm and decibel expressions respectively as follows:

$$EIRP = G_T \cdot P_T \ [W] \quad and \quad EIRP = [G_T] + [P_T] \ [dBW] \tag{4.19.}$$

The EIRP value is an important parameter in evaluating the transmitting performance of an MES terminal including an antenna. However, the EIRP amount (dBW) is defined by the sum of the antenna gain (dB) and the output power of HPA (dBW), taking account of feed losses such as feed lines, cable and a diplexer.

4.3. Classification of Mobile Satellite Antennas (MSA)

In many respects the mobile satellite antennas currently available for MSC applications constitute the weakest links of the system. If the mobile antenna has a high gain, it has to track the satellite, following both mobiles and satellite orbital motions. Thus, sometimes this is difficult and expensive to synchronize. Therefore, if the vehicular antenna has low gain, it does not need to perform tracking but the capacity of the communications link is limited. In general, according to the transmission direction, there are three types of MSA: **1)** transmitting and receiving or so-called transceiving, as a part of all types of MES; **2)** only receiving is part of the special Inmarsat EGC receiver and **3)** only transmitting is built in satellite beacon antennas for maritime, land and aeronautical applications.

The Inmarsat, Eutelsat, ESA, Cospas-Sarsat, Iridium, Globalstar, ICO and other GEO and Non-GEO current and forthcoming mobile satellite providers have conducted research on all network segments, including different types of MSA and their future development and improvements. On the other hand, the Engineering Test Satellite-V (ETS/V) experiments conducted in Japan for the transmission of voice, video and different data rate digital communications between ships, land vehicles and aircraft were successful. Moreover, a test of low-speed data transmission by using briefcase-size transportable equipment, onto which two small printed antennas were mounted, was among the experiments.

4.3.1. Shipborne MSA

The different types of shipborne satellite antenna systems were developed for installation on board ocean-going ships and inland sailing vessels, on sea platforms and other offshore infrastructures. In general, these antennas must have strong and rugged constructions, with corresponding mechanical and electrical particulars. The Inmarsat-A SES is the inheritor of the first generation of Marisat, similar to SES and was the first Inmarsat operating standard of MMSC. In fact, this analog standard was started in 1982, using the Inmarsat standard-A transceiving antenna system known as Above Deck Equipment (ADU). In the meantime Inmarsat-C and EGC were developed with small omnidirectional antennas. In addition, Inmarsat-B digital standard started to be in service on ships at the end of 1993, using the second generation of Inmarsat satellites. This standard, compatible with Inmarsat-A, uses the same antenna specification. After employing more powerful satellite constellations, the Inmarsat system developed Inmarsat-M, mini-M and D+ standards with special shipborne antenna systems. All the above-mentioned MSA are transceiving antennas, except for the Inmarsat-EGC receiver, which can also use a receiving antenna. At the same time, Inmarsat developed the Inmarsat-E L-band satellite EPIRB, with a built-in shaped or hemispherical beam transmitting antenna of 0 dB nominal gain. On the other hand, the Cospas-Sarsat system developed shipborne EPIRB, with small built in VHF and UHF antennas. Thus, rugged MSA with added-value system functions compatible with shipborne operation for Inmarsat, Iridium, Globalstar and other systems and compact antennas for small lifeboat operation are some of the fascinating requirements for antenna designers in the future.

4.3.2. Vehicleborne MSA

The vehicleborne MSA were developed for installation on road and rail vehicles, such as trucks, trailers, buses, cars and trains. Thus, the transceiving Inmarsat-C, D, M, mini-M and other MSA are used on board land vehicles. Similar to the maritime application, the Cospas-Sarsat system has developed personal or vehicleborne PLB handheld satellite beacons with small built in VHF and UHF band transmitting antennas.

4.3.3. Airborne MSA

The airborne MSA were developed for installation on board aircraft, such as airplanes and helicopters. There are several types of low and high-gain airborne satellite transceiving antennas for Inmarsat-C, H, H+, I, L, mini-M and other MSS. Parallel to the maritime and land distress applications, the Cospas-Sarsat system developed airborne ELT satellite beacons with small built-in VHF and UHF band transmitting antennas. Because of the high speed of aircraft, the aerodynamic constraints are significant and antennas for both radio and satellite systems must conform to minimum drag and reliability requirements.

4.3.4. Transportable MSA

Inmarsat developed special transceiving satellite antenna systems, which can be integrated into mobile Transportable Earth Stations (TES) of Inmarsat-B, C, M and mini-M terminals.

4.3.5. MSA for Personal Satellite Terminals

The new GEO and Non-GEO MSC systems developed small satellite antennas for GMPSC terminals, such as handheld antennas, antennas for mobiles and roof antennas for indoor and outdoor satellite terminals. Some of these systems provide mobile tracking, so the tracking antenna has to be combined or integrated with the communications antenna.

4.3.6. Other Types of MSA

The other solutions for MSA are receiving broadcast radio and TV MSA, receiving GNSS MSA, transceiving broadband MSA and antennas for military satellite communications and navigation equipment.

4.4. Low-Gain Omnidirectional Antennas

As mentioned, the antenna systems for MSC are classified into omnidirectional and directional. The gain of omnidirectional antennas is low and generally from 0 to 4 dB in the L-band, which does not require the capability of satellite tracking. There are three types of low-gain omnidirectional antennas, which are very attractive for all mobile applications owing to the small size, light weight and circular polarization properties. These antennas are also used as elements of directional antennas for special configurations.

4.4.1. Quadrifilar Helix Antenna (QHA)

The QHA low-gain model is composed of four identical helixes wound, equally spaced, on a cylindrical surface. The helix elements are fed with signals equal in amplitude and 0, 90, 180 and 270^0 in relative phase. This antenna can easily generate circular polarized waves without a balloon or a 3 dB power divider, which are required to excite a balanced fed dipole and circular polarized cross-dipoles. It can also be operated on a wide frequency bandwidth of up to 200% because it is a traveling-wave-type antenna. The components of QHA are ground plane **(g)**, pitch **(p)**, pitch angle **(a)**, length **(l)** and diameter **(d)**, presented in **Figure 4.2. (A)**. The diameter of the ground plane is usually selected to be larger than one wavelength and the number of turns is N = l/p. However, it is well known that the parameters for (a) are about 12 to 15 and the circumference of the helix (πd) is about 0.75 to 1.25 wavelengths. Circular polarized waves with good axial ratios can be transmitted along the (z) axis direction (axial mode). The gain of a helical antenna depends on the number of (N) turns and typical gain and half-power beam width are about 8 dBi and 50^0 when N = 12 ~ 12 but is usually about 3 dBi. This antenna is employed as a receiving antenna for GPS and as a transceiver antenna for L-band Inmarsat-C SES, VES and AES (low air drag Aero-C) applications covered by different kinds of radomes. All three types of antennas can also be combined with GPS receiving. This antenna is also a component of AMSC, MSAT, MSAT-x, Iridium, Globalstar, Mobilesat ETS-V, forthcoming ICO and other MSC terminals.

In general, QHA, as a mobile antenna is the best solution and has two advantages over a conventional unifilar helical antenna. The first is an increase in bandwidth, namely, it can generate axial mode circular polarized waves in the frequency range from 0.4 to 2.0 wavelengths of the helix circumference. The second is lowered frequency for axial mode operation. The principle disadvantage is an increase in the complexity of the feed system. The area of the ground plane is usually about 3 times the diameter of the helix.

4.4.2. Crossed-Drooping Dipole Antenna (CDDA)

A dipole antenna with a half-wavelength (λ/2) is the most widely used and it is also the most popular, having been used in antenna systems such as the parabolic antenna for MSC.

Figure 4.2. Types of Low-Gain Omnidirectional Antennas

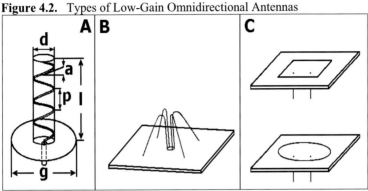

Courtesy of Book: "Mobile Antenna Systems Handbook" by K. Fujimoto and J.R. James

A half-wavelength dipole is a linear antenna whose current amplitude varies one-half of a sine wave, with a maximum at the centre. As a dipole antenna radiates linearly polarized waves, two crossed-dipole antennas have been used in order to generate circular polarized waves. The two dipoles are geometrically orthogonal and equal amplitude signals are fed to them with π/2 in-phase difference. In order to optimize the radiation pattern, a set of dipole antennas is bent toward the ground, as shown in **Figure 4.2. (B)** and for that reason it is called a drooping dipole antenna. Otherwise, the CDDA serves as a transceiver antenna for L-band Inmarsat-C SES and VES applications mounted inside a radome.

The CDDA is the most interesting for LMSC, where required angular coverage is narrow in elevation and is almost constant in azimuth angle. By varying the separation between the dipole elements and the ground plane, the elevation pattern can be adjusted for optimum coverage for the region of interest. The general characteristics of this antenna are: gain is 4 dBi minimums, axial ratio is 6 dB maximum and the height of the antenna is about 15 cm. This antenna has a maximum gain in the boresight direction.

4.4.3. Microstrip Patch Antenna (MPA)

A microstrip disc (patch) antenna is very low profile and has mechanical strength, so it is considered to be the best type for mobiles such as cars and especially in aircraft at the hybrid L to Ku-band, which requires low air drag. In general, a circular disk antenna element has a circular metallic disc supported by a dielectric substrate material and printed on a thin dielectric substrate with a ground plane. In order to produce a circularly polarized wave, a patch antenna is excited at two points orthogonal to each other and fed with signals equal in amplitude and 0 and 90° in relative phase. Thus, a higher mode patch antenna can also be designed to have a similar radiation pattern to the drooping dipole. To produce conical radiation patterns (null on axis) suitable for land mobile satellite applications, the antenna is excited at higher mode orders. In **Figure 4.2. (C)** is illustrated the basic configuration for a circular patch antenna (above it is shown square patch with the same characteristics), which has two feed points to generate circular polarized waves. The resonant frequency excited by basic mode and given as:

$$f = 1,84c/2\pi a \sqrt{\epsilon_r}$$ (4.20.)

Figure 4.3. Types of Aperture Antennas

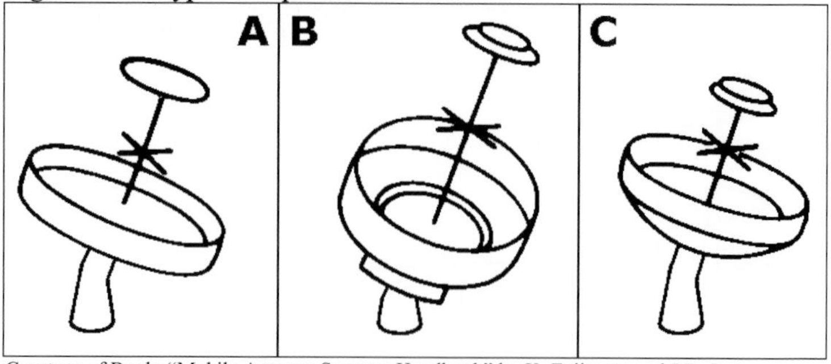

Courtesy of Book: "Mobile Antenna Systems Handbook" by K. Fujimoto and J.R. James

where (a), (c) and ($\mathcal{E}_r$) are the radius of circular disc, the velocity of light in free space and the relative dielectric constant of the substrate, respectively. In LMSC, a MPA antenna with higher order excitation is considered better because it can optimize the gain in elevation angle to the satellite in the same way as a CDDA. In fact, the area of a higher mode circular MPA is about 1.7 times larger in radius on the gain is about 6 to 8 dBi. The circular patch is also suitable as a satellite navigation-receiving antenna for GPS receivers.

4.5. Medium-Gain Directional Antennas

The medium-gain directional MSA are solutions with a typical gain between 12 and 15 dBi, although some antennas can have even bigger gains. These MSC antenna systems can provide voice, Fax and HSD for Inmarsat-M shipborne and vehicleborne applications and for Inmarsat airborne standards, including other systems developed by ESTEC.

4.5.1. Aperture Reflector Antennas

The aperture reflector antennas are good solutions with medium-gain characteristics used in MSC, with three basic representatives such as SBF, modified SBF and improved SBF antennas, illustrated in **Figure 4.3. (A)**, **(B)** and **(C)**, respectively. The main characteristics of these three antennas are shown in **Table 4.2.**
Moreover, due to the excellent radiation characteristics of SBF antennas, all three types of aperture antenna with half-power beam width of about 34° have been in their time proposed for shipborne antenna of Inmarsat-M. The SBF antennas consist in the stabilized platform with two gyroscopes for azimuth and elevation angles, diplexer, HPA and LNA, which are enclosed under the protective cupola of a radome. In order to stabilize the antenna, two gyro wheels rotate in opposite directions on a platform.

4.5.1.1. Short Backfire (SBF) Plane Reflector Antenna

The SBF plane reflector antenna that was developed experimentally by H.W. Ehrenspeck in the 1960s is well known as a highly efficient antenna of distinctly simple and compact construction. Its high directivity and low sidelobe characteristics make it a single antenna with high, even values, which is applicable to MSC, tracking and telemetry. Therefore, an SBF antenna is very attractive for gains in the order of 13 to 15 dBi peak RHCP and can be

Table 4.2. Particulars of Aperture Types of Antennas

Characteristics	SBF Antenna	Modified SBF Antenna	Improved SBF Antenna
Effective Gain	14.5 dB	15 dB	15 dB
Half-Power Bandwidth	34°	34°	34°
Directive Gain	14.8 dB	15,5 dB	15.5 dB
First Sidelobe Level	−21 dB	−22 dB	−22.5 dB
Axial Ratio	−1.3 dB	−1.1 dB	−1.1 dB
Aperture Efficiency:			
Effective - Directive Gain	65% − 75%	75% − 80%	76% − 85%
RF/VSWR Bandwidth, under 1,5	3%	7%	9%
Diameter of Large Reflector (D_R):			
Bigger (D_{R1})	40 cm (2,05λ)	40 cm (2,05λ)	40 cm (2.05λ)
Smaller (D_{R2})	-	27 cm (1.38λ)	-
Diameter of Small Reflector (D_r):			
Bigger (D_{r1})		-	
Smaller (D_{r2})	9 cm (0.46λ)	9.5 cm (0.48λ)	9 cm (0.46λ)
Width of a Rim	-	8.5 cm (0.43λ)	8 cm (0.41λ)
Distance Between (D_R) and (D_r)	4.9 (0.25 λ)		4.9 (0.25 λ)
Distance Between Exciter & D_r	9.7 cm (0.49λ)	19.5 cm (0.99λ)	12.9 cm (0.66λ)
Distance Between (D_{r1}) & (D_{r2})	4.9 cm (0.25λ)	-	5.7 cm (0.29λ)
Slanting Angle of a (D_R)	-	-	1.8 cm (0.09λ)
	0°	-	15°

mounted primarily on small but on any size of ships. Otherwise, this type of antenna consists in two circular planar reflectors of different diameter, separated generally by about one-half wavelength, forming a shallow leaky cavity resonator with a radiation beam normal to the small reflector. Namely, the antenna is fed by a dipole at around the midpoint between two reflectors and it has almost a quarter-wavelength rim on the larger reflector. It has the problem of a narrow bandwidth of about 3% because of its leaky cavity operation. The Rx terminal G/T is −12 dBK and the EIRP of the Tx terminal is 28 dBW.

The basic configuration of the SBF antenna consists in a cross-dipole element, which is required to generate a circularly polarized wave, large and small reflectors and a circular metallic rim. The antenna has the strong directivity normal to the reflector and its performance is superior to that of other types of mobile antennas with the same diameter, however, it has the problem of narrow frequency band characteristics. This antenna has many beneficial characteristics, such as efficiency and the simplicity of construction and is also considered a favorite option for a compact and high-efficiency shipboard antenna. It is produced by many world manufacturers approved by Inmarsat or ESTEC. For instance, one of the most known manufacturers of Inmarsat-M MES with antenna is Thrane and Thrane with their partners, while the manufacturer of ESTEC SBF antenna is G&C McMichael.

4.5.1.2. Modified SBF Plane Reflector Antenna

A modified SBF antenna differs from the conventional SBF antenna in that there is either an additional step on the large reflector or a change in the shape of the large reflector from a circular to a conical plate in order to improve the gain characteristics and the frequency bandwidth of the VSWR. The dual reflector improves the input impedance characteristics covering the frequency range between transmitting and receiving sides.

The conventional SBF model is a resonant-type antenna, producing input impedance characteristics that are narrow in bandwidth, so wider bandwidth is required to cover the 1.6/1.5 GHz range for MES of the Inmarsat system. In effect, the improvement in the input impedance is greatly dependent on the size and the separation of the small reflectors. The VSWR can be reduced from 1.7 and 1.5 (at 1.54 and 1.64 GHz) to below 1.2 for each RF.

4.5.1.3. Improved SBF Conical Reflector Antenna

The main research activities of the ETS-V program in MSC have been focused on studying the reduction of fading, using compact and high-efficiency antennas with a gain of around 15 dBi, so the electrical characteristics of a simple SBF antenna have been improved by changing its main reflector from a flat disk to a conical or a step plate and by adding a second small reflector. The gain is improved by 1 dB without changing sidelobe levels. Comparisons of electrical and other parameters of three types of SBF antennas are shown in **Table 4.2.** Stabilization of the antenna is obtained by a two-axis stabilized method and satellite pointing is carried out by a tracking program using output signals from the ship's gyroscope. It is also considered to be a suitable option for mounting aboard ships.

4.5.2. Wire Antennas

The wire antenna systems are monosyllabic construction or combinations of elements, such as different shapes of wire spirals and helixes, dipoles and patches. These types of antennas have a very simple construction, with any reflector specified for medium-gain directional antennas and, with some modification, responds well to the demands of MSC applications.

4.5.2.1. Helical Wire Antennas

Since an axial mode helical antenna has good circular polarization characteristics over a wide frequency range, it has been put into practical use as a single wire antenna or as an array element. With respect to the structure, this antenna can be considered a compromise between the dipole and the loop antennas and the radiation mode varies with the pitch angle and the circumference of the helix. In particular, a helix with a pitch angle of 12 to 15° and a circumference of about 1 λ, has a sharp directivity towards the axial direction of the antenna. This radiation mode is called the axial mode, which is the most important mode in helical antennas. Several studies have been carried out on the properties of the axial mode helical antenna with a finite reflector. The current induced on the helix is composed of four major waves, which are two rapidly attenuating waves and two uniform waves along the helical wire. These waves include the traveling wave and the reflected wave. Thus, in a conventional helical antenna, the uniform traveling wave will be dominant when the antenna length is fairly large, with typical versions such as a conical helix shown in **Figure 4.4. (A)** and a cylindrical helix in **Figure 4.4. (B).** A conical helix is interesting for L-band MSS enabling HPBW in the order of 100° and circular polarization without hybrid gain of 4 to 7 dBi. Cylindrical antennas can be monofilar or multifilar, also suited for L-band MSS, while in a short-cut cylindrical helix antenna, the rapidly attenuating traveling wave will be dominant, especially in a two-turn (N = 2) helical antenna.

1. Conical Helix Antenna – This antenna can be regarded as a low-gain development of the cylindrical helix antenna and is suitable for wide-beam width applications with good efficiency. Thus, with suitable choices of cone angle and turn spacing, it is possible to achieve a beam width in the order of 100°. This type of antenna can also achieve an input VSWR of 1.5:1 or better than 5% frequency bandwidth merely by incorporating a simple quarter-wavelength transformer. The typical size for an L-band application is in the order of 15 cm in length and the ground plane is about 20 cm in diameter. The resultant gain is approximately 4 to 7 dBi, which is between low and medium-gain requirements.

Figure 4.4. Types of Wire Antennas

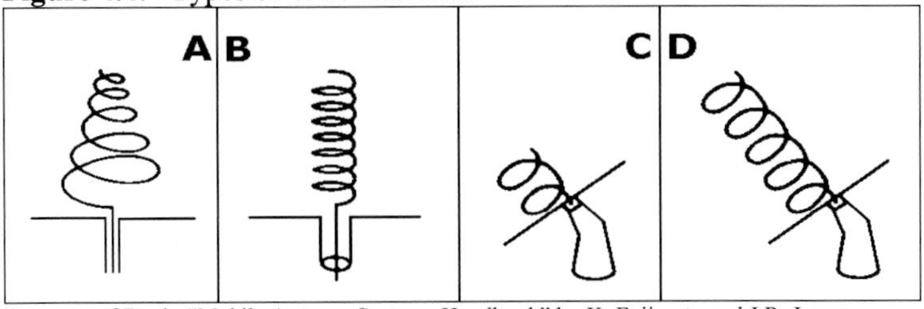

Courtesy of Book: "Mobile Antenna Systems Handbook" by K. Fujimoto and J.R. James

2. Two-Turn Cylindrical Helix Antenna – This antenna has two-turns of wires, forming a simple helical antenna solution with reflector, illustrated in **Figure 4.4. (C).** This model has relatively high antenna gain and excellent polarization characteristics for its size. Radiation patterns characteristically are calculated with respect to (E_0) and (E_ϕ) planes. The gain of this antenna is 9 dBi and the axial ratio is about 1 dB, with reflector diameter (d) around 1λ. Such types of antenna have comparatively high performance in spite of their small size and compact construction. From the above-mentioned considerations, a highly efficient antenna for the Inmarsat-M MES can be realized by applying this antenna to elements of an array antenna.

3. Five-Turn Cylindrical Helix Antenna – This antenna solution is illustrated in **Figure 4.4. (D).** The main electrical characteristics are: gain is 12.5 dBi of peak RHCP for Tx and 11.5 dBi for Rx; sidelobe level has value of about –13 dB; axial ratio is 3 dB; beam width of 3 dB has angle of –47°; terminal G/T has –16 dBK and terminal EIRP has 29 dBW. This antenna solution is designed and developed by the European research institution ESTEC. In addition, stabilization of this antenna is obtained by gravity elevation on double-gimbaled suspension. Thus, the pendulum aligns itself with the vertical when not subject to other acceleration. However because the centre of rotation of the pendulum is distant from that of the ship pitch and roll movement induces horizontal acceleration to which the pendulum is sensitive. In order to limit perturbations, the resonant frequency of the pendulum must be low with respect to the excitation frequencies in pitch and roll and the damping (friction) must be minimum. Low resonance frequency is achieved by minimizing the distance between the centre of gravity of the rotating part and its centre of rotation. This also reduces torque due to horizontal acceleration but at the same time reduces the stabilizing torque due to gravity.

4.5.2.2. Inverted V-Form Cross Dipole Antenna

The inverted V-type Crossed Dipole Antenna is an advanced circularly polarized antenna with tick V-elements, shown in **Figure 4.5. (A).** The resonance of this antenna is obtained when the length is somewhat shorter than a free-space half wavelength. Thus, as the thickness is increased, the resonant length is reduced. Circular polarization can be produced by a pair of orthogonally positioned dipoles driven in quadrature phase with equal amplitudes. The crossed-dipole antenna arrangement cannot provide a good axial ratio off boresight because the radiation patterns for the straight dipole are different in both principal planes, called the H and E-planes. This shortcoming can be improved by modifying the straight dipoles to a nonstraight version, such as the V and U-forms. The improved dipoles

Figure 4.5. Types of Wire Antennas

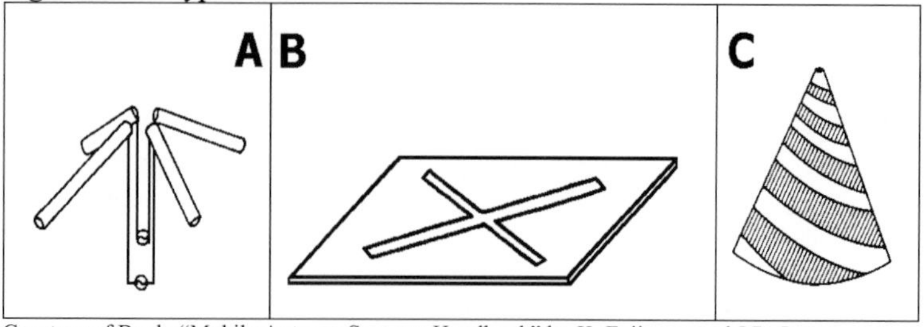

Courtesy of Book: "Mobile Antenna Systems Handbook" by K. Fujimoto and J.R. James

are called V and U-type dipoles. According to some conducted measurements, the U-type provides better electrical performance than the V-type, though the V-type is simpler in mechanical structure and is less complex. The crossed-dipole can also produce circular polarization without using any external circuits, such as the hybrid component. Thus, the condition to excite the circularly polarized waves can be established by a balun and the self-phasing of four radiating elements. Two of the elements are at a 0° phase angle and the other two are at an 180° phase angle. The desired 90° phase difference is obtained by designing the orthogonal elements such that one is larger relative to making it inductive, while the other is smaller to make it capacitive. This type of antenna is a good model for Ku-band aeronautical satellite communications.

4.5.2.3. Crossed-Slot Antenna

These antennas are useful for L-band aeronautical satellite communications on high-speed aircraft because they are low-profile in structure and suitable for a flush-mounting application, which is shown in **Figure 4.5. (B).** The slot antenna is circularly polarized and is complementary with the corresponding dipole antenna, so that the radiation pattern is the same as that for the horizontal dipole. There are only two differences: first is the property that the electric and magnetic fields are interchanged and second, is that the slot electric field component normal to the perfectly conducting sheet is discontinuous from one side of the sheet to the other because the direction of the field reverses. In this case, the tangential component of the magnetic field is, likewise, discontinuous. This antenna can be also complementary with the corresponding crossed-dipole antenna, although the feeding method for the circular polarization is more complicated. Thus, on a model of this antenna known as a cavity-backed it needs one 90° hybrid to produce the circular polarization. This feed technique is effective not only to suppress undesired coupling between the cross slots but also to match the input impedance over a wider frequency band.

4.5.2.4. Conical Spiral Antenna

This type of antenna has spiral wire elements on a cone with circular polarization and is suitable for L-band LMSC and GPS applications, while the bifilar version is also used in Ku-band satellite communications, see **Figure 4.5. (C).** In comparison with a conical helix antenna, this type of antenna provides better performance and is more versatile, though the geometry is somewhat complex. Therefore, such an antenna is independent of frequency and its geometry can be presented mathematically in spherical coordinates (r, θ, φ) as:

$$r = e^{a\phi}g(\theta) \tag{4.21.}$$

where (a) and $g(\theta)$ are an arbitrary constant and angular function, respectively. Its radiation mechanism can be understood by regarding the two spirals as a transmission line. When two conductor arms are fed in antiphase at the cone apex, waves travel out from the feed point and propagate along the spirals without radiating until a resonant length has been traversed. Strong radiation occurs at that point and very little energy is reflected by the outer limits of the spiral. Conveniently, two conductor arms can also be fed directly at the centre point or apex from a coaxial cable bonded to one of the spiral arms without any external baluns because the spiral arm can itself act as a balun. In this case, a dummy cable may be bonded to another arm to maintain the symmetrical performance. If the width of arm is decreased to a narrow constant value, the arms can be formed by the cable alone.

4.5.2.5. Planar Spiral Antennas

Cavity-backed planar spiral antennas are commonly divided into three main categories: equiangular, logarithmic and Archimedean spiral antennas. These types of antennas are well suited for flush-mounting on aircraft for L to Ku-band satellite communications. In general, this antenna has been fed by using the external balun but it can also be fed at the centre point, or apex, from a coaxial cable bonded to one of the arms, without any external baluns, like the conical spiral antenna.

1. Equiangular Spiral Antenna – The geometry of this antenna corresponds to the special case of the conical spiral antenna, bifilar with logarithmic period, cavity-backed and can be obtained by substituting a $\pi/2$ into θ_0 to give:

$$r_1 = r_0 e^{a\phi}; \quad r_2 = r_0 e^{a(\phi-\Delta)}; \quad r_3 = r_0 e^{a(\phi\pm\pi)}; \quad r_4 = r_0 e^{a(\phi-\Delta\pm\pi)} \tag{4.22.}$$

This antenna needs no external hybrid circuits to produce circular polarization and the example shown in **Figure 4.6. (A)** can radiate LHCP waves outward from the page and RHCP waves into the page when the pair of spirals is excited in antiphase at the centre. Otherwise, according to experimental measurements, the axial ratio is near unity and the HPBW is in the order of $90°$ over a decade bandwidth or even more. As for the input impedance the resistive part on the thickness of the antenna elements and thin elements lead to high impedance values. This implies that the impedance depends on the arm width when the structure is planar. If the angular extent (Δ) is chosen to be $90°$, the geometries of the arm and the space between arms are identical, except for a rotation of $90°$ around an axis. This structure is defined as self-complementary, just like the conical spiral antenna but it should be noted that the planar spiral antenna has a constant impedance of 60π [Ω] for the two arm configurations.

2. Logarithmic Spiral Antenna – This bifilar antenna design with logarithmic period and cavity-backed, shown in **Figure 4.6. (B)**, can be presented mathematically by:

$$r_1 = a^{\phi}; \quad r_2 = a^{(\phi-\Delta)}; \quad r_3 = a^{(\phi\pm\pi)}; \quad r_4 = a^{(\phi-\Delta\pm\pi)} \tag{4.23.}$$

This antenna can radiate RHCP waves outward from the page and LHCP waves into the page without any external hybrid circuits, as a pair of spirals is excited with an antiphase at the centre.

Figure 4.6. Types of Wire Antennas

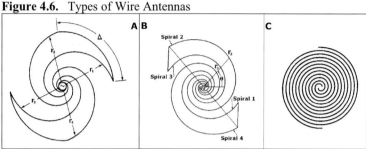

Courtesy of Book: "Mobile Antenna Systems Handbook" by K. Fujimoto and J.R. James

3. Archimedean Spiral Antenna – The Archimedean spiral thin-wire bifilar cavity-backed antenna, shown in **Figure 4.6. (C)**, is another geometry of the planar spiral. This antenna has superior bandwidth properties when fully optimized and typically consists in a pair of thin wire arms, of which the geometry can be presented by:

$$r_1 = r_0\phi; \quad r_2 = r_0(\phi - \pi) \tag{4.24.}$$

This antenna also needs no external hybrid circuits to produce circular polarization and can radiate RHCP waves outward from the page and the LHCP waves into the page if the pair of thin wire arms is excited in antiphase at the centre. It is a broadband antenna and has properties similar to the standard planar spiral antenna, although it is not theoretically a frequency independent structure. When placed in a quarter-wave cavity, this antenna can achieve near-octave bandwidth, even when the cavity consists in a metal-based cylinder without any absorber. Thus, if an absorber-loaded cylinder is employed in the cavity, a greater-than decade bandwidth may be achieved, although about half the power is dissipated into heat by the absorber. A typical Archimedean spiral antenna has an octave bandwidth for a VSWR less then 2, an axial ratio of less then 2 dB and a beam width of about 70°, while a gain of 7 to 8 dBi is achieved without an absorber. The structure has several mechanical advantages: it is compact and fairly simple to construct and the spiral arms can be easily fed, using a suitable impendence-transforming balun.

4.5.3. Array Antennas

Several different type of antenna can be arrayed in space to make a directional pattern or one with a desired radiation pattern. This type of integrated and combined antenna is called an array antenna consisting in more than two elements, such as microstrip, cross-slot, cross-dipole, helixes or other wire elements and is suitable for MSC. Each element of an array antenna is excited by equal amplitude and phase and its radiation pattern is fixed.

4.5.3.1. Microstrip Array Antenna

The microstrip array antenna (MAA) is a nine-element flat antenna, disposed in three lines spaced at 94 mm, namely about a half wavelength at 1.6/1.5 GHz and whose antenna volume is about 300 x 300 x 10 mm, see **Figure 4.7. (A)**. As shown in this figure, the element arrangements of the MAA solutions are 3 x 3 rows square arrays in order to obtain

Figure 4.7. Types of Array Antennas

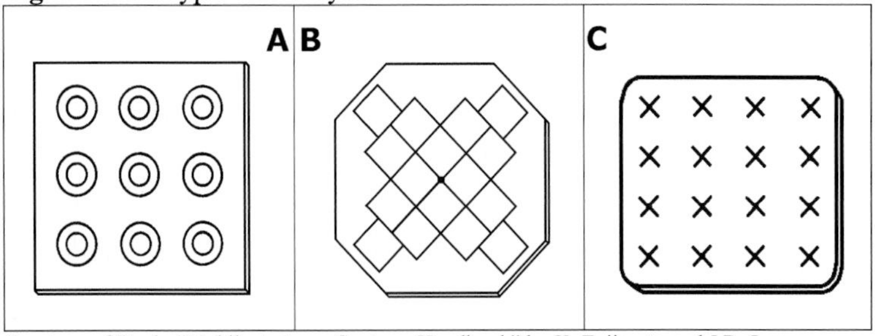

similar radiation patterns in different cut planes. The MAA beam scanning is performed by controlling four-bit variable phase shifters attached to each antenna element. This type of antenna is very applicable for the MES Inmarsat-M and the Inmarsat-Aero standard.

4.5.3.2. Cross-Slot Array Antenna

The cross-slot array antennas (XSA) are a 16-element solution with 97 mm spacing and their volume is about 560 x 560 x 20 mm, shown in **Figure 4.7. (B)**. Evident is the element arrangement of the XSA, which is a modified 4 x 4 square array in order to obtain similar radiation patterns in different cut planes. The XSA antenna beam scanning is carried into effect to control four-bit variable phase shifters associated to each antenna element. Otherwise, this type of antenna is also suitable for the MES Inmarsat-M maritime and land, including Inmarsat-Aero solutions.

4.5.3.3. Cross-Dipole Array Antenna

The cross-dipole array antenna is composed of 16 crossed-dipoles fed in phase with a peak gain of 17 dBi and with the feeding circuit behind the radiating aperture, shown in **Figure 4.7. (C)**. The main electrical characteristics of this antenna are: gain is 15 to 17 dBi with peak RHCP transmit; axial ratio has a value of 0.7 dB; beam width of 3 dB is $-34°$; terminal G/T is -9.5 dBK and EIRP terminal value is 32 dBW. Otherwise, the antenna system consists in a stabilization mechanism for tracking, flat antenna array, diplexer, HPA and LNA, which are all protected by a plastic radome. Stabilization of the antenna is obtained by a single-wheel gyroscope, when the azimuth pointing is controlled by the output from the ship's gyrocompass. Similar to the previous two models, this antenna is also suitable for maritime and aeronautical applications.

4.5.3.4. Four-Element Array Antennas

There have been several four-element antenna models developed, such as Yagi-Uda, Quad-Helix and four elements SBF array.

1. Yagi-Uda Crossed-Dipole Array Antenna – This array antenna has been developed for use on board ships and is protected with a radome, shown in **Figure 4.8. (A)**. The feeder of this antenna is a simple formation of four in-line crossed-dipoles fixed in the middle of the reflector. This endfire array has circular polarization and the gain is between 8 and 15 dBi.

Figure 4.8. Types of Array Antennas

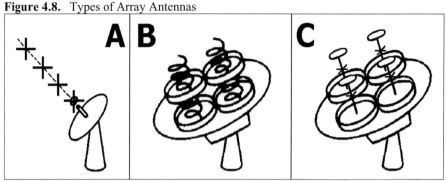

Courtesy of Book: "Mobile Antenna Systems Handbook" by K. Fujimoto and J.R. James

2. Quad-Helix Array Antenna – The quad-helix array antenna solution is composed of four identical two-turn helical wire antennas in the shape of a square and whose elements are oriented in the manner illustrated in **Figure 4.8. (B)**. According to previous studies, the effect of mutual coupling between each element of this antenna is not negligible and this mutual coupling mainly degrades the axial ratio. The axial ratio of a single helical antenna is about 1 dB but this value is degraded to about 4.5 dB in the case of the array antenna with an array spacing of 0.7λ. However, the best properties of antenna gain and axial ratio can be obtained at a rim height of about 0.25λ. The antenna gain is improved by 0,4 dB and the axial ratio is also improved by 3.5 dB, compared to that of the quad-helix array antenna without rims. The performance characteristics of this small antenna are essentially, gain is about 13 dB (HPBW is 38°) and aperture efficiency is about 100%. It appears that this type of helix-integrated antenna is also well suited for the shipborne Inmarsat-M standards.

3. Four-Element SBF Array – This antenna is developed on the basis of a conventional SBF antenna as an integrated array with four SBF elements, see **Figure 4.8. (C)**. The antenna provides high aperture efficiency, circular polarization and almost high-performance gain between 18 and 20 dBi. Because of the high gain characteristics, this array is very suitable for maritime applications as a shipborne antenna.

4.5.3.5. Spiral Array Antenna

Directional antennas for LMSC have been expected to provide voice and HSD links not only for long haul tracks but also for private cars. From that point of view, cost is an important factor to be taken into account in designing antenna systems. In the early stage of LMSC, a mechanical steering antenna system was considered the best candidate for vehicles, however, it will be replaced by a phased array antenna in the near future because it has many attractive advantages, such as low profile, high-speed tracking and potential low cost.

The mechanical steering antenna with eight spiral elements and with adopted closed loop tracking method gives about 15 dBi in system gain, shown in **Figure 4.9. (A)**. The antenna is 30 cm in radius, 35 cm in height and 1.5 kg in weight. The array consists in 2 x 4 spiral elements and it forms a fan beam with a half-power beam width of 21° in the azimuth and 39° in the elevation plane at L-band. Its peak gain is about 15 dBi, including the feeder losses, and is suitable to track the satellite for MSC, because elevation angles to the satellite

Figure 4.9. Types of Array Antennas

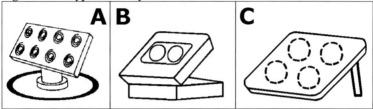

Courtesy of Book: "Mobile Antenna Systems Handbook" by K. Fujimoto and J.R. James

are not as varied as those of the azimuth angles. In effect, the antenna beam direction can be shifted in two azimuth directions, from the E or W side, by switching the pin diode phase shifters. Consequently, the difference between the received signals in both directions is used to drive the antenna system towards the satellite. The beam-shifting angle is set to approximately 4°.

4.5.3.6. Patch Array Antennas

The main feature of the future MSC will be portability, which means that a person can directly access the satellite to establish a link using a very small TES transceiver with antenna system. Even in the present Inmarsat L-band system, great efforts have been made to develop transportable and portable terminals with corresponding antennas.

1. Two-Patch Array Antenna – This antenna for TES and briefcase portable terminals is developed in the Inmarsat and ETS-V programs, see **Figure 4.9. (B).** This antenna has two microstrip patch elements (one for Rx and another for Tx), gain is 6 dBi, EIRP is 6 dBW and G/T is –21 dBK. The reason for adopting separate Rx and Tx antennas is to eliminate a diplexer, which is too large and heavy for a compact and lightweight terminal. The antenna beam width on the lid is wide enough to point to a satellite by manual tracking.

Two microstrip patch array antennas mounted on the lid of the briefcase TES transceiver for low-speed data transmission serves the Inmarsat and ETS-V TES terminals.

2. Four-Patch Array Antenna – Several four-element patch array transportable antennas were developed for universal Inmarsat-C and other TES terminals. On the other hand, JPL and NASA designed similar L-band types of mobile antennas, mainly for LMSC for regional utilities in the USA. These include a mechanically steered, tilted 1 x 4 patch array and two electrically steered planar-phase array antennas. Otherwise, the mechanically steered four-square patch arrays can be fixed in one line, similar to two-patch array, or can have the shape of a four-circular patch array manually steered antenna, which arrangement is shown in **Figure 4.9. (C).** All three of these medium-gain antennas feature beams that are narrow in azimuth angle, hence, they require azimuth steering to keep the beam pointed toward the desired satellite as a mobile changes its azimuth orientation. They provide 9 to 12 dBi gain, reject multipath signals outside their beam pattern and allow two satellites separated by 30° in a GEO arc to reuse the same frequency to cover the continental US region. A dither-tracking four-element, circular, polarized array for AMSC/MSAT terminals has been designed, 10.16 cm high, 50.8 cm in diameter, with 20 – 60° elevation coverage and with a minimum of 10 dBi gain. In fact, this antenna employs a kind of closed-loop for tracking the satellite in azimuth. The rotating antenna platform is mounted on the fixed platform that includes the motor drive and pointing system hardware.

Figure 4.10. Types of Array Antennas

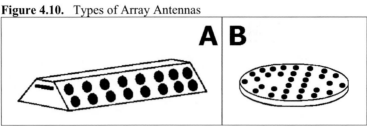

Courtesy of Book: "Mobile Antenna Systems Handbook" by K. Fujimoto and J.R. James

4.5.3.7. Phased Array Antenna

Phased array antennas were developed for maritime, land and aeronautical applications to provide a design for a thin antenna that can be installed on land vehicles and aircraft. Otherwise, these mobile antennas were developed initially for aircraft and are well known for their complexity and high cost. As a result, emphasis was placed on the selection of manufacturing techniques, materials and component types, in addition to meeting the RF and pointing requirements and keeping the cost down.

As mentioned previously, the radiation pattern of an array antenna is fixed, however, the radiation pattern can be scanned in space by controlling the phase of the exciting current in each element of the array. This type of antenna is called a phased array antenna, which has many advantages in terms of MSC applications such as compactness, light weight, high-speed tracking performance and potentially, low cost.

1. Phased-Array Antenna for AMSC – A directional medium-gain antenna is considered a key technology in AMSC. This type of antenna was developed and tested by CRL for the ETS-V program. At the same time, Inmarsat approved all aeronautical antenna standards for installations on board commercial and military aircraft. Taking account of the electrical and mechanical requirements of AMSC, a phased array with low-profile antenna elements was chosen for a directional main antenna, while a microstrip antenna was chosen as an antenna element because of its very low profile, very light weight and mechanical strength, which satisfy the requirements for airborne antenna. However, one disadvantage is the very narrow frequency bandwidth, usually 2 to 3%. The antenna adopted is a two-frequency resonant element because it provides a compact array and a simple feed line configuration. On the other hand, this type of antenna has very poor axial ratio values. The problem was overcome by using the sequential-array technique, where a thin substrate with high dielectric constant is used over a wide frequency bandwidth with excellent axial ratios. The microstrip phased-array antenna is mounted on top of the fuselage and has two planes with 16 circular patch elements, 2 of which are in elevation and 8 in azimuth, see **Figure 4.10. (A)**. In practice, the required coverage angles are as narrow as $+20°$, so the beams are not steered in elevation directions and the array plane is set at $65°$ to the horizon in order to optimize the beam coverage on the flight routes. Thus, by controlling eight 4-bit digital phase shifters, the antenna beam scans in a $4°$ step with $+60°$ with respect to a line perpendicular to the axis of the aircraft. In addition, a step track method was adopted to track the satellite. Therefore, this array has the following characteristics: 2 L-band Tx and Rx frequencies; polarization is LHCP; gain is 14.7 dBi for Tx and 13.5 dBi for Rx; EIRP is about 30 dBW, G/T -10.8 dBK, the axial ratio is about 2 dB, volume is 760 (l) x 320 (w) x 180 (h) mm and the weight is 18 kg.

Figure 4.11. Types of Array Antennas

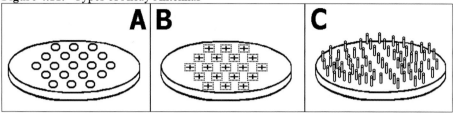

Courtesy of Book: "Mobile Antenna Systems Handbook" by K. Fujimoto and J.R. James

2. Phased-Array Antenna for MMSC – The new shipborne phased-array antenna has been developed for utilization on board big ocean-going vessels. Besides other directional types of shipborne antenna, this unit is designed to serve as ADE of the Nera F77 transceiver or other brand SES for new Inmarsat Fleet 77 service. The Nera F77 antenna is a mechanically steered circular disc with 32 low-profile radiating elements arranged in two circles with 16 and 12 elements and 4 elements are located in the middle of the disc, which is illustrated in **Figure 4.10. (B).** The ADE unit can be mounted on a mast or directly on the deck and is covered by a radome. The dimensions of antenna mast mount are 180 x 132 cm with a weight of 65 kg and deck-mounted antennas are 108 x 91 cm and 50 kg in weight. The gain of the antenna is about 15 dBi in both transmission satellite links.

3. Breadboard Phased-Array Antenna for LMSC – The breadboard mobile phased-array antennas have been developed in the USA to meet the L-band frequency requirements, 20° to 60° elevation and full 360° azimuth angles of special coverage, gain of 10 dBi above 30° in elevation and 8 dBi at 20° in elevation, half-power bandwidth of 25° in azimuth and 35° in elevation, intersatellite signal isolation of 20 dB between two GEO separated by circa 35° and beam pointing accuracy of $+5^{\circ}$. Two phased-array antennas were developed for LMSC: Ball and Teledyne designs illustrated in **Figure 4.11. (A)** and **(B)**, respectively. Each antenna consists in 19 low-profile radiating elements, with 18 3-bit diode phase shifters. The Ball model uses dual resonant stacked circular microstrip elements to cover both the Tx and Rx bands, while the Teledyne model employs stripling crossed-slot radiators. For the antenna's beam-pointing system, the initial acquisition of the satellite is accomplished by a full azimuth search for the strongest received signal. An angular rate sensor is used to establish an inertial reference point when the acquisition is performed while the mobile is turning. Tests show that the antenna can acquire a reference pilot signal in two seconds from a random spatial position. After the desired satellite signal has been acquired, the antenna tracks the satellite by a closed-loop sequential lobing technique. In the event of severe signal fade due to shadowing, the sequential lobing can no longer function properly. In this case, the open-loop angular rate sensor takes over the pointing for a 10 sec period, until the sensor drifts away. This antenna with all the radiated elements, hybrids, diode phase shifters, pin diode drivers and power/combiner driver (except diplexer) can be mounted on the roof of a land vehicle cabin. Thus, based on the research experiments of the airborne phased-array antenna, a new phased-array car antenna has been proposed by CRL. The antenna was installed on a test van and tracks the ETS-V satellite at an elevation angle of 47°. The receiving signal from the satellite was almost constant except for shadowing and blocking effects. The gain of this antenna is 10 and 18 dBi for elevations of 30° and 90°, respectively, the system temperature is about 200°K, axial ratio is 4 dB for 30° of elevation, the volume is 60 x 4 cm and weight is 5 kg.

4.5.3.8. Adaptive Array Antennas

The antenna for the MSAT program of LMSC was developed by the Canadian CRC. Linearly and circularly polarized arrays with a gain values of 9–11 dBi and 10–13 dBi, respectively, have been designed and evaluated in trials using the ex-Marecs-B satellite. Both antennas use L-band with spatial coverage of 15° to 50° elevation and 360° azimuth angles. The satellite tracking is initially acquired with a closed-loop method of stepping through 16 azimuth beam positions and selecting the beam with the strongest signal. In the event that the signal falls below a given threshold, the acquisition sequence is again initiated until the signal is required. The speed of operation is determined by the terminal C/N_0 ratio and the value of S/N requirements in the control loop bandwidth, which takes 0.1 sec to acquire the satellite after an initial phase look. Otherwise, tracking of the satellite is performed by periodically switching on either side of the current beam position and selecting the beam with the strongest signal. A number of algorithms have been devised to minimize any perturbation of the communications signal to less than 1% of the time.

The main characteristics of this type of antenna are as follows: operating RF is from 1.530 to 1.660 GHz; spatial coverage is 15° to 50° elevation and 360° azimuth; gain is 9 to 11 dBi for linear and 10 to 13 dBi for circular model and the size of both models is about 61 cm diameter, while the linear is 6.3 cm and circular is 20.3 cm high, respectively.

The maximum phase transients in azimuth can be kept to less than ± 10° over the required angular coverage and operating frequency band.

1. Linearly Polarized Adaptive Array Antenna – This antenna consists essentially in a driven quarter-wave monopole surrounded by concentric rings of parasitic elements all mounted on a ground plane of finite size. The parasitic elements are connected to the ground through pin diodes. With the application of suitable biasing voltages, the desired parasitic elements can be activated and made highly reflective. The directivity and pointing of the antenna beam can be controlled both in the elevation and azimuth planes using high speed digital switching techniques. The use of a circular polarizer in the linearly polarized design can realize an increase in gain at the expense of an increase in antenna height. The polarizer has an elliptical cross-section, a diameter of 40 cm and a height of 20 cm. It consists in a number of conformal scattering matrices to achieve the 90° differential phase shift between two orthogonal polarizations. Part of a five-ring linearly polarized antenna is shown in **Figure 4.11. (C)**. The antenna incorporates sufficient electronics to control the radiation patterns and pointing on command. It is designed to serve MSAT and AMSC transceivers and to be mounted on the metallic roof of a vehicle, where the effective ground plane can significantly enhance antenna gain at low elevation angles.

2. Circularly Polarized Adaptive Array Antenna – This type of adaptive array is similar to linear adaptive array and hence, is obtained by adding the linearly polarized array.

4.6. High-Gain Directional Aperture Antennas

High-gain directional aperture antennas are more powerful transmission reflectors and panels used for Inmarsat maritime and transportable applications. The typical gain of these antennas is more than 20 dBi, EIRP is a maximum of 33 dBW and G/T is about –4 dBK. There are two basic types of directional parabolic antennas: dish and umbrella and the third new solution for Inmarsat transportable units is the Quad flat panel antenna designed by the South African-based company OmniPless and introduced in the next section.

Figure 4.12. High-Gain Directional Aperture Antennas

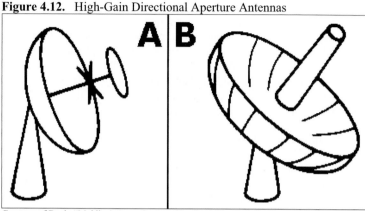

Courtesy of Book: "Mobile Antenna Systems Handbook" by K. Fujimoto and J.R. James

4.6.1. Parabolic Dish Antenna

The first generation of parabolic dish antennas used a reflector in diameter of max 1.2 m, whereas on newer models it is likely to have reduced in size to approximately 0.7 to 0.8 m, see **Figure 4.12. (A).** Because a large proportion of the Rx signal gain and Tx EIRP is produced by the antenna, the area of the dish can only be reduced if the transmitting power from the satellite transponder is increased, when the receive preamplifier gain can be increased without an appreciable increase in noise. The parabolic reflector is most often used for high directivity for radio signals traveling in straight lines, as do light rays. They can also be focused and reflected just as light rays can, namely, a microwave source can be placed at focal point of antenna reflector. The field leaves this antenna as a spherical wave front. As each part of the wave front reaches the reflecting surface it is phase-shifted 180°. Each part is then sent outward at an angle that results in all parts of the field traveling in parallel paths. Due to the special shape of a parabolic surface, all paths from the focus to the reflector and back into space line are the same length. When the parts of the field are reflected from the parabolic surface, they travel to the space line in the same amount of time. This antenna is a large microwave parabolic consisting in the reflector of dish shape, feeder structure, waveguide assembly, servo and drive system and protective radome.

1. Stable Platform – This is the antenna support assembly, which must remain perfectly stable when the ship is pitching and rolling in extremely bad weather conditions. Namely, it is essential that the stable platform holds the reflector in its A/E angular positions despite movement of the ship. The platform usually consists in a large solid bed mounted in such a way that four gyro compasses are able to sense movement and correct any errors detected, holding the platform level. In practice, it is a form of electronic gimbal.

2. Tracking System – The antenna is controlled in A/E angles by stepping motors, which in turn are electronically controlled in a simple feedback system. This electromechanical antenna arrangement enables the dish to maintain a lock on a satellite despite navigation course changes. In such a way, as the ship changes course, both A/E control corrections must be made automatically.

3. Computer Control – The antenna unit processor controls all ADE functions, which include satellite tracking and electronic control.

4. RF Electronics – This segment contains the Tx HPA and the Rx RF front-end LNA stages, plus all the critical Bandpass signal filter stage.

5. Multiplexer Unit – In modern equipment it is common practice to reduce the number of cables between ADE and BDE. Hence, this is achieved by multiplexing up/down signals or commands between ADE and BDE onto the one coaxial feeder.

Therefore, this antenna is connected to BDE transceivers both incorporated in an Inmarsat-A and B SES terminal. Each manufacturer inevitably produces a different design of ADE and BDE, however they do in fact perform the same functions.

4.6.2. Parabolic Umbrella Antenna

An umbrella-type antenna is a deployable, compact and lightweight parabolic type suitable for transportable Inmarsat-A and B transceivers, illustrated in **Figure 4.12. (B).** Otherwise, this antenna has almost all the same technical characteristics as a parabolic dish antenna.

4.7. Antenna Systems for Particular MSC

Typical MSC systems are Inmarsat, AMSC, MSAT, Optus and others in which the L-band 1.6/1.5 GHz MSA are used for the service link. The Big LEO systems use S and L-band (2.5/1.6) and Little LEO use VHF band. The only exceptions are US-based OmniTRACS and EutelTRACS in which the MSA antenna is serviced in the 14/12 GHz Ku-band, the Japanese N-Star service link uses 2.6/2.5 GHz S-band and Teledesic and new advanced MSS will operate on 30/18 K and Ka-bands. A typical satellite navigation system GPS uses 1.6 and 1.3 L-band for transmission in satellite-to-Earth direction. In addition, an antenna system for ships, land vehicles and aircraft has been developed to receive TV programs.

4.7.1. Antennas for Inmarsat System

In general, the biggest mobile satellite operator, Inmarsat, developed four types of MSA for maritime, land, aeronautical, transportable and fixed applications, while in particular, some maritime MSA are used or transformed for land vehicles and off/onshore installations.

4.7.1.1. Ship Earth Station (SES) Antennas

The SES antennas are more sophisticated than other MSA owing to the many ship's motions and rugged environment. The MSA systems are designed according to SES standards, which have been developed since 1981, see four types of MSA in **Figure 4.13**.

1. Inmarsat-A/B MSA – Standard-A was introduced in 1982 as a first operating Inmarsat analog SES, while digital standard-B was started in 1993, using the 2nd generation Inmarsat satellites. Both standards are compatible and suitable for SES mounting on large ships, sea platforms and on-shore infrastructures. They use same high-gain parabolic steerable dish with the same specifications: gain is from 20 to 24 dBi and G/T is –4 dBK. The EIRP of Inmarsat-B is less than that of Inmarsat-A by 3 dB, due to the slightly shaped global beam antenna adopted in the Inmarsat-2 satellites. Considering the desired EIRP and the noise performance of current LNA, the Inmarsat-B EIRP of 33 dBW can be obtained by the antenna with a gain of about 20 dBi. Thus, for both standards, an aperture antenna such as a directional parabolic dish is suitable because of the high aperture efficiency.

Figure 4.13. Inmarsat Antenna Systems for SES

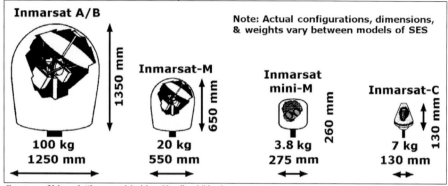

Courtesy of Manual: "Inmarsat Maritime Handbook" by Inmarsat

This antenna is especially desirable because it is simple in its structure and can be designed flexibly with respect to gain values larger than 20 dBi. Satellite tracking is an essential capability for ship's motions and the small half-power beam width of about 10°. The four-axis (X-Y-A-E) stabilizer is the one most commonly used, when a fixed horizontal plane is obtained by controlling motion about the X and Y axes and an antenna system installed on the X-Y plane can be stabilized by controlling the A and E axes. Otherwise, the antenna is directed to the satellite by controlling the A and E axes. It needs antenna pedestal control circuits with servomotors to control the axes and it also needs some sensors, such as an accelerometer, a rate sensor and a level sensor, to provide information about ship motions. A flywheel stabilizer has sometimes been used in order to avoid the need for antenna control sensors and electronic circuits. This kind of stabilizer makes use of the inertial force generated by one or two rapidly rotating flywheels. Thus, a three-axes stabilizer was developed and used in some recent standard-A and B, while the present standards have generally used a closed-loop tracking system because of their very simple configuration. The most popular open-loop tracking is a step track method, which drives in elevation and azimuth directions alternatively by a step angle of 0.5° in such a way as to keep the received signal level as high as possible.

2. Inmarsat-M MSA – This standard was initiated in 1993 as a solution for small, lightweight, simple in configuration and low-cost SES for installation on small ships, land vehicles and rural infrastructures. The antenna applicable to this system can be a single medium-gain directional parabolic SBF dish or a phased array with an antenna gain that ranges from 13 to 16 dBi. For the medium-gain parabolic dish antenna it will be difficult to illuminate the parabolic efficiency with the primary radiator because the diameter of this antenna is smaller than a high-gain antenna. In fact, this type of antenna is not adequate for the Inmarsat-M compared to other types of antennas with high aperture efficiency, such as SBF antenna. Traveling-wave antennas, such as the cross-Yagi/Uda array, a single helical antenna and a log-periodic antenna have high efficiency for medium-gain performances. Since those types of antennas are comparatively long in their axial direction, the volume of the radome becomes fairly large and conflicts with the requirements of Inmarsat-M. This antenna needs a suitable mount/steering system on board ships.

The SBF antenna, with typical gain of 15 dBi, is one of the favorite standard-M antennas for SES. Although the SBF antenna is a compact, simple configuration and high-efficiency shipborne antenna, it has a narrow frequency bandwidth of about 3% (about 5% less than required). The electrical characteristics of the conventional SBF have been improved in an upgraded SBF antenna by changing the main reflector from a flat disc to a conical or a step plate and by adding a small second reflector. These improvements give better performance with an aperture efficiency of about 80% and an RF bandwidth of 20% for VSWR under 1.5 and gain is also improved by about 1 dB without changing sidelobe levels.

3. Inmarsat mini-M MSA – The Inmarsat mini-M SES introduced in 1997 offers the same service as Inmarsat-M but in smaller, more lightweight and compact transceiver and antenna units for mounting on small ships and land vehicles. This SES can be made smaller because it operates only in the spot-beam coverage of the Inmarsat-3 beams, of which a typical representative is TT-3064A Capsat SES, designed by Thrane & Thrane. The mini-M antenna is a compact, lightweight, 3-axis stabilized platform with a minimum of moveable parts and all motors, sensors and electronics are mounted in the bottom bowl. Antenna pointing is controlled by various sensors and a step-tracking algorithm for signal level peaking. However, it does not require cable unwrapping but can track the satellite in any position above the horizon and it is connected to the transceiver through a single coaxial cable to ensure simple installation. The sensors in the antenna are:

a) 3-D compass (magnetometer) which measures the Earth's magnetic field vector;

b) 2-D inclinometer which measures the vessel's roll and pitch angles and a

c) gyro which measures the vessel's yaw, pitch and roll rates.

The sensor stabilization of the antenna pointing compensates for the fast movements of the ship. In case of a no signal situation the sensors are able to point the antenna independently but in all other situations a step-tracking algorithm maximizes the strength of the received signal. The antenna has step-tracking and sky-scan modes of normal operation:

1) During step-tracking mode the platform tracks the satellite using the sensors and the step-tracking algorithm. If synchronism with the satellite is lost for more than 130 seconds the sky-scan mode is entered to re-acquire synchronism with the satellite.

2) The sky-scan mode is automatically entered during the initial acquisition of a satellite after power-up, when a different satellite is selected by the user or the signal has been lost for more than 130 sec. During a sky-scan the antenna pointing is moved over the sky from horizon to zenith in a smooth spiral movement and stabilized by the sensors, searching for a particular satellite. Therefore, when the satellite is found, the antenna exits the sky-scan mode and enters step-tracking mode. Sensor stabilized platforms work with directional RHCP antennas. Maximum pointing error ±10° and beam width ±30°. The value of G/T is minimum –17 dBK and EIRP is from 8 to14 dBW in 2 dB steps.

4. Inmarsat-C MSA – This MSA system is the simplest and most compact configuration without mount and tracking/pointing systems, introduced in 1990 as the second Inmarsat SES standard. The omnidirectional low-gain antennas are the most suitable for this mobile standard, such as cross-drooping dipole, quadrifilar helix and the microstrip patch antennas. In fact, the quadrifilar antenna is well known as a shipborne antenna because of its good performance and axial ratios in wide angular coverage, with about 4 dBi in gain, maximum 3 dB in axial ratio, –24 dBK in G/T and 14 dBW EIRP. The Inmarsat-C MSA is suitable for installation on board medium and small ships, land vehicles and rural infrastructures.

5. Inmarsat-D and D+ MSA – The Inmarsat-D (receiver) and D+ (transceiver) standards were developed in 1996 and 1997, respectively. Both standards use omnidirectional small low-gain antenna similar to that introduced for Inmarsat-C, with the only difference that is smaller, more compact and with the G/T value of –25 dBK. Besides, this antenna can be a

Figure 4.14. Inmarsat Antenna Systems for VES

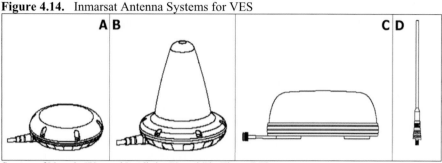

Courtesy of Manuals: "User and Installation Manuals" by Thrane & Thrane

separate unit in the smallest radome for Inmarsat-D+ transceiver made by JRC or the antenna can be built in a small box together with the Inmarsat-D+ transceiver and GPS Rx antenna made by Canadian-based SkyWave. These antennas are serving equipment similar to paging systems for vessels/vehicle messaging and SCADA Rx/transceivers equipment.

4.7.1.2. Vehicle Earth Station (VES) Antennas

Land vehicles can generally use all shipborne terminals, except Inmarsat-A and B because of the large size of their antennas and indoor equipment. The most popular standards for VES are Inmarsat-C, M and Mini-M sometimes with modified antennas and terminals. In the next section three of the most known VES antennas, made by the Denmark-based manufacturer Thrane & Thrane, will be introduced.

1. Inmarsat-C/GPS VES MSA – The most suitable and smallest all-in-one antenna in the world for land vehicle mounting is the TT-3002L omnidirectional Inmarsat-C/GSP RHC polarized antenna with built-in LNA/HPA electronics designed to operate in an aggressive environment, shown in **Figure 4.14. (A).** The antenna housing is sealed and contains no user serviceable parts, with dimensions of 122 x 35.5 mm and a weight of 0.5 kg. The antenna is very compact and is designed to operate in a corrosive environment and in extreme weather conditions without any service. It has a modular construction that allows easy exchange of antenna elements and is designed to work with the VES Capsat transceiver TT-3022C and GPS receiver in compliance with the Inmarsat-C specifications. Thus, this antenna can handle 32 Kbytes transmission length, with up to 100 metres of coax cable and it is designed to operate when the satellite is visible over the horizon in coverage down to 20° of elevation and no signal path blockage is present. The antenna gain is about 4 dBi, axial ratio is about 3 dB, G/T is –23 dBK and EIRP is 14 dBK at a 5° elevation angle. This antenna is suitable for magnetic base mounting, especially on cars and on all land vehicles for message and data transmissions.

The larger version of this antenna has only difference in dimension of 122 x 125 mm and in weight of 0.6 kg, see **Figure 4.14. (B).** This antenna is designed to work with the Capsat Thrane & Thrane transceivers TT-3020C, TT-3022D and TT-3022C and is suitable to be installed on any kind of land vehicle.

2. Inmarsat mini-M VES Plate MSA – The most suitable VES for installation in a large variety of vehicles, ranging from long haul trucks to off-roaders and trains is the TT-3007F low profile antenna built on an aluminum base plate and covered by a radome, illustrated in

Figure 4.14. (C). This transceiving unit consists in an HPA/LNA board (TT-3010G) which amplifies received/transmitted voice and data radio signals. Otherwise, a controller board calculates tracking-information based on input from a rate sensor and the received signal strength. Thus, the TT-3007F low profile antenna is an automatic tracking antenna designed to work with the TT-3062D Capsat Compact Carphone. When it has locked onto a satellite, it tracks by turning the antenna element via a stepper motor. Elevation is adjustable by switching between two different aerials; one Helix aerial for low elevation and a Patch for high elevation, both with an RHCP characteristic. The helix aerial has elevation coverage of 13° to 40°; elevation beam of 37° and azimuth beam of 73°. While the patch aerial has elevation coverage of 30° to 70°; elevation beam of 62° and azimuth beam of 67°. Thus, if driving in areas where the elevation is greater than 70°, the user might need to choose another satellite with a smaller elevation angle. However, to switch between the two aerials a cable has to be manually moved from one connector to another, which requires the radome to be removed and this is a disadvantage. Perhaps the problem can be solved with some kind of simple switch or diplexer. Moreover, the antenna comes with magnetic feet for roof mounting on a vehicle. It is also possible to install the unit with bolts, for a more secure and permanent installation. The value of gain is 12 to 15 dBi, G/T is a minimum of –17 dBK and EIRP is from 8 to14 dBW in 2 dB steps. The dimensions of the antenna are 285 x 90 mm and the weight is 2.7 kg.

3. Inmarsat mini-M VES Whip MSA – The TT-3007B rod antenna is a passive whip omnidirectional unit for use on cars and all kinds of land vehicles, see **Figure 4.14. (D)**. The only movable parts are the lower part of the unit, where adjustments for correct land zone are carried out. The rod is rigid but will bend down to an angle of approximately 30° at the base if necessary, e.g.,, when passing under obstacles. It must always be in its upright position when the system is in operational mode. To gain high receiver performance, the unit is connected to the antenna Front-End unit via a low-loss coaxial cable. The elevation angle is set according to the geographical position and is determined by the maps provided. In fact, manual readjustment of the elevation is necessary when driving over long distances. Hence, the zones on the body of the antenna set the elevation angle as follows: Zone 1 = 25–32.5°; 2 = 32.5–40°; 3 = 40–47.5°; 4 = 47.5–55°; 5 = 55–62.5° and Zone 6: 62.5–70°. Tuned all the way down to maximum (beyond zone 6), the antenna will work up to 90° because of the way the beam reforms. Tuned all the way up to the highest point it will turn and the antenna will work down to 15°. The antenna has an elevation beam-width of approx 15°. This relatively narrow angular aperture is necessary to obtain sufficient gain. Due to this, short signal drop-outs might appear if driving in very rugged terrain. To avoid water ingression to the antenna, a rubber "Raincoat", i.e., a rubber tube is rolled down over the lower part of the antenna. This antenna has the same electrical characteristics as the one previously discussed; dimensions are only 930 x 22 mm and the weight is 0.63 kg.

The TT-3010A antenna front-end unit contains a high power amplifier, a duplex circuitry, an LNA and supervisory control circuitry. Data communication between the main unit and the antenna is accomplished by serial communication through the single coaxial cable, which also carries the RF-signal.

4.7.1.3. Aircraft Earth Station (AES) Antennas

The commercial aeronautical MSC services worldwide have been provided by the Inmarsat network since 1990 with Aero-H and L standards. Hence, aircraft can easily use Inmarsat-C

Figure 4.15.   Inmarsat Antenna Systems for AES

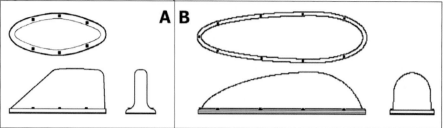

Courtesy of Manuals: "User and Installation Manuals" by Thrane & Thrane

and mini-M shipborne terminals but because of the high speed of aircraft and small space in the cockpit, all AES terminals and antennas design are modified more aerodynamically and in a very flat shape. At this point, in 1992 and 1998 Aero-C and L standards were developed, respectively and finally the Aero mini-M standard for AES mounts was recently developed. In fact, there are many different categorizations of Aero standards but the most reasonable today is classified into three main groups: AES with Low-Gain Antenna (LGA), Intermediate Gain Antenna (IGA) and High-Gain Antenna (HGA). All kinds of aircraft antenna provide simultaneous full duplex and two bands of transceiving operations.

1. LGA System – The Low-Gain Antenna system is a small omnidirectional low gain and not a tracking satellite antenna designed to support both aeronautical Inmarsat-C/GPS and L transceivers for aeronautical automatic low bit rate data reporting and message transfer of position reports, performance data and operational messages on a global basis, from sea level to about 20 km and all the way from 70° N to 70° S. This system consists in an antenna element, a diplexer, an LNA and a C-class HPA. Its gain is 0 dBi, EPIRB is about 12 dBW, G/T is –32 dBW and its radiation pattern is omnidirectional, to cover over 85% of the upper hemisphere above an elevation angle of 5°. A typical example of this antenna is the aeronautical Capsat antenna system, manufactured by Sensor Systems Inc., consisting in a TT-3002A MSA (Jet Blade) and a TT-3001F HPA/LNA Pack. This antenna is connected to the Thrane & Thrane TT-3024A Aero Inmarsat-C/GPS Capsat Transceiver. Dimensions of the antenna are 116 x 297.4 x 108 mm and the weight is 680 g. Working ambient temperature is between –55°C and +85°C at an altitude of about 21 km. These particulars are for Aero-C standards and the assembly drawing is presented in **Figure 4.15. (A)**, although some of the Aero-L standards have the same shape of antenna. The Aero-L antenna provides EIRP of minimum 16.5 dBW, G/T of –13 dBK, 93% coverage and can serve as the reversionary antenna below 7 dBW of EIRP for Aero-H configurations. The LGA system is suitable for aircraft that do not need voice or HSD communications.

2. IGA System – The Intermediate Gain Antenna system is a medium 2-axis mechanically steered phased array intermediate gain antenna designed to support both aeronautical Inmarsat-I and M. A typical example of IGA is the integrated TT-5006A innovative solution for Aero-I and M installations on small and medium-sized business aircraft, which can be used for both multi-channel Aero-I and single channel Aero-M applications, shown in **Figure 4.15. (B)**. It is the smallest aero-satellite antenna which interfaces to the TT-5000 and TT-3000M series of Thrane & Thrane transceivers for Aero-I and M respectively, available for voice, fax prints, data transfer or E-mail messages. This antenna contains its built-in Navigational Reference System, including the GPS antenna with magnetic and altitude sensing for stand-alone operations, angular sensors and electronics for performing

Figure 4.16. Types of Phased Array Antenna

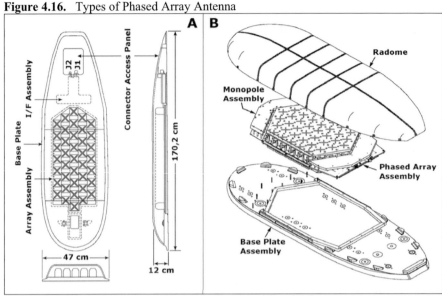

Courtesy of Manuals: "High Gain Antenna" by A) Canadian Marconi Co and B) Tecom

this task. All these functions are contained within the unit. The electrical characteristics of TT-5006A MSA are as follows: gain is 6 dBi, EIRP is 14 dBW, G/T is –17dBK and axial ratio of minimum 6 dB with coverage volume of over 85% above 5° elevation angles. Dimensions of this antenna system are 560 x 150 x 124mm and the weight is 2.2 kg. The antenna is the smallest combined NRS and MSC antenna in this class with an ultra light construction, which offers very low drag (less than 3 pounds drag at Mach 0.8 on a small jet). The latest state-of-the-art technology from the well-proven Thrane & Thrane IGA products provides ideal solutions for all short/medium haul business and military aviation.

3. HGA System – The High Gain Antenna system is a medium 2-axis electronically steered phased array high gain antenna, which is designed to support aero Inmarsat-H, H+ and HSD+ standards. The two typical representatives of HGA products are as follows:

a) The solution of HGA to support Inmarsat Aero-H and H+ standards with which both provide channel rates up to 10.5 Kb/s, supporting multichannel voice, Fax and data, while the Aero-H+ system optionally can also use the higher power of the Inmarsat-3 satellites when operating within the spot-beam coverage area.

b) The combination of a HGA TT-5000HSD+ Thrane & Thrane and the new Inmarsat GAN or Swift 64 technology, provides Aero-H+ capability together with a dedicated HSD channel for up to 64 Kb/s of Mobile Packet Data Service (MPDS).

The HGA is usually a planar phased array of a number of identical crossed-dipole elements arranged in a grid. For instance, the CMA-2102 HGA of the Canadian Marconi Company has 39 crossed-dipole array elements, presented in **Figure 4.16. (A)** and the T-4000 HGA array system manufactured by US-based Company Tecom has 44 crossed-dipoles, shown in **Figure 4.16. (B)**. Hence, both antennas provide simultaneous transmission and reception of full-duplex satellite signals, namely the Tx and Rx band. Beam steering of the HGA is performed via serial transmission of phase-shifter data from the Beam Steering Unit (BSU).

The BSU converts tracking and pointing coordinates (relative azimuth and elevation of antenna) from the Satellite Data Unit (SDU) into a signal needed to select the antenna array elements in combinations that point the antenna beam in the desired direction. This form of beam steering is known as open-loop steering, which requires the SDU to receive accurate and timely navigation information. At any rate, all steering operations are transparent to the user since the antenna BSU/SDU combination ensures that the beam automatically points in the desired direction.

Diplexer/Low Noise Amplifiers (DIP/LNA) enable an RF signal to be sent and received. The 3 RF ports: antenna, transmit and receive of the DIP/LNA provide signal routing and filtering functions. Signals in the Rx band are routed from the antenna port to the Rx port; while Tx band signals are routed from the transmit port to the antenna port. The Rx path filters the Tx signal and other out-of-band signals to prevent the LNA and other Rx side components from being driven into nonlinear operation. The Tx path filters receive band signals so that noise and spurious signals from HPA do not increase the noise floor of the Rx. This unit establishes the noise floor by boosting the signals and noise received from the antenna to a level much greater than the noise level of subsequent components in the receive path. The DIP/LNA provides at least 55 dB gain and noise figure of less than 1.8 dB. Since any transmission line losses between the antenna and the DIP/LNA decrease the signal strength and increase the noise floor, cable lengths between the antenna and DIP/LNA should be minimized, with the maximum loss allowed being 0.3 dB.

The base plate of the antenna is a mount for the phased array antenna and the radome and provides a mating interface to the aircraft airframe. The phased array antenna is mounted to the antenna system base plate. The radome provides physical protection for the electrical components. It is of multilayer fibreglass construction and is painted with Teflon paint. This paint prevents ice build-up and resists common chemicals normally used around aircraft, such as de-icing fluid. The radome is outfitted with metallic strips, grounded to the base plate and fuselage, which serve to divert lighting and eliminate the disturbances of any antenna or avionics electrical components.

The HGA's low profile, aerodynamically designed radome minimizes drag and eliminates icing. Hence, its design has been optimized for installation on long-/medium-haul passenger aircraft. Since it can be mounted anywhere along the top of the fuselage, it is suitable for a wide range of commercial and military airframe types. Otherwise, the typical performance characteristics of the HGA are as follows: gain is between 12 and 17 dBi over 90% of the Inmarsat Hemisphere (IH), i.e., coverage or minimum of 9,5 dBi over 100% of IH; EIRP is about 25.5 dBW over 92.6% of IH, it assumes 40 W HPA and 2.5 dB of cable loss; G/T is about −13 dBK over 90% of IH and axial ratio is about 6 dB over 96.5% of the industry (Inmarsat) hemisphere. Practically HGA is known as a high-gain antenna for aeronautical application but actually, this type of MSA is a medium-gain antenna.

4.7.1.4. Transportable Earth Station (TES) Antennas

Recently, after the introduction of Inmarsat-A and C standards, Inmarsat started to provide a new transportable and portable telephone and facsimile communication service in rural areas and in regions with out of range or damaged terrestrial telecommunication networks. At first, Inmarsat-A transportable transceivers were developed by Magnavox Corporation, EB Communications, JRC and other manufacturers of MSC equipment, with directional aperture dish and umbrella parabolic antennas. Soon after Nera, Thrane & Thrane, Toshiba

Figure 4.17. Inmarsat Antenna Systems for TES

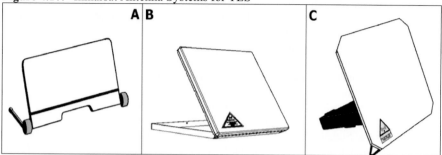

Courtesy of Manuals: "User and Installation Manuals" by Thrane & Thrane

and other equipment manufacturers designed the TES Inmarsat-C transceiver with small patch antennas. Today, the most eminent manufacturer of TES units is Thrane & Thrane and for TES antennas, OmniPless from South Africa.

1. Inmarsat-C TES Patch Antenna – This antenna was developed for the Inmarsat-C portable Capsat TT-3026A transceiver designed by Thrane & Thrane, shown in **Figure 4.17. (A).** A transceiver unit was integrated into a small laptop PC as a portable terminal, providing the same service as other Inmarsat-C terminals in rural and remote areas.

2. Inmarsat mini-M TES Patch Antenna - The mini-M TT-3007A patch antenna resides on top of the Capsat TT3060A Mini-M Electronics Unit (EU), where a short coaxial cable of 0.15 m connects the antenna and the EU, see **Figure 4.17. (B).** It is possible to place the patch antenna at a distance from the laptop-sized EU when the short antenna cable can be replaced with the supplied cable of 5 m or even longer up to 70 m. The patch antenna has a stand, which in unfolded condition can hold it at the correct angle for transmission; alternatively, when the stand is folded up it can be fastened to a bracket. It is completely watertight and sealed for outdoors operation and has an acoustic signal strength indicator, which beeps with an increased repetition rate as the signal received via the antenna becomes stronger. Generally, this antenna meets or exceeds current and proposed Inmarsat specifications for Inmarsat phone spot-beam operations. It has a directional RHCP Patch Array with ±15° horizontal and ±15° vertical beam width and electrical characteristics as follows: gain is about 14 to 15 dBi, EIRP is from 11 to 17 dBW in 2 dB steps and G/T is a minimum of –17 dBK. Dimensions of the antenna are 52 x 270 x 200 mm and the weight is 2.2 kg (including handset, battery pack and antenna). The label attached to the TT3007A antenna gives a warning that the antenna radiates microwave signals and that the operator has to keep a minimum of 0.5 m safety distance. All types of flat panel antennas have similar warning labels. The new version of this antenna should still have one panel or combination of two panels and will support the M4 system multimedia modem service and will links users to the GAN via Inmarsat-3 satellites providing mobile HSD, ISDN line, voice, Fax or combinations of services from virtually anywhere and anytime.

3. Inmarsat mini-M Small Dish TES Antenna – The Small Dish TT-3008D antenna is just one of the many antennas available for the Thrane & Thrane M4 service, see **Figure 4.17. (C).** The directional patch array antenna combined with Capsat Messenger Terminal is the perfect choice for a combined fixed and mobile solution at remote and rural sites. It offers both portability and an easy to set-up fixed solution. The antenna has a robust frame mount construction that also gives the option of installation on a pole or tripod. In addition,

Figure 4.18. Inmarsat Antenna Systems for TES

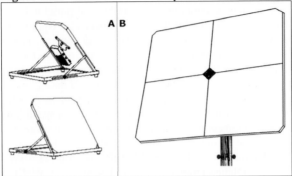

Courtesy of Webpage: "Antenna Products" by OmniPless

built in compass, audible signal strength and clear elevation indication aid in locking onto the satellite. The antenna can be stored in a separate and sturdy carry bag. The transceiver is placed in a bracket inside the antenna, making it one unit; the smallest ISDN units only weigh 5.2 kg. The 64 Kb/s mobile ISDN connections give the user access to the Internet, high quality voice calls, Fax, video conferencing and many other applications. In addition, this antenna also enables access to the inexpensive low-speed Mini-M services for voice, fax and data. Should a 128 Kb/s speed solution be required, then this is made possible by combining two M4 terminals with two small dish antennas to the ISDN interfaces of the laptop. In this case it is possible to use the GAN service offered by Inmarsat, which covers both high-speed, and low-speed data needs. The electric characteristics of the antennas are: gain is from 17 to 19 dBi, EIRP is 8 to 14 dBW for mini-M and 19 to 25 dBW for HSD and G/T is –7 dBK. Dimensions of the antenna panel are 397 x 397 x 75 mm and the weight is 3.1 kg. The new antenna version is combined with three connected (folded-over) panels.

4. Inmarsat mini-M Big Dish TES Antenna – The Inmarsat-M Remote Antenna is a proven product designed by OmniPless for use with satellite phones that do not have a remote function, see **Figure 4.18. (A).** Due to its high gain, the antenna can be positioned more than 20 metres from the terminal without any degradation of the signal, depending on the type of terminal and quality of cable used. Thus, this allows the transceiver to be used indoors even when no clear line-of-sight to the satellite is available. This antenna is fully sealed and can be mounted outdoors even in the harshest weather conditions. It can be mounted on a wall and remotely controlled from up to 70 m from the main unit. It is also available with a dedicated pole mount option, which allows simple installation for fixed or semi-fixed applications, such as rural public payphones, roadside emergency telephones and SCADA applications. This kind of antenna will support the new Inmarsat GAN service with 64 Kb/s ISDN HSD and all Internet solutions. This antenna is also ideal for roof mounting of indoor transceivers in rural offices or remote business solutions in gas/oil pipe or mining industries. Besides, the antenna enables access to the inexpensive low speed Mini-M services for voice, Fax and data transmissions. The polarization of the dish antenna is RHCP with manual adjustable elevation setting and with a beam width of approximately 20°. The electrical characteristics of this antenna are as follows: gain value is about 19 dBi, EIRP is from 8 to 14 dBW in 2 dB steps, G/T is a minimum of –7 dBK and the axial ratio is 2 dB. The dimensions of the antenna are 558 x 550 x 63 mm and the weight is 3.5 kg.

5. Inmarsat-B Quad Flat Panel TES Antenna – The TES flat panel quadrant antenna is an innovative solution for new generation transportable and portable Inmarsat-B terminals, see **Figure 4.18. (B)**. Designed by OmniPless for compact TES equipment, it provides ease of deployment and small stowage size. It consists in four separate identical panels and a base plate. Proprietary attachments allow the panels to be quickly mounted onto the base plate, ideal for integration with the devices of the deployment structure and RF electronics. A retaining clamp is available, which locks the panels to the base plate for permanent and semi-permanent installations. This antenna enables transmission of voice, data and Fax calls at speeds of 9.6 Kb/s; advanced ISDN capabilities for high speed Internet/LAN access and broadcast uplinks. By connecting a standard Fax and a PC, one has a complete office at a remote or rural location, with optional HSD of 56/64 Kb/s. The electrical characteristics are: Tx Gain is 21.4 dBi and Rx gain >21.3 dBi, EIRP is 25 to 33 dBW, G/T is about –4 dBK, axial ratio is 1 dB, antenna temperature is 114°K and with RHCP polarization. The dimensions of the antenna low panel stowage is 390 x 387 x 62 mm (4 panels), deployed panel size in total is 774 x 774 x 15.5 mm (about 0.6 m^2) and the weight is 2.4 kg.

4.7.2. Antennas for the Cospas-Sarsat System

The Cospas-Sarsat system uses few types of antennas built in or mounted on the topside of EPIRB, PLB and ELT units, for maritime, land and aeronautical applications, respectively, as illustrated in **Figure 4.19**. In general, the antenna for Distress Satellite Beacons (DSB) may have been designed to transmit signals, such as 121.5; 243 and 406.025 MHz. Some antenna for handheld DSB have to receive signals emitted on 243 MHz and to conduct power to a strobe light mounted above the antenna. It is possible that the radiated signal can be composed of an unknown ratio of vertical and horizontal polarizations. For this reason, some consideration shall be given to the type of antenna and its radiated field. The results shall encompass all wave polarizations. Therefore, the antenna pattern and field strength measurements should provide sufficient data to evaluate the antenna characteristics. There are many manufacturers of DSB for maritime, land and aeronautical applications but in this section only three types of DSB antennas will be introduced, manufactured by Jotron Electronics AS from Norway.

1. Antenna for Handheld DSB – The handheld floating Tron 1E MK II VHF 121.5 MHz emergency beacon transmitter is designed for manual operation as a waterproof floating device for ships, and can also be used for land and aeronautical applications, shown in **Figure 4.19. (A)**. A rugged, flexible, helical-type antenna is fastened to the top part of the DSB electronic unit. The DSB unit floats horizontally in water and the user must therefore keep the antenna above the water. If the antenna is kept under the water, the searching craft will not be able to receive the emergency signals. Otherwise, there are similar types of handheld DSB designed by other manufacturers. One kind of this model is a stainless steel telescopic transmitting dual band antenna at 121.5 and 243 MHz mounted on top of the handheld unit. The next important sample is 3 band rot antenna mounted on top of PLB MR 509 handheld emergency equipment, used for example in aircraft or vessels to locate crew members and passengers in distress and as a communication transceiver. This antenna uses emergency transmit frequencies: F1 (VHF) = 121.5 MHz; F2 (UHF) = 243 MHz and F3 (UHF) = 406.025 MHz and training transmit frequencies 122.5, 245.1 and 406.025 MHz (training code), respectively. Thus, the Rx frequency is 243 MHz and the training Rx frequency is 245.1 MHz.

Figure 4.19. Antennas for Emergency Satellite Beacons

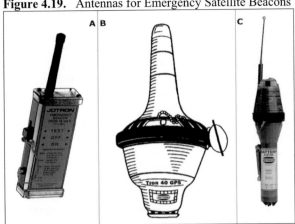

Courtesy of Pamphlets: "Satellite Emergency Beacons" by Jotron

2. Antenna for Floating EPIRB/GPS – The floating EPIRB Tron 40 GPS set is a GMDSS DSB with buoy shape and is designed for manual or automatic operation and designated to enhance crew and passenger lifesaving on board ocean vessels, as is shown in **Figure 4.19. (B).** This DSB unit has two built in antennas: the first is an omnidirectional 406.025 MHz antenna for Tx distress signals via LEOSAR and GEOSAR Cospas-Sarsat satellites and the second is a receiving GPS patch antenna providing accurate position improvement of the EPIRB from a radius of 5 km to an amazing 100 m. This unit can be used as ELT for aircraft in an emergency on the sea's surface. Similar top-mounted or built-in antennas have also been designed for ELT aeronautical applications with different body shapes.

3. Antenna for Floating EPIRB – The floating EPIRB Tron 45S is designed for manual or automatic seawater activation on board vessels in distress, see **Figure 4.19. (C).** An omnidirectional whip transmitting 121.5 and 406 MHz antenna with a length of 180 mm is mounted on top of the electronic unit. Moreover, similar top-mounted or built-in antennas have also been designed for ELT and PLB units, for land and aeronautical mobile applications, respectively, with different body shapes.

4.7.3. Antenna Systems for GMPSC

Many emerging GMPSC networks offer services based on small handheld terminals. The design of these terminals is based upon the services provided by the system, so the electronics and the antenna should follow certain guidelines to fulfill the system, network, service and market requirements. Briefly, the requirements are for the antenna system to have a shaped pattern in elevation (depending on the satellite antenna, orbital parameters and satellite system statistics); circular polarization, an adequate G/T, (depending on the link budget parameters, the mobile's environment and handling of the handset by the user); an adequate bandwidth (depending on the system specifications) and finally, an appropriate size (dictated by the handset size and aesthetic constraints but also depending on all the other requirements). Hence, the QHA seems to be the most promising antenna for such an application because it offers an elevationally shaped pattern, whose shape can be changed by small structural changes and circular polarization within the main lobe.

The G/T of these antennas has to be measured in conjunction with the environment and the operating system applied, while the instantaneous bandwidth for VSWR is better than at 2:1 – 3 to 5% for slim antennas and goes up to 10 – 15% for wider structures. Otherwise, the pattern shape bandwidth goes up to 15% for all structures.

The required pattern shape was determined based on an analysis of the G/T for different antenna patterns in different environments and for a system with the constellation statistics of GMPSC, such as Iridium, Globalstar and others. It is assumed that the user cooperates by pointing the antenna towards the zenith.

The main lobe should have the maximum gain at about 60° above the horizon, with a half transmission power beam width starting from the maximum gain angle and ending at the minimum elevation angle.

The minimum elevation angle was defined by the system to be 10° above the horizon but by taking into account the satellite statistics of the system, an 18° minimum elevation angle can be selected, offering a 100% coverage of most of the Earth's surface and a 98% coverage in latitudes between 3 and 5° and above 70° North and South. A small gain dip at the zenith is desirable, since this optimizes coverage at low elevation angles where extra path and fading losses are encountered. The antenna sizes for handheld terminals were designed to be as small as possible for the operational L and S-bands. At this point, the antennas chosen are relatively short QHA with very good pattern shapes and can be produced using simple design instructions.

4.7.3.1. Wire Quadrifilar Helix Antenna (WQHA) for Handheld Terminals

The WQHA is a highly resonant antenna invented by Kilgus in the 1970s. This antenna consists in four helices placed at $90°$ to each other. The four helical elements are connected by radial parts and are fed in $90°$ phase differences ($0°$, $90°$, $180°$ and $270°$). The analysis by Kilgus for the resonant QHA (RQHA) is based on the assumption that the QHA consists in four helical and four radial parts. At this point, the current distribution on the helical elements is assumed to be sinusoidal, with magnitude maxima at the feed and the distal end for each 1/2 length. The current in the radials is approximated by a uniform distribution. Another analysis approach to the RQHA is based on the assumption that a QHA consists in two Bifilar Helices (BH) placed at $90°$ angular distance and fed in phase quadrature. The radials in the distant end can be shorted or open-circuited giving only changes to the input impedance. One of the major disadvantages of the QHA is the complex feed network required. In addition, one approach is to feed each BH with the assistance of a balun. Most of these configurations need a $90°$ phase difference hybrid and two baluns are needed to feed both BH. The exceptions to that rule are the Self-Phased configuration and the Keen balun, which can offer the phase difference without a hybrid and with the use of only one balun. The other way is to separately feed each one of the four helical elements, with $90°$ phase difference, using three hybrids. This method does not need a balun to feed the QHA but can be used only in an end-fire configuration, for constructional reasons.

Otherwise, the QHA can operate in both the satellite mode (hemispherical pattern) and terrestrial mode (toroidal pattern). There are three possible ways to accommodate a QHA on handheld terminals, illustrated in **Figure 4.20.** The first one **(A)** is a bottom fed, end-fire QHA with shorted end radials and normal QHA. The second one **(B)** is the same antenna but uplifted to be above the head, known as a moved-up QHA and finally, the third one **(C)** is a top fed, backfire QHA with open bottom radials, i.e., with open radial QHA.

Figure 4.20. Three Solutions of QHA

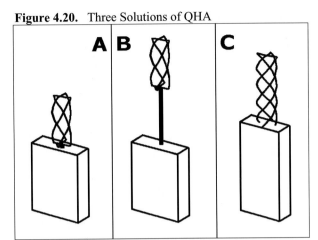

Courtesy of Paper: "QHA" by A.A. Agius and other

4.7.3.2. MSA Solutions for Handheld Terminals

The new GEO and Non-GEO MSC systems are employed as personal handheld terminals with small whip or telescope antennas similar to cellular systems. In this context, several different manufacturers in the world have designed types of antenna for top mount on the handheld terminals. These systems also provide service for all three mobile applications, with specific antenna solutions for installation on board ships, vehicles and aircraft.

The Japanese firm Kyocera developed a handheld terminal with a patch telescopic antenna for the Iridium system, see **Figure 4.21. (A).** It is omnidirectional providing a spherical coverage with maximum reception regardless of the angle at which the handset is held.

Figure 4.21. Three Possibilities of MSA for Handhelds

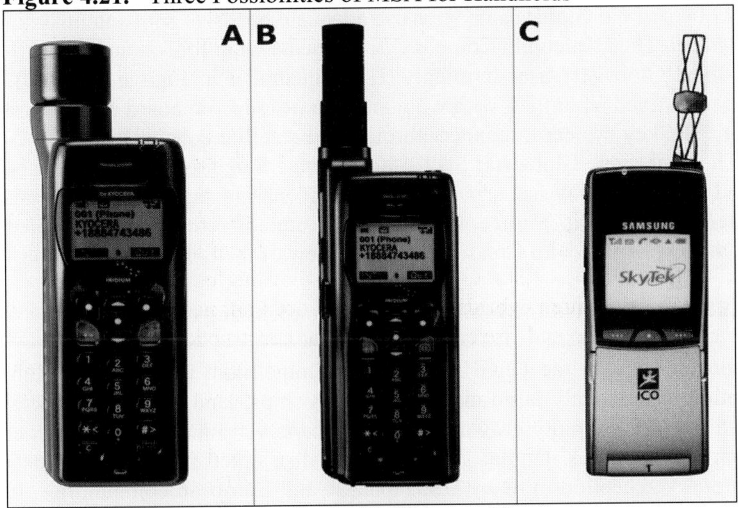

Courtesy of Prospects: "Product Specifications" by Kyocera and Samsung

Figure 4.22. Three Possibilities of Marine Antennas for SES

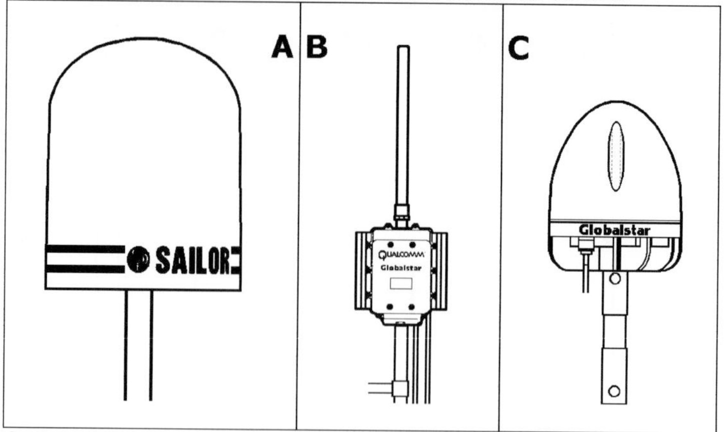

Courtesy of Prospects: "Product Specifications" by S.P. Radio A/S, Qualcomm and Telit/ICS

The same company has developed another, similar SS-66K model of handheld terminal with helical antenna, see **Figure 4.21. (B).** The antenna has omnidirectional radiation with swivel, providing maximum reception regardless of the angle at which the handset is held. The Korean-based company Samsung designed a handheld terminal for the forthcoming non-GEO ICO system, with a hypothetical antenna proposed by ICO, see **Figure 4.21. (C).** The Denmark-based manufacturer of maritime radio and MES known as S.P. Radio A/S, have developed a helical omnidirectional antenna for the Iridium single-channel transceiver Sailor SC4000 for MMSC, see **Figure 4.22. (A).** The antenna has $210°$ or 3 dB bandwidth and a link margin of 16 dB. Four such separate helical omnidirectional aerial constructions can be mounted separated by $90°$ on board ships and with an adequate PBX switchboard, can make it possible to run four simultaneous satellite links. These four antennas are part of a multichannel Sailor transceiver, MC4000, which uses the Iridium satellite network.

The subscribers of the Globalstar system use adequate handsets with antennas similar to the Iridium system. The Qualcomm Company developed a whip roof-mounted satellite antenna for semi-fixed Globalstar user terminals (UT), illustrated in **Figure 4.22. (B).** For large vessels, a semi-fixed mount UT with whip antenna installed on board enables a ship's fixed phones or PBX/key system to connect through the satellite constellation of the Globalstar system. The dual mode Globalstar SAT 550X Marine Satellite Terminal designed by Telit (Italy) and ICS (UK) may use separate waterproof marine satellite antennas and optional GSM antennas, shown in **Figure 4.22. (C).** The omnidirectional marine antenna can be mounted in any position with a reasonably clear view of the horizon, without obstacles.

4.7.4. Antenna Systems for Mobile Satellite Broadcasting (MSB)

Satellite broadcasting at the 12 GHz band has become more widespread and the need for TV reception by relatively large mobiles, such as ships, land vehicles (trucks, buses and trains) and aircraft is rapidly growing. This service is very attractive for both commercial and military applications. Digital TV and audio distributed by satellite is becoming the standard for TV reception in many US, Japanese and European communities. In the USA waters, services are currently performed by: DirecTV, USSB, Dish Network and AlphaStar

Broadcast emissions. Europe and the Middle East use the services of Astra, Eutelsat and Orbit, while in Japan, the NHK (Japanese Broadcasting Corporation) studied, designed and developed two types of mobile Direct Broadcasting Satellite (DBS) antenna systems: with a combination of mechanical and electrical tracking.

Since a satellite broadcasting system covers a wide area, it is inherently suitable for mobile reception of two new services known as Direct Audio Broadcasting (DAB) and Direct Broadcasting Satellite (DBS). This reception by mobiles has advantages in propagation conditions compared to its terrestrial TV broadcast counterparts because the former suffers less interference due to multipath propagation as commonly experienced in the latter.

However, in order to receive satellite broadcasts with good quality, the following electrical, mechanical and structural requirements are imposed on mobile antennas for MSB:

1) Accurate Beam Pointing – In the existing MSB systems, the total gain of the receiving antenna is required to be minimum 32 dBi, hence, the beam must be quite narrow, i.e., about $2°$. Therefore, an accurate beam pointing capability is required.

2) High-Speed Tracking – Since a mobile moves fast, especially in the case of airplanes and rapid trains and changes its direction very frequently, the mobile receiver must track the satellite very quickly, not only for beam steering but also for compensation of the vehicle's vibration. In this sense, electronic scanning may be promising.

3) Countermeasure for Satellite Signal Interruption – In the monopulse mobile tracking system the receiving mobile antenna can point its beam to the satellite as long as the signal from the satellite is being received. Hence, if the signal is interrupted by surrounding obstacles, tracking becomes impossible. For that reason, in order to cope with satellite signal blocking, several techniques are required: space diversity, holding the antenna direction by using a gyroscope or a magnetic sensor and high-speed mechanical searching for the satellite signal for recapture.

4) Height Limitation – The receiving MSC configurations as a whole must be compact, lightweight and especially, not too high, to prevent accommodation on the mobile. Usually, the antenna unit on the mobile should be less than a foot high. In order to satisfy these requirements, several kinds of MSA for MSB reception were developed and some of these are used on intercity fast trains, such as Shinkansen in Japan.

4.7.4.1. Mechanically Tracking Antennas for Land MSB

An example of a mechanical tracking antenna is the planar-array, illustrated in **Figure 4.23. (A).** This MSA is a microstrip patch array with very high gain of about 40 dBi. Otherwise, this antenna works on 12 GHz frequency band for TV and audio reception on mobiles. Japanese designers developed another two planar-arrays with mechanically controlled beams in both azimuth and elevation directions. Thus, the first consists in four planar antennas mounted on 88 x 88 cm aluminum honeycomb plate. In order to make the antenna height lower, to reduce wind pressure and keep the beam to the elevation of the GEO satellite, the antenna was designed to provide a beam tilt of $30°$. More recently, these antenna elements have been replaced by a waveguide slot array because this array can realize a large beam tilt and consequently, a lower aerodynamic profile antenna can be obtained. The antenna consists in four sub arrays, each of which is a 16 x 24-element cross-slot array. Moreover, the overall dimension of the antenna is 662 x 646 mm^2 and the tilt angle of the beam is $52°$. The gain of the antenna is 35 dB at 11.95 GHz, RHCP, operating at 25 to $58°$ of elevation and $±120°$ in azimuth range.

Figure 4.23. MSB TV Solutions for Land Mobiles

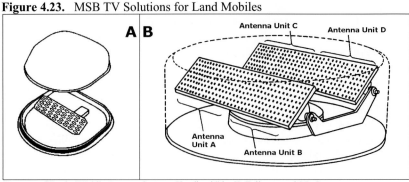

Each antenna has an LNA to reduce C/N degradation due to feeder loss. The phase shifters connected behind the LNA are switched by its driver from 100 to 400 Hz to swing the antenna beam by up to 0.2° sequentially, in the four main directions. In the case of train applications, especially the bullet train, a radome is used to protect the antenna from wind pressure of about 800 kg/m². This type of antenna with radome is mounted on the Japanese bullet train Shinkansen. That means that this type of antenna can also be mounted on buses.

4.7.4.2. Mechanical and Electrical Tracking Antennas for Land MSB

Since land mobiles such as cars and trains move fast and change their direction frequently, electrical tracking is preferable but in order to fully cover 360° in azimuth, a facility for mechanical tracking is used. Such an antenna, developed by NHK in Japan, uses a gyroscope that is always calibrated with a phase comparison monopulse method for mechanical tracking and with an adaptive beam-forming method of electrical tracking, illustrated in **Figure 4.23. (B).** Four flat panel antenna units are divided into two pairs to reduce the system's height. The system is housed in a cylindrical sandwich-structure radome of 350 mm in height and 862 mm in diameter, whose weight is about 50 kg. The flat panel antenna unit consists in an 8 x 16-element circularly polarized printed array. Each element is a rectangular microstrip patch with dual feeds on a two-layer structure of the substrate. The two-layer structure makes an optimum system design possible for both the feed line and the radiation element separately and it provides a low-loss and wide-band microstrip antenna. Otherwise, this receiving MSB antenna employs mechanical steering in both elevation and azimuth range for coarse tracking and electrical beam steering by the adaptive technique for fine tracking. The antenna units A and B, which are mounted on the turntable and rotate around the azimuth axis, produce a phase difference corresponding to the direction of the incident wave. Using the phase difference as the error signal of the phase comparison monopulse, it is possible to form a feedback loop. On the other hand, the two pairs of this antenna unit, A/B and C/D, rotate around the different elevation axes. The phase difference between the pairs is independent of the elevation angle. This means that the elevation mechanism, controlled by the phase comparison monopulse, cannot form a feedback loop. The phase difference relates to the elevation direction of the satellite. It is possible to determine the elevation angle of the two antenna panels using the information of the phase difference between them, by directing the beams of the panels to the satellite.

Figure 4.24. TV MSB Solutions for Ships and Aircraft

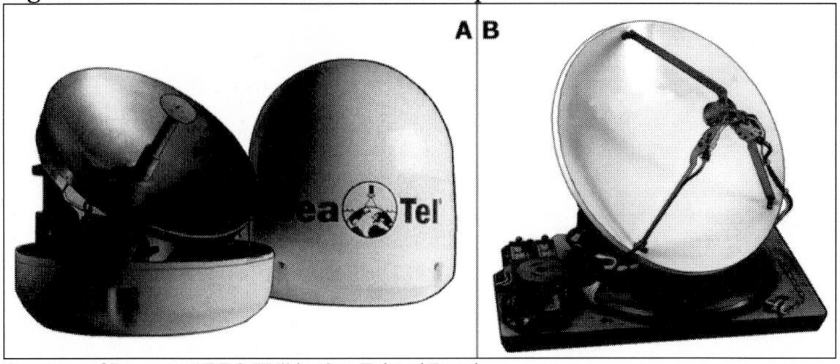

Courtesy of Prospect: "DBS TV" by Sea Tel and Racal

4.7.4.3. Antennas for Maritime MSB

The SeaTel Company is a well-known manufacturer of Marine Stabilized Antenna Systems for direct TV reception on board commercial and military vessels and for fishing and pleasure boats. This company has developed several models of MSA, presented in **Table 4.3.** Almost all these antennas have US and European dual and linear circular polarization, respectively; type of stabilization is 3-axis servo; roll and pitch stabilization of ±30°; roll and pitch response rate of more than 30°/sec; turn rate of 12°/sec, azimuth range of 540° continuous and elevation range of 25 to 60° allows full clearance for roll and pitch of the vessel on the high seas at a minimum elevation angle. However, the inertial stabilization sensors and proprietary control software completely eliminate the need for expensive and complex gyro or fluxgate compass references in acquiring and reacquiring the satellite, even in the roughest sea conditions. With a sophisticated new tracking antenna controller, getting that digital TV picture on the set could not be simpler: it is only necessary to select a TV service provider and to enter the ship position, or let the GPS do it. The controller then takes over to point the antenna at the satellite and start tracking. From that point on, the TV set on board the ship is ready to receive a TV program. The smallest 1898 SeaTel model of MSA for MSB is shown in **Figure 4.24. (A).** The WeSat versions of 1898, 3294, 4094 and 4894 have the possibility for GEO weather tracking and receiving equipment.

Table 4.3. Classification of SeaTel Antennas for MSB

Antenna Characteristics	Antenna Models					
	1898	**2094**	**2494**	**3294**	**4094**	**4894**
Dish Diameter/cm	45.72	50.80	60.96	81.28	101.60	121.92
Radome Diam./cm	50.80	60.96	78.74	101.60	121.92	152.40
Weight/kg	14.97	21.77	29.48	54.43	81.65	95.26
Gain/dBi	31.0	33.0	35.5	37.5	39.5	41.0
EIRP/dBW	52-53	50-51	48-49	46-47	44-46	42-43.5

4.7.4.4. Antennas for Aeronautical MSB

Racal Avionics is a manufacturer of MSA for aircraft and, in addition, is a major supplier and integrator of DBS TV antennas, which are optimized for the business multi-channel Jet VIP operator, shown in **Figure 4.24. (B)**. This antenna can support simultaneously up to four Airshow DBS TV and audio channels. Hence, the antenna system, which incorporates

Figure 4.25. Components of Ship Motion

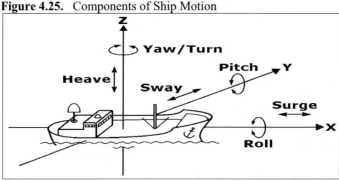

state-of-art technology, comprises two main elements: **(1)** Mechanically steered antenna dish and **(2)** Antenna Control Unit (ACU), housed in the Multichannel Receiver Decoder Unit (MRDU). In fact, the mechanically steered antenna is optimized for separate fin-top installation or can be co-located with a fin-mounted MSA unit. Polarization of the antenna is RHCP/RHCP and it is capable of horizontal and vertical linear polarization. The volume of the antenna reflector is 29.21 cm in diameter and weight is 6.8 kg.

4.8. Antenna Mount and Tracking Systems

The communication MSA for mobiles are relatively large and heavy, especially shipborne directional Inmarsat-A and B antennas. Over the past two decades, the directional antenna system, which comprises the mechanical assembly, the control electronics and gyroscope, the microwave electronic package and the antenna assembly (dish, array or similar), has reduced considerably in both physical size and weight. These reductions, brought about by greater EIRP from satellite transponders coupled with GaAs-FET technology at the front end of the receiver, leading to higher G/N RF amplifiers, has made the fitting of shipborne antennas on tracks and airplanes a reality.

4.8.1. Antenna Mount Systems

The MSA system is generally mounted on a platform, which has two horizontally stabilized axes (X and Y), achieved by using a gyrostabilizer or sensors, such as accelerometers or gyrocompasses. The stabilized platform provides a horizontal plane independently of mobile motion, such as roll or pitch. For example, all mobiles have some kind of motions but ship motion has seven components during navigation, such as: roll, pitch, yaw, surge, sway, heave and turn, shown in **Figure 4.25**. Turn means a change in the ship's heading, which is intentional motion, not caused by wave motion and the other six components are caused by wave motion. Surge, sway and heave are also caused by acceleration.

4.8.1.1. Two-Axis Mount System (E/A and Y/X)

An antenna mount is a mechanically moving system that can maintain the antenna beam in a fixed direction. In MMSS, the mount must have the capability to point in any direction on

Figure 4.26.   Two-axis and Four-axis Mount Systems

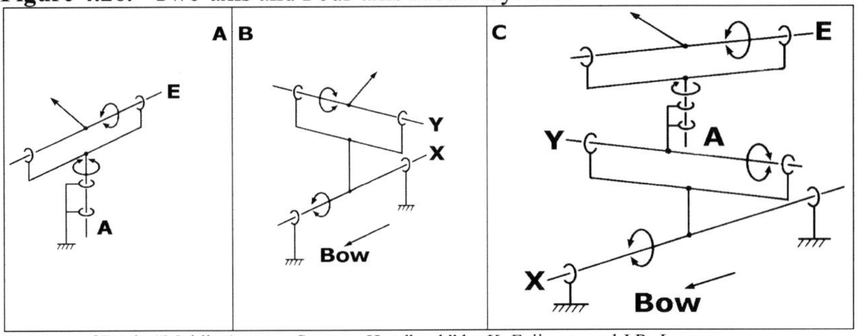

Courtesy of Book: "Mobile Antenna Systems Handbook" by K. Fujimoto and J.R. James

the celestial hemisphere because ships have to sail across heavy seas. It is well known that the two-axis configuration is the simplest mount providing such functions. There are two typical mounts of the two-axis configuration: one is the E/A (elevation/azimuth) mount and the other is the Y/X mount. Simplified stick diagrams of both mounts are illustrated in Figure **4.26. (A)** and **(B)**, respectively. In the E/A mount, a fully-steerable function can be obtained by choosing the rotation range of the azimuth axis (A-axis) from 0 to 90°. In the Y/X mount a fully steerable function is achieved by permitting the rotation angle to be from –90° to +90° to both the X and Y-axis. This is the basic configuration for the ship's utility, so a special function required for its antenna mount system is to compensate the ship's motions due to sailing and ocean waves and to keep the antenna beam in a nearly fixed direction in space. In the case of pointing/tracking under ship's motions, the required rotation angle range of each axis is from 0° to more than 360° for the A-axis and from –25° to +120° for the E-axis with respect to the deck, assuming that the operational elevation angle is de facto restricted above 5°. Both mount types have several disadvantages.

4.8.1.2.   Three-Axis Mount System (E/A/X, E'/E/A and X'/Y/X)

The three-axis mount system is considered to be a modified two-axis mount, which has one additional axis. The three-axis mount of an E/A/X type as shown in **Figure 4.27. (A),** is the E/A mount with one additional X-axis. The function of the X-axis is to eliminate the rapid motion of the two-axis mount due to roll. However, in this system, the possibility of gimbal lock for pitch is still left near the zenith, when the E-axis is parallel to the X-axis. The three-axis mount of an E'/E/A type, shown in **Figure 4.27. (B)** is the E/A mount with an additional cross-elevation axis, E. In the mount system, the change of the azimuth angle is tracked by rotating the A-axis and the change of the azimuth angle is tracked by a combined action of the E and E' axes. Hence, the E and E' axes allow movements in two directions at a right angle. With an approximate axial control, this mount is free from the gimbal lock problem both near the zenith and the horizon. The three-axis mount of an X'/Y/X type is the two-axis Y/X mount system with the X'-axis on it to obviate the gimbal lock at the horizon, presented in **Figure 4.27. (C)**. When the satellite is near the horizon, the X-axis takes out the rapid motion due to yaw and turn. In this sense, the X'-axis rotates within ±120°, so the X'-axis can only eliminate the rapid motion within the angular range. In general, this axis mount is rather more complex than that of the four-axis mount because steering and stabilization interact with each other.

Figure 4.27. Three-axis Mount System

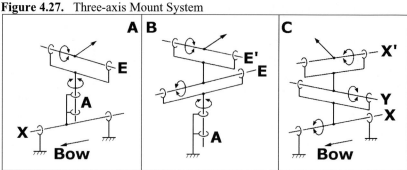

4.8.1.3. Four-Axis Mount (E/A/Y/X)

The stabilized platform is made by the X/Y-axis to take out roll and pitch and a two-axis mount of the E/A type is settled on the stabilized platform. This is the four-axis mount configuration, shown in **Figure 4.26. (C).** The tracking accuracy of this mount is the best solution because the stabilization function is separated from the steering function and at any rate, four major components such as roll, pitch, azimuth and elevation angle are controlled by its own axis, individually. The four-axis mount has been adopted in many SES of the current Inmarsat-A and B standards.

4.8.2. Antenna Tracking and Pointing Systems

The tracking and pointing system is another important function required of the antenna mount system. It should be noted that the primary requirement for SES tracking systems are that they be economical, simple and reliable. Tracking performance is a secondary requirement when an antenna beam width is broad.

1) Manual Tracking – This is the simplest method, wherein an operator controls the antenna beam to maximize the received signal level. At first, the operator acquires the signal and moves the antenna around one axis of the mount. If the signal level increases, the operator continues to move the antenna in the same direction. If the signal decreases, the operator reverses the direction and continues to move the antenna until the signal level is maximized. The same process is repeated around the second axis and the antenna is held both axes when the received signal level decreases. This method is suitable for LMSC and especially for portable and fly-away communication terminals.

2. Step Tracking – Among various auto track systems, the step track system has recently been recognized as a suitable tracking mode for SES because of its simplicity for moderate tracking accuracy. The recent development of integrated circuits and microprocessors has brought a remarkable cost reduction to the step track system, whose principle is the same as that of the manual track. The only difference is that an electric controller plays the role of an operator in the manual track. The schematic block diagram of the step tracking system is shown in **Figure 4.28. (A).** Sample-hold circuits are used to hold the signal levels, which are compared before and after the antenna have been moved by an angular step. If the level increases, the antenna is moved in the same direction and vice versa, if the level decreases, the direction will be reversed. In the other words, this process will be carried out alternately

Figure 4.28. Functional Block Diagrams of Step and Program Tracking

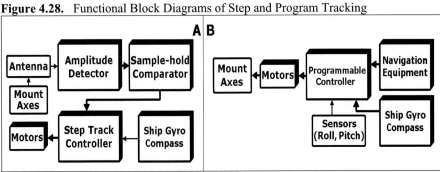

Courtesy of Book: "Mobile Antenna Systems Handbook" by K. Fujimoto and J.R. James

between two axes, whose good accuracy depends on the sensitivity of the comparators. As a result, the beam centre is maintained in the vicinity of the satellite direction. Wrong decisions on the comparison of levels generally arise from the S/N ratio and level changes due to multipath fading and stabilization error.

3. Program Tracking – The concept of the program tracking system is based on the open-loop control slaved to the automatic navigation equipment, such as a gyrocompass, GPS, the Omega and Loran-C systems. Namely, in program tracking, the antenna is steered to the point of the calculated direction based on the positional data of the navigation equipment. Since the satellite direction changes because of roll, pitch and turn, a function to remove these rapid motions is required in the program track, whose block diagram is illustrated in **Figure 4.28. (B)**. Thus, the error of the navigation equipment is negligibly small for the program track system, while its error mainly depends on the accuracy of sensors for roll, pitch and turn, which is the stabilization error. At this point, an adequate sensor for the antenna program track system is a vertical gyro because it is hardly affected by the lateral acceleration. Accordingly, when the stabilization requirement is lenient, a conventional level sensor, such as inclinometer, a pendulum and a level may be used with careful choice of the sensor's location. The controller calculates the direction of the satellite to compensate for the ship's motions, which is affected by all components.

In any event, the simpler the axis configuration of the antenna mounts, the more complex the program calculation procedure becomes. More exactly, since the program controller has to execute calculations of many trigonometric functions, a microprocessor is a candidate for the controller. However, the program tracking system is also applicable to the four-axis mount. A combination with the step track system is more desirable because the error of the program track system can be compensated by the step track system and its error due to the rapid ship motions can be compensated by the program tracking system.

4.8.3. Omnidirectional Shipborne Antenna Mounting

When installing MSA it is necessary to find a location on board ship that is as free from obstruction as possible. Thus, it is also important to maintain a certain distance from other communication antennas, especially radar installations. Normally, the best place for the MSA would be above the radar scanning antennas or far a way from them. Otherwise, a minimum safe distance should be maintained to HF antennas of 5 m, to the VHF antennas 4 m and to the magnetic compass, 3 m.

Figure 4.29.   Safe Distance of Inmarsat-C Antenna from Obstructions

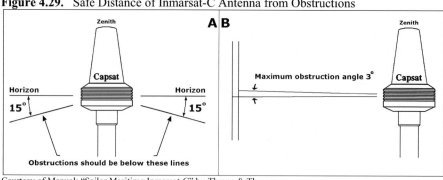

Courtesy of Manual: "Sailor Maritime Inmarsat-C" by Thrane & Thrane

The omnidirectional antenna is designed to provide satellite coverage even when the vessel has pitch and roll movement up to $15°$. Thus, to maintain this coverage the antenna should be free from obstructions in the area down to $15°$ below the horizon, as illustrated in **Figure 4.29. (A)**. Since this may not be possible in the fore and aft directions of the vessel, the clear area can be reduced to $5°$ below the horizon in the fore and aft directions and $15°$ below the horizon in the port and starboard directions. Otherwise, any compromise in this recommendation will degrade performance. If an obstruction, such as a pole or a funnel is unavoidable, the distance to these objects should be large enough, so that the obstruction only covers $3°$. For instance, if the diameter of the obstructing object is 0.1 m, the safe distance should be about 2 m, as shown in **Figure 4.29. (B)**.

The safety levels for the Thrane & Thrane Capsat-C Antenna Unit and similar Inmarsat-C aerials are based on the ANSI standard C95.1-1982. Namely, this standard recommends that the maximum power density at 1.6 GHz exposed to human beings should not exceed 5 mW/cm^2. Therefore, at maximum radiation output, the power from an Inmarsat-C antenna of 16 dBW EIRP corresponds to a minimum safety distance of about 30 cm. The future standard from the European Telecommunication Standard Institute (ETSI) concerning 1.5/1.6 GHz MES the recommendation will be maximum 8 W/m^2 (0.8 mW/cm^2) with a minimum safety distance of 62 cm at 16 dBW of EIRP.

4.8.4. Directional Shipborne Antenna Mounting and Steering

The Above Deck Equipment (ADE) consists in an antenna unit mounted on a pedestal, an RF unit, power and control unit, all covered by a radome. Ideally, the antenna should have free optical sight in all directions above an elevation angle of $5°$. The antenna must be placed as high as possible, in the best position on board ship to avoid blind spots with degradation or loss of the communications link.

4.8.4.1. Placing and Position of SES Antenna Unit

The directional antenna has a beam width of $10°$ and ideally requires a free line-of-sight in all directions above an elevation angle of $5°$. Possible obstructions will cause blind spots, with the result of degradation or even loss of the communications link with the spacecraft. So, complete freedom from degradation of the propagation is only accomplished by placing

Figure 4.30. Theoretical Antenna Installation

Courtesy of Brochure: "SES Design and Installation Guidelines" by Inmarsat

the antenna above the level of any possible obstructions. This is normally not feasible and a compromise must be made to reduce the amount of blind spots. The degree of degradation of the communication depends on the size of the obstructions as seen from the antenna, hence the distances to them must also be considered. However, it should be remembered that the antenna RF beam of energy possesses a width of $12°$ angle cone and consequently, objects within 10 m of the radome, which cause a shadowing sector greater than $6°$, are not likely to degrade the equipment significantly. Preferably, all obstructions within 3 m of the antenna system should be avoided. Obstructions less than 15 cm in diameter can be ignored beyond this distance. Knowing the route that the ship normally sails allows a preferable sector of free sight to be established, thus facilitating the location of the antenna unit. The azimuth and elevation angles, relative to any satellite of the Inmarsat Network, can be derived from azimuth and elevation maps, shown in **Figure 2.4. (A)** and **(B)**, respectively.

The antenna beam must be capable of being steered in the direction of any GEO satellite of the Inmarsat constellation, whose orbital inclination does not exceed $3°$ and whose longitudinal excursions do not exceed $±0.5°$. Therefore, means must be provided to point the antenna beam automatically towards the satellite with sufficient accuracy to ensure that the G/T and EIRP requirements, namely receive and transmit signal levels, are satisfied continuously under operational conditions.

Careful and important consideration should be given to the placing of an Inmarsat-A or B ADE with radome. Essentially, the focal point of the parabolic antenna must be pointing directly at the GEO satellite being tracked without any interruption of the microwave beam, which may be caused by any obstruction on the ship. Inmarsat specify that there should be no obstacle that is likely to downgrade the performance of the equipment in any angle of azimuth down to an elevation of $-5°$, which is not easy to achieve. Thus, the SES design and installation guidelines of Inmarsat give a theoretical antenna installation instructions mode satisfying this advice but with the disadvantage that the antenna is very high above the vessel's deck and would be impossible to install in such a way, see **Figure 4.30.** This type of installation is not practical because a ship's antenna would certainly be adversely affected by strong, gusty wind and vibration and it would be difficult to gain access for maintenance purposes.

If ship's structures do interrupt the antenna beam, blind sectors will be caused, leading to degraded communications over some arc of azimuth travel. Otherwise, if it is like that and as is often the case, it is impossible to find a mounting position free from all obstructions; the identified blind sectors should be recorded. It may be possible for the operator, when in an area served by two satellites, to select the satellite whose azimuth and elevation angles with respect to the ship's position are outside the blind sector. Obviously, this method is not enough practical because the satellite overlapping sectors inside of Inmarsat's four ocean regions cover relatively small areas. The best solution to avoid all blind sectors is to place the antenna unit on top of the radar mast or on a specially designed mast.

4.8.4.2. Antenna Mast and Stabilizing Platform

The mast has to be designed to carry the weight of the antenna unit, maximum 300 kg, depending on model design or manufacturer, presented in **Figure 4.31. (A).** It must also be able to withstand the forces imposed by severe winds up to 120 knots on the radome and strong vibrations due to very rough seas on the whole ADE construction. The top end of the mast should be fitted with a flange with holes matching the bolts extending from the bottom of the radome. The flange must not be so large as to interfere with the hatch in the bottom of the antenna unit. The holes through the mast flange must be positioned symmetrically around the ship's longitudinal axis. If the height of the mast makes it necessary to climb up to the antenna unit, a ladder must be provided on the mast column. A guard rail must be attached to the upper section for safety purposes. If the height of the mast exceeds approximately 4.5 m, an access platform should be attached to the mast about 1.5 m below the radome bottom.

Figure 4.31.   ADE Mast and Stabilized Platform

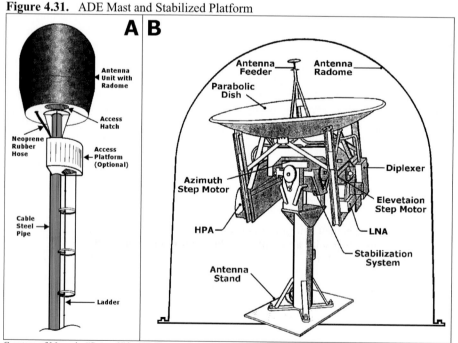

Courtesy of Manuals: "Saturn 3" by EB Communications and "Maritime Communications" by Inmarsat

The radome completely encloses the periphery of the base plate assembly to protect the electronic and mechanical components from corrosion and weather. It is usually fabricated from high-gloss fibreglass and is electronically transparent to RF signals in the assigned frequency band. The radome is secured to the circular or square antenna stand (base plate) with several screws and can be removed easily without special tools.

The antenna stabilized platform is housed inside the radome and consists in the electrical and mechanical elements, presented in **Figure 4.31. (B).** There are two antenna control stepper motors. First, there is the azimuth step motor, which controls the position of the antenna reflector in the horizontal A plane (azimuth) and second, is the elevation step motor that controls the vertical E plane (elevation). Each motor has four phase-inputs coming from the drive circuit in the BDE Control Board and a supply voltage from the master power supply located on the antenna stand. As a stepper motor turns the antenna, it also adjusts the setting of the relevant sensor potentiometer. The sensor voltage supply is the reference voltage for the A/D converter on the Control Board. Therefore, two stepper motors move the antenna in both azimuth and elevation angles and move the relevant sensor potentiometers, which provide feedback information on the position of the antenna. The stabilization system, or gyroscope with two gyro motors, stabilizes the platform for the antenna against the roll and pitch of the ship. A two-turn solenoid clamps the antenna platform to the gyroscope assembly. A diplexer passes the Rx signal from the antenna to the LNA and the Tx signal from the transceiver assembly to the antenna. The LNA amplifies the Rx signal and the HPA amplifies the Tx signal. The antenna, a parabolic dish, radiates EM energy to and from the antenna feeder in Rx or Tx direction, respectively. Finally, the antenna assembly is mounted below the radome for protection purposes.

4.8.4.3. Antenna Location Aboard Ship

The ship's antenna unit should be located at a distance of at least 4.5 m from the magnetic steering compass. At this point, it is not recommended to locate the antenna close to any interference sources or in such a position that sources such as the radar antenna, lie within the antenna's beam width of $10°$ when it points at the satellite. The ADE should also be separated as far as possible from the HF antenna and preferably by at least 5 m from the antennas of other communications or navigation equipment, such as the antenna of the satellite navigator or the VHF and NAVTEX antennas. In addition, it is not practical to place the antenna behind the funnel, as smoke deposits will eventually degrade antenna performance. Regardless of the location chosen for the antenna, it should be oriented to point forwards in parallel with the ship's longitudinal axis when in the middle of its azimuth range, which will correspond to zero degrees on the azimuth indicator.

The EM RF signals are known to be hazardous to health at high radiation levels. Thus, it is inadvisable to permit human beings to stand very close to the radome of an SES when the system is communicating with a satellite at a low elevation angle. In this case, Inmarsat recommends that the radiation levels in the vicinity of the antenna should be measured. The crew members and passengers should not be admitted to areas closer than 10 m away from the antenna unit at desk level above 2 m, measured beneath the lowest point of the radome, as illustrated in **Figure 4.32. (A)**. No restrictions, therefore, are required when the antenna radome is installed at least 2 m above the highest point accessible to crew and passengers. Hence, authorized personnel should not remain close to the antenna system for periods exceeding 1 hour per day without switching off the RF transmitter. However, radiation plan

Figure 4.32. Antenna Radiation Precautions and Azimuth Limit

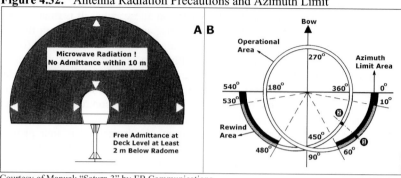

Courtesy of Manual: "Saturn 3" by EB Communications

diagrams may be produced and located near the antenna as a warning for crew members, passengers and ship's visitors, or distances from the antenna may be physically labeled at the relevant place.

4.8.4.4. Satellite Determination and Antenna Azimuth Limit

An Inmarsat-A, B and M MSA must be capable of locating and continuously tracking the GEO satellite available or selected for communication, namely if the ship has in view only one satellite or if the ship is in an overlapping position, respectively. Thus, Inmarsat-C has an omnidirectional antenna and does not need a tracking system. Locating and tracking may be done automatically, as in the case of an SES, or manually, as with a portable MES. In fact, it is common practice to believe that the GEO satellites are fixed and that once the link has been established it will remain so as long as the mobile does not move. However, ships or other mobiles are always moving during operational management of voyages and satellites are under the influence of a number of variable astrophysical parameters, which cause it to move around its station by up to several degrees. Therefore, an ADE tracking system must counteract this by repositioning the SES antenna at regular intervals and in case of need. Namely, the carrier signal is monitored continuously and, if a reduction in its amplitude is detected, a close programmed search is initiated until the carrier strength is again at maximum. No loss of signal occurs during this process, which is automatically initiated. Obviously, the greatest tracking problem will arise when the SES is moving at speed with respect to the satellite. An Inmarsat-A and B MSA may be moved through any angle in azimuth and elevation as the vessel moves along its course. In this case, it is essential that electronic control of the antenna is provided. In practice, antenna control may be achieved by manual or simple electronic feedback methods.

a) Manual Commands – When the radio operator has selected manual control, elevation is commanded by up and down keys, whereas azimuth positioning is controlled clockwise and counter-clockwise keys. A command would be used when the relative positions of both the vessel and the satellite are known. Azimuth and elevation angles can be derived and input to the equipment, by using the two A and E charts of Inmarsat network coverage. Once the antenna starts to detect a satellite signal, the operator display indicates signal strength. Fine positioning can now be achieved by moving the antenna in A and E in 1/6th degree increments until maximum field strength is achieved.

b) Automatic Control – Once geostationary satellite lock has been achieved, the system will automatically monitor signal strength and apply A/E corrections as required in order to maintain this lock as the vessel changes course.

c) Automatic Search – An automatic antenna search routine commences 1.5 minutes after switching on the equipment, or it may be initiated by the operator. Therefore, the elevation motor is caused to search between $5°$ and $85°$ limits, whereas the azimuth motor is stepped through $10°$ segments. If the assigned common signaling channel signal is identified during this search the step tracking system takes over to switch the antenna above/below and to each side of the signal location searching for maximum signal strength.

d) Gyroscopic Control – Lock is maintained irrespective of changes to the vessel's course by sensing signal changes in the ship's gyro repeater circuitry. Satellite signal strength is monitored and if necessary, the A/E stepper motors are commanded to search for maximum signal strength.

e) Antenna Rewind – The antenna in the ADE is provided on a central mast and is coupled by various control and signal cables to a stationary stable platform. Thus, if the antenna was permitted to rotate continuously in the same direction, the feeder cables would eventually become so tightly wrapped around the central support that they would either prevent the antenna from moving or they would fracture. To prevent this happening, a sequence known as antenna rewind is necessary, as is shown in **Figure 4.32. (B)**. In fact, an antenna has three areas with rewind time of approximately 30 seconds plus stabilizing time, giving a total of about 1.5 minutes:

1. Operational Area is the antenna-rotating limit in the azimuth plane. In fact, the antenna can rotate a total of $540°$, which is shown as a white area in **Figure 4.32. (B)**. Normally, the antenna will operate in the operational area, which is between $60°$ and $480°$.

2. Rewind Area is necessary for the following reasons: if the antenna moves into one of the rewind areas, i.e., $10°$ to $60°$ or $480°$ to $530°$ (antenna azimuth lamp lights) and if no traffic is in progress, the antenna will automatically rewind $360°$ to get into the operational area and still be pointed at the satellite, which is illustrated as a dotted area in **Figure 4.32. (B)**. For example, the antenna moves from position 1 to 2 and the rewind lamp lights. If the SES is occupied with a call and the ship turns so that the antenna enters the rewind area, no rewind will take place unless the antenna comes into the azimuth limit area. If this happens, rewind will take place and the call will be lost. The azimuth warning indicator on the operator display will light to indicate that antenna rewinding is in progress.

3. Azimuth Limit Area is an important factor because when the antenna is in this area the azimuth limit lamp lights. If the antenna moves into the outer part of the azimuth limit area, i.e., $0°$ to $10°$ or $530°$ to $540°$, rewind will start automatically, despite traffic in progress.

4.8.4.5. Antenna Pointing and Tracking

The directional reflector antenna is highly directive and must be pointed accurately at the satellite to achieve optimum receiving and transmitting conditions. In normal operation the antenna is kept pointed at the satellite by the auto tracking system of, for example, Saturn 3 SES. Before the auto tracking can take over, the antenna must be brought within a certain angle in relation to the satellite. This can be obtained using the command "find" or by manually setting the antenna using the front push buttons on the terminal or via teleprinter command. For manual pointing it is necessary to provide the ship's plotted position, ship's heading by gyro, azimuth angle map and elevation angle map of the satellite.

Figure 4.33. Antenna Pointing

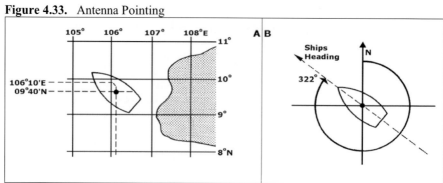

Courtesy of Manual: "Saturn 3" by EB Communications

1. Ship's Plotted Position – The plotted position is needed to decide which satellite can be used, namely which Inmarsat network area can be tuned: Indian Ocean Region (IOR), Pacific Ocean Region (POR), Atlantic Ocean Region-West (AORW) or Atlantic Ocean Region-East (AORE), depending on the ship's actual position, as presented in **Figure 4.33. (A)**. Sometimes, the ship can be in an overlapping area covered by two or even three Inmarsat satellites. In this case it will be important to choose convenient CES and to point the antenna towards one of overlapping ocean regions.

2. Ship's Heading by Gyrocompass – The permanent heading of the ship determined by gyrocompass is needed for the antenna auto-tracking system, shown in **Figure 4.33. (B)**.

3. Azimuth Angle – The azimuth is the angle between North line and horizontal satellite direction as seen from the ship, as is shown by example of 259°, in **Figure 4.34. (A)**. Thus, the actual azimuth angle for the various satellites due to the ship's plotted position can be found on the map, illustrated in **Figure 2.4. (A)**.

4. Elevation Angle – The elevation angle is the satellite height above the horizon as seen from the ship, as is shown by the example of 38°, in **Figure 4.34. (B)**. In this case, the actual elevation angle for the various satellites due to the ship's plotted position can be found on the map in **Figure 2.4. (B)**.

Figure 4.34. Azimuth and Elevation Angle

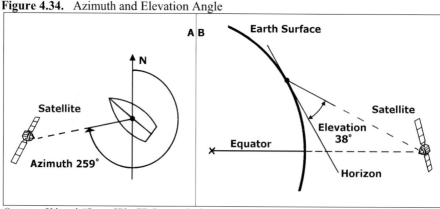

Courtesy of Manual: "Saturn 3" by EB Communications

Figure 4.35. Top-Mounted Airborne Antenna

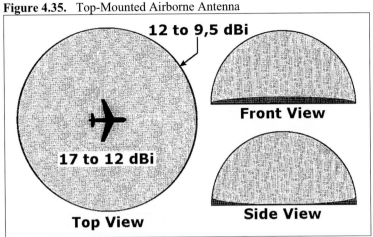

Courtesy of Manual: "CMA-2102 HGA Technical Description" by Canadian Marconi Company

4.8.5. Airborne Antenna Mounting and Steering

Gain is the fundamental parameter controlling antenna operation. The HGA CMA-2102 of the Canadian Marconi Company provides very good gain coverage, exceeding 12 dBi in 90% and 9.5 dBi over 100% of the Inmarsat hemisphere, which extends from $5°$ upward over the horizon in level flight. This antenna provides very high gain values of 14 to 17 dBi within 50% of the hemisphere and has profound implications for the economics of the end user. With superior receive gain, less satellite power is required for nominal service and more satellite channels can therefore be supported. High transmit gain improves multi-channel operation capabilities by ensuring robust operation and the continuity of six-channel operation over a maximum coverage zone. The HGA CMA-2102 also exhibits good coverage below the Inmarsat hemisphere. Measured gain exceeds 9 dBi when looking down to $-3°$ in the roll plane, while trials conducted in cooperation with Inmarsat verified that HGA has 10 dBi gains at $0°$ elevation and can maintain a voice call to at least $-2°$.

4.8.5.1. AES Antenna Mounting

The airborne antenna assembly can be mounted on top of the aircraft or on the sides of the fuselage. Ideally, the best position for placing the antenna is a location on the centreline of the aircraft coordinate system, as is shown in **Figure 4.35**. In this case, the antenna assembly location should be moved forward or aft along the centreline from the ideal position to maintain proper separation from other antenna systems. All L-band antennas must be separated by at least 50.8 cm, with 1.5 m preferred. The top mount provides satellite coverage independently of the direction of aircraft travel. Namely, the top view is 100% clear, while front and side views have a very small shadowing on the lower levels of the hemisphere. This heading-independent antenna gain coverage is a very significant advantage, which may, in the future, become an important condition for all type of aircraft before being permitted to use the optimal flight paths made possible by satellite-based Air Traffic Management (ATM).

Figure 4.36. Typical Side-Mounted Airborne Antenna

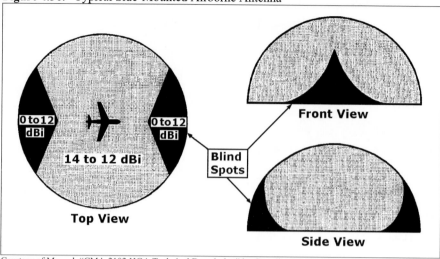

Courtesy of Manual: "CMA-2102 HGA Technical Description" by Canadian Marconi Company

On the other hand, typical side-mounted antennas have an inherent coverage deficiency due to their limited scan range. Thus, nose and tail bearings and shallow elevation angles to the satellite can reduce HGA gain values below 12 dBi, sometimes all the way to zero. Thus, these triangular areas, called blind-spots or "keyholes", occur in an aircraft's fore and aft directions, extending horizontally about 45° to each side at an elevation angle to the GEO satellite of 5° and rising to a point about 45° above the nose and tail, shown in **Figure 4.36.** Recent operational experience by a number of airlines has confirmed that the blind-spots cause problems with satellite signal reception, especially in certain areas of the world, such as the Pacific Rim. Actually, these blind-spots are the frequent cause of problems, including intermittent black-outs, dropped calls and sometimes zero communications, even when the aircraft is well within the Inmarsat coverage zone, particularly at high latitudes with low elevation angles. In any case, these operational problems have been important factors in the switch from side-mounted to top-mounted. Side-mounted installations also require two feeds from the HPA, as well as a high-power relay, adding to system losses.

The top-mounted antenna can typically be installed with up to 1 dB less loss than permitted in Arinc-741, resulting in better communications and greater channel availability. Hence, **Table 4.4.** compares the top-mounted with a typical side-mounted antenna design.

Table 4.4. Comparison of Top-Mount and Side-Mount Airborne Antennas

Gain Characteristics	Top-Mount	Side-Mount
Minimum Gain over 75% Inmarsat Hemisphere	13 dB	12 dB
Average Gain over 75% Hemisphere	15 dB	<13 dB
Minimum Gain over 100% Hemisphere	>9,5 dB	0 dB

Therefore, as illustrated in **Figure 4.36.**, the top view of a side-mounted airborne antenna is reduced on the nose and tail with shadowing of about 12% in total, while both front and side views have shadowing spots of about 15% each.

4.8.5.2. Performance at High Latitudes and Low Elevation Angles

The MSC performance can be limited at high latitudes and low elevation angles, regardless of which HGA is used. For a GEO satellite orbiting at an altitude of about 35.786 km, the highest latitude that can be viewed is $81°$ 18', corresponding to an elevation angle of $0°$. For aircraft flying above this latitude, the satellite is below the horizon and unavailable for communications. Hence, when the aircraft's longitude differs from that of the satellite, the maximum latitude is even less; at longitudes midway between two satellites, it can be as low as $71°$. Even before the $0°$ elevation angle is reached, MSC signals suffer from several sources of increased loss at low elevation angles. These factors combine to reduce the number of available voice channels at $5°$ elevation compared to $20°$ elevation.

High gain below the horizon can actually degrade an HGA's low-elevation performance because it increases the interference from signals reflected from the ground. Thus, side-mounted antennas can meet these limits only over uneven ground or rough seas, where multi-path signals are diffused. At low elevation angles, the multi-path rejection of HGA is further enhanced by its top-mounted design, since the antenna's predominantly vertical polarization ellipse helps to cancel out the effects of multi-path reflections, mostly horizontally polarized.

The major source of antenna noise results from HGA "looking" at portions of the terrain, which are "hotter" than the sky. In any event, the top-mounted antenna's controlled gain roll-off below the horizon, at all azimuth angles, giving much lower noise characteristics than side-mounted antennas, especially at low satellite elevation angles. In reality, the noise of this antenna is almost independent of elevation angle, whereas patterns of side-mounted antenna vary widely around the horizon.

At low elevation angles, it is critical that MSA can seamlessly transfer to the next satellite as soon as it has a higher elevation angle. The algorithm normally used to determine when hand-over should occur assumes that the satellite with the highest elevation angle is the optimal one but this is not necessarily true of side-mounted antennas, since the higher satellite may be in a blind-spot fore or aft of the aircraft. This can cause delays in satellite selection and could lead to a temporary communications black-out. However, the HGA top-mounted unit has no blind-spots and will perform better than side-mounted during satellite hand-over.

4.8.5.3. Beam Steering Performances

Steering of the HGA is controlled by the Beam Steering Unit (BSU), which transmits steering commands to the HGA over an asynchronous EIA-RS-422A serial data link. The HGA interprets the steering command and steers the antenna beam to the desired position. Thus, the BSU translates antenna beam position data and beam change commands received from the Satellite Data Unit (SDU) in a standard digital format into signals needed to select phasing of the antenna elements that result in the antenna beam pointing at the desired satellite. The BSU is housed in an ARINC-600 rack-mounted 2-MCU package.

During subsystem operation, open-loop steering words and control words are received from the SDU. The steering words contain azimuth and elevation information for the direction in which the beam will be steered, using the coordinate system, which is illustrated in **Figure 4.37.** In the proper manner, after receiving an open-loop steering word, the BSU calculates the beam number (one of more than 1,000) and associated phase-shifter settings, which are

Figure 4.37. HGA Coordinate System

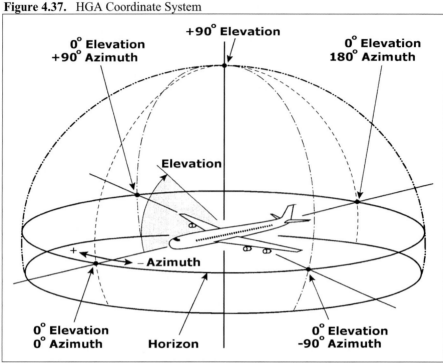

Courtesy of Manual: "CMA-2102 HGA Technical Description" by Canadian Marconi Company

then up linked to the HGA at a maximum rate of 20 Hz, via an RS-422A serial data links. However, should the rate of transmission of the open-loop steering words exceed 20 Hz, the BSU buffers the incoming words unit it has up linked and strobes in the phase-shifter settings for current azimuth/elevation value. If multiple steering words have been received during the processing time, only the most recent value is retained. Control words will be processed as they are received. Another function of the BSU includes reporting of the selected antenna beam and gain to the SDU.

Moreover, BSU's operational software meets the requirements of Level 2 Flight Essential category. In addition to operating software, the BSU terminal contains diagnostic software to assist ground maintenance personnel in fault isolation. The diagnostic software is capable of identifying faults to a more detailed level than required by ARINC 741.

5

PROPAGATION AND
INTERFERENCE CONSIDERATION

Propagation and interference characteristics are very important for providing quality and reliability of MSC propagation channels. The Quality of Service (QoS) can be expressed in terms of the Bit Error Rate (BER) performance, which depends on the Carrier-to-Noise C/N_0 density ratio; meanwhile, the service reliability is manifested in the relation of service availability. The intervening medium between MES within the network and satellites is termed a transmission channel. The fixed satellite services have two constant channels between a minimum of two FES using the same spacecraft, while in the GMSC network there are two types of channels to be considered: the variable channel between the MES and satellite or service link and the constant channel between the LES (Gateway) and the satellite or feeder link. These two channels have many different characteristics, which need to be taken into account during the system design examination. The more critical are the variable channels used by MES, since transmitter power, receiver gain and satellite visibility are sometimes quite restricted in comparison to the constant link. In this case the propagation path profile between all kind of MES and satellites varies continuously whenever the mobiles (ships, land vehicles or airplanes) are in motion. These variations are most significant and frequent in the case of maritime and aeronautical environments.

The common satellite channel environment affects radiowave propagation in changeless ways. The different parameters influenced are mainly path attenuation, polarization and noise. The factors to be considered are gaseous absorption in the atmosphere, absorption and scattering by clouds, fog, all precipitation, atmospheric turbulence and ionospheric effects. In this sense, several measurement techniques serve to quantify these effects in order to improve reliability in the system design. Because these factors are random events, GMSC system designers usually use a statistical process in modeling their effects on radiowave propagation. To design an effective GMSC model it is necessary to consider the quantum of all propagation characteristics, such as signal lost in normal environment, path depolarization causes, transionospheric contribution, propagation effects important for mobile systems, including reflection from the Earth's surface, fading due to sea and land reflection, signal blockage and to the different local environmental interferences for all mobile and handheld applications. At any rate, the local propagation characteristics on the determinate geographical position have very specific statistical proprieties and results for ships, vehicles and aircraft.

5.1. Radiowave Propagation

Radiowaves travel in space as electrons at the speed of light approximately 300,000,000 m/sec. When radiowaves are propagated from a transmitter antenna, they form three modes of propagation path: surface, sky and space wave propagation, shown in **Figure 5.1. (A)**. Radiowaves are generally transmitted from an antenna omnidirectionally but transmission can be also modified in a directional path by using a directional dish or antenna arrays.

Figure 5.1. Radiowave Modes of Propagation and Free Space Loss

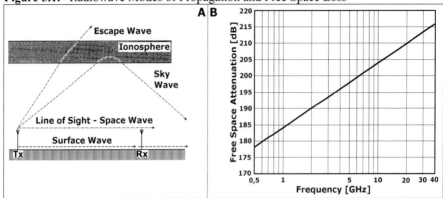

The surface wave is a radiowave that is modified by the nature of the terrain over which it travels. The surface wave propagation predominates at all radio frequencies up to 3 MHz for VLF, LF and MF bands. Hence, sky waves are severely influenced by the action of free electrons, called ions, in the upper atmosphere, known as the ionosphere and are caused to be attenuated and reflected and possibly returned to Earth. In fact, sky wave propagation predominates in the HF band between 3 and 30 MHz.

Above 30 MHz, the predominant mode of propagation of radiowaves is by the space wave. This wave, when propagated between 30 and 300 MHz into the troposphere by an Earth VHF or UHF radio station, is subject to deflection by variations in the refractive index structure of the air through which it passes, which will cause the radiowave to follow the Earth's curvature for short distances up to about 100 km. However, space waves above 300 MHz propagated upwards, away from the troposphere, may be termed free space waves and are primarily used for satellite communications.

Whilst all transmitting antenna systems produce one or more of the three main modes of propagation, one of the modes will predominate. The predominant mode may be equated to the frequency used, as all other constraints remain constant. Therefore, for the purposes of this explanation of propagated radiowaves, it is assumed that the mode of propagation is dependent upon the frequency used in an adequate system, for that is the only parameter, which may be changed by an operator.

The simplest situation involving the transmission of information by EM waves is when the Tx antenna transmits directly to the Rx antenna, without any obstacles in the path through which the EM waves must travel. This situation occurs when both antennas are located at a relatively short distance on the Earth's surface with full mutual visibility. As the distance between the two antennas increases due to the Earth's curvature, the EM waves are no longer in a direct path (line-of-sight). In order to increase the path length the antenna can be fitted on a higher mast but only to the extent that the costs do not become prohibitive. In this way, the problem can be solved by means of forthcoming stratospheric platforms or communications satellite payloads. Since these satellites are placed at very high altitudes, their reach is far superior to that of the higher towers that can be built on the Earth's surface, see altitudes and other parameters in **Table 5.1**. When two antennas situated on the

Table 5.1. Altitude Location of Stratospheric Platforms and Spacecraft

System	Altitude (h) in Km	Angle in degrees	Distance (d) in Km
International Flight	10	3.313	368.4
Stratospheric Platforms	30	5.549	617.1
Amateur Satellites	150	12.320	136.2
LEO Satellites/Low	780	27.008	3003.6
LEO Satellites/High	2000	40.438	4497.1
MEO Satellites	10000	67.095	7461.7
GEO Satellites	35600	81.268	9037.8

Earth are very far apart and there is no visibility between them, stratospheric platforms or satellites can be used to relay the signals. There is otherwise visibility between the Tx antenna and the spacecraft (platforms) and vice versa, between spacecraft (platforms) and Rx antenna. At this point, the ways these signals will travel on their line-of-sight depend on propagation characteristics and conditions of interference in the hypothetical environment. Because communications satellites are located at higher altitudes relative to the Earth's surface than stratospheric platforms and masts, they cover a wider area, which is beneficial for ocean spaces, continents and for countries with large dimensions.

5.2. Propagation Loss in Free Space

It is necessary to define the loss between Tx and Rx antenna separated by a distance from the transmission medium, assumed to be a vacuum and the antenna is isotropic. It may serve either as Tx or Rx antenna in the radiation field where it is situated.

Considering that the isotropic antenna radiates signals in all directions of its spherical surface with total power flow P_T in Watts, including receiving antenna gain, at a sufficiently large distance (r) in metres away from the centre to the surface of the sphere, the power flux per unit area through any portion of the spherical surface must be as follows:

$$P_f = P_T/4\pi r^2 \tag{5.1.}$$

The amount of power that the receiving antenna absorbs in relation to the RF power density of the EM field is determined by its effective aperture, which is defined as the area of the incident EM wave front that has a power flux equal to the power dissipated in the load connected to the receive antenna output terminals. Following the previous (5.1.) equation the receiving power at a receiving antenna can be expressed by:

$$P_r = P_T (\lambda/4\pi r)^2 = P_T/L_f \tag{5.2.}$$

where λ = wavelength and L_f = free space transmission loss, which can be given in dB as:

$$L_f = 10 \log_{10} (P_T/P_r = 32,40 + 20 \log f_{MHz} + 20 \log r_{km} \tag{5.3.}$$

where f = frequency of the emitted radio field in dB and r = distance in km between Tx and Rx antenna. Thus, free space loss is related to operating frequency and transmission distance. **Figure 5.1. (B)** shows free space loss as a function of frequency in GHz for a distance of 36,000 km.

5.3. Atmospheric Effects on Propagation

This study deals with propagation effects in all regions of the atmosphere and free space, including the Earth's ionosphere. Namely, most of the Earth's weather precipitation and hydrometeors occur in the troposphere, which is the non-ionized region from the Earth's surface up to a height of about 15 km above the surface at the Equator. The thickness of the troposphere decreases towards the poles. Therefore, propagation effects in the troposphere tend to increase in importance as the RF increases above 1 GHz. For GMSC systems, the effects of reflection from the Earth's surface are critically important at even lower RF. At frequencies below about 1 GHz, the most important region of the Earth's atmosphere is the ionosphere, the ionized region above the stratosphere, within which low RF propagation effects are quite strong. Propagation effects within the ionosphere have an influence on the terrestrial and Earth space paths from VLF to SHF bands. However, it should be borne in mind that for an RF band range below a certain number of GHz, the ionospheric propagation effects can be quite important and for frequencies above a certain number of GHz, tropospheric effects may be negligible. At this point, radiowave signals passing through the atmosphere or over the surface of the Earth begin to lose strength. The decrease of signal due to the medium through which it passes is called attenuation.

5.3.1. Propagation Effects of the Troposphere

The troposphere extends upwards from the Earth's surface to a height of approximately 15 km, where it meets the stratosphere. More exactly, at the boundary between the two spaces there is a region called the tropopause, which possesses a different refractive index to each neighboring layer. The effect exhibited by the tropopause on a space wave is to produce a downward bending action, causing it to follow the Earth's curvature. Hence, the bending radius of the radiowave is not as severe as the curvature of the Earth; nevertheless, the space wave will propagate beyond the visual horizon. In practice, the radio horizon exceeds the visual horizon by approximately 15%. This effect is known to navigators whereby the surface radar range of a maritime PPI radar picture extends slightly beyond the horizon.

5.3.1.1. Attenuation due to Atmospheric Gases

The different types of gases present in the atmosphere may attenuate the electromagnetic waves, which is caused by the molecular absorption of the atmospheric constituents and is strongly frequency dependent. The main contributors to this attenuation below 70 GHz are water vapor and oxygen. Thus, absorption increases as elevation angle is reduced, which at any frequency is a function of temperature, pressure, humidity of the atmosphere and the elevation angle of the satellite.

At sufficiently high frequencies, EM waves interact with the molecules of atmospheric gases to cause attenuation. These interactions occur at resonance radio frequencies and are apparent in plots of zenith ($90°$ elevation angle) attenuation versus frequency, as shown in **Figure 5.2. (A).** This is the theoretical total estimated one-way attenuation for vertical paths through the atmosphere, where solid curves are for moderately humid atmosphere, dashed curves represent the limits for 0% and 100% relative humidity, V is vertical polarization and A denotes limits of uncertainty, for a vertical Earth space path as a function between 1 and 200 GHz at $45°$ North latitude using US standard atmosphere.

Figure 5.2. Theoretical One-Way and Rainfall Attenuations

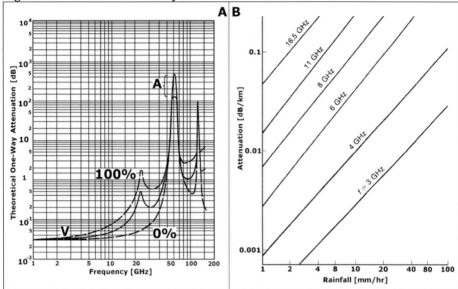

Courtesy of Books: "Satellite Communications Systems" by Richharia and "Satellite Communications" by R. Gagliardi

It may be noted that the specific frequency bands where the absorption is high are about 22.23 GHz caused by vapor (left peak in the curve) and centered between 53.50 and 65.20 GHz caused by oxygen (right peak in the curve). Meanwhile, in the frequency range of current interest about 1–18 GHz the zenith one-way absorption is in the range of 0.03–0.20 dB and 0.35–2.30 at 5° elevation. The corresponding upper limit for 100% humidity is 0.7 dB at zenith and 8 dB at 5° elevation. Therefore, these effects are not very important for GMSC systems because the influence of attenuation due to atmospheric gases can be effectively ignored at frequencies below 10 GHz.

5.3.1.2. Attenuation by Precipitation and Hydrometeors

Hydrometeors are condensed water vapors existing in the atmosphere, such as cloud, fog, rain, hail and snow. The last three forms of atmospheric water are called precipitation. They produce transmission impairments and attenuation by absorbing and scattering the energy of radiowaves.

1. Clouds and Fog – Cloud and fog may cause attenuation but much less than that caused by light rain of about 10 mm/h. Clouds and fog are suspended water droplets, usually less than about 0.10 mm in diameter, whose effect is significant only for systems operating above 10 GHz. It is important to note that attenuation to radiowaves depends on the liquid water content of the atmosphere along the propagation path. Clouds have liquid contents from 0.05 to 5 g/m^3 and their formation shape can be due to a variety of atmospheric processes, which result in cloud layer types for each of three cloud heights: low, middle and high. Thunderstorm cumulo-nimbus high clouds cause the maximum attenuation. The fog liquid content is in the order of about 0.40 g/m^3 and typically fog extends from 2 to 8 km. The attenuation from fog is negligible for satellite communications.

Figure 5.3. Equivalent Rain Cell of Rainfall Rate and Rain Attenuation Statistics

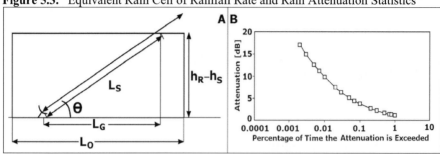

Courtesy of Handbook: "Radiowave Propagation Information for Predictions for Earth-to-Space Path
Communications" by ITU

2. Rain – Rain precipitation is the most significant contribution to atmospheric attenuation, which is caused by radiowave absorption and scattering from raindrops. First of all, to evaluate this additional rain loss, it is necessary to obtain the expected rainfall rate in mm/hr for the region of the communication link. For this reason, in **Figure 5.2. (B)** are presented six curves, which can be used to read off dB losses per EM path length at the operating frequencies. Namely, these curves are generated from a combination of empirical data and mathematical models that fit this data. Rainfall rate distribution is inhomogeneous in space and time and can impair a satellite communications link at frequencies above 10 GHz, as well as increase the noise temperature and impairment, or cross-polarization discrimination. Rain gauge records show short intervals of higher rain rates occurring in longer periods of lighter rain. Therefore, various precipitation methods have been proposed for predicting rain attenuation statistics calculations from rainfall rate measurements near the signal path. The slant path rain attenuation prediction is based on the estimation of the attenuation exceeded at 0.01% of the time $A_{0.01}$ from the rainfall rate $R_{0.01}$ (mm/h), exceeding at the same time percentage. The effective path length is given by the formula:

$$A_{0.01} = \gamma_R \cdot L_e = \gamma_R \cdot L_s \cdot r_{0.01} \quad [dB] \tag{5.4.}$$

Where γ_R (dB/km) = specific attenuation, L_e is the effective path length, L_s = slant path length and $r_{0.01}$ = path reduction factor. The spatial structure of rain can be modeled by an equivalent rain cell of uniform rainfall rate with a rectangular cross-section of equivalent length L_0 and effective height h_R-h_S in the plane of the path, see **Figure 5.3. (A)**.
The relation for slant path length for elevation angles above 5° is as follows:

$$L_s = h_R\text{-}h_S/\sin\theta \quad [km] \tag{5.5.}$$

and for elevation angles less than 5° is:

$$L_s = 2(h_R - h_S)/ \left[\sin^2\theta + 2(h_R\text{-}h_S)/R_e\right]^{\frac{1}{2}} + \sin\theta \quad [km] \tag{5.6.}$$

where θ = path elevation angle and R_e = effective radius of the Earth (8,500 km). Using simple trigonometry relations the horizontal projection is given by the formula:

$$L_G = L_s \cos\theta \quad [km] \tag{5.7.}$$

The reduction factor has the following relation:

$$r_{0.01} = 1/1 + L_G / L_0 \tag{5.8.}$$

For example, the specific attenuation using the frequency-dependent coefficients k and α gives the following expression:

$$\gamma_R = k(R_{0.01})^\alpha \quad [dB/km] \tag{5.9.}$$

Values of the frequency-dependent coefficients are shown in **Table 5.2.** Values not given there can be obtained by interpolation using logarithmic scales for frequency and k and a linear scale for α. For linear and circular polarization, these parameters can also be obtained by the following equations:

$$k = [k_h + k_v + (k_h - k_v) \cos^2\theta \cos 2\tau] / 2 \tag{5.10.}$$

$$\alpha = [k_h\alpha_h + k_v\alpha_v + (k_h\alpha_h + k_v\alpha_v) \cos^2\theta \cos 2\tau] / 2k$$

where θ = path elevation angle and τ = polarization tilt angle relative to the horizon (τ = 45° for circular polarization). The attenuation value predicted for a percentage in the range of 0.001 to 1% can be obtained by relation:

$$A_p = A_{0.01} \, 0,12 \cdot p^{-(0,546+0,043\log P)} \tag{5.11.}$$

For percentages of the time outside the 0.001 to 0.1% is a relatively complicated procedure, while the predicted attenuation exceeded for 0.01% for an average year is $A_{0.01} = 9.8$ dB. Finally, the estimated attenuation exceeded for any other percentages of the time P may be calculated from $A_{0.01}$ by the expression:

$$A_p = A_{0.01} (P/0.01)^{-\alpha} \quad [dB] \tag{5.12.}$$

The cumulative distribution, for example, of the rain attenuation statistics is illustrated in **Figure 5.3. (B)**.

3. Hail and Snow – Hail depolarization often precedes or follows rain and the presence of hail above rain is one factor that causes wide variations in the instantaneous value of Cross Polarization Discrimination (XPD), which may accompany a given attenuation. Hail, with its ice meteors, causes a small loss due to scattering but not due to absorption.

Dry snow likewise does not absorb radiowave energy, however, the water content of wet snow dropping slowly through the ray path causes more absorption than would arise if the same amount of water were falling as rain.

5.3.1.3. Sand and Dust Effects

The Earth's atmosphere contains a variety of solid particles in suspension, whose size range from aerosols of approximately 50 μm to coarse dust about 1 mm. On the other hand, if the size of particles is larger, they fall out of suspension more quickly but aerosols can remain aloft for days and sometimes can be carried high up into the stratosphere.

Otherwise, aerosol and small solid particles do not significantly affect the propagation of EM until the optical frequencies are approached and their extinction cross-section becomes appreciable. The effect of fine and coarse dust on the radiowave path can be noticeable at much lower frequencies and so in many cases, the impact is usually difficult to separate from the meteorological phenomena that often accompany the dust storms. Then, enhanced humidity or even rain that might accompany a severe convective activity in an arid region might cause the dust particles to attenuate more strongly than would be the case in a dry atmosphere due to water up-take into the crystal structure of the particles themselves.

In other cases, the strong winds that generate a dust storm may cause antenna depointing that will lead to a loss of radio signal strength, which is difficult to distinguish from the attenuating effect of dust particles.

An intensive study of the sand and dust particle effects on 4/6 and 11/14 GHz of satellite paths made a clear distinction between dust and sand. Namely, sand particles have limits from 10 to 100 μm and are usually restricted to within 10 m of the Earth's surface and in this case their impact on the satellite link due to radio propagation effects is usually insignificant at elevation angles above 5°. On the other hand, dust particle sizes are below 10 μm and can come up to 1,000 m in the atmosphere. Sand particles can absorb about 1% and dust particles can absorb 6 to 9% of air humidity. Thus, severe dust storms occur in eight desert areas with average wind speeds from 6 to 22 m/s, gust speed is from 15 to 50 m/s, width and length of affected area is maximum 8,000 km; dust can spread up to 12 km high in the atmosphere, expected visibility is from 0 to 1,000 m in a severe storm and the duration of events is from a few hours to 2 weeks.

The location and direction of the dust storms are in the vicinity of big desert areas, such as the southwest part of the USA, central and southern part of the South American Continent, northern and southern African Continent, around the Caspian Sea and Persian Gulf including central Asia and regions of the northwest and east coast of Australia. The main sources of heavy dust storms are big deserts such as the Sahara, the Kalahari etc, including nearby areas affected by Haboob storms. Namely, the Haboob storm is a direct result of a thunderstorm generated by a strong and cold wind, known as a cold dust storm. The severe turbulence and gusting in this storm area leads to very high suspended dust densities, which give rise to a significant reduction of optical visibility. For this reason, optical visibility has been used as a classification mechanism for dust storm effects. The principal criterion applied to defining the occurrence of a dust storm is when visibility drops below 1 km and when the duration of storm activities lasts from a few hours to 174 days in the various eight desert regions on Earth.

At this point, if parameters k_r and k_i are the real and imaginary parts of the complex relative dielectric constant of the dust particles, respectively, the specific attenuation (Flock, 1987) can be given by:

$$A_s = 189r/V\lambda \ [3k_i/(k_r + 2)^2 + k_i^2] \ \ [dB/km] \qquad (5.13.)$$

where r = radius [m]; V = visibility 2 m above the ground [km] and λ = wavelength [m].

Table 5.2. Specific Attenuation in dB/km for a Visibility of 10 m at the Given Frequency

Frequency:	1 GHz		3 GHz		10 GHz		30 GHz	
g H$_2$O/g soil:	0.30%	10%	0.30%	10%	0.30%	10%	0.30%	10%
A$_s$ [dB/km]	0.001	0.004	0.003	0.02	0.01	0.07	0.03	0.60

Table 5.2. gives typical results of attenuation predictions for nominated frequencies and where circular dust particles are assumed to have a uniform distribution. In any event, the specific attenuation is significant when Ka-band frequencies are employed on 20/30 GHz.

Even then, it is only in relatively humid air with about 10% water uptake that the path attenuation approaches 2 dB. The value of XPD for depolarization is almost benign until the visibility becomes very small, about 2 m, and the elevation angle is low. However, if the air becomes more humid and the dust particles absorb the moisture, the XPD values are worsened by about 5 dB. On some paths, for elevation angles below 10° and for some frequencies below 10 GHz, a severe dust storm that contains a large proportion of moist particles could cause the XPD to fall below 12 dB.

In general, however, for satellite communications link paths, the radiowave impairments caused by sand and dust storms are significantly less of a problem than the mechanical difficulties they may cause, such as severe wind gusts and dust in the antenna feeds and auxiliary equipment.

5.3.1.4. Site Diversity Factors

Path diversity in satellite systems involves the provision of alternate propagation paths for signal transmission, with the capability to select the least-impaired path when conditions warrant. In this case, implementation of path diversity requires the deployment of two or more interconnected Earth terminals at spatially separated sites, hence the use of the term "site diversity". This combination is considerable when an LES is being used for a commercially or militarily important purpose, where an outage is considered more serious, e.g., as the head end of a cable system, the probability of an outage can be reduced by having a second LES a few kilometres away and using whichever has the better signal at moment. Therefore, this is because heavy rain, in particular, tends to be quite localized and would probably not impact both stations at the same time.

5.3.1.5. Other Diversity Schemes

Without an ability to increase the margin (e.g.,, through the application of power control via additional transmit gain and power, or increased resource allocation in the TDMA frame), there are basically three types of diversity schemes that can be used by satellite systems to overcome impairments at a given LES.

1. Orbital Diversity – This diversity is different from site diversity in that only one LES site is used. To achieve a measure of diversity, the LES uses two antennas that can access different satellites simultaneously and is not necessarily the diversity interconnecting link between sites that is required for site diversity. However, to obtain significant decorrelation of concurrent attenuations along the two paths, the angle between the two paths at the LES must be large. Inasmuch as this angle is large, at least one or even both of the links will be at a relatively low elevation angle and therefore encounter a greater degree of impairment than at higher elevation angles. In general, the achievable diversity gain is very small, with values of about 2 to 3 dB in the 14/11 GHz bands.

2. Frequency Diversity – Path losses caused by particulates on the path increase as the frequency increases, particularly for rain. In effect, at 6/4 GHz, attenuation due to rain is negligible; at 14/11 GHz it can be significant in high rainfall regions of the world; at 30/20 GHz it is the dominant link impairment nearly everywhere. At this point, if it is possible to switch communication transmissions to a lower-frequency band, significant increases in availability might be achieved. This capability requires that both frequency bands (the higher, impaired one and the lower one to which the communication channels are to be switched) be simultaneously available at the LES in question. Furthermore, there must be spare capacity available in the lower frequency spectrum whenever needed, implying that significant spare capacity must be provided if the link is a high-capacity channel and that the complete network need to be under dynamic control. Namely, both elements require significant investment. Should such dynamic network control features be in place and the additional capacity in the lower frequency band be available on-call, frequency diversity can therefore undoubtedly provide large increases in availability.

3. Time Diversity – Severe rain events do not usually last long at a given location. Thus, this characteristic can be used in any communication RF link that does not require interaction between the caller and the receiver. The service can be said to be acceptable if, for example a Fax is successfully sent without any error within a two-hour period. The delay in sending the Fax can be considered a form of time diversity. This feature could also be used with advantage to determine the capacity requirement of a given RF link for optimal economic performance. Thus, if a link is sized for a maximum anticipated capacity it will have excess capacity for most of the time. If some transmissions can be delayed and sent, for example, at off-peak times, the capacity requirements can be reduced. The time delay could therefore be used either at times of peak capacity (i.e., the equivalent of call-blocking) or when the LES is undergoing a severe rain event.

5.3.2. Clear-Sky Effects on Atmospheric Propagation

Radio signals traveling through the atmosphere layers suffer attenuation even during fine weather. In any event, the clear-sky attenuation is mainly the result of absorption of energy from the transmission by water vapor and oxygen molecules; although there are other modes of clear-sky effect that have influences on propagation.

1. Defocusing and Wave-Front Incoherence Contribution – Several expressions have already been evaluated and are provided in No 2.3.2. of ITU-R P.618 recommendation to estimate the defocusing (beam-spreading) losses on paths at very low elevation angles. Otherwise, the loss is implicitly accounted for in the prediction methods for low-angle fading found in articles No 2.4.2. and 2.4.3. of the Recommendation. Hence, small-scale irregularities of the refractive index structure of the atmosphere cause incoherence in the wave front at the receiving antenna. In any case, this will result in both rapid radio signal fluctuations and an antenna-to-medium coupling loss that can be described as a decrease of the antenna gain. In practice, signal loss due to wave-front incoherence is therefore probably only significant for large-aperture antennas, high frequencies and elevation angles below 5°. Measurements made in Japan with 22 m antenna suggest that at 5° elevation angle the loss is about 0.2 to 0.4 dB at 6/4 GHz, while measurements with a 7 m antenna at 15.5 GHz and 31.6 GHz gave losses of 0.3 and 0.6 dB, respectively, at a 5° elevation angle.

2. Scintillation and Multipath Influence - Small-scale irregularities in the atmospheric refractive index cause rapid amplitude variations. Thus, tropospheric effects in the absence of precipitation are unlikely to produce serious fading in space telecommunication systems operating at a frequency below about 10 GHz and at elevation angles above 10°. Besides, at low elevation angles and at a frequency spectrum above about 10 GHz, tropospheric scintillations can, on occasion, cause serious degradations in performance. The atmosphere scintillation measure model includes frequency, elevation angle and antenna diameter, but also including meteorological parameters, can be used to account for regional and seasonal dependencies.

3. Propagation Delays – Additional propagation delays superimposed on the delay due to free space propagation are produced by refraction through the troposphere precipitation and the ionosphere. Therefore, at a frequency above 10 GHz, the ionospheric time delay is generally less than that for the troposphere.

4. Angle of Arrival Values – The gradient of the refractive index of the atmosphere causes a bending of the radio ray and the angle of arrival varies from that calculated based on the geometry of the path. Since the relative index varies largely with altitude, the angle-of-arrival variation is much greater in the elevation than in the azimuth angle. In addition, turbulent irregularities of the refractive index can give rise to angle-of-arrival scintillations. Both of these effects decrease markedly with elevation angle and are generally insignificant for elevation angles above 10°. The effects are independent of frequency.

5.3.3. Transionospheric Propagation

Radiowaves at frequencies of VHF and above are capable of penetrating the ionosphere and therefore, they provide transionospheric telecommunications. The ionosphere consists in a layer somewhere between 80 and 150 km altitude, where the density of the atmosphere is very low. Radiation from the Sun ionizes some molecules and it takes a long time for them to be neutralized by other ions. The concentration of ions varies with height, time of day, the season and in what part of its 11-year sunspot cycle the Sun happens to be.

5.3.3.1. Faraday Rotation and Group Delay

Ionospheric effects are significant for frequencies up to about 10 GHz and are particularly important for GEO and Non-GEO satellite constellations operating below 3 GHz. At the frequencies used for satellite transmission, signals pass right through and are subject to negligible refraction, less than 0.01° at 30° elevation.

The Total Electron Content (TEC) accumulated through the transionospheric transmission path results in the rotation of the linear polarization of the signal carrier and a time delay in addition to the anticipated propagation path delay. Given knowledge of the TEC, Faraday rotation and group delay can be estimated for communication applications. Therefore, this delay is known as the group delay, while the rotation of the linear polarization of the carrier is known as Faraday rotation. The TEC, denoted by N_T can be evaluated by the formula:

$$N_T = \int_s N_e(s)\, ds \quad [\text{electrons/m}^2] \qquad (5.14.)$$

where N_e = electron density [electrons/m²] and s = propagation path length through the ionosphere [m]. Typically, N_T varies from 1 to 200 TEC units (1 TEC unit = 10^{16} el/m²). N_T has typical values in the range of 10^{16} and 10^{18} el/m². Even when the precise propagation path is known, the elevation of N_T is difficult to determine because N_e is highly variable in space and time. When propagating through the ionosphere, a linearly polarized wave will suffer a gradual rotation of its plane of polarization due to the presence of the geomagnetic field and the anisotropy of the plasma medium. Namely, this trend slows down the signal because the Earth's magnetic field penetrates the ionosphere when ions (charged particles),

Figure 5.4. Faraday Rotation as a Function of TEC and RF

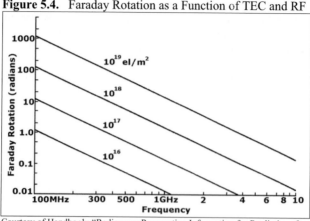

Courtesy of Handbook: "Radiowave Propagation Information for Predictions for
Earth-to-Space Path Communications" by ITU

subject to the alternating electric field of a signal, tend to gyrate around the local line of
force. The magnitude of Faraday rotation will depend on the frequency of the radiowave,
the geomagnetic field strength and the electronic density (concentration) of the plasma as:

$$\Phi = N_T \, (KM/f^2) = 2.36 \times 10^2 \, B_E N_T f^2 \quad [\text{radians}] \qquad (5.15.)$$

where $K = 2.36 \times 10^4$ [MKS units]; M = value of (B_E secϕ) at 420 km of height; B_E =
longitudinal component of the Earth's magnetic induction along the ray path [Tesla]; ϕ =
zenith angle of the ray and f = frequency [Hz]. Typical values of Φ as a function of
frequency for representative TEC values are shown in **Figure 5.4.** Hence, the occurrence of
Faraday is well understood and can be predicted with a high degree of accuracy and
compensated for by adjusting the polarization tilt angle at the LES. The GPS and similar
satellite navigation systems, which use the 1 to 2 GHz frequency spectrum and depend on
measuring the travel time of EM signals, has to correct for this effect.

The presence of charged particles in the ionosphere slows down the propagation of radio
signals along the path and produces a phase advance. Thus, the time delay in excess of the
propagation time in free space is called the group delay and is given by:

$$T_g = 1,34 \times 10^{-7} \, N_T/f^2 \quad [\text{sec}] \qquad (5.16.)$$

Accordingly, the time delay with reference to propagation in vacuum is an important factor
to be considered for digital communication and navigation positioning systems.

5.3.3.2. Ionospheric Scintillation

Ionospheric effects are important at frequencies below 1 GHz, although they may even be
important at frequencies above 1 GHz and are dependent on location, season, solar activity
(sunspots) and local time. Thus, ionospheric scintillation occurs as short-term, rapid signal
fluctuations and is mainly caused by irregularities in the ionosphere ranging from altitudes
of 200 to 600 km. In fact, the frequency-dependence depends on the ionospheric conditions

but the attenuation varies approximately at the same rate as the square of the wavelengths. The effect is greater for lower frequencies and at lower latitudes; while high latitude areas near the Arctic polar region bounded between ± 20° are susceptible to intense scintillation activity. In the L and S-band, this effect can be ignored at medium latitudes except during periods of solar activity. When the Sun is very active, L-band enhancement and fading of 6 dB and –36 dB, respectively, were observed even at 37° latitude. Scintillation activity is at a maximum during the night, lasting from 30 minutes to a number of hours.

5.3.3.3. Other Ionospheric Effects

1. Dispersion – When transionospheric radio signals occupy a significant bandwidth the propagation delay, being a function of frequency, introduces dispersion. The differential delay across the bandwidth is proportional to the integrated electron density along the ray path. Hence, for an integrated electron content of 5×10^{17} el/m^2, a signal with a pulse length of 1 μs will sustain a differential delay of 0.02 μs at 200 MHz, while at 600 MHz delay would be only 0.00074 μs.

2. Refraction – When radiowaves propagate obliquely through the ionosphere layer they undergo refraction, which produces a change in the direction of arrival of the ray.

3. Absorption – For equatorial and mid-latitude regions, radiowaves of frequencies above 70 MHz will assure penetration of the ionosphere without significant absorption, while for frequencies below 70 MHz the ionospheric absorption loss is significant.

4. Doppler Frequency Shift – The special effects of frequency change due to the temporal variability of the ionosphere layer upon the apparent frequency of the carrier (the Doppler shifted carrier). For example, at f = 1.6 GHz (GPS system), the observed frequency change Δf at high latitude is: $\Delta f/f < 10^{-9}$.

5.4. Sky Noise Temperature Contributions

The mechanism that causes absorption of energy from a wave passing between space and the Earth also causes the emission of thermal noise at RF. In fact, some radio noise is added to the emission reaching the receiver, whereas the Earth itself radiates noise, which can enter the transmission path via satellite or the LES receiving antennas. Therefore, the radio noise emitted by all matter, while used as a source of information in radio astronomy and remote sensing, may be a limiting factor in communication services. Otherwise, sources of radio noise of interest on Earth-to-space paths are the atmosphere, clouds, rain, extraterrestrial sources and noise from the surface of the Earth. Prediction methods are given in the ITU-R P.372 Recommendation. The thermal noise power N, available from a black-body having a noise temperature of the source T [K], measured in bandwidth B [Hz], is given by the form:

$$N = kTB \quad [W] \tag{5.17.}$$

where k = Boltzmann's Constant. The special power density N_0 of noise from source is:

$$N_0 = N/B = kT \quad [WHz^{-1}] \tag{5.18.}$$

Thus, in considering the level of noise received at an LES or satellite from sources external to the environment, it is convenient to identify a brightness temperature T_B for each separate source and a coefficient (η), which represents the efficiency, with which the receiving antenna captures noise from that source. Then t, the noise temperature component due to the identified source, is given by:

$$t = \eta T_B \quad [K] \tag{5.19.}$$

Therefore, the total noise entering the system from all of these sources, expressed as a noise temperature, can be obtained by summing all the component noise temperatures.

5.4.1. Environmental Noise Temperature Sources

The LES antennas in GEO satellite infrastructures are typically designed and sited so that the main lobe does not intersect the local terrain or obstructions, such as mountains or large buildings. Side lobes are also minimized to reduce the effect of the Earth's temperatures on the system performance. However, in LMSC systems, the antenna beam may pass through vegetation and be obstructed by buildings or mountain terrain. Measurements suggest that the impact of the additional terrestrial noise is greater when the antenna has a low internal noise temperature, that is, for less directive antennas. Although these obstructions will raise the noise temperature seen by the antenna, they will also cause shadowing or multipath effects, which are likely to be more significant in the total link performance.

Industrial man-made sources of noise affect VHF and UHF frequencies for all but the quietest rural areas. Unlike other noise effects, there is a polarization dependence in that the vertical component is higher that the horizontal. In general, the median level of noise will decrease linearly with log(f). There are significant variations with location and time and little data are available to develop models to predict levels.

Thus, considering the noise received at an LES from the ground, the sea etc. and buildings nearby, T_B typically lies between 100 and 250K, approaching the lower limit at sea at low angles of elevation angles and the upper limit on land. The relation for this noise temperature component t_{gr} is same as for t. If the angle of elevation of the main beam and the gain of the antenna are both high and the antenna design is good, with well-suppressed side lobes and little sub reflector spillover, the corresponding value of η will be small and t_{gr} may be no more than 20K. If, however, the gain of the antenna is very low, as is typical of LES, η may reach 0.5 and t_{gr} may exceed 100K.

On the other hand, satellite antennas directed towards the Earth, having sufficient gain for the main lobe to be filled by the Earth, will also receive terrestrial noise with $\eta \approx 1$, while T_B will be about 210K if land occupies a large fraction of the beam footprint. However, except in the atmospheric absorption bands, T will be somewhat less, perhaps as little as 160K, if the sea occupies a large part of the footprint.

5.4.2. Atmospheric Noise Temperature Elements

The noise temperature of a satellite-based antenna is dominated by the high temperature emitted by the Earth, which fills, or mostly fills, the main beam of the antenna. Additional noise from precipitation or other variables is insignificant in this case. For a global beam, the noise temperatures are dependent both on frequency and on the position of the satellite with relation to the major land masses of the Earth. The ground-based antenna observes the relatively cool sky and therefore, the presence of clouds and rain can significantly raise the noise temperature of the antenna. In general, the brightness temperature of the atmosphere due to the permanent gases and rain if it is present and seen by an antenna, is given as:

$$T_B = T_m (1 - 10^{-A/10}) \quad [K] \tag{5.20.}$$

where T_m = effective temperature of the attenuating medium (atmosphere, clouds, rain), typically about 270K and A = total attenuation due to the medium. The effect of rain on a satellite downlink is not just the attenuation but the decrease in C/N due to the higher noise temperature seen in rainy conditions, compared to clear sky conditions. In some cases, the noise temperature increase can have more effect on the link than the attenuation itself.

5.4.3. Galactic and Other Interplanetary Noise Effects

Noise from interplanetary sources, particularly the Sun, the Moon and from the galactic background, is well understood and the effect on the total extraterrestrial noise temperature of a system can be calculated with the following relation:

$$T_B = T_g x \, 10^{-A/10} \quad [K] \tag{5.21.}$$

where T_g = temperature of any interplanetary radio sources, including background Galactic noise (about 3K above 3 GHz). Thus, the brightness temperature of the Sun decreases with increasing frequency, from about 10^6 at 30 MHz to 10^4 at 10 GHz under quiet conditions. At 20 GHz, an antenna of 2 m diameter and a beam width of about $0.5°$ would have an increase in noise temperature of about 8.100K with a quiet Sun. The Sun and Moon each sub tend an angle of about $0.5°$, so that if the antenna beam is significantly larger than that, the effect t of the Sun or Moon is averaged with a larger portion of relatively cool sky.

5.5. Path Depolarization Causes

The atmosphere behaves as an anisotropic medium for radio propagation. Consequently power from one polarization is coupled to its orthogonal component, causing interference between the channels of a dual polarized system.

In this sense, depolarization or cross-polarization may occur when EM waves propagate through media that are anisotropic, namely asymmetrical with respect to the incident of polarization. Meanwhile, depolarization in the form of Faraday rotation of the plane of linear polarization occurs in the ionosphere because of presence of the Earth's magnetic fields. At this point, the resulting impairments are typically circumvented by using circular polarization at frequencies below 10 GHz, for which the effect can be significant. Depolarization is often the most significant path impairment for 6/4 GHz satellite systems and can be the limiting performance factor for some 14/11 GHz satellite paths, especially at lower path elevation angles in moderate rain climates.

On the other hand, depolarization in precipitation is caused by differential attenuation and phase shifts that are induced between orthogonal components of an incident wave by anisotropic hydrometeors. Orthogonally polarized radiowaves propagating in a medium that causes only differential phase shift are depolarized but maintain orthogonality. If the medium induces differential attenuation, the waves are also deorthogonalized.

Figure 5.5. Generalized Elliptical Waveform

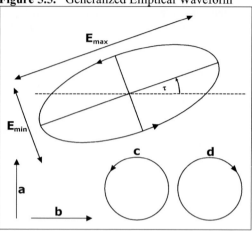

Courtesy of Book: "Mobile Satellite Communication Networks"
by R.E. Sheriff and Y.F. Hu

5.5.1. Depolarization and Polarization Components

The importance of depolarization for satellite communications systems depends on a few components: f = frequency signal, geometry of path (θ = elevation angle and τ = tilt angle of the received polarization), local climatic factors (severity of the rain) and sensitivity to cross-polar interference (whether the system employs frequency reuse).

The EM waves comprise both the electric and magnetic field vectors. Therefore, these two components travel in the direction of the transmission path and are orthogonal, while the orientation of the electric field vector defines the polarization of the transmitted waves. In general, as the wave progresses in time the tip of the electric vector traces an ellipse in a plane perpendicular to the propagation direction. A representative polarization ellipse of representative elliptically polarized radiowave is displayed in **Figure 5.5.** Two important parameters in this figure are the axial ratio and τ = the inclination tilted angle with respect to the reference axis. The polarization ellipse may be tilted at an angle (τ) with respect to the particular coordinate frame. Thus, the general form of a polarized wave, when viewed perpendicular to the direction of travel, is elliptical in shape.

The polarization state of a wave is completely specified by its polarization ellipse, i.e., the amplitudes of the major axis (E_{max}), the minor axis (E_{min}) and the sense of rotation of the vector also defines the axial ratio by using the following expression:

$$A_R = 20\log (E_{max}/ E_{min}) \quad [dB] \tag{5.22.}$$

In satellite communications four types of polarization are employed, shown in **Figure 5.5.**:
(a) Vertical Linear Polarization (VLP); **(b)** Horizontal Linear Polarization (HLP); **(c)** Left-Hand Circular Polarization (LHCP) and **(d)** Right-Hand Circular Polarization (RHCP). The direction of the travel is symbolical "into the paper". Horizontal and vertical polarizations are defined with respect to the horizon, LHCP has an anticlockwise and RHCP has a clockwise rotation when viewed from the antenna in the direction of travel.

Therefore, if E_{max} and E_{min} are equal in magnitude, the polarization state can be RHCP or LHCP, depending on the sense of rotation and in the case where E_{max} is nonzero and E_{min} is zero, the value of electric vector maintains a constant orientation defined by E_{max} and the polarization state is said to be linear.

The polarization quantity of interest for frequency re-use communication systems is the Cross-Polar Isolation (XPI), defined as the decibel ratio of the (desired) co-polar power received in a channel to the (undesired) cross-polar power received in that same channel. In practice, however, XPI is difficult to measure because the cross-polarized components cannot be distinguished from noise in the co-polar channel. The quantity usually measured is the Cross-Polar Discrimination (XPD), defined as the ratio of the co-polarized power received in one channel to the cross-polarized power detected in the orthogonal channel, both arising from the same transmitted signal. Moreover, theory predicts that XPD and XPI components are equivalent for most practical situations.

In fact, a polarized wave will comprise the wanted polarization together with some energy transmitted on the orthogonal polarization. The degree of the coupling of energy between polarizations is given by:

$$XPD = 20\log |E_{cpr}/E_{xpr}| \quad [dB] \qquad (5.23.)$$

where parameters E_{cpr} = received co-polarized electric field strength and E_{xpr} = received cross-polarized electric field strength. For the case of coexisting circular polarized waves the XPD can be determined from the axial ratio by the expression:

$$XPD = 20\log (A_R + 1/A_R - 1) \quad [dB] \qquad (5.24.)$$

5.5.2. Relation between Depolarization and Attenuation

The ITU provides a step-by-step method for the calculation of hydrometer-induced cross-polarization, which is valid for frequencies within the range 8 GHz < f < 35 GHz and for elevation angles less than 60°. In addition to operating frequency and elevation angle, the attenuation due to rain exceeded for the required percentage of time p and the polarization tilt angle τ, with respect to the horizontal, also needs to be known. The XPD due to rain is given by the following equation:

$$XPD_{rain} = C_f - C_A + C_\tau + C_\theta + C_\sigma \quad [dB] \qquad (5.25.)$$

where C_f = frequency-dependent term or 30 logf [GHz]; C_A = rain-dependent term or V(f) logA_p; C_τ = polarization improvement factor or $-10\log [1-0.484 (1+\cos4\tau)]$; C_θ = elevation angle-dependent term or $-10\log (\cos\theta)$ and C_σ = canting angle term or 0.0052σ. In the above, the canting angle refers to the angle at which a falling raindrop arrives at the Earth with respect to the local horizon. Terms τ, θ and σ are expressed in degrees, while σ has value of 0°, 5°, 10° and 15° for 1, 0.1, 0.01 and 0.001% of the time, respectively. Taking ice into account, the XPD not exceeded for p% of time is given by:

$$XPD_p = XPD_{rain} - C_{ice} \quad [dB] \qquad (5.26.)$$

where C_{ice} = ice depolarization or XPD_{rain} x (0.3+0.1 logp)/2.

5.6. Propagation Effects Important for GMSC

The MES operates in a dynamic and often unsuitable environment in which propagation conditions are constantly changing. Namely, satellite transmission path profile in GMSC varies continuously while the mobiles are in motion. The MES use relatively broad beam antenna systems, which have only a limited discrimination against signals reflected from scattered objects and surfaces. Because of the inherent random nature of disturbances, radio signals are usually characterized statistically. In general, a signal arriving at the antenna of a MES consists in the vector sum of a direct component and diffused components arising from multipath reflection. The resultant effects of these additional considerations are:

1) Signals suffer attenuation whenever the satellite path to mobiles is shadowed.

2) Signals fluctuate randomly because reflected and scattered random signal components arriving at the mobile antenna are picked up, which is known as the multipath.

3) The power spectral density of multipath noise is a function of the mobile's speed and the environmental conditions.

Depending on the environment in which a mobile operates, the satellite channels in GMSC may be categorized as maritime, land and aeronautically based. Each category has its typical channel characteristics because of the different propagation environments. Three different categories of MES: SES, VES and AES, with their own distinctive channels, need to be considered separately. Therefore, the local operation variable environment has a significant impact on the achievable quality of transmission service in GMSC.

On the contrary, the LES or Gateway terminals serving in a mobile satellite network can be optimally located in fixed positions with constant channel characteristics to guarantee good visibility to the satellite at all times, eliminating or reducing the negative and undesirable effect of the local environment to a minimum.

5.6.1. Propagation in MMSC

Maritime GMSC requirements have specific tasks and problems for shipping designers and operators. In fact, the modern ship is constructed of steel, which whilst floating on salt seawater becomes a very effective electromagnetic shield capable of rejecting and reflecting part of radiowaves. Meanwhile, modern ocean-going vessels have reduced or removed many very heavy traditional deck structures, which can affect transmission and reception. Hence, ships are sailing sometimes on very rough seas, which causes the motion and vibration of satellite antennas. In this case, the mobile satellite antenna goes quickly out of satellite focus and causes interrupted or intermediated transmission.

Afterwards, multipath fading caused by reflections from the sea surface can significantly impair MMSC channels, especially at the low elevation angles of an antenna system on board ships. Fading characteristics depend on the antenna gain of the SES, elevation angle, sea surface conditions (smooth or rough) and so on. A number of multipath fading reduction techniques, such as polarization shaping, antenna pattern shaping, multielement antenna systems and error-correction coding have to be considered.

Furthermore, MMSC systems may operate on a worldwide basis, including paths with low elevation angles. Due to the use of L-band frequencies for the Inmarsat system, the effect of ionospheric scintillation is not negligible, particularly in equatorial regions during years of high solar activity on an 11-year cycle. On the other hand, tropospheric effects, such as rain attenuation and scintillation, will be negligible for the said frequency bands. But signal level attenuation due to blocking by the ship's superstructure is a considerable problem, for which there are some adequate solutions and modifications.

In any event, the necessity of communications to ships via satellite is greater every year. Due to safety regulations of the IMO and GMDSS, communication channels to vessels need to be specified to a high degree of reliability. A typical voyage on board a ship comprises the following phases:

1. The ship, after sailing, usually uses coastal navigation to a convenient position on the open sea ready to take-over the shortest course or orthodrome towards the next port of call.

2. Celestial and satellite navigation in a constant open sea environment.

3. Going over to coastal navigation and maneuvering the ship at anchorage or in port.

4. The ship is moored in port under merchant operations.

Each of the above phases can be considered to have particular channel characteristics. For instance, while the ship is in the port, an LMSC channel type environment could be envisaged, where it is subject to sporadic shadowing due to buildings, cranes, other ships and obstacles. However, during coastal navigation ship's satellite antennas could be under the influence of ground refraction and diffuse reflected components. With regard to effects of multipath reflection from the sea, this will be explained later in this chapter.

Clearly, the MMSC network offers a very different propagation scenario to that in the LMSC case. Whereas with the LMSC channel, the modeling of the specular ground reflection is largely ignored, in the instance of MMSC reflections from the surface of the sea provide the major propagation impairment. At this point, such impairments are especially severe when using antennas of wide beam width, whilst operating at a low elevation angle to the satellite. Such a scenario is not untypical in the MMSC environment.

5.6.2. Propagation in LMSC

Land GMSC requirements have different problems to those for MMSC application because of a more specific local environment. Namely, when the direct satellite signal is obstructed by roadside trees, utility poles, buildings, tunnels, hilly terrain, mountains or other kinds of blockage, communication outages will occur. If the disturbance lasts for a long period of time, the communication will be disconnected. At all events, it is necessary to know the statistical characteristics of signal attenuation and its frequency as well as the duration of outages in given service areas, such as urban, suburban, rural, vegetation and mountain areas. However, in comparison to the modeling of the MMSC propagation channel, results for the LMSC channel are relatively few.

Radiowaves reflected from mountains, hills and artificial structures, such as buildings and bridges, cause interference to the direct signal from a satellite and cause amplitude and phase fluctuations in the received signal. Namely, such multipath fading may degrade the BER performance for digital modulation. The received LMSC signal is combined with three components: the direct line-of-sight wave, the diffuse wave and the specular ground reflection. The direct wave arrives at the receiver without reflection from the surrounding environment. The only L and S-band propagation impairments that significantly affect the direct component are free space loss and shadowing. The diffuse component comprises multipath reflected signals from the different nearby surrounding environment, such as buildings, mountains, trees and telegraph poles. Without consideration of LMSC, networks rely on multipath propagation, where multipath has only a minor effect on mobile satellite links in most practical operating environments. The specular ground components are results of the reflected signal reception from the ground near to the mobile. Antenna of low gain and wide beam width operating through satellites with low elevation angle are particularly susceptible to this form of impairment. In fact, such a scenario could also include GMPSC systems operating via LEO spacecraft. The first step towards designing an LMSC channel is to define and divide certain transmission environments into typical categories as follows:

1. Urban areas are characterized by almost complete obstruction of the direct wave.

2. Suburban, mountains and tree shadowing environments are affected when intermittent partial obstructions of the direct wave occur.

3. Open and rural areas are not affected by obstruction of the direct wave.

The last two of the above environments are of particular interest for LMSS. However, in urban environments, it is difficult to guarantee visibility to the satellite, resulting in the multipath component dominating reception. Currently, it is not very economical and practical to use LMSC system in urban areas because it is possible to employ landline infrastructure or dual cellular/GMPSC networks, such as Globalstar and Iridium.

In open and rural areas, the multipath phenomenon is the most dominant link impairment. The multipath components can either add constructively, with enhancement of signal, or destructively, with fade caused to the direct wave component. This results in the received MES transmissions being subject to significant fluctuations in signal power.

In suburban areas the major contribution to signal degradation is caused by buildings, trees and other man-made obstacles. These obstacles manifest as shadowing of the direct signals, which results in attenuation of the received signal. The MES motion through suburban areas results in continuous variation of the received signal strength and variation in the received phase.

5.6.3. Propagation in AMSC

Aeronautical GMSC requirements and propagation conditions are superior to those in MMSC and LMSC because there are no difficulties and obstacles between a satellite and AES. However, at low elevation angles and when a low-gain antenna is used, multipath fading caused by reflection from the sea or ground surface occurs, although it is less than in the maritime case. Otherwise, in an AMSC an aircraft demodulator must track the received signal and remove the Doppler effect due to the high speed of flight, using digital signal processing or using a pilot signal.

At any rate, the ability of communications to aircraft via satellite is becoming increasingly important. Due to the safety regulations of the ICAO, communication channels to an aircraft need to be specified to a high degree of reliability. A typical flight on board an aircraft comprises the following phases:

1. The aircraft taxies to a position on the runway, ready to take-off.

2. Take-off and ascension to cruise altitude at a constant height above the cloud layer.

3. Descent from cruise altitude to a landing on the runway.

4. Taxiing from the runway to a stand-by place in the airport for a new flight.

Each of the above phases can be considered to have particular channel characteristics. For example, while the aircraft is in the airport, an LMSC channel type environment could be subject to sporadic shadowing due to adjacent buildings, airport structures, other aircraft and close obstacles. In effect, the AMSC channel is further complicated by the maneuvers performed by an aircraft during the course of a flight, which could result in the aircraft's structure blocking the line-of-sight to the satellite. The body of the airplane is also a source of multipath reflection, which also needs to be considered. In fact, as was discussed earlier, the speed of an aircraft introduces large Doppler spreads as an additional effect.

The effect of multipath reflections from the sea for circular polarized L-band transmissions will be explained later, together with the maritime case. Thus, ITU provides a methodology similar to that for the maritime link, for determining the multipath power resulting from specular reflection from the sea. Thus, the ITU Recommendation includes a methodology to derive the mean multipath power as a function of elevation angle and mobile antenna gain. In this sense, by applying this method for an aircraft position 10 km above the sea and for a minimum elevation angle of $10°$, the relative multipath power will be in the range of approximately -10 to -17 dB, for antenna gains varying from 0 to 18 dBi, respectively.

Presently, AMSC systems are served by the L-band. Due to the bandwidth restrictions at this spectar, services are limited to voice and low data rate applications. Inmarsat recently improved speed of transmission with a new generation of satellite constellation. The need to provide broadband multimedia satellite services, akin to those envisaged by new satellite UMTS/IMT-2000, will require the move up in frequency to the next suitable bandwidth, the K and Ka-bands. At these frequencies, tropospheric effects will have an impact on link availability during the time when the aircraft is below the cloud layer. The channel characteristics for a K-band satellite system have been investigated in Europe and the USA. Thus, here a Rice-factor of 34 dB is reported for line-of-sight operation, while shadowing introduced by the aircraft's wing during a turning maneuver resulted in a fade of 15 dB.

5.6.4. Surface Reflection and Local Environmental Effects

Surface reflections and local environmental effects are important for GMSC because such factors generally tend to impair the performance of satellite communications links, although signal enhancements are also occasionally observed. Local environmental effects include shadowing and blockage from objects and vegetation near the MES.

At this point, surface reflections are generated either in the immediate vicinity of the MES terminals or from distant reflectors, such as mountains and large industrial infrastructures. The reflected transmission signal can interfere with the direct signal from the satellite to produce unacceptable levels of signal degradation. In addition to fading, signal degradations can include intersymbol interference, arising from delayed replicas.

The impact of the impairments depends on the specific application, namely in the case of typical LMSC links, all measurements and theoretical analysis indicate that the specular reflection component is usually negligible for path elevation angles above $20°$. Moreover, for handheld terminals, specular reflections may be important as the low antenna directivity increases the potential for significant specular reflection effects. For GMSC system links, design reflection multipath fading, in combination with possible shadowing and blockage of the direct signal from the satellite, is generally the dominant system impairment.

5.6.4.1. Reflection from the Earth's Surface

Prediction of the propagation impairments caused by reflections from the Earth's surface and from different objects (buildings, hills, vegetation) on the surface is difficult because the possible impairment scenarios are quite numerous, complex and often cannot be easily quantified. For example, the degree of shadowing in LMSC satellite links frequently cannot be precisely specified.

Therefore, impairment prediction models for some complicated situations, especially for LMSC links, tend to be primarily empirical, while more analytical models, such as those used to predict sea reflection fading, have restricted regions of applicability. Nevertheless, the basic features of surface reflections and the resultant effects on propagating signals can be understood in terms of the general theory of surface reflections, as summarized in the following classification:

1. Specular Reflection from a Plane Earth – Here, the specular reflection coefficient for vertical polarization is less than or equal to the coefficient for horizontal polarization. Thus the polarization of the reflected waves will be different from the polarization of the incident wave if the incident polarization is not purely horizontal or purely vertical. For example, a circularly polarized incident wave becomes elliptically polarized after reflection.

2. Specular Reflection from a Smooth Spherical Earth – Here, the incident grazing angle is equal to the angle of reflection. The amplitude of the reflected signal is equal to the amplitude of the incident signal multiplied by the modules of the reflection coefficient.

3. Divergence Factor – When rays are specularly reflected from a spherical surface, there is an effective reduction in the reflection coefficient, which is actually a geometrical effect arising from the divergence of the rays.

4. Reflection from Rough Surface – In many practical cases, the surface of the Earth is not smooth. Namely, when the surface is rough, the reflected signal has two components: one is a specular component, which is coherent with the incident signal, while the other is a diffuse component, which fluctuates in amplitude and phase with a Rayleigh distribution.

5. Total Reflected Field – The total field above a reflecting surface is a result of the direct field, the coherent specular component and the random diffuse component.

6. Reflection Multipath – Owing to the existence of surface reflection phenomena signals may arrive at a receiver from multiple apparent sources. Thus, the combination of the direct signal (line-of-sight) with specular and diffusely reflected waves causes signal fading at the receiver. The resultant multipath fading, in combination with varying levels of shadowing and blockage of the line-of-sight components, can cause the received signal power to fade severely and rapidly for MES and is really the dominant impairment in the GMSC service.

5.6.4.2. Fading in MMSC and AMSC Systems Due to Sea Surface Reflection

Multipath fading due to sea reflection is caused by interference between direct and reflected radiowaves. The reflected radiowaves are composed of coherent and incoherent components, namely specular and diffuse reflections, respectively, that fluctuate with time due to the motion of sea waves. The coherent component is predominant under calm sea conditions and at low elevation angles, whereas the incoherent becomes significant in rough sea conditions. If the intensity of the coherent component and the variance of the incoherent component are both known, the cumulative time distribution of the signal intensity can be determined by statistical consideration.

In any event, a prediction model for multipath fading due to sea reflection, however, was first developed for MMSC systems at a frequency near 1.5 GHz. Although the mechanism of sea reflection is common for MMSC and AMSC systems, only with the difference that fading characteristics for AMSC are expected to differ from those for MMSC, this is because the speed and altitude of aircraft so much greater than those of ships. At this point, the effects of refractions and scattering by the sea surface become quite severe in case of MMSC and AMSC, particularly where antennas with wide beam widths are used.

The most common parameter used to describe sea condition is the significant wave height (H), defined as the average value of the peak-to-trough heights of the highest one-third of all waves. Empirically, H is related to the r.m.s. height (h_o) by:

$$H = 4h_0 \qquad\qquad (5.27.)$$

Hence, at 1.5 GHz the smaller-scale waves can be neglected and the r.m.s. value of the sea surface slopes appear to fall between 0.04 to 0.07 in the case of wave heights less than 4 m. Thus, with diminishing satellite elevation angle, the propagation path increases, causing a decrease of signal power at the Rx side. The noise level is initially constant; however, upon reaching some critical value of the elevation angle, sea-reflection signals appear at the Rx input, which begins to affect the C/N value. To include the effect of multipath interference caused by sea-refracted signals, the reception quality would be more properly described by the C/N plus M, where M is an interfering sea-reflected signal acting as a disturbance. Thus, sea-reflected signals differ in structure and can be divided into two categories:
1. Radio signals with the rapid continuous fluctuations of amplitudes and phases and with a possible frequency shift due to the motion of small portions of the specular cross-section relative to the source of signals (noise or diffused components).
2. Radiowaves with relatively slowly changing phase close to the phase of the basic signal and with an amplitude correlating with that of the basic signal (specular component).
Consequently, within the overall specular cross-section, an angle of arrival reflected radio signals relative to the horizontal plane may be regarded as constant and can be described by the following expression:

$$\alpha = 90^{\circ} - \gamma \qquad\qquad (5.28.)$$

where α = angle of radio signals arrival in accordance with **Figure 5.6.** and γ = reflection angle. The modulus of sea reflection factor for L-band signals is within 0.8 and 0.9, which means that the amplitude of the specular reflected signal is nearly the same as that of the direct signal. As measurements have shown, the noise component depends only upon an elevation angle and a wave height. Decreasing the elevation angle and increasing the wave height result in an increase in the total amplitude of the noise, which includes the noise component. At elevation angles below 5°, the amplitude component reaches a peak value and is no longer affected by the wave height. Now an increase of the wave height causes primarily more frequent variations in the noise component. The corresponding deviation of C.N measured in 1 kHz bandwidth amounts from 4.5 to 5 dB.
The specular component that appears at the Rx input together with the direct signal causes fading in the direct signal due to both the minor difference between their phases and the slow change of the parameters of the reflected signals. The ratio of the direct to specular reflected signal can be described as:

$$C/M = (C + G_{\varepsilon}) - [C - G_{(\alpha + \varepsilon)}] \qquad\qquad (5.29.)$$

where C = direct signal; M = specular reflected signal power from the sea; G_{ε} = maximum gain of the receive SES antenna pointing towards the satellite; ε = elevation angle and α, as is shown in **Figure 5.6.** In addition, keeping accuracy sufficient for practical purposes, the previous relation gives:

Figure 5.6. Geometry of Sea Reflection of Satellite Radio Signals

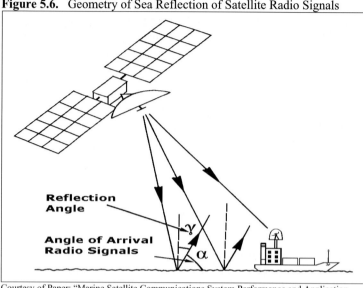

Courtesy of Paper: "Marine Satellite Communications System Performance and Application near the System Operational Boundaries" by V.I. Volodin and R.V. Zharnovetsky

$$C/M = \beta_{C/N} + [G_\varepsilon - G_{(\alpha + \varepsilon)}] \qquad\qquad (5.30.)$$

where $\beta_{C/N}$ = deviation of C/N ratio. With decreasing elevation angle, the C/M diminishes monotonically, except for the elevation angle range of 5° to 8°, within which a rise in C/N is observed. This is obviously due to the fact that at the said angles the difference in path between the direct and the specular signal becomes negligible, so that conditions appears close to the summation of the similar signals at the receiver input. An increase of the C/N plus M ratio is observed simultaneously due to reaching a peak value of amplitude in the noise (diffused) component. In fact, experimental measurements show that as the elevation angle decreases from 10° to 1°, the mean C/N plus M diminishes from 22–24 dB to 17–18 dB, with the deviation increasing from 1.5–2 dB to 4.5–5 dB.

5.6.4.3. Multipath Fading Calculation Model for Reflection from the Sea

The amplitude of the resultant signal at the SES terminal, being the sum of the direct wave component, the coherent and the incoherent reflection components, has a Nakagami-Rice distribution (see ITU R P.1057). The cumulative distribution of fading depends on the coherent-to-incoherent signal intensities. For example, in the case of rough sea conditions at 1.5 GHz, the coherent reflection from the sea is virtually non-existent and the coherent signal is composed only of the direct component. Therefore, the fading is determined by the Carrier-to-Multipath ratio (C/M), i.e., the power ratio of the direct signal and multipath component caused by incoherent reflection. The maximum fade depth (Φ_{max}) occurs when the coherent multipath signal is in anti-phase with the direct signal, given by:

$$\Phi_{max} = - 20 \log (1 - A_r) \quad [dB] \qquad\qquad (5.31.)$$

Figure 5.7. Estimates of Coherent Reflection and Multipath Power

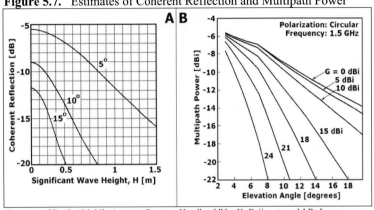

Courtesy of Book: "Mobile Antenna Systems Handbook" by K. Fujimoto and J.R. James

where A_r = amplitude of the coherently reflected component. The value decreases rapidly with increasing wave height, elevation angle and RF. In practice, due to the vertical motion of the ship antenna relative to average sea surface height, the maximum fade value will seldom occur. By adding Φ_{max} and $\Phi_i(p)$ as signal fade due to the incoherent component in the function of time percentage (p), a practical estimate of the combined fading effects of the coherent and incoherent multipath signal for sea conditions is obtained:

$$\Phi_c = \Phi_{max} + \Phi_i(p) \hspace{6cm} (5.32.)$$

The maximum fade value due to the coherent component will not occur constantly because of the vertical motion of the ship antenna relative to average sea surface height; therefore, the estimate using this equation seems to give the worst-case value. In practice, for low elevation angles (les than 10°) at around L-band frequencies, the maximum fading occurs when the significant sea wave height is between 1.5 and 3 m, where the coherent reflected component is negligible. Accordingly, the dependence of fading depth on wave height in this range is relatively small.

The amplitude level of the coherent component decreases rapidly with increasing sea wave height, elevation angle and frequency. **Figure 5.7. (A)** illustrates the relationship between coherent reflection and significant wave height. Namely, estimates of amplitude of the coherent component for an omnidirectional antenna as a function of a significant wave height for low elevation angles are illustrated; the frequency is 1.5 GHz and polarization is circular. Thus, the incoherent component is random in both amplitude and phase, since it originates from a large number of reflecting facets on the sea's waves. The amplitude of this component follows Rayleigh distribution and the phase has a uniform distribution.

Since the theoretical model concerning the incoherent components is not suitable for engineering computations using a small calculator, simpler prediction models are useful for the approximate calculation of fading. Such simple methods for predicting multipath power or fading depth have been recently developed. Thus, in **Figure 5.7. (B)** is presented the relationship between multipath power and elevation angle for different antenna gains. Although fading depth depends slightly on sea surface conditions, even if the incoherent is dominant, the simple model is useful for a rough estimate of fading depth.

Figure 5.8.   Estimates of Fading Depth and Spectral Bandwidth

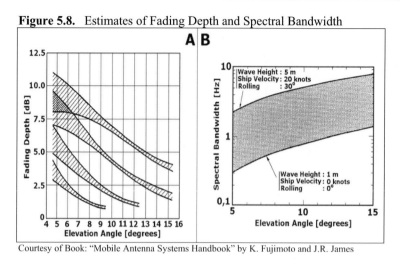

Courtesy of Book: "Mobile Antenna Systems Handbook" by K. Fujimoto and J.R. James

Fading depth, which is a scale of intensity of fading, is usually defined by the difference in decibels between the direct wave signal level and the signal level for 99% of the time. The fading depth can be approximated by a 50% to 99% value for fading where the incoherent component is fully developed. Large fading depths usually appear in rough sea conditions, where the incoherent component is dominant. Thus, **Figure 5.8. (A)** shows the fading depth estimated by the simple method for antenna not exceeding for 99% of the time and the corresponding C/M ratio for circular polarization at 1.5 GHz band under the condition of significant wave heights from 1.5 to 3 m. The antenna gains of 24, 20, 15 and 8 dB are functions of elevation angle with a fully-developed incoherent component. The calculation is based on the theoretical method, where the shaded area covers the practical range of the sea wave slope, which depends on fading depth in rough sea conditions. Values estimated by this simple method give the mean values of those given in **Figure 5.8. (B)**.

On the other hand, as the theoretical model is not suitable for engineering computations using a small calculator, these simple prediction models are really useful for the approximate calculation of fading or interference. Such simple methods for predicting multipath power or fading depth have been developed by Sandrin and Fang [1986] and by Karasawa and Shiokawa [1988] for MMSS and Karasawa [1990] for AMSS.

Furthermore, the frequency spectral bandwidth of temporal amplitude variations enlarges with increasing wave height and elevation angle. **Figure 5.8. (B)** shows the probable range of –10 dB spectral bandwidth (which is defined by the frequency corresponding to the spectral power density of –10 dB relative to the flat portion of power spectrum) of L-band multipath fading obtained by the theoretical fading model as a function of the elevation angle under the usual conditions of MMSC; namely, significant wave height of 1 m to 5 m, ship speed of 0 to 20 knots and rolling conditions of 0 to 30°.

5.6.4.4.   Other Estimations of Fading for MMSC and AMSC Systems

The error pattern in digital transmission systems affected by multipath fading is usually of the burst type. Accordingly, a firm understanding of the fade duration statistics of burst type fading is required. Mean value of fade duration (Φ_D) and fade occurrence interval (Φ_o)

Figure 5.9. Scattergram and Altitude Dependence of Fading Depth

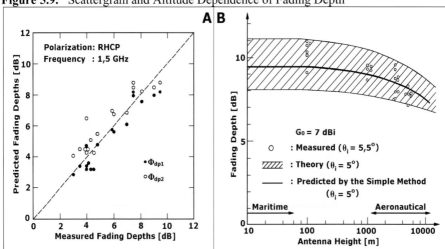

Courtesy of Handbook: "Radiowave Propagation Information for Predictions for Earth-to-Space Path Communications" by ITU

for a given threshold level as a function of time percentage, can be estimated from the fading spectrum. A simple method for predicting the mean value from the -10 dB spectral bandwidth is available as a theoretical fading model. Predicted values of (Φ_D) and (Φ_o) for 99% of the time at an elevation angle from 5 to $10°$ are 0.05 to 0.4 sec for (Φ_D) and 5 to 40 sec for (Φ_o). The probability density function of (Φ_D) and (Φ_o) at any percentages ranging from 50% to 99% approximates an exponential distribution.

1. Simple Prediction Method of Fading Depth – According to theoretical analysis and experimental results made by the mentioned researchers in Japan, the lowest elevation angle Earth-to-space path at 1.5 GHz RF band satisfies the energy conservation law: [Power of coherent component] + [Average power of incoherent component] ~ Constant. If this expression is satisfied, the maximum incoherent power can be estimated easily by calculating the coherent power at u = 0. Otherwise, for a more accurate estimation, small modifications of some parameter dependencies are necessary. The modified procedure has been adopted in P.680 for MMSC and P.682 ITU-R Recommendations for AMSC. **Figure 5.9. (A)** shows a scattergram of measured and predicted fading depths (i.e., fade for 99% of the time relative to that for 50%) in the case of MMSC systems between measured data and predicted values derived from the simple calculation method with the same conditions. In this figure, Φ_{dp2} are values from the method set out in ITU-R P.680, while Φ_{dp1} are those from an alternative procedure of the prediction method for scattering angles. It is evident that the values given by these methods agree well with the experimental values although the methods are rather approximate. In **Figure 5.9. (B)** is shown the altitude dependence of signal fade depth not exceeded for 99% of the time vs. antenna height on board ships or aircraft. This experiment was obtained from measurements with a helicopter together with the calculated values from the simple estimation method of the solid line and the theoretical model of the shaded region in the figure. From the figure it can be seen that the simple prediction method agrees well with both the theoretical model and measured data even in the case of the AMSC system.

Figure 5.10. Typical LMSC Propagation Environment

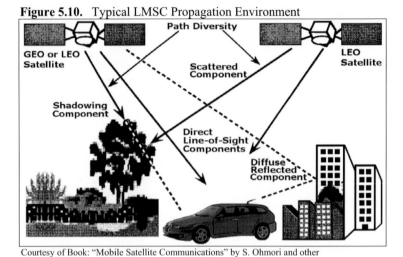

Courtesy of Book: "Mobile Satellite Communications" by S. Ohmori and other

2. Fading Spectrum – In system design, particularly for digital transmission systems, it is important not only to estimate the fading depth but also to know the properties of temporal variation, such as the frequency power spectrum. For MMSC systems, theoretical analyses were carried out in Japan and all parameters affecting the spectrum such as wave height, wave direction, ship's direction and velocity, path elevation angle and antenna height variations due to ship's motion (rolling and pitching) were taken into account. In general, spectrum bandwidth is broader with increasing wave height, elevation angle, ship velocity and the relative motion of the ship borne antenna. The dependence of the spectral shape on antenna polarization and gain is usually very small. Moreover, since the speed of aircraft is significantly higher than that of ships, the fluctuation speed of multipath fading in AMSC is much faster than that in MMSC, depending on the flight elevation angle measured from the horizontal plane. The calculated –10 dB spectral bandwidth is between 20 and 200 Hz for elevation angles of 5° to 20°, for flight elevation angles 0° to 5° at a speed of 1,000 km/h.

5.6.4.5. Fading in LMSC System due to Signal Blockage and Shadowing

Recently, in the USA, Canada, Australia, Japan and Europe, domestic LMSC and GMPSC services have started. The main purpose of these systems is to extend MSC voice services to rural/remote areas where terrestrial/cellular services are not provided. In a typical urban environment, line-of-sight for cellular systems sometimes is not available due to blockage by buildings and other structures, when a MES can receive many waves reflected from these structures and conduct communication link using these signals. The LMSC system can be expected to use the direct line-of-sight signal from a satellite because of the high elevation angles. When the line-of-sight is blocked by any obstacle, MSC is not available, but using path diversity the link is available because of overlapping of two or more signal from adjacent satellites. At present, path diversity from separate satellites is rarely used for GEO MSS, but Non-GEO MSS have an inherent capability to exploit diversity, because the number of satellites is large providing path diversity. At this point, **Figure 5.10.** presents a typical propagation environment for scattering, multipath fading, shadowing, diffusion, ect.

Figure 5.11. Loss due to Knife-Edge Diffraction

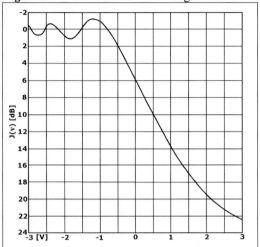

Courtesy of Handbook: "Radiowave Propagation Information for
Predictions for Earth-to-Space Path Communications" by ITU

Therefore, to design LMSC system, one needs information about the propagation statistics of multipath fading and shadowing. A vehicle runs at a distance of 5 to 20 m from roadside obstacles using an omnidirectional antenna, which has azimuthally uniform gain but elevation directivity, or a medium or high-gain antenna with automatic tracking capability. Thus, signal blockage and shadowing effects occur when an obstacle, such as roadside trees, overpasses, bridges, tunnels, utility poles, high buildings, hills or mountains, impedes visibility to the focus of satellite. This results in the attenuation of the received signal to such an extent that transmissions meeting a certain quality of service may not be possible. At any rate, in the shadowing environments the presence of the trees will result in the random attenuation of the strength of the direct path signal. Hence, the depth of the fade is dependent on a number of parameters including tree type, height, as well as season due to the leaf density on the trees. Whether a VES is transmitting on the left or right-hand side of the road could also have a bearing on the depth of the fade, due to the line-of-sight path length variation through the tree canopy being different for each side of the road. In fact, fades of up to 20 dB at the L-band may be presented due to shadowing caused by roadside trees. This shadowing by roadside trees cannot occur on modern highways because they are free of trees, only sometimes can shadowing appear by tunnels, very big constructions, bridges and mountains or hills in narrow passages.

1. Tree Shadowing – Attenuation due to trees nearby the roads arises from absorption by leaves and blockage by trunks and branches. Absorption by leaves is a function of the type and size of leaves and the water content therein. Blockage due to trunks is primarily a function of their size. In addition to attenuation of the direct signal, trees also cause an incoherent component due to signals reflected and diffracted off the tree surfaces.

The overall attenuation from different types of fully foliated trees varies from 10.6 to 14.3 dB and the attenuation coefficient is from 1.3 to 1.8 dB/m. Measurements were conducted with MES Rx in a rural environment. Based on these average values, a frequency scaling law for the attenuation coefficient has been derived [Goldhirsh and Vogel] by:

Figure 5.12. Geometry of Knife-edge Diffraction Phenomena

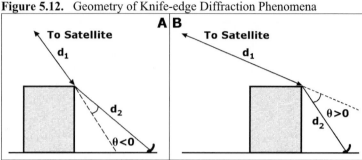

Courtesy of Handbook: "Radiowave Propagation Information for Predictions for Earth-to-Space
Path Communications" by ITU

$$a_1 = a_0 \sqrt{f_1/f_0} \quad [dB/m] \tag{5.33.}$$

where a_1 and a_0 = attenuation coefficients at frequencies f_1 and f_0 [GHz], respectively. Hence, the range of variation for a_1 at 1.5 GHz is from 0.5 to 1.7 dB/m. Moreover, trees without foliage attenuate less and the reduction in attenuation appears to be proportional to the total attenuation experienced when the tree is fully foliated. The received strength of the direct signal behind a tree will depend on the orientation of the signal path with respect to the tree. Therefore, the amount of absorbing matter lying along the path will determine the degree of attenuation and hence, on average, the length of the signal path through the tree shield can be considered a major factor in determining the signal level. The path length is a function of the elevation angle and the distance between the receiver and the tree. At this point, the average attenuation behind an isolated tree can be estimated as the product of the attenuation coefficient and the path length through the tree. Thus, the path length through the tree canopy will depend on its shape and the orientation of the signal path within the canopy. Depending on the type being considered, the tree canopy may be modeled as any one of the shapes. Otherwise, for the intermediate elevation angle (20° to 50°), attenuation is almost independent of elevation and dependence becomes important only at the higher and lower ends of the elevation angle range. By considering the path length variability as a statistical parameter, however, a tree can be modeled as giving an average attenuation and a distribution around it. Both the coherent and incoherent components will vary with the receiver position and complete decorrelation of the signal is expected over distances in the order of a few wavelengths.

2. Building Shadowing – Signal reception behind buildings takes place mainly through diffraction and reflection. A direct line-of-sight component does not usually exist and therefore shadowing cannot be defined unambiguously, as in the case of trees. However, shadowing may be loosely defined as the power ratio between the average signal levels to the unshadowed direct signal level. Otherwise, diffractions from buildings can be studied using knife-edge diffraction theory, which gives reasonable estimates. A concept view of knife-edge diffraction phenomena is shown in **Figure 5.11.** for all losses caused by the presence of the obstacles as a function of a dimensionless parameter v and in **Figure 5.12.**, which illustrates the geometry of the path for both the illuminated **(A)** and shadowed cases **(B)**, in order to calculate the parameter v, by using elevation and wavelength as follows:

$$v = \varepsilon \sqrt{2}/[\lambda(1/d_1 + 1/d_2)] \quad \text{but } d_1 \gg d_2, \text{ so:} \quad v = \varepsilon \sqrt{2}/\lambda \, (1/d_2) = \varepsilon \sqrt{2d_2}/\lambda \tag{5.34.}$$

The signal strength at the shadow boundary is 6 dB below the line-of-sight level. In the illuminated region, the signal fluctuations are experienced due to interference between the direct and the diffracted components. Hence, once inside the shadowed region, the shadow increases rapidly. An experimental investigation into building shadowing loss conducted by Yoshikawa and Kagohara in 1989 confirmed the applicability of the knife-edge diffraction theory. Measured signal strength behind a building at various distances was found to follow the prediction made, assuming a single diffraction edge. However, where the building is narrow compared with its height, there may be significantly less shadowing than predicted by the above procedure. When the direct signal path is blocked by a building, diffractions of the buildings are not expected to play a dominant role in establishing the communication link, unless MES is close to the shadow boundary. Reflections may play a useful role in such situations, as happens in cellular systems. Building penetration depends on the type of exterior material of the building and the location inside the building. Thus, the loss through the outer structure, known as the penetration loss, is defined as the difference in median signal levels between that measured immediately outside the building at 1.5 m above the ground and that immediately inside the buildings at some reference level on the floor of interest. Measurements made at 940 MHz in a medium-size city in the USA indicate that on the ground floor of typical steel-concrete-stone office buildings, the average penetration is about 10 dB with a standard deviation of about 7 dB. While another set of measurements in a large city resulted in average ground floor penetration loss of 18 dB with a standard deviation of 7.7 dB. However, the overall decrease of penetration loss with height was about 1.9 dB per floor. The biggest average attenuations are 12 dB and 7 dB and standard deviations are 4 db and 1 dB for metal and concrete, respectively. Attenuation through glass ranges from about 2 to 6 dB depending on the type of glass, i.e., plain glass produces less attenuation compared to tinted or coated glass, containing metallic components. Otherwise, the smallest average attenuation is through office furnishings, aluminum and wood/brick of about 1, 2 and 3 dB, respectively. Losses within a building are both of distance from the exterior wall blocking the signal path, as well as the interior layout. Measurements have resulted in an inverse distance power law coefficient ranging from 2 to 4. The forthcoming ICO system has conducted experiments with satellite-borne signals whose final target is to improve and even to eliminate building shadowing.

5.6.4.6. Fading in AMSC System due to Land Reflection

An experiment aboard a helicopter over land was carried out by receiving right-hand circularly polarized 1.5 GHz beacon signals from an IOR Marisat satellite at an elevation angle of 10°. Fading depths measured over plains such as paddy fields were fairly large (about 5 dB), nearly equal to that for sea reflection. However, fade depths measured over mountainous and urban areas were less than 2 dB. In the case of mountains, reflected waves are more likely to be shadowed or diffused by the mountains. As to urban areas, the shadowing and diffusing effects of reflected wave by buildings are also large. For this reason, the ground reflected multipath fading in these cases is not generally significant.
Measurements of Sea-Reflection Multipath Effect – A study of multipath propagation at 1.5 GHz was performed with KC-135 aircraft and the NASA ATS-6 satellite. Otherwise, the signal characteristics were measured with a two-element waveguide array in the aircraft noise radome, with 1 dB beam width of 20° in azimuth and 50° in elevation. Namely, data was collected over the ocean and over land at a normal aircraft altitude of 9.1 km and with

Figure 5.13. Ocean Mean-Square Scatter Coefficients vs. Elevation

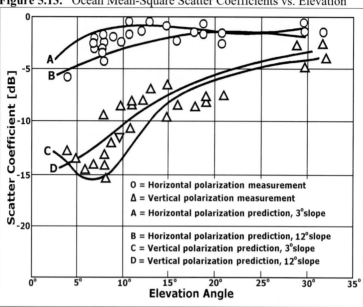

Courtesy of Handbook: "Radiowave Propagation Information for Predictions for Earth-to-Space
Path Communications" by ITU

a nominal speed of 740 km/h. Coefficients for horizontal and vertical antenna polarization
were measured in an ATS-6 experiment, where values for r.m.s. sea surface slopes of 3°
and 12° were plotted versus elevation angle, in **Figure 5.13.**, along with predictions derived
from a physical optics model. Sea slope was found to have a minor effect for elevation
angles above about 10°. The agreement between measured coefficients and those predicted
for a smooth flat Earth as modified by the spherical Earth divergence factor increased as
sea slope decreased. The relationship between r.m.s. sea surface and wave height is
complex but conversion can be performed. Namely, for most aeronautical systems, circular
polarization will be of greater interest than linear. For the simplified case of reflection from
a smooth Earth, which should be a good assumption for elevation angles above 10°,
circular co-polar and cross-polar scatter coefficients (S_c and S_x), respectively, can be
expressed in terms of the horizontal and vertical coefficients (S_h and S_v), respectively by:

$$S_c = (S_h + S_v)/2 \quad \text{and} \quad S_x + (S_h - S_v)/2 \tag{5.35.}$$

For either incident RHCP or LHCP is employed. Thus in general, the horizontal and
vertical coefficients are complex values and phase information is required to apply the last
equitations to the curve, in **Figure 5.13**.

5.6.5. Interference from Adjacent Satellite Systems

In GMSC systems for ships, vehicles and aircraft, small mobile antennas are essential for
operational and economic reasons. As a result, a number of low G/T value MES terminals

Figure 5.14. Basic Model for Intersatellite Interferences Phenomena

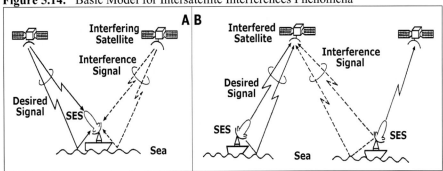

Courtesy of Handbook: "Radiowave Propagation Information for Predictions for Earth-to-Space Path Communications" by ITU

with smaller antennas have been developed. However, such antenna systems are subject to the restriction of frequency utilization efficiency, or coexistence between two or more satellite systems in the same frequency band and/or an overlap area where both satellites are visible. For coordination between two different systems in the same frequency band, a highly reliable interference evaluation model covering both interfering and interfered with conditions is required. Investigation into this area has been undertaken in particular by ITU-R Study Group 8. Advancement of such a model is an urgent matter for the ITU-R considering the number of MSC systems that are being developed in the meantime.

In GMSC systems, the desired signal from the satellite and the interfering signal from an adjacent satellite independently experience amplitude fluctuations due to multipath fading, necessitating a different treatment from that for fixed satellite systems. The main technical requirement is a formulation for the statistics of differential fading, which is the difference between the amplitude of the two satellite signals. At this point, the method given in No 5 of ITU-R P.680 Recommendation therefore presents a practical prediction method for signal-to-interference ratio where the effect of thermal noise and noise-like interference is taken into account; assuming that the amplitudes of both the desired and interference signal affected by the sea reflected multipath fading follow Nakagami–Rice distributions. In fact, this situation is quite probable in MMSC systems.

The basic assumptions of the intersatellite model are shown in **Figure 5.14.**, as an example of interference between adjacent satellite systems, where **(A)** is downlink interference on the MES side and **(B)** is uplink interference on the satellite side. This applies to multiple systems sharing the same frequency band. It is anticipated that the interference causes an especially severe problem when the interfering satellite is at a low elevation angle viewed from the ship presented in this figure because the maximum level of interference signal suffered from multipath fading increases with decreasing elevation angle. Another situation is interference between beams in multi-spot-beam operation, where the same frequency is repeatedly allocated.

5.6.6. Specific Local Environmental Influence in MMSC

Local environmental influence is important for SES equipped with beam width antenna. Many factors, with different kinds of noise sources, tend to make disturbances in MMSC

Figure 5.15.   Geometry of Blocking

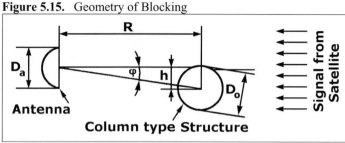

Courtesy of Handbook: "Radiowave Propagation Information for Predictions for
Earth-to-Space Path Communications" by ITU

channels. Another factor that affects communication links is RF emission from different
noise sources in the local environment. Specific local ship environmental factors can be
noise contributions from various sources in the vicinity of the SES and the influence of the
ship's superstructure in the operation of maritime mobile terminals. However, some of
these local environmental factors can affect SES when a ship is passing nearby the coast
and, some of these are permanent noise sources. More exactly, these environmental sources
include broadband noise sources, such as electrical equipment and motor vehicles and out-
of-band emission from powerful transmitters such as radars and ships HF transmitters.

5.6.6.1.   Noise Contribution of Local Ships' Environment

Some of the noise contributions from the local ships' environment are as follows:
1. Atmospheric Noise from Absorption – Absorbing atmospheric media, such as water
vapor, precipitation particles and oxygen emit thermal noise that can be described in terms
of antenna noise temperature. These effects were discussed at the beginning of this chapter.
2. Industrial Noise – Heavy electrical equipment tends to generate broadband noise that
can interfere with sensitive receivers. Therefore, a high percentage of this noise originates
as broadband impulsive noise from ignition circuits. Namely, the noise varies in magnitude
by as much as 20 dB, depending on whether it is measured on a normal working day or on
weekends and holidays when it is lower in magnitude.
3. Out of Band Emission from Radar – Ship borne and surveillance radars operating in
pulse mode can generate out of band emission that can interfere with SES receivers. In
general, such emissions can be suppressed by inserting waveguide or coaxial filters at the
radar transmitter output.
4. Interference from High Power Communication Transmitters – High power ships and
terrestrial transmitters, for example HF ship radio transceivers; HF radio diffusion and TV
broadcasting can interfere with SES.
5. Interference from Vehicles – Under certain operational conditions, RF emissions from
vehicles may impair Rx sensitivity. Accordingly, in one measurement the noise emanating
from heavy traffic has to be about –150dB (mW/Hz) within the frequency band 1.535 to
1.660 MHz.
6. Shipyard Noise – Extremely high peak amplitudes of noise of –141 dB (mW/Hz) were
recorded from Boston Navy Yard, which was in full operation at that time. Thus, this noise
is also a combination of city ambient noise and broadband electromagnetic noise from
industrial equipment.

Figure 5.16.   Estimated Attenuation due to Blocking

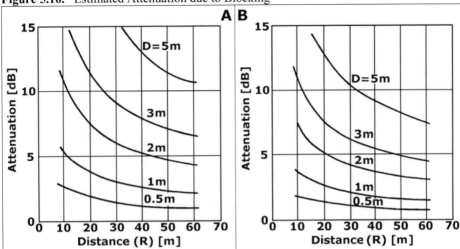

Courtesy of Handbook: "Radiowave Propagation Information for Predictions for Earth-to-Space Path
Communications" by ITU

5.6.6.2. Blockages Caused by Ship Superstructures

Ship's superstructures can produce both reflection multipath and blockage in the direction
of the satellite. For the most part, reflections from the ship's superstructure located on the
deck can be considered coherent with the direct signal. The fading depth due to these
reflections depends on a number of construction parameters including shape of the ship,
location of the ship's antenna, antenna directivity and sidelobe level, axial ratio and
orientation of the polarization ellipse, azimuth and elevation angles towards the satellite,
etc. Antenna gain has a significant influence on the fading depth. In this case, low gain
antennas with broader beam widths will collect more of the reflected radio signals,
producing deeper fades.

Blockage is caused by ship superstructures, such as the mast and various types of antennas
deployed on the ship. The geometry of blockage by a mast is presented in **Figure 5.15**.
Signal attenuation depends on several parameters including diameter of column, size of
antenna and distance between antenna and column. Accordingly, estimated attenuation due
to blocking by a column type structure is shown in **Figure 5.16**. for antenna gains of 20 dB
(A) and 14 dB **(B)**, respectively.

5.6.6.3. Motion of Ship's Antenna

The motion of mobile satellite antennas is an important consideration in the design of
MMSC systems. The received signal level is affected by the antenna off-beam gain because
the antenna motion is influenced by the ship's motion. The random ship motion must be
compensated by a suitable stabilizing mechanism to keep the antenna properly pointed
towards the satellite. This is normally achieved either through a passive gravity stabilized
platform or an active antenna tracking system. In either case, the residual antenna pointing
error can be significant enough to warrant its inclusion in the overall link calculation.

Figure 5.17. Measured Stabilized Antenna Motion

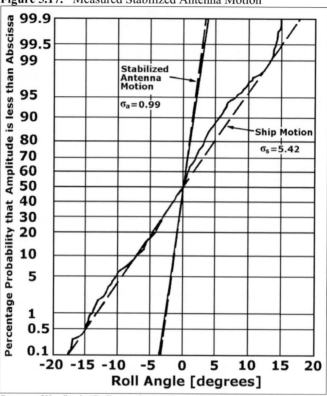

Courtesy of Handbook: "Radiowave Propagation Information for Predictions for
Earth-to-Space Path Communications" by ITU

Earlier experimental evidence suggests that the roll motion of a ship follows a zero mean
Gaussian distribution over the short-term of the sea waves. The standard deviation of the
distribution (σ_s) is a function of the vessel characteristics and the sea state of the wave
height. In **Figure 5.17.** is illustrated the distribution of the instantaneous roll angle of a ship
under moderate to rough sea conditions. The distribution of the ship motion approximates
to a Gaussian standard deviation of distribution with $\sigma_s = 5.42$ value.

Also shown in the figure is the distribution of roll angle of a passively stabilized antenna
under the same conditions, which also follows a zero mean Gaussian distribution with a
quantum of $\sigma_s = 0.99$. Solid curves in **Figure 5.17.** represent measured values and dashed
curves show calculated values for stabilized antenna motion over the sea conditions with
wave heights of approximately 5 m.

Otherwise, the relation between the standard deviations of the two distributions depends on
the design of the passive stabilizer. Although the ship's antenna motion is much reduced,
depending on the antenna beam width, the residual pointing error may be large enough to
produce appreciable signal fluctuations. Over a long period of sea waves time σ_s varies as a
function of the sea surface conditions and its distribution can be approximately by either a
log normal distribution or a Weibull distribution.

6

INMARSAT GEO GMSC SYSTEM

6.1. Inmarsat System and Structure

 Inmarsat Ltd is a 75 member-country internationally owned cooperative, which operates and maintains the Inmarsat Ground Network (IGN) constellation of nine GEO satellites and many Earth stations in mobile service (MES and LES). Founded in 1979, the company has over 22 years experience in designing, implementing and operating innovative satellite communication networks and MSC configurations. Inmarsat delivers its services through an IGN of approximately 260 partners in over 80 countries including some of the world's largest telecommunications companies. Thus, Inmarsat offers a portfolio of visionary MSS and FSS for governments and enterprises requiring reliable voice, data and video communication on land, at sea or in the air, with over 98% of the Earth's surface (excepting the polar regions). Inmarsat also provides connectivity in regions of developed and developing countries were there is no TTN service available.

Inmarsat was the world's first international and nongovernmental GMSC operator and is still the only one to offer a mature range of modern communications services to maritime, land, aeronautical and other mobile or semi-fixed users. Formed as a commercial shipping and fishing maritime-focused intergovernmental organization, Inmarsat has been a limited company since 1999, serving a broad range of markets. Starting initially with a user base of 900 ships in the early 1980s, it now supports more than 250,000 ships, vehicles, aircraft, portable and semi-fixed terminals and that number is growing at several thousand per month. Inmarsat is a subsidiary of the Inmarsat Ventures PLC holding company and has designed a new generation of GEO satellite constellations to enhance MSC all over the world. The satellites are controlled from Inmarsat's headquarters in London, which is also home to Inmarsat Ventures as well as the small IGO created to supervise the company's public service duties for the maritime community, known as GMDSS, implemented with IMO, and aviation ATC and CNS communications implemented with ICAO. The keystone of the strategy is the new Inmarsat I-4 satellite system, which from 2004 will support the Inmarsat Broadband Global Area Network (BGAN), mobile data communications at up to 432 Kb/s for Internet access, mobile multimedia and many other advanced applications.

6.2 Inmarsat Space Segment

For the first decade of Inmarsat's operation, the space segment has been leased from Comsat (series of three Marisat satellites F1, F2 and F3), from ESA (series of two Marecs satellites A and B2) and from Intelsat (series of three Intelsat V-MCS A, B and D). These satellites were initially configured in three ocean regions: AOR, IOR and POR, each with an operational satellite and at least one spare in-orbit. In fact, this satellite constellation was known as the first generation of the Inmarsat network, when Inmarsat was not responsible for TT&C. However, operations were controlled by Inmarsat NCC in London.

6.2.1. Current Satellite Constellation

Inmarsat's primary space segment consists in four Inmarsat-3 and Inmarsat-2 satellites in GEO constellation. Between them, however, the main global beam of the satellites provides overlapping coverage of the whole surface of the Earth apart from the Poles and in this way it is possible to extend the reach of terrestrial wired and cellular networks to almost anywhere on Earth. A call from an Inmarsat MES goes directly to the satellite overhead, which routes it back down to a Gateway on the ground called an LES. From there, the calls and messages are passed into the TTN public phone, data and ISDN networks.

The Inmarsat-3 satellites are backed up by a fifth Inmarsat-3 and four previous generations of Inmarsat-2, also in GEO. A key advantage of the Inmarsat-3 over their predecessors is their ability to generate a number of spot-beams as well as large global beams. Spot-beams concentrate extra power in areas of high demand as well as making it possible to supply standard services to smaller and simpler terminals. In **Table 6.1.** is shown a list of the Inmarsat spacecraft of 1998/99 deployed to cover the entire globe except the Poles.

Table 6.1. Ocean Regions and Satellite Longitude

Satellite Status	Atlantic Ocean Regions			
	Atlantic/AOR-W	Atlantic/AOR-E	Indian/IOR	Pacific/POR
1st Operational Position	Inmarsat-2 F4 54° W	Inmarsat-3 F2 15.5° W	Inmarsat-3 F1 63.9° E	Inmarsat-3 F3 178.1° E
2nd Operational Position	Inmarsat-3 F4 54° W	–	–	–
In-Orbit Spare Position	Inmarsat-2 F2 55° W	–	Inmarsat-2 F3 65° E	Inmarsat-2 F1 179° E

Figure 6.1. View of Inmarsat GEO Satellites

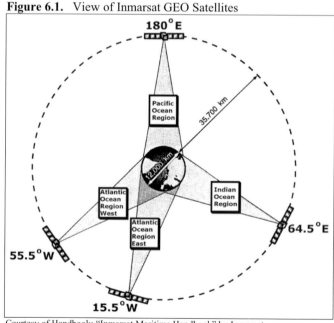

Courtesy of Handbook: "Inmarsat Maritime Handbook" by Inmarsat

Figure 6.2.   The Position of four Inmarsat Satellite Ocean Regions

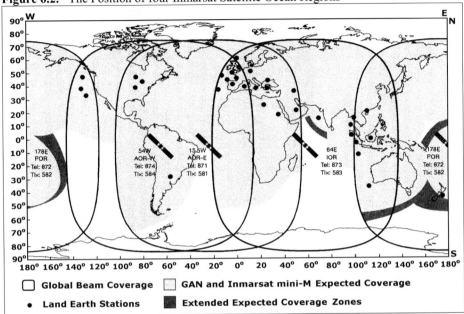

Courtesy of Prospect: "Inmarsat Coverage Map" by Inmarsat

Moreover, the current list of Inmarsat GEO satellite constellations is presented in the new Inmarsat Maritime Communications Handbook (Issue 4 of 2002) and gives the same 1st positions for operational satellites but in the event of a satellite failure the values change as follows:

AOR-W at 98° W; AOR-E at 25° E; POR at 179° E; IOR (For Inmarsat-A, B, C and M) at 109° E and IOR (For Inmarsat-C, mini-M and Fleet) at 25° E.

The Inmarsat organization bases its Earth coverage on a constellation of four prime GEO satellites covering four ocean regions with four overlapping, illustrated in **Figure 6.1.**

The coverage area for any satellite is defined as the area on the Earth's surface (sea, land or air), within which line-of-sight communication can be made with the satellite. That means, if an Inmarsat MES terminal is located anywhere within a particular satellite coverage area and an antenna of MES is directed towards that satellite, it will be possible to communicate via that satellite with any LES that is also pointed at the same particular satellite.

In **Figure 6.2.** is illustrated the footprints projected onto the surface of the Earth from the four Inmarsat GEO satellites currently in use. It should be noted that the recommended limit of latitudinal coverage is within the area between 75° North and South. However, as a large percentage of the Earth's MSC requirements lies within this roaming area, the system is considered to possess a global coverage pattern. As well as a "global" beam covering a complete hemisphere, each satellite generates up to seven spot beams designed to increase the amount of communications capacity available in areas of high demand. Namely, the MSC services are delivered to ships, land vehicles, aircraft, transportable and semi-fixed mobile terminals through the spot beams of the four Inmarsat 3 satellites. In **Figure 6.3.** is presented the spot-beam coverage of each Inmarsat-3 satellite for all ocean regions.

Figure 6.3. Inmarsat Spot-Beam Coverage

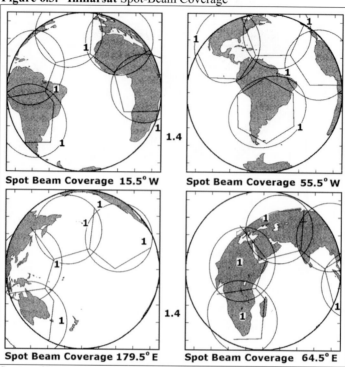

Courtesy of Prospect: "Inmarsat Coverage Map" by Inmarsat

In this sense, with its own GEO satellite constellation the Inmarsat network provides the GMSC service, which in Article-3 of the Inmarsat convention is stated as: "The purpose of Inmarsat is to make provision for the space segment necessary for improving mobile maritime and, as practicable, aeronautical and land communications and as well as MSC on waters not part of the marine environment, thereby assisting in improving GMSC for distress and safety of life, communications for ATS, the efficiency and management of transportation by sea, on land and in air and other mobile public correspondence services and radio determination capabilities".

In such a way, the present spot-beam coverage concentrates extra power in areas of high demand as well as making it possible to supply standard service to smaller, simpler and less powerful mobile terminals. At this point, a key advantage of the Inmarsat-3 satellites over their predecessors is their ability to generate a number of spot-beams as well as single large global beams.

6.2.1.1. Second Generation of Inmarsat-2 Satellites

Inmarsat operates a total of four 2nd generation Inmarsat-2 birds launched in 1990/92 with a capacity equivalent to about 250 Inmarsat-A voice circuits. These four Inmarsat-2 satellites were built to Inmarsat specifications by an international consortium headed by the Space and Communications division of British Aerospace (now Matra Marconi Space).

Figure 6.4. Inmarsat-2 and Inmarsat-3 Spacecraft

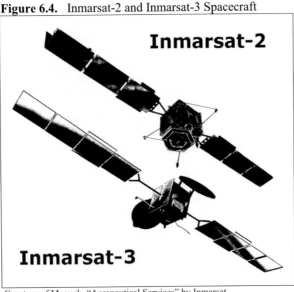

Courtesy of Manual: "Aeronautical Services" by Inmarsat

The Inmarsat-2 satellite design is based on the Eurostar three-axis-stabilized satellite platform with 10 years lifetime. At launch, each weighed 1,300 kg and had an initial in-orbit mass of 800 kg and 1,200 watts of available power. Each communications payload has two transponders, which provide outbound (C to L-band) and inbound (L to C-band) links with MES in the 6.4/1.5 and 1.6/3.6 GHz bands, respectively. The L-band EIRP is a minimum of 39 dBW, while G/T for L-band is about –6 dB/K for global coverage. Each satellite's global beam covers roughly one-third of the Earth's surface.

6.2.1.2. Third Generation of Inmarsat-3 Satellites

The Lockheed Martin Astro Space US-based company built the spacecraft bus, based on the GE Astro Space Series 4000, 2.5 m high and with a 3.2 radial envelope centered on a thrust cone. Matra Marconi Space built the communications payload, antenna systems, repeater and other communications electronics. Payload and solar arrays are mounted on N and S-facing panels, while L-band Rx and Tx reflectors, mounted on E and W panels, are fed by an array of cup-shaped elements. Furthermore, the navigation antenna is located on the Earth-facing panel.

The tremendous advantage of the Inmarsat-3 satellites is their ability to concentrate power on particular areas of high traffic within the footprint. Each satellite utilizes a maximum of seven spot beams and one global beam. Thus, the number of spot beams will be chosen according to traffic demands. In addition, these satellites can re-use portions of the L-band frequency for non-adjacent spot beams, effectively doubling the capacity of the satellite. Each satellite weighs about 2,066 kg at launch, compared to 1,300 kg for an Inmarsat-2 satellite, shown in **Figure 6.4**. Thus, the satellites produce up to 48 dBW of EIRP, a measure of how much signal strength a satellite can concentrate on its service area. All other parameters of Inmarsat-3 F1 satellite are presented in **Table 6.2**.

Table 6.2.   Orbital Parameters of Inmarsat-3 Spacecraft

Background Owner/Operator: Inmarsat Organization Present status: Operational Orbital location: 64° East Altitude: About 36,000 km Type of orbit: Inclined GEO Inclination angle: ±2.7° Number of satellites: 1 operational & 1 spare Number of spot beams: 5 Coverage: IOR Additional information: Other Ocean regions POR, AOR-W and AOR-E have 1 operational and 1 or 2 spare Inmarsat-2 or 3 satellites **Spacecraft** Name of satellite: Inmarsat-3 F1 Launch date: 4 April, 1996 Launch vehicle: Atlas IIA Typical users: Maritime, Land and Aeronautical Cost/Lease information: Nil Prime contractors: Lockheed Martin Other contractors: Matra Marconi Type of satellite: GE Astro Series 4000 Stabilization: 3-Axis Design lifetime: 13 years	Mass in orbit: 860 kg Launch weight: 2,066 kg Dimensions deployed: 2 x 7 x 20 m, Electric power: 2.8 kW SSPA power: C-band 1 @ 15 W; L-band 1 @ 490 W **Communications Payload** Frequency bands: a) Communications: L-band (Service Link) 1.6/1.5 GHz C-band (Feeder Link) 6.4/3.6 GHz b) Navigation: L1 1.5 & C-band 6.4/3.6 GHz Multiple access: TDM/TDMA Modulation: BPSK, O-QPSK, FEC Transponder type: L-C/C-L & L1-C-band Number of transponders: 1 L & C-band Channel bit rate: From 600 b/s to 24 Kb/s Channel capacity: About 2000 voice circuits Channel bandwidth: L-C/C-L 34 MHz; Navigation 2.2 MHz; L-L 1 MHz; C-C 9 MHz Channel polarization: L-band RHCP; C-band LHCP & RHCP EIRP: L-band Global 44 dBW & Spot 48 dBW; C-band 27.5 dBW G/T: L-band Global –6.5 & Spot –2.5 dB/K

The main mission spacecraft payloads of Inmarsat-3 are the communication transponders on both C and L-band and are the frequency-translated by the transponders for the downlink within the same band. The uplink signal is rebroadcast to users within ocean coverage and spot beam areas. As a secondary payload, Inmarsat-3 has the navigation transponders that provide the WAAS capability of Augmentation CNS capabilities. Two frequencies, L1 on 1.57542 GHz and C-band on 3.6 GHz, are used to allow correction of ionospheric delay. The WAAS signal will be broadcast to the users at L1 frequency. For additional integrity purposes and for checking the data received by the satellite, the data information being broadcast to mobile users is also down linked back to the control site in the C-band. In such a manner, the 6.4 GHz L-band repeater is power-limited to ensure that the navigation signal can never interfere with the GPS or GLONASS signals.

6.2.1.3. Fourth Generation of Inmarsat-4 Satellites

Responding to the growing demands from corporate mobile satellite users of high-speed Internet access and multimedia connectivity, Inmarsat is now building its fourth generation of satellites as a Gateway for the new mobile broadband network. Inmarsat has awarded European manufacturer Astrium a 700 million US$ contract to build three Inmarsat I-4 satellites, which will support the new Broadband Global Area Network (BGAN). The BGAN will be introduced in 2004 to deliver Internet and Intranet content and solutions, video on demand, videoconferencing, fax, E-mail, phone and LAN access at speeds up to 432 Kb/s worldwide and it will be compatible with third generation (3G) cellular systems.

Three Inmarsat I-4 satellites to be launched in 2004 will have the advanced technology to reduce service costs by 75%, compared to existing Inmarsat-M4 charges. They will be 100 times more powerful than the present generation and BGAN will provide at least 10 times as much capacity as today's network. The spacecraft will be built in the UK; the bus will be assembled in Stevenage and the payload in Portsmouth. The two sections will be united in France, together with the US-built antenna and German-built solar arrays.

6.2.2. Inmarsat MSC Link Budget

A link budget analysis forms the cornerstone of the space system design. Link budgets are performed in order to analyse the critical factors in the transmission chain and to optimise the performance characteristics, such as transmission power, bit rate and so on, in order to ensure that a given target quality of service can be achieved.

The sample a of maritime link budget is courtesy of Inmarsat for the link MES-to-GEO at 1.64 GHz and from the LES-to-GEO at 6.42 GHz and in the reverse direction for the link GEO-to-LES at 4.2 GHz and GEO-to-MES at 1.5 GHz is presented in **Table 6.3**.

Table 6.3. Maritime Mobile Link Budget

Parameter	MES-to-GEO	LES-to-GEO
MES/LES EIRP Carrier	36 dBW	58 dBW
Absorption & FSL at 1.6/6.42 GHz of 5° Elevation	189.4 dB	201.3 dB
Satellite Rx G/T	−13.0 dBK	−14.0 dBK
Uplink C/N_o	62.2 dBHz	17.3 dBHz
Total Satellite EIRP	16.0 dBW	33.0 dBW
Intermodulation Noise Power Ratio	15.0 dB	9.0 dB
Transponder Bandwidth (7.5 MHz)	68.8 dBHz	68.8 dBHz
Satellite EIRP/Carrier	−5.0 dBW	18.0 dBW
Satellite C/N_o	62.8 dBHz	62.8 dBHz
Parameter	**GEO-to-LES**	**GEO-to-MES**
GEO Satellite EIRP Carrier	−5.0 dBW	18.0 dBW
Atmospheric & FSL at 4.2/1.5 GHz of 5° Elevation	197.6 dB	188.9 dB
LES/SES Rx G/T	32.0 dBK	−4.0 dBK
Downlink C/N_o	58.0 dBHz	53.7 dBHz
Satellite Link C/N_o (Up/Intermodulation/Downlink)	55.7 dBHz	53.1dBHz
Intersystem Interference C/I_o	64.4 dBHz	61.8 dBHz
Overall C/N_o	55.2 dBHz	52.6 dBHz
Required C/N_o	52.5 dBHz	52.5 dBHz
Margin	2.7 dB	0.1 dB

The MES terminal is considered to be an Inmarsat standard-A installation, which G/T is −4 dB/K. The up and downlink budget for C/N_o are fairly standard except can be noted that: $C/N_o = EIRP − FSL − L_f + G/T − K$, where FSL = Free Space Loss; L_f = fixed losses made up of antenna misalignment of the receiver; and K = Boltzmann Constant in logarithmic form (−228.6 dBW/Hz/K). The intermodulation on board the spacecraft is given as follows: the total intermodulation noise in the 7.5 MHz (68.8 dBHz) transponder is 24 dBW. The subtraction 24 − 68.8 = −44.8 dBW/Hz (intermodulation noise density), and in such a way, the carrier to intermodulation noise $(C/N_o)_{IM}$ = 18 dBW −44.8 dBW/Hz = 62.8dB-Hz. The required (C/N_o) has to be 52.5 dB-Hz.

6.3. Inmarsat Ground Segment and Networks

The ground segment comprises a network of LES, which are managed by LES operators, Network Coordination Stations (NCS) and Network Operations Centre (NOC). However, the major part of the ground segment and network are mobile subscribers or MES. Each LES operator provides a transmission link between satellite network and TTN, capable of handling many types of calls to and from MES terminals simultaneously over the Inmarsat networks.

6.3.1. Inmarsat Mobile Earth Station (MES) Solutions

An MES is an RF device installed on board different mobiles, such as SES, VES, AES and TES, or it can be installed in a fixed location in maritime or land-based environments, indoor or outdoor, public payphones, on board mobiles and in suburban, rural and remote locations including SCADA applications. Inmarsat does not manufacture such equipment itself but permits manufacturers to produce models, which are type-approved to standards that have been set by Inmarsat and other international bodies, such as IMO, ICAO and the International Electrotechnical Commission (IEC). Therefore, only type-approved terminals are permitted to communicate via Inmarsat's space and ground segments. At this point, all types of MES provide different communication services in both mobile-to-ground and ground-to-mobile direction and inter-mobile communications.

The list of MES terminals, types of service and access codes and the countries in which they are registered are given in Inmarsat Operational Handbooks, in the Admiralty List of Radio and Satellite Services, in the ITU list of Ship Stations, and in SITA and ARINC list of Aircraft Stations.

6.3.2. Inmarsat Land Earth Stations (LES)

The LES terminal is a powerful land-based receiving and transmitting station serving in a GMSC system. Because LES infrastructure is fixed and, it can serve but cannot be part of FSS. In a more precise sense, every LES is a part of MSS network, although it has a fixed location and can provide FSS. Some LES, such as Goonhilly, provide widely fixed links for FSS and consequently, the Inmarsat MSC network is a small part of the overall LES providing service for Inmarsat-A, B/M, mini-M, D+ and Aero MES. Each LES in the IGN is owned and operated by an Inmarsat Signatory with the mission to provide a range of services to all types of MES. There are more than 40 LES terminals located in 30 countries around the globe but usually in the Northern Hemisphere, which is a small anomaly. The MES operator and shore subscribers can choose the most suitable LES, as long as they are within the same Ocean Region.

The fundamental requirement for each Inmarsat LES is that it be capable of communicating reliably with all MES terminals. There are two major types of LES: CES for maritime and land mobile applications, providing service to all standards of SES and VES, and GES for aeronautical mobile applications providing service to all standards of only AES terminals. The list of LES terminals, types of service and access codes, the countries in which they are based and Ocean Region of operation are given in Inmarsat Operational Handbooks, in the Admiralty List of Radio and Satellite Services, in the ITU list of Coast Stations, and in the SITA and ARINC documentations.

Figure 6.5. Typical Block Diagram of LES Terminal

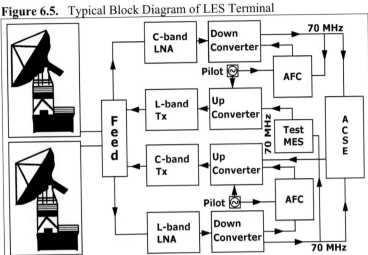

Courtesy of Book: "An Introduction to Satellite Communications" by D.I. Dalgleish

The technical side of typical mobile LES (CES and GES) consists in three main features: the antenna system (left), the communication RF equipment (between Feed and ACSE) and ACSE unit (right), as illustrated in **Figure 6.5.**

1. Antenna System – A typical LES antenna system for an entire IGN would be a Cassegrain structure with a dish reflector of about 14 m diameter. At this point, each LES can have a minimum of one operational and one spare antenna system in order to continue transmissions during maintenance. Some stations have more than two antennas, which depends on the Ocean Region covered and the services provided. In a more general sense, the antenna system operates in both the L and C-band to and from the satellite, with gain requirements of 50.5 dBi and 29.5 dBi, respectively. This antenna is designed to withstand high wind speeds up to 60 m/h in its operational attitude and 120 m/h when stowed at 90° and the parabolic dish is steerable ±135° in azimuth and 0 to 90° in elevation angle. Tracking is either by automatic program control or operator initiated. An antenna tracking accuracy of 0.01° r.m.s. and a repositioning velocity of 1o s^{-1} would be typical parameters for such a dish. RF and base band processing hardware design varies greatly with LES design and requirements. In the other words, a single antenna may be used to transmit and receive L-band as well as the C-band signals or the employment of a separate L-band antenna avoids the need for a relatively complex feed system (to combine and separate the outgoing and incoming L-band and C-band signals) but this advantage must be weighed against the cost of procuring and installing a second antenna.

2. Communication RF Equipment – The equipment is situated inside the LES building and must be able to operate in Tx and Rx L-band links to monitor the MSC L-band channel and respond to requests for frequency allocations by the NCS; to verify signal performance by loop testing between satellite and LES and to receive the C-to-L Automatic Frequency Control (AFC). The AFC provides Tx and Rx direction control, which helps to keep MES as simple and as cheap as is practicable. A complete test of the LES equipment can be carried out without the cooperation of the MES because a separate test terminal is provided at each LES for this purpose.

Figure 6.6. Multiplexing Six Channels onto a Single Satellite Access

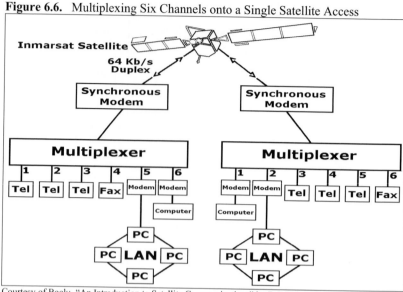

Courtesy of Book: "An Introduction to Satellite Communications" by D.I. Dalgleish

The Inmarsat system requires AFC to correct for Doppler shift (caused by inclination of the GEO) and errors in frequency translation in the satellite and LES. The total frequency shift from this cause without AFC could be more than 50 kHz and thus is of the same order as the spacing between the NBFM channels and would be enough to cause failure of the system. The AFC reduces the RF shift to a few hundred hertz by comparing pilot carriers transmitted via the satellite with reference oscillators at the LES and using the difference signals to control the RF of the local oscillator associated with the up and down converter. A pilot transmitted at C and received at L-band is used to control the up converter and thus offset the frequencies of the operational carriers to compensate for Doppler shift and satellite frequency translation errors in the ground-to-mobile direction. Similarly, a pilot transmitted at L and received at C-band controls the down converter and corrects for Doppler shift and errors in translation in the mobile-to-ground direction. All RF errors are corrected except those arising from the frequency instability of the MES up and down converter and Doppler shift resulting from the relative velocities of satellite and MES.

3. ACSE – The Antenna Control and Signaling Equipment (ACSE) is part of LES, whose principal purpose is to recognize requests for calls sent by MES, to set and release. In such a mode, this requires response to and initiation of in-band and out-band signaling over the satellite and the terrestrial path. The next ACSE tasks are to recognize distress calls (usually from SES – AES) and preempt channels for them when necessary; to check that MES are on the list of authorized users and to bar calls (except all distress calls) from or to unauthorized MES; to switch voice circuits between TTN circuit and the LES FM channel modem; to switch Tlx circuits between TTN channels and the TDM/TDMA time slots; to determine Tx/Rx frequencies used by the FM channel unit in accordance with the channel allocations made by the NCS; to allocate TDM and TDMA time slots and to collect statistics for billings, international accountings (for transit calls), traffic analyses and management and maintenance purposes, etc.

Figure 6.7. Configuration of CES for MMSC and LMSC

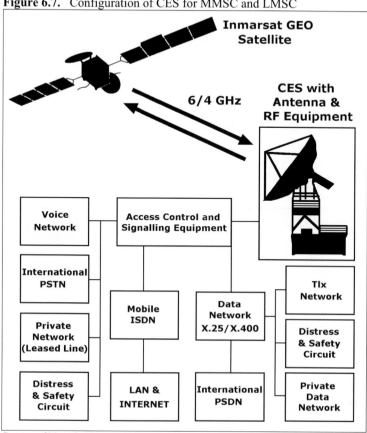

Courtesy of Manual: "Aeronautical Services" by Inmarsat

In this scenario, the ACSE block is sometimes specified to include all the communication equipment of LES terminal other than RF and IF equipment. For instance, this devices are such as: modulators and demodulators, data and voice channels, RF assignments to MES, line control subsystem, system control processor, etc.

The MSC services offered by an LES vary depending upon the complexity of the station selected. For example, a typical LES could offer a wide range of services from and to the MES located in convenient ocean regions, such as two-way voice including Fax/Paging, Tlx, all data rate, Video, GAN/Internet and Mobile Emergency services (Distress, Urgency, Safety and Medical assistance calls). Multiplexing as a number of communication channels onto a single satellite link becomes possible by using duplex HSD, such as multiplexing six communication channels onto a single satellite connection using the Inmarsat duplex HSD service, which is shown in **Figure 6.6**.

The term CES is included in the generic name LES, which applies to Earth stations used for either maritime or land-based MSC, which is illustrated in the configuration diagram **Figure 6.7**. At this point, there are numbers of Inmarsat CES terminals worldwide that can provide a communications service to SES, VES and TES standards.

Figure 6.8. Inmarsat-A/B Network for MMSC Application

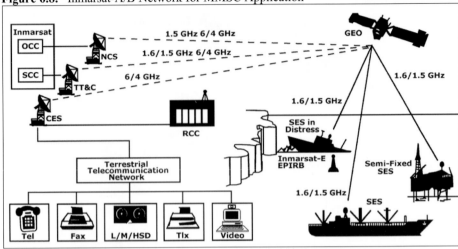

Courtesy of Manual: "Technical Summary of Inmarsat-M and B Standards" by Inmarsat

6.3.2.1. Inmarsat Coast Earth Stations (CES) for Maritime and Land Applications

1. CES-A – The CES-A standard has been the workhorse of MMSC since February 1982, when Inmarsat started to provide GMSC for shipping, fishery and offshore industries. It provides analog MSC two-way direct-deal phone, Fax, Tlx, E-mail, data at rates up to 9.6 Kb/s and HSD up to 64 Kb/s for SES and TES, which is presented in **Figure 6.8**.
2. CES-B/M – The Inmarsat CES-B/M supports both Inmarsat-B and M terminals. The Inmarsat-B digital system was introduced in 1994 to provide high quality voice, Fax, Tlx, E-mail and all rates data transmission, with MMSC network infrastructure identical to Inmarsat-A configuration, as is shown in **Figure 6.8**. The Inmarsat-M system entered into exploitation in 1993 to complement the existing Inmarsat-A service by providing global inexpensive voice/Fax and data communications, which is shown in **Figure 6.9**.
Inmarsat-B HSD terminals can make or receive data and voice calls on a direct dial basis. The connection is made via one of the IGN GEO satellites to an LES, which relays the call to and from the international TTN. There are more than 17 LES operators around the world capable of handling Inmarsat-B HSD. Transmissions at 64 Kb/s require connection via leased or switched ISDN links. Furthermore, the M4 service, including new Fleet and GAN, is supported by Eik CES in Norway and other large CES worldwide.
3. CES-mini-M – The Inmarsat mini-M unit was launched in January 1997 with the same service as Inmarsat standard-M but with a smaller, more lightweight and compact terminal because it operates only in the spot-beam coverage of Inmarsat-3 satellites, which is shown in **Figure 6.9**. Some mini-M CES can also support Aero mini-M.
4. CES-C – The CES-C terminals were introduced in 1991 to complement Inmarsat-A by providing global low-cost two-way data communications for all type of MES installed on board vessels, fishing boats, yachts, supply craft, land vehicles, small aircraft and remote TES for rural and SCADA services.
4. CES-D/D+ – The Inmarsat CES-D/D+ supports MES and SCADA very small terminals with one-way (D) and two-way (D+) data messaging.

Figure 6.9. Inmarsat-M and mini-M IGN for MMSC and LMSC

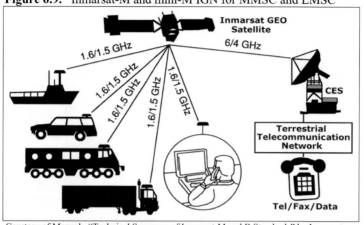

Courtesy of Manual: "Technical Summary of Inmarsat-M and B Standards" by Inmarsat

6.3.2.2. Inmarsat Ground Earth Stations (GES) for Aeronautical Application

The GES is a fixed satellite station in mobile aero service capable of communicating with aircraft via GEO satellite, see **Figure 6.10.** Thus usually, they consist in a dish transceiving antenna typically 10 m diameter, RF system and ACSE. The GES operating in the Inmarsat AMSC system is compatible with a wide variety of voice and data terminals. Digital voice coding and decoding units (codecs) are employed at the GES to convert ground-to-air voice signals into digital code for efficient error-free transmission. Otherwise, codecs are also used to translate air-to-ground digital code back into clearly intelligible toll quality speech, which is better than VHF quality. Data communications are supported through interfaces with public and private data networks. In this sense, the interfaces conform to CCITT recommendations X.25 and X.75, which define packet-data parameters and will support ISO-8208 compatible data communications. This means that the system can accommodate applications such as PC links between the aircraft and a computer or database TTN. There are many Inmarsat GES worldwide, which today operate and support all Aero standards including the new Aero-HSD and Swift64 AES. To obtain Aero services, users must contact and work through Inmarsat service providers and GES operators. Thus, a number of GES operators have formed consortia, such as ARINC, SITA, Avicom Japan, Satellite Aircom, Skyphone and Skyways Alliance, to offer an aeronautical service worldwide.

6.3.3. Inmarsat Ground Network (IGN)

A general overview of IGN is illustrated in **Figure 6.11.** This maritime configuration is applied to each of the four Inmarsat Ocean Regions for setting up MSC channels for a ship-to-shore call. The same scheme can be implemented for any other MES configuration of mobile-to-ground calls. Each MES has always to be tuned to the Common Signaling Channel (CSC), to listen for assignments Requesting Channel (RC), when not engaged in passing traffic, namely MES is an idle state, while each LES also watches the CSC to receive their channel assignments. The CSC is also referred to as TDM0 and is the origin of all traffic. The IGN is interfaced to the TTN as a Gateway to all fixed subscribers.

Figure 6.10. Configuration of GES for AMSC

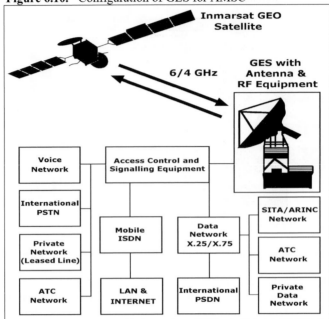

Courtesy of Manual: "Aeronautical Services" by Inmarsat

6.3.3.1. Network Coordination Stations (NCS)

The Inmarsat system uses four NCS, one in each Ocean Region separate for each standard, to monitor and control MSC traffic within the region. Usually some LES perform dual services and when required to be specifically identified, the LES serving as an NCS or Standby NCS will be referred as a collocated station. Hence, the NCS is involved in monitoring and control functions and in setting up calls between MES and LES, which is illustrated in **Figure 6.11.** The illustration shows in general terms how the NCS responds to a request from an MES (SES) for a communication channel, by assigning a channel to which both the MES and LES (CES) operator must tune for the call to proceed.

Therefore, an LES serving as the NCS or Standby NCS shall comply with all the technical requirements applicable to any Inmarsat standard LES and shall normally process its own calls in the same manner as a normal LES. In addition, the station serving as the NCS shall perform the following condition functions for the IGN as follows:

a) Transmits continuously on a special channel known as a CSC at 6 GHz L-band;
b) Accepts Tlg assignment MSG from all LES and re-broadcasts it to MES on the CSC;
c) Accepts Tel and HSD channel request-for-assignment messages from all LES in IGN and makes Tel and HSD channel assignment via the common TDM channel;
d) Maintains a Tel and HSD channel-activity list that indicates which channels are in use as well as the LES and MES using each channel;
e) Determines if an addressed MES is busy with another call;
f) Clears a telephone call in progress if necessary to service an SOS priority request; and
g) Maintains a record of RC, CSC and Tel channel used for IGN analysis purposes.

Figure 6.11. Inmarsat Ground Network

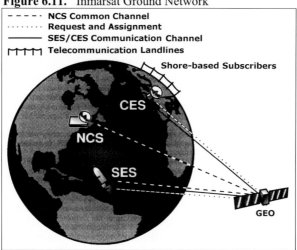

Courtesy of Manual: "Inmarsat-A Maritime User's Manual" by Inmarsat

Furthermore, NCS shall also change to the Alternative Common TDM RF of 6 GHz for Inmarsat satellite, Tx transponder load control carriers and facilitating measurements of RF signals at both C and L-band from the satellite. For these functions to be performed, an MES Rx must initially be synchronized to the NCS common channel and logged-in to the NCS for its Ocean Region, either automatically or manually at the MES.

6.3.3.2. Network Control Centre (NCC)

The Inmarsat NCC is located in the Inmarsat Headquarters building in London. It monitors, coordinates and controls the operational activities of all satellites (payload and antennas) and makes it possible to transfer operational information throughout the Network and via worldwide TTN routes, data between the NCC in each of the 4 Ocean Regions. The NCC station can send Inmarsat system messages via one or all of the NCS to inform the MES in their Ocean Regions of news relevant to any Inmarsat standard. It controls characteristics of the space segment throughout TT&C stations located in different countries; realizes all plans for new technical solutions and conducts development of the entire system; controls functions of current and newly introduced MES and LES and provides information about all MES, LES, NCS and the working condition of the entire Inmarsat system.

6.3.3.3. Satellite Control Centre (SCC)

Whereas the NCC is crucial to the MSC service management, the SCC located in London at Inmarsat House is crucial to spacecraft management and functions of station-keeping and TT&C. All data to and from the SCC is routed over worldwide TTN or tracking stations, which also provide backup capacity if required. The TT&C LES terminals are equipped with VHF, C and L-band for controlling spacecraft in all four Ocean Regions. Data on the status of the nine Inmarsat satellites is supplied to the SCC by four TT&C stations located at Fucino (Italy), Beijing (China), Lake Cowichan and Pennant Point in western and eastern

Canada and there is also a back-up station at Eik in Norway. Thus, this service provides TT&C, i.e., operational status of spacecraft subsystems and payload, such as transponder signals; decoders and converters; temperature of all equipment and surface; diagnostics on all electrical functions; satellite orientation in space; situation of attitude control fuel; telemetry of process decoder and Rx beacons and provides tracking and control of all parameters during launch of satellite.

6.3.3.4. Rescue Coordination Centres (RCC)

As the name implies, RCC are used to assist with SAR in distress situations for maritime and aeronautical applications. Namely, extensive MSC links provide end-to-end connection between the vessel or airplane in distress and competent rescue authorities. Because of the very high priority status accorded to distress alerts and the use of automatic signaling systems, this direct connection linking is rapidly established, usually within only a few seconds. Comprehensive MSC systems link an individual RCC and can have either LES, MCC or LUT. When an RCC receives an original distress alert (SOS or MAYDAY) via one of these stations, it will relay details of the alert to SAR units and to other ships (if the distress is at sea) within the general area of the reported distress. Hence this relayed message should provide the vessel or aircraft in distress with identification, its position and any other relevant information of practical use in rescue operations. The RCC, which initially receives a distress alert, appropriately called First RCC, assumes responsibility for all further coordination of subsequent SAR operations. However, this initial responsibility may be transferred to another RCC, which may be in a better position to coordinate rescue efforts. The RCC stations are also generally involved in subsequent SAR coordinating communications, such as between the designated On-scene Commander or Coordinating Surface Search (ship or helicopter), who are on board SAR units within the general area of the distress incident.

6.3.3.5. Terrestrial Telecommunications Network (TTN)

The TTN operators are usually Inmarsat Signatories and can be PTT or any government or private TTN providing landline public Tel and Tlx service. The TTN operators enable interface of IGN on their landline infrastructure for voice, Fax, Tlx, data (LSD, MSD and HSD) and video services, as illustrated in **Figure 6.8.** The newest Inmarsat HSD service enables connections to the TTN and other infrastructures, such as ISDN, PSDN, PSTN, Leased Lines, Data Network (X.25, X.75 and X.400), Private Data Networks, ATC Network, SITA/ARINC Networks and so on.

6.4. Inmarsat Mobile Earth Stations (MES)

The MES terminal is electronic equipment consisting in an antenna and transceiver with peripheral devices usually installed on board mobiles or sea-platforms and can be mounted in fixed indoor or outdoor sites of remote and rural areas. In particular, MES can contain only Rx or Tx terminals with mandatory or optional specific equipment in MSC service. All Inmarsat MES terminals use working Tx frequencies between 1626.5 and 1645.5 MHz and Rx frequencies between 1530.0 and 1545.0 MHz in four Inmarsat satellite regions, through about 40 LES located around the globe, as is shown in **Figure. 6.2.**

In **Table 6.4.** is presented a comparison table of parameters between SES and AES MES.

Table 6.4. Comparison of Parameters between Maritime and Aeronautical MES

Parameter	Standard				
	A	**B**	**M**	**C**	**Aero**
Digital Data Services	-	9.6/16 Kb/s	2.4 Kb/s	600 b/s	10.5 Kb/s
SCPC Assignment	NCS	NCS	NCS	NCS/CES	GES
TDM/TDMA	CES	CES	-	-	GES
SCPS RF Assignment	Paired	Unpaired	Unpaired	Unpaired	Unpaired
Spot-beam Identification	-	Yes	Yes	Yes	Yes
Access Channel	As Tlx	TDM-BPSC	TDM-BPSC	TDM-BPSC	TDM-BPSC
Return Request Channel	Aloha BPSC (BCH) 4800 b/s	Aloha O-QPSK (1/2-FEC) 24 Kb/s	Slotted Aloha BPSC (1/2-FEC) 3 Kb/s	Aloha BPSC (1/2-FEC) 600 b/s	Aloha BPSC (1/2-FEC) 0.6-24 Kb/s
Signal Unit Size	-	96 bits	96 bits	Variable	96 bits
MES G/T	–4 dBK	–4 dBK	–10/–12 dBK	–23 dBK	–13 dBK
MES EIRP	36 dBW	25-33 dBW	19-27 dBW	11-16 dBW	13.5-25.5 dBW
MES HPA Class-C (typical)	30-40 W	25 W	20 W	10 W	60 W
Tx/Rx Chains	2	1	1	1	1
Synthesizer Step Size	25 kHz	10 KHz	5 kHz	5 kHz	17.5 kHz
Voice Channel Coding/-Modulation	NBFM - 2:1 Companding	16 Kb/s APC O-QPSK	4.8 Kb/s IMBE code O-QPSK	No Voice	9.6 Kb/s APC O-QPSK
Voice Channel FEC - Rate	-	3/4-24 Kb/s	3/4-8Kb/s	No Voice	1/2-21 Kb/s
Voice Chan. Bandwidth-Rate	50kHz-9.6KB/s	20kHz-2.4Kb/s	10 kHz	No Voice	17kHz-0.3Kb/s
Tlx Chan.Modulation/Coding (Forward-Return)	TDM BPSK - TDMA BPSK	TDM BPSK - TDMA BPSK (1/2-FEC)	-	BPSK - BPSK (1/2-FEC)	-
Tlx Chan. Bandwidth/Rate (Forward-Return)	25kHz-1200b/s 25kHz-4800b/s	10kHz-6Kb/s 20kHz-24Kb/s	-	2.5kHz-600b/s 2.5kHz-600b/s	-
Tlx Capacity per Carrier	22	56	-	Variable	-
Satellite L-band EIRP	17.5 dBW	16 dBW	19 dBW	-	23 dBW

6.4.1. Inmarsat Maritime Mobile Ship Earth Station (SES)

Inmarsat started operations with only Inmarsat-A maritime service and for that reason devised a synonym: INternational MARitime SATellite (INMARSAT) organization. The SES terminal is electronic equipment consisting in antenna and transceiver with peripheral devices usually installed on board ships or sea-platforms. Later, Inmarsat developed other SES standards, shown in **Figure 6.12.** with mandatory and obligatory equipment.

Figure 6.12. Maritime SES Standard Configurations

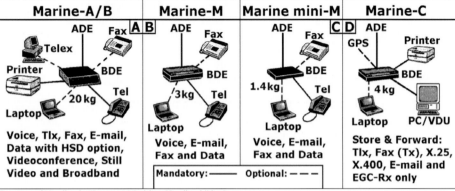

Courtesy of Handbook: "Inmarsat Maritime Handbook" by Inmarsat

Figure 6.13. Maritime ADU and BDU Configuration

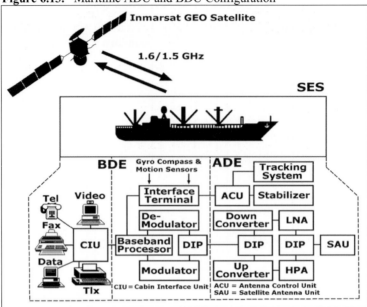

Courtesy of Handbook: "Inmarsat Maritime Handbook" by Inmarsat

In **Figure 6.13.** are presented ADE and BDE general block diagrams of electronic units for SES maritime terminals.

6.4.1.1. SES-A

The Maritime Inmarsat-A is the first analog standard for SES and TES applications and is more than 20 years old. Even though it is almost out of production, this does not mean that the infrastructure is not capable of fulfilling modern needs well into the 21st century. On the contrary, the service of this standard will be supported until at least the years 2005/2006. The Inmarsat standard-A SES provides two-way direct-dial phone (high quality voice), Fax, Tlx, E-mail and different data rates, shown in **Figure 6.12. (A)**.
This standard supports data of 9.6 Kb/s and 64 Kb/s HSD depending upon the different elements of end-to-end connection. These applications include the simultaneous transfer of two-way HSD applications, such as high-speed file transfer, broadband, transmission of still and compressed video pictures, high-quality 15 kHz audio and videoconferencing.
This SES can interface with a range of options and value added services, such as data modems, PABX that route voice, Fax and data calls from the bridge, radio room and crew cabin; that can support LAN and Internet facilities via a server that interfaces with MSD and HSD ports for sophisticated remote office operations and provides a variety of cordless DECT, encryption and other middleware, which can be deployed for specific benefits.
The MSD transfer requires modem, PC and communications software via RJ45 interface; while HSD in place of the modem needs a digital encoder and modulator via the PC including CCITT V.35, RS-422 and the ISDN-type CCITT V.36 and for Fax data it is necessary to connect a Fax machine that has been recommended by the SES manufacturer.

Figure 6.14. Maritime SES Standard-B Configuration

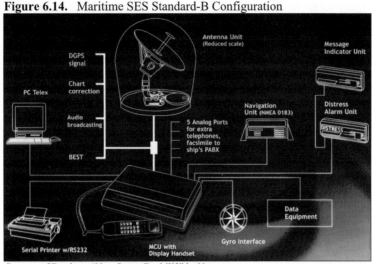

Courtesy of Brochure: "Nera Saturn Bm MK2" by Nera

6.4.1.2. SES-B and B/HSD

The Inmarsat-B SES new digital standard is seen as the successor to the highly successful Inmarsat-A analog system. Compared with Inmarsat-A, this standard makes improved use of satellite power and bandwidth, enabling lower charges, high quality and reliability. A basic Inmarsat-B SES can provide all the communications of a well-equipped mobile office with services, such as direct-dial high quality voice, Group 3 Fax, Tlx, 9.6 Kb/s data and 56/64 Kb/s HSD facilities, as is shown in **Figure 6.12. (A).** Enhanced and modified SES terminals are available from some manufacturers for fixed, multiple-channels and other special applications, such as compressed or delayed video transmission using HSD and special modems, such as the model Nera Saturn Bm MK2 SES, shown in **Figure 6.14.** It interfaces LAN with the communications security of an encryption service, provides broadcast facilities, electronic chart corrections and Lease Gateway CN17 including Bandwidth Efficient Satellite Transport (BEST) optional service. The ship's GPS can be connected to the NMEA 0183-port at the CIU (main control unit) to download Differential GPS (DGPS) transmissions from the Inmarsat-B system and to provide navigation information even if is out of range of land-based DGPS stations. Furthermore, it provides ship engine monitoring, important access to medical and maritime assistance, navigation/WX hazard warnings and meteorological forecast transmissions.

The Inmarsat-B/HSD services are suitable for applications such as high-speed file transfer, store-and-forward video, high-quality PC-based videoconferencing and audio transmission, broadband networking and multiplexed channels combining voice, Fax and data. This MES service is interfaced to worldwide TTN subscribers with shore-based LAN via ISDN, as well as dedicated audio transmission circuits for broadcasters.

The IMO has certified Inmarsat-A/B as satisfying the requirements for its GMDSS mission, giving safety coverage for virtually all of the world's navigable ocean waters. A call from an Inmarsat-B is routed via the Inmarsat-3 satellite global or spot-beams to a CES in each ocean region to the TTN and RCC distress and safety mission.

Figure 6.15. Maritime SES Capsat GMDSS Dual Mode Configuration

Courtesy of Brochure: "Maritime Systems" by Thrane & Thrane

6.4.1.3. SES-M

The Inmarsat-M system is designed to provide cost effective digital MSC for SES, VES and TES. It is the world's first portable MSC phone making possible voice (4.8 Kb/s), Group 3 Fax and data calls (2.4 Kb/s) from briefcase-sized terminals. The Inmarsat-M SES provides, in real-time mode, voice, data (X.25 and X.400), Internet and E-mail. For MSD transmission this standard has an internal data modem, which is built into the electronic unit. This system provides the group calling: simultaneous transmission of a message to a certain group of customers or according to their geographical location but it does not meet the GMDSS requirements. The Inmarsat-M set may include a 70 cm phased array antenna with a radome, main electronic unit, handset with a display, standard PC, printer, distress button and power supply unit, as presented in **Figure 6.12. (B)**. The whole control over the MES is performed by the software by choosing commands from the menu displayed on the handset or the main electronic unit and PC, while entry of commands and text is carried out via the handset or PC keyboard.

6.4.1.4. SES Mini-M

With 100,000 registered units worldwide and sales doubling in the last two years, mini-M has consolidated its position as the world's biggest selling mobile satellite phone. It offers portable, maritime, land vehicles, corporate jets and recently, Inmarsat launched payphone versions. It provides Advanced Multi-band Excitation (AMBE) voice, Fax, data and E-mail services, using a voice coding rate of 4.8 Kb/s including error detection/correction.

The mini-M SES offers global and spot-beam coverage of the Inmarsat-3 satellites, as is shown in **Figure 6.12. (C)**. The unit can work on batteries or via AC/DC adapter/charger, which is very important for mobiles. The Thrane & Thrane Capsat GMDSS SES dual mode Inmarsat-C/mini-M solution is illustrated in **Figure 6.15.** This integrated SES configuration complies with an Inmarsat CN 114 specification and SOLAS/GMDSS requirements and is one of the best solutions for integrated commercial and distress combinations. In the event of a distress alert, excepting Standard-C alert via Tlx, the system can also initiate a phone call to an RCC, in order to optimize the efficiency of the SAR operation. In addition, the Capsat dual mode solution can optionally be connected to a PC, Remote Alarm/EGC Printer, Fax machine, Integrated GPS, NMEA port, Black box, etc. A Dual Mode cradle with handset serves for mini-M Tel (voice) and relevant Inmarsat-C, indication including a flashing distress indicator and audio alarm to ensure that the SOS is both seen and heard. The Remote Alarm/EGC printer is using to initiate a distress alert when the red button is pressed, while it can also obtain printouts of all incoming messages.

The latest, most compact mini-M is aimed at business and remote-site customers. Features include Subscriber Identity Module (SIM) card capability, which enables a number of subscribers to use the mini-M service without having to create complex billing arrangements. A SIM card also protects the user from fraud because the information stored on it, user identity and billing details are encrypted, making it very difficult to copy. Moreover, if a card is lost or stolen, it can be canceled and replaced very quickly, without any need for re-programming. These terminals are also available in land vehicles, coastal vessels and rural phone versions, with the latter being fitted with an 80 cm dish antenna. They feature gyro-stabilized antenna platforms and maritime BDE units, which are ideal for all ships, coastal and fishing vessels to deep-sea commercial operations, offshore oilrigs, boats and yachts. Similarly, the antennas of land vehicular versions are readily roof-mounted on cars, trucks and trains. A large antenna utilizes the extra power to the mini-M terminal and is ideal for semi-fixed or fixed sites and the public payphone variant.

6.4.1.5. SES-C/Mini-C

The Inmarsat-C and mini-C are the smallest terminals with Inmarsat standards suitable for all mobile, semi-fixed and transportable applications for transmission of two-way data and telex messages at an information rate of 600 b/sec on L-band, while facsimile and E-mail messages are transmitted only in ship-to-shore direction.

The Inmarsat-C is second standard developed by Inmarsat dedicated at first for commercial and distress maritime application on merchant and even military fleets. The typical SES-C has a small and compact omnidirectional antenna as an ADE which because of its light weight and simplicity, can be easily mounted on all type of ships, yachts, fishing boats and offshore platforms. The ADE can be as a single SES-C or combined Inmarsat-C/GPS omnidirectional antenna. The BDU equipment can be an Inmarsat/C transceiver or combined with a built-in GPS receiver installed on board ships in the radio station or on the navigating bridge with mandatory and optional devices, shown in **Figure 6.12. (D)**. Some terminals have built-in message preparation and display facilities; others come with a standard RS-232 port so that users can connect their PC or other data equipment. The power requirements of Inmarsat-C terminals can be met from a ship's mains or battery sources. Inmarsat-C terminals can be also programmed to receive only multiple-address messages from shore offices via CES known as Enhanced Group Calls (EGC). The best solution for all types of vessels is the integrated Capsat GMDSS Dual Mode Thrane & Thrane SES-C and mini-M, shown in **Figure 6.15.**

The TT-3026LM Mini-C comprises antenna, Inmarsat transceiver and a 12-channel GPS Rx in one single device with a total weight of 1.1 kg and a size of 15 cm. It provides data, E-mail, position reporting/polling, Fax, Tlx, X.25, intership communication, SCADA, etc.

6.4.1.6. SES-D/D+

Inmarsat-D offers global one-way and Inmarsat-D+ two-way data communications utilizing equipment no bigger than a personal CD player. Complete with integrated GPS, Inmarsat SES-D+ systems are ideally suited for tracking, tracing, short data messaging and SCADA applications. Inmarsat-D+ may be used in the point-to-multipoint broadcast of information, typically for financial data, such as exchange rates and stock-exchange prices, credit-card listings and disaster alerts. The D+ units can store and display up to 40 messages of up to 128 characters each. Subscribers can receive tone, numeric and alphanumeric messages, as well as clear data. In addition, D+ terminals will be able to transmit position information derived from integral GPS and short messages. The service provides the capability to send a message to a group of MES users, which will require a group member Pager Identity (PID) in addition to the individual PID. Mobile subscriber equipment will not generate an acknowledgement to a group call. Any action required by the MES in response to a group call message must be controlled via the end-to-end application. All messages sent to an MES will be numbered to enable the subscriber to identify any lost messages. Repeated messages will be sent with the same message number to allow repeated call indication. The D service provides the capability to transmit from the mobile subscriber to the base: **a)** Acknowledgement Burst, **b)** Short Data Burst and **c)** Long Data Burst.

6.4.1.7. Inmarsat-C EGC Receiver

Reception of EGC messages may be via an existing Standard-A and B and either built into the electronics or provided as a separate unit and connected through an IF interface. In both cases, EGC would be made with the existing antenna, LNA and diplexer. The simplest Rx could be part of Standard-C SES with its omnidirectional antenna. It could be dedicated to the reception of EGC messages, distress alerts and other MSI. A basic EGC only Rx would consist in a decoder, demodulator, processor and built-in printer, it can use SES-C antenna and can be installed on the ship's bridge. Operation of the EGC receiver is extremely simple, with interface limited to a simple telephone-type keypad. The basic operation control would permit the selection of message types to be received, such as meteorological, navigational, etc. It would be possible to deselect ALL SHIPS messages such as distress alert and other MSI such as SafetyNET and FleetNET transmissions.

6.4.1.8. Inmarsat-E EPIRB Transmitter

The Inmarsat-E system provides global maritime distress alerting via Inmarsat satellites to dedicated Rx equipment located at four CES infrastructures: Raisting, Germany (T-Mobil); Niles Canyon, USA (Stratos); Perth, Australia (Telstra) and BT Atlantic, UK (BT). The distress alerts transmitted by an EPIRB will be received by two CES in each ocean region, giving 100% duplication in case of failures or outages associated with any of the CES.
Following reception of the distress alert, it is immediately forwarded to RCC via an X.25 TTN connection within 5 minutes, so that appropriate action can be taken. Therefore, the Inmarsat-E distress system supports Float Free L-band EPIRB, which incorporates the features: GPS positioning, which is accurate to within 200 metres. Automatic activation when the EPIRB is released by "floating free". Remote activation and information input from vessels' bridge or other manned situation. Optional SAR Radar Transponder (SART) and 121.5MHz beacon and High intensity, low duty cycle flashing light, see **Figure 6.8**.

Technical particulars of L-band EPIRB are as follows: the size of the EPIRB is between 22 cm and 70 cm high and weighs about 1.2 kg, modulation is non-coherent binary FSK, Tx frequency is between 1.6455 and 1.6465 GHz, Tx power/gain is nominally 1 W/0 dBi, frame length data/synchronization/party bits are 100, 20 and 40 bits respectively; Code is NRZ-L; Modulation rate is 32 bauds, total and number of transmissions are 40 min/4.

This EPIRB may be activated from a remote control position on the bridge or the conning position of the vessel; manually by using a switch on the side of the equipment if the EPIRB has been carried into the survival craft; or automatically as soon as the EPIRB has been released by immersion in water (hydrostatic release). Carriage of a satellite EPIRB is required by IMO GMDSS, which came into effect on 1 August 1993 for all vessels over 300 GRT and passenger ships engaged in international voyages. Inmarsat-E EPIRB models are also available for small craft and pleasure boats, so all RCC stations hold details of all registered Inmarsat-E EPIRB terminals.

6.4.2. Inmarsat Land Mobile Vehicle Earth Station (VES)

The VES terminals have been designed for installation and operation on road and railway vehicles. The vehicular antenna is normally placed on the roof and is typically of low or medium gain, omnidirectionally or directionally radiated, respectively. The RF equipment is usually mounted in the cabin nearby the driver controls. The VES standards are technically almost the same as SES. The only difference is that VES Above Haul Device (AHD) or antenna system and Below Haul Device (BHD) or RF unit with peripherals are almost all smaller and more compact because of the reduced space, especially installations in road vehicles. Officially there are four Inmarsat VES standards derived from SES such as VES-M, VES mini-M, VES-C and VES-D. All land mobile classes of MES terminals are divided into land mobile VES and Transportable TES.

6.4.3. Inmarsat Aeronautical Mobile Aircraft Earth Station (AES)

The Inmarsat AMSC provides two-way voice, Fax, variable rate data and video service for aircraft operating virtually anywhere in the world. It has been developed by Inmarsat and the aviation industry and complies with the Standard and Recommended Practices (SARP) for AMSC systems developed by the ICAO. The Inmarsat Aero system is supported by the civil aviation communities, including such organizations as ARINC, SITA and others. Satellite equipment designed and built to these standards may operate worldwide without any restriction and can be fitted on long-haul aircraft. This promises a revolution in oceanic ATC, with pilots and controllers talking to each other as often as they need and aircraft automatically reporting accurate positions at regular intervals throughout the flight. Therefore, for the first time, they know exactly where the aircraft are and can reach them quickly and controllers have the flexibility to issue new routes in mid-flight, in response to changes in the winds. The resulting fuel savings are expected to be worth many millions of dollars in the years to come. In the airline industry, aircraft downtime means dollars down the drain. If an Inmarsat-equipped airliner develops a fault in-flight, its crew can alert destination staff hours ahead of landing so that they can organize the labor and resources needed to fix the problem quickly and get the aircraft flying and earning again. Should the fault have safety implications, the airline's dispatch office has all the data it needs to work with air traffic control towards a safe outcome.

Figure 6.16. Aeronautical AES Standard Configurations

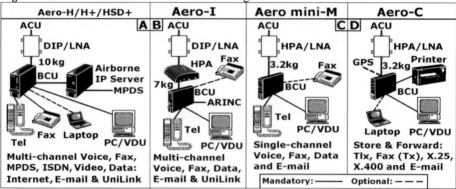

Courtesy of Manual: "Aeronautical Services" by Inmarsat

The AES are aeronautically installed satellite terminals capable of communicating with GES in the Inmarsat Network for access to TTN. The Inmarsat has developed several standards of AES for installation on board civil and military airplanes and helicopters, which are   presented in **Figure 6.16.** The AES terminal receives and processes RF signals from the satellite, then formats and transmits RF signals to the satellite at L-band (1.6/1.5 GHz). The AES interfaces on board systems, such as duplex Tel, Fax, data and video equipment for aircraft, which standards meets the requirements of the ICAO and industry standards, such as ARINC Characteristics 741 as well as Inmarsat standards. In this case, Inmarsat is not concerned with packaging and only requires compliance to its SDM for system access. In fact, ARINC 741 describes one physical implementation of the Inmarsat system, X Characteristics 741 for AES, which comprises ACU and BCU units presented in **Figure 6.17**.

1. Satellite Data Unit (SDU) – The SDU is the heart of the AES. It interfaces with other on-board avionics including the aircraft navigation system and performs most of the protocol, data-handling, modulation/coding and demodulation/decoding AES functions.

2. RF Unit (RFU) – The RFU converts IF inputs from the SDU into L-band RF signals, which are sent to HPA for transmission. It also receives the L-band RF signal from the satellite via the LNA, converts them to IF and passes them to the SDU.

3. High Power Amplifier (HPA) – The HPA amplifies the transmitted RF signal from the RFU to the appropriate power level required to maintain the air-to-ground MSC link.

4. Diplexer/Low Noise Amplifier (DIP/LNA) – The diplexer provides separation of Rx and Tx signals, while the LNA amplifies the RF signals received by the antenna to compensate for system signal losses and forwards them to the RFU.

5. Beam Steering Unit (BSU) – The BSU is otherwise known as the Antenna Control Unit (ACU) and when used with a mechanically steered antenna, controls the pointing of the airborne antenna. Namely, it receives instructions from the SDU on where to point the antenna beam. The instructions are converted into steering commands, either electronic or electro-mechanical, to point the antenna beam towards the desired satellite.

6. Satellite Antenna Unit (SAU) – The antenna as a part of the ACU is the component that is mounted to the exterior of the aircraft enabling the system to transmit/receive RF signals to/from the satellite. Thus, three kinds of antennas have already been presented: LGA, IGA and HGA, specified for use in the Inmarsat AMSC system.

Figure 6.17. Aeronautical ACU and BCU Configuration

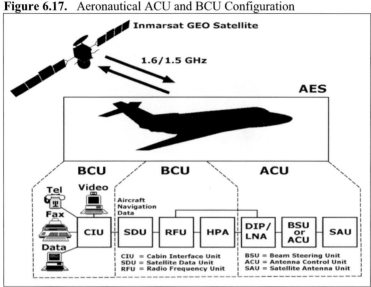

Courtesy of Manual: "Aeronautical Services" by Inmarsat

6.4.3.1. AES-H/H+/HSD+

The first Inmarsat Aero-H MSC service provides simultaneous two-way digital voice, Fax Group 3 at 4.8 Kb/s and real-time packet mode data transfer at 10,5 Kb/s anywhere in the global beam. It has been developed to meet the MSC requirements of the flight crew, cabin crew and passengers, for social, operational, administrative and safety applications.

Aero-H+ is an evolution of the Aero-H service that uses the higher power of the Inmarsat-3 spot beam or operates using the global beam. This Aero standard supports the same services as Aero-H. The Aero-H+ interfaces with international X.25/PSTN/PSDN networks and is compatible with the ISO 8208 internetwork standard, when the circuit-mode data option supports user-defined protocols. This standard is fully compliant with the ICAO requirements to support CNS/ATM in oceanic and remote air space, enables safety AMSC and automatic position reporting for ATC/SCN, including all operations management and ACARS/AIRCOM-type messaging worldwide. This system also employs pilot/controller voice and data link satellite communications and incorporates a satellite-based ADSS. Additional applications of the H+ are real-time aircraft engine and airframe monitoring and reporting; maintenances and fuel requests; weather and flight plans updates, NOTAMS, a point-to-multipoint data broadcast, catering information and crew scheduling.

The latest Inmarsat Aero-HSD+ or Swift64 is a 64 Kb/s two-way Mobile ISDN and MPDS such as the Thrane & Thrane TT-5000HSD+ AES, which supports the full range of ISDN compatible communications and TCP-IP Internet connectivity, presented in **Figure 6.18**. Inmarsat-compatible MSC equipment standards must be installed on board an aircraft in order to access the Inmarsat Aero-H/H+/HSD+ service. Otherwise, this equipment usually comprises a steerable high gain antenna, suitable avionics and other peripheral devices, such as telephone sets, facsimile machines, laptop and PC with peripherals, see **Figure 6.16. (A)**. In fact, the range of services available depends on the type of equipment chosen.

Figure 6.18. Aero-M4 HSD+ (Complete Swift64) 64 Kb/s Solution

Courtesy of Brochure: "Aeronautical Satellite Communications" by Thrane & Thrane

6.4.3.2. AES-I

The Aero-I standard uses an intermediate-gain terminal exploiting the higher power of the Inmarsat 3 satellites. Aero-I allows aircraft flying within spot beam coverage to receive multi-channel voice, Fax and circuit mode data services through smaller, cheaper terminals. However, packet data services are available virtually worldwide in the global beams, which include ACARS/data link and ATN. At any rate, this standard is a good solution to the MSC needs to support safety, flight operation, administrative and crew and passenger voice and data communications. Otherwise, this AES is ideally suited to short and medium haul aircraft but its characteristics make it well-suited for installation on a wide range of aircraft types as follows: commercial airliners; cargo aircraft; corporate and general aviation and military transport aircraft. The reductions in aircraft equipment size, weight and cost mean that Aero-I is also proving to be very attractive to some operators of wide-body aircraft.

In building aircraft devices that incorporate the new codec, manufacturers have also had the opportunity to repackage and update previous designs, again allowing more compact and lightweight Aero-I aircraft equipment that offers identical or improved performance, which configuration is presented in **Figure 6.16. (B)**. The latest Inmarsat Aero-I terminal is Thrane & Thrane TT-5000 Aero-I configuration, illustrated in **Figure 6.19**.

The Aero-I offers the following services:

Figure 6.19. Aero-I Solution

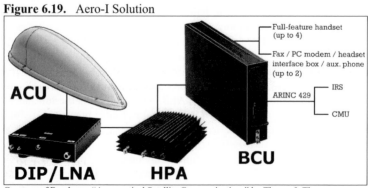

Courtesy of Brochure: "Aeronautical Satellite Communications" by Thrane & Thrane

Figure 6.20. Aero mini-M and Aero-C Solutions

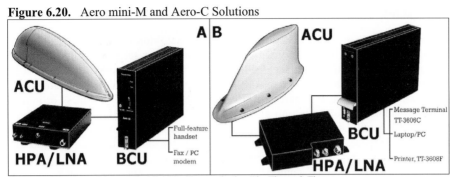

Courtesy of Brochure: "Aeronautical Satellite Communications" by Thrane & Thrane

1. Cockpit and passenger voice, Fax, PC data and E-mail are available in spot beams.
2. Packet mode ISO 8208 data communications from 600 b/s to 4.8 Kb/s and emergency (distress) voice communications in global beams.
3. Multi-channel operation from 1 to 7 channels per AES and packet data integrated with In-Flight Entertainment systems to provide on-line access to ground-based information sources and services.

Aero-I also complies with the ICAO's proposed revisions to their SARP for AMSS and may be used as part of the wider developments in CNS/ATM. This provides further benefits for aircraft operators and justification for the installation of Aero-I. Thus, flight crew and ATC Centres may also exploit other proven features of Inmarsat's aeronautical system, which includes secure voice access via short dialing codes to major ATC centres.

The additional power radiated within the Inmarsat-3 spot beams has allowed Aero-I to be smaller, lighter and cheaper than previous AES systems and to use a smaller IGA antenna, providing 6 dB of gain, compared with 12dB required for the Aero-H service. The second important development for Aero-I is the adoption of state-of-the-art voice codec technology that allows the same voice quality to be communicated using half the previous MSC bandwidth. Aero-I uses a new 4.8 Kb/s voice codec compared with the 9.6 Kb/s codec used in the Aero-H service. This allows the same or improved voice to be carried using less than half the satellite resource, allowing Inmarsat to provide MSC at lower cost to users.

6.4.3.3. AES Mini-M

The Aero mini-M AES provides voice, Fax and data communications for small corporate aircraft and general aviation users. Planned to take advantage of the spot beam capability of the Inmarsat 3 satellites, Aero mini-M is designed based upon technology developed for maritime-based MSC users and features affordable lightweight aircraft equipment offering a single channel such as Tel, Fax and PC data configurations, shown in **Figure 6.16. (C).** A mini-M requires an externally mounted antenna compatible with an IGA Aero-I antenna. Thus, the smaller aircraft for which the Aero mini-M service is intended do not operate in airspace using MSC for air traffic management, as discussed, this service is performed by Aero-I and Aero-H applications. The Aero mini-M service offers a single channel for Tel calls via PSTN; circuit mode data for Fax or PC data with SIM card capability and encrypted voice using STU-III technology if required. The latest Inmarsat Aero mini-M terminal is the compact Thrane & Thrane TT-3000 configuration, see **Figure 6.20. (A).**

6.4.3.4. AES-C

The Aero-C version of low-rate data transmission allows store-and-forward text or data messaging, one-way Fax/E-mail, polling and data reporting non safety-related MSC service at a rate of 600 b/s, which interfaces with X.25/PSTN, telex networks. The Aero-C has been developed to meet highly reliable AMSC capabilities do not envisage operating in the ICAO CNS/ATM system. It is particularly suitable for smaller sized business or military aircraft and helicopters operating on a regional basis or in remote regions, which do not need the full Tel and data capability services. The additional potential service applications of this standard are weather and flight plan updates, maintenance and fuel requests, position reporting, business and airways corporate communications and en-route and destination weather updates. The Aero-C comprises an antenna, a diplexer and a transceiver with optional GPS, using an Aero-C antenna. The automatic Doppler shift compensation for subsonic speeds handles messages of up to 32,000 characters. Finally, the transceiver requires an interconnection to cockpit text-based data terminal equipment and/or a laptop type PC, as is presented in **Figure 6.16. (D)**. Otherwise, the optional printer can also be connected to the system for hard-copy printouts. The latest Inmarsat Aero-C terminal is the compact Thrane & Thrane TT-3024A configuration, shown in **Figure 6.20. (B)**.

6.4.3.5. AES-L

The Inmarsat Aero-L MSC service provides commercial or military aircraft and helicopters with a real-time, two-way data communications capability at 600/1200 b/s and interfaces with X.25/PSTN/PSDN TTN. This service has been developed to match the needs of aircraft operators for their flight crew, cabin crew and passengers. It is fully compliant with ICAO requirements and with the ISO 8208 internetwork standard for PSDN services to support safety, CNS/ATM, worldwide automatic position reporting/polling for ATC, operations management, communication and CARS/AIRCOM-type messaging. Thus, this system employs pilot/controller voice and data link MSC and incorporates a satellite-based ADSS and consists in low gain antenna, avionics and data terminal equipment similar to Aero-C, see **Figure 16.6. (D)**. The service provisions of Aero-L are as follows: real-time aircraft engine and airframe monitoring and reporting; maintenance and fuel requests; weather and flight plans updates, NOTAMS, point-to-multipoint data broadcasting, catering information and crew scheduling.

6.4.4. Inmarsat Semi-fixed and Transportable Earth Station (TES)

Many of Inmarsat's users are on land and this proportion is growing rapidly as customers discover the potential of a high quality communications service that operates anywhere, without the need for any support infrastructure. Suitcase-sized transportable terminals with removable antenna offering the latest in compact design are already in widespread use and other terminals are available for specialized applications. The Inmarsat TES terminals have the same standards and characteristics of transceivers and antenna units as those introduced for Inmarsat SES or VES terminals. The TES-A/B is a rugged terminal with directional high-gain quad flat panel antenna or umbrella dish reflector. TES-M, mini-M and TES-C are ideal for travelers by car or fixed locations in remote areas for outdoor or indoor installations. Mini-M with a Big dish antenna is suitable for rural public payphones.

6.5. Inmarsat Commercial GMSC Service

Each year more and more business is conducted on the move aboard vessels, on long-distance vehicles, in aircraft and at locations far beyond the reach of ground-based fixed and mobile communications. However, mobile workers see no reason why, wherever they go, they should not enjoy the same facilities of voice, video, data, Fax, Tlx, E-mail and web via Inmarsat, as they are accustomed to in their home offices. A notebook-sized satellite unit will meet the needs out on site, at a remote and rural office. For the passenger who loves the sea but hates the isolation, today's Inmarsat-equipped ship offers in-cabin phones, Internet access, news bulletins, and videoconferencing, etc. Particularly on the international airlines, travelers who need to work during long flights can make use the same facilities.

The Inmarsat CNS solution is one of the keys to managing maritime transport, land-based transport by road, rail or inland waterways and for aeronautical transport. It will increase the capacity and the safety of transportation industries everywhere. Actually, managers need to know where their ships, vehicles or planes are at all times, as well as to improve safety at sea, on the ground or in air. Namely, they will be able to know exactly when a consignment has been held up and its exact location.

A comparison of service facilities between SES and AES is presented in **Table 6.5**.

Table 6.5. Comparison of Service Facilities between Maritime and Aeronautical MES

Services	Inmarsat Standards				
	A	**B**	**M**	**C**	**Aero**
Start Service	1982	1993	1993	1991	1990
Voice	Yes	Yes	Yes	No	Yes
Tlx	Yes	Yes	No	Yes	No
Group 3 Fax	9.6 Kb/s	9.6 Kb/s	2.4 Kb/s	No	9.6 Kb/s
Data	9.6 Kb/s	16 Kb/s	2.4 Kb/s	0.6 Kb/s	0.6/ 9.6 Kb/s
X.25	Yes	Yes	Yes	Yes	Yes
X.75	No	No	No	No	Yes
X.400	Yes	Yes	Yes	Yes	No
HSD	56/64 Kb/s	56/64 Kb/s	No	No	56/64 Kb/s
HSD File Transfer	56/64 Kb/s	56/64 Kb/s	No	No	56/64 Kb/s
HSD Store & Forward Video	56/64 Kb/s	56/64 Kb/s	No	No	56/64 Kb/s
Videoconference	56/64 Kb/s	56/64 Kb/s	No	No	56/64 Kb/s
Mobile Broadband	56/64 Kb/s	56/64 Kb/s	No	No	56/64 Kb/s
HSD Multiplexed Channels	56/64 Kb/s	56/64 Kb/s	No	No	56/64 Kb/s
Short Data/Position Reports	No	No	No	Yes	No
Group Calls	Yes	Yes	Yes	Yes	Only Maritime
SafetyNET/FleetNET	Yes with EGC	Yes with EGC	Yes with EGC	Yes	Only Maritime
GMDSS/Distress Button	Yes	Yes	Yes	Yes	Only Maritime

6.5.1. Maritime Onboard Applications and Services

The Inmarsat Maritime systems offer the numerous MSC applications designed to meet the communication requirements of the navigation-bridge or cabin crew and passengers.

1. Bridge and Cabin Crew Service – These onboard applications include traffic for ship's operation and management purposes during navigation, initiated from the sea or the shore:

a). Two-way voice, Tlx, Fax or data between ship's Master and shipping company, agents, port authorities, pilots, medical clearance offices, ship chandlers, ship repairers and others.

b). Direct voice contact with company operations or engineering for immediate advice and consultation for on board administration, operational or engineering problems.

c). Position reporting to AMVER, AUSREP, JASREP and other on shore organizations for providing safe navigation in seas under their supervision.

Figure 6.21. Comparison of Antennas between Fleet Applications

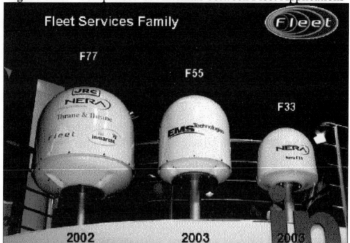

Courtesy of Webpage: "Fleet Service Family" from Internet

d). Meteorological data reporting (OBS) to shore-based weather centres.
e). Reception of ship-routing messages.
f). Other ship's traffic could include Tel or Tlx medical advice to masters of ships dealing with unwell crewmembers and passengers and all kinds of private crew communications.
2. Cabin Passenger Service – The cabin communication includes telephones employing all of the standard signals, dialing tone, ringing signal, etc., by cash, credit and prepaid card activated handsets, like normal public telephones. The wall-mounted payphones, similar to urban public phones, can also be installed in passages or other public spaces.
3. Ship borne Communication Office – Additional Inmarsat equipment usually installed in ship borne offices or former radio stations can offer telephone, PC VDU, Tlx and Fax machines similar to those found in offices on shore; connection of PC to shore network and databases; hotel and rental car reservations; teletext type news and so on. This office can also be a shipping bureau with PC/LAN facilities in connection with conducting cargo, administration and ship management.

6.5.1.1. Maritime Fleet System

The Fleet system and range of service is a development of the Inmarsat mini-M and GAN systems, offering both ocean-going and coastal vessels with comprehensive voice, fax and data communications. Fleet's high-speed Mobile ISDN and cost-effective IP-based MPDS offer unparalleled connections, including access to the Internet/E-mail, weather updates, videoconferencing and an advanced voice distress safety system (Fleet F77 only). The two new members of the family, Fleet F55 and Fleet F33, also offer global ship-to-shore communications but with the benefit of reduced-size antennas and above-deck equipment, making them ideal for medium to smaller vessels, see **Table 6.6.** and **Figure 6.21.**
The new Fleet F77 solution provides high-quality MMSC, while the new Fleet F55 and F33 launched in 2003 allow smaller vessels to benefit from voice, E-mail, secure Internet and Intranet access, plus high-quality Fax services, sea and weather charts, etc.

Table 6.6. Comparison of Service Facilities between three Fleet Applications

Service	Coverage	Voice	Data - Circuit Switched	Data - Packet Mode	Fax	GMDSS	Antenna Size
Fleet F77	Global	Global (Digital)	64K ISDN Euro Standard	MPDS Standard Fit	2.4K/9.6K (option) G4 64K	Voice IMO	75-90 cm (diameter)
Fleet F55	Global Voice Spot Data & Fax	Global (Digital)	64K ISDN Euro Standard	MPDS Standard Fit	9.6K (option) G4 64K	Not Applicable	50-60 cm (diameter)
Fleet F33	Global Voice Spot Data & Fax	Global (Digital)	9.6K Data Variant in 2003	MPDS Data Variant in 2003	9.6K Fax with 9.6K data variant	Not Applicable	30-40 cm (diameter)

1. Fleet F77 – The Inmarsat F77 provides full global networking coverage for the maritime community, offering new Inmarsat services based on Mobile ISDN or the MPDS available to mariners anywhere in the world. Benefits from the extended coverage provided by Fleet F77 enable craft operating in the deep ocean waters, far away from land, to take advantage of reliable, high-quality MSC and transform them into an "Office at sea", as is shown in the basic Thrane & Thrane Capsat Fleet77 package in **Figure 6.22**. This package consists in a BDE transceiver (TT-3038C), an ADE sensor stabilized platform with directional RHCP antenna (TT-3008), a cradle for the handset (TT-3622B) and a handset (TT-3620F).

The other illustrated equipment enable as follows: LAN Web/E-mail access, Large File Transfer and Photo transmission, Videoconferencing/Video Phone over IP (VCoIP/VPoIP) using the MPDS network, which charges only for the transferred amount of data sent and received in MG/s; Video Phones, Secure Telephone Equipment (STE), Audio Recorder and Group-4 Fax via Mobile ISDN and Low speed voice line/cordless Tel, mini-M voice, switchboards and Group-3 Fax via a RJ11 modular jack. On the other hand, the Fleet SES configuration for F55 and F33 will be somewhat similar.

Seafarers have never been in safer hands than with F77 because they can benefit from greater security, such as built-in pre-emption and voice prioritization. This standard feature ensures that non-essential and low priority regular MSC can be interrupted, so that essential safety or any emergency calls can get through instantly, and SAR forces can have seamless

Figure 6.22. Capsat Fleet77 GMSC Package

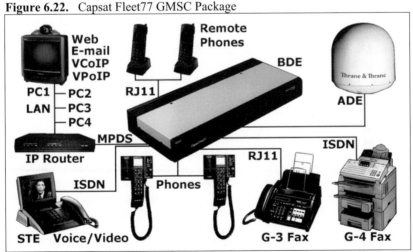

Courtesy of Brochure: "Global Maritime Connectivity via Inmarsat Fleet F77" by Thrane & Thrane

contact with the relevant vessels. It will always work in an hierarchical way: distress calls will pre-empt all other MSC; urgency calls will pre-empt both safety and routine calls and safety calls will pre-empt routine calls. This development is the result of a decision by the IMO to enhance the GMDSS mission after the m/v Achille Lauro disaster in 1994.

Fleet F77's digital system allows everyone on board to benefit from voice service around the world. Additional SIM card functionality on Fleet F77 also makes it possible to offer capabilities such as Universal Crew Calling, a range of communications solutions designed to take social calling away from the bridge and allow crew to call home in greater privacy. In addition to the standard voice service (4.8K AMBE), a high quality 3.1 KHz audio channel is also provided. The Fleet F77 service allows fleet managers to select either or both of two communication methods: Mobile ISDN or MPDS, suiting different types of operational needs, gives on-board access to a wide range of useful applications, as follows: ship management, technical support, telemetry and remote ship operations; telemedicine and teleeducation; Internet/Intranet/E-commerce and banking; telemetry and monitoring; technical support; electronic charts; weather routing and digital image transfer.

2. Fleet F55 – The benefits of the new Fleet F55 include digital voice in 4 Inmarsat Ocean Regions, while two distinct services are available in the Inmarsat-3 spot-beam coverage: 64 Kb/s Mobile ISDN, providing high-quality voice, Group-4 Fax and high-speed voice-band data and MPDS as "always on" IP, charges by the amount of data sent and received, not the time spent online. This new system is very attractive for smaller volumes of data or interactive data. Fleet F55 has been designed for ships needing a smaller antenna and low-cost BDE and hardware installation. Offering spot-beam data and global voice coverage, it is ideal for medium to large vessels, medium-sized craft such as merchant or patrol vessels. Similarly to Fleet F77, an additional SIM card functionality will also be on Fleet F55.

3. Fleet F33 – The new Fleet F33 service allows users to select either or both of two MSC channels: an integrated data service within the spot-beam or MPDS. Otherwise, there are some fundamental differences between the two forms of communication, suiting different types of operational needs. Currently, Fleet F33 offers an integrated data service within the spot-beam, delivering a data stream at speeds up to 9.6 Kb/s. The best will be used for Fax transfer and sending batched transmissions, such as file transfers via E-mail. During 2004, the MPDS is scheduled to be introduced for Fleet F33. This will provide always-on service, where charges are made for the amount of data sent and received, rather than the time spent online, which will be ideal for E-mail and Web browsing. Small and lightweight antenna, similar to mini-M, will offer increased ease of supply and installation. This system will use an additional SIM card functionality in the same way as the previous two solutions.

6.5.1.2. Mobile ISDN and MPDS Fleet Solutions

As discussed previously, Inmarsat Fleet is a development of the mini-M and GAN systems. The voice service is the same as the mini-M AMBE voice rate of 4.8 Kb/s and HSD is the same as the HSD defined for the GAN circuit-switched 64 Kb/s channel and associated services, as well as the mandatory Inmarsat Packet Data Service (IPDS). An asynchronous data service will be provided via this channel, which is the same as a Group-3 Fax at rate of 2.4 Kb/s. Inmarsat Fleet does not support the hybrid class SES.

1. Mobile ISDN – The Integrated Services Digital Network (ISDN) is the ITU-T, formerly CCITT, term for the digital public telecommunications network. It is offered in two system packages: Basic Rate (2B+D) provides two 64 Kb/s (B) data channels and one 16 Kb/s (D) signaling channel for individual users and LAN data links or high quality audio feeds for broadcast applications, and Prime Rate provides up to 30 x 64 Kb/s (B) data channels and 1 x 64 Kb/s (D) signaling channel for high band width business, such as videoconferencing and high capacity on-demand LAN bridge/router links. Because the Inmarsat-B service operates at only 64 Kb/s it is normally used with the Basic Rate ISDN service but can be also used on Primary Rate with applications that can operate on a single B-channel, such as videoconferencing. There is a possibility that two separate Inmarsat-B HSD can use ISDN single channel TA simultaneously, which is less expensive than a 2B+D unit.

Therefore, the 64 Kb/s data service supports applications between ISDN terminals using ISDN protocols such as V.120 or X.75. It will support any 64 Kb/s data stream and is the service used for implementing ISDN mobile applications, such as videoconferencing, LAN routing, file transfer, broadcast-quality audio transmissions and secure telephony. Thus, the service is accessed primarily through the RJ-45 connector on the SES and therefore up to eight multiple ISDN devices can be attached to the SES. A Point-to-Point Protocol (PPP) modem data service, suitable for data file transfer, E-mail or Internet access, may also be available via an RS-232, USB or infrared port. With Inmarsat Mobile ISDN, the customer uses a dedicated line or channel between the mobile equipment and the satellite. This channel provides up to 64 Kb/s of bandwidth. Users are charged by the length of time this dedicated channel is allocated. Hence, an ISDN call typically takes less than five seconds to connect, which is something to take into account when the call is over a MSC system.

Because of the global growth of ISDN, a whole range of telecommunications applications that were once the domain of large corporations have now become cost-effective and easily available to even the smallest of businesses. Dial-up networking using ISDN enables any number of LAN to be quickly and easily linked. Other services available through the Mobile ISDN are videoconferencing and broadcast-quality audio. On the other hand, with the introduction of the Inmarsat Mobile ISDN service, there is no longer any reason why people working in remote and rural locations should not enjoy the sophisticated IT and other solutions that are taken for granted in today's suburban office.

2. Mobile Packet Data Service (MPDS) – When using a computer/PC on a network, the information is not constantly being transmitted on the network in both directions. In fact, it is being sent/received in bursts, with gaps in between the bursts. The reason for this is because most applications use the so-called "query/response" mechanism, which burst of information contains "queries sent" with wanted information and "responses received" as a confirmation of received information. In fact, these bursts of information are called simply, packets of data. Since this is the method of transferring data on the Internet, it is called Internet Protocol (IP). Each of the packets sent contains both the sender's and the receiver's Internet address. Because a message is divided into a number of packets, each packet can be sent by a different route across the Internet. Packets can arrive in a different order than they were sent. The IP just delivers them, although it is another protocol known as the Transmission Control Protocol (TCP) that puts each packet back in the right order.

The Inmarsat MPDS has been developed to provide transfer of packet data over Inmarsat Networks, thereby giving users more efficient and flexible data transmission models. It operates on 64 Kb/s satellite channel, in both the to-ship and from-ship directions. These channels are allocated depending on the level of generated traffic. The individual terminal sends and receives data instantly but during the quiet periods, when customers are listing and reading a WebPage or typing an E-mail, the channels are free to be used by other SES. Short maintenance bursts are sent to keep the system informed of the ship's status.

The MPDS users are only charged by the amount of data they send and receive, rather than by how long the application takes or how long they are connected. The data is packaged in such a way that allows it to be sent through a channel simultaneously shared with other applications or users' data being transmitted under the same satellite in that spot-beam. Because the bandwidth of each channel is fixed, the more subscribers connect the more the available bandwidth gets reduced and therefore the speed decreases. Namely, this way of operating is based on a "best effort" or Undefined Bit Rate (UBR) foundation. In future, Inmarsat will look to provide more Constant Bit Rate (CBR) service, where the user will be guaranteed a minimum service level.

The MPDS-enabled terminal becomes simply a device connected to the Internet. When using MPDS, the CES operator is effectively acting as Internet Service Provider (ISP) as well. In this sense, mobile IP is a perfect solution for many applications such as Web browsing, interactive E-mail sessions, database enquiries, Web mail, IP/LAN connectivity, Intranet access, etc. Speed or throughput can be irrelevant for a specific data size in some applications. The Inmarsat IP network can be configured to route packets for certain addresses over one type of network and packets for all other addresses over another type of network. Users should be aware that any application using Public Network Access (PNA) is not necessarily secure, due to the very nature of the Internet. Encryption would always be recommended and where highly confidential information is being transferred, Private Network Access would be the best solution. Private networks can be accessed through Inmarsat Mobile IP by setting up a Virtual Private Network (VPN), which maintains privacy through the use of a tunneling protocol and security procedures. Using a VPN involves encrypting data before sending it through the PNA and decrypting it at the receiving end. Otherwise, an additional level of security involves encrypting not only the data but also the originating and receiving network address.

6.5.1.3.   First Generation of Inmarsat Data Services

One-way ship-to-shore demand-assigning 56/64 Kb/s service has been available since 1991 by using the Inmarsat-A SES type approved EB Saturn 3S-90. Duplex HSD or Inmarsat-A64 demand-assigned service became operational in 1992 with the HSD/A64, supported with adequate modem versions: Inmarsat HSD File Transfer; Inmarsat HSD Store-and-Forward Video; Inmarsat HSD High Quality Audio Broadcast; Inmarsat-A64 Multiplexed Channels and Inmarsat-A64 Multimedia Communications.

1. Inmarsat-A Data Services – To send and receive data messages through the Inmarsat-A network requires a PC via an external modem connected to one of the telephone ports on the Inmarsat-A SES, as shown in **Figure 6.23. (A)**. When an SES has dual ID numbers, it is recommended to connect a PC with the appropriate software to the second ID. It is possible to operate reliable data communications at 9.6 Kb/s via the Inmarsat-A system and faster speeds may be achieved by optimizing the faster modem and terminal settings. The Inmarsat-B does not need a modem and can be connected directly to the PC.

4. Inmarsat-C Data Service – As discussed earlier, the Inmarsat standard-C network offers a unique facility of store-and-forward data mode and messaging service at a rate of 600 b/s. As well as messages being delivered via the conventional means of Tlx or Fax, they can also be delivered via different data services, as is presented in **Figure 6.23. (B)**.

Many CES terminals operate most of the service on a subscription basis, so SES should be configured to use a Special Access Code (SAC) or two-digit access code.

Figure 6.23. Inmarsat-A and C Data Service Terminals

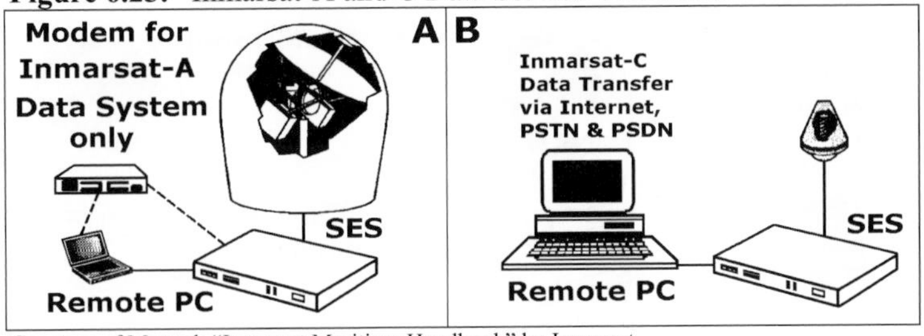

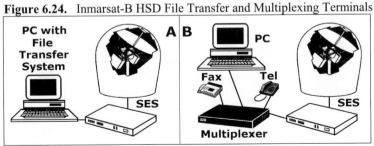

Courtesy of Manual: "Inmarsat Maritime Handbook" by Inmarsat

6.5.1.4. Inmarsat-B HSD Service

As discussed, both Inmarsat-A and B SES provide an HSD service operating at either 56 or 64 Kb/s. These systems offer the following three services and six systems:

a). Telepresence can bring assistance to the most remote sites, avoiding the expense of specialists and can support applications such as the repair of ship's engines at sea.

b). Telemedicine via Inmarsat-B HSD on board ships can provide rapid access to shared and remote medical expertise, using interactive audio-visual and data communications.

c). Teleeducation can support the training of staff on board ships, platforms, in remote and rural locations, in suburban installations, including different maintenance and emergency procedures. The resulting savings in expenses and travel can be significant.

1. HSD File Transfer – This service enables a PC-based file to be transferred from one location (SES or TES) to another, such as head office, at speeds of 56/64 Kb/s, as is shown in **Figure 6.24. (A).** Namely, it is ideal for users who send high volumes of data, such as seismic vessel surveyors, oil and gas exploration platforms firms and cruise liners.

2. HSD Multiplexing – A HSD 64 Kb/s channel can be used to carry up to six multiplexed or combined Tel, Fax and MSD circuits. Namely, SES is connected to PC, Fax and Tel via multiplexer, see **Figure 6.24. (B).** This multi-channel capability is also suitable for big ships, seismic surveyors and oil/gas exploration firms. This application will also interest banks and other firms with numbers of staff working in remote areas, where it is difficult to obtain Tel lines and can allow national authorities to quickly extend their networks or to restore their communications infrastructure in the event of a disaster. Without doubt, cruise liner ships have the greatest need for this service, enabling them to lower their own communications costs while increasing the profit margin on passenger traffic.

Figure 6.24. Inmarsat-B HSD File Transfer and Multiplexing Terminals

Courtesy of Manual: "Inmarsat Maritime Handbook" by Inmarsat

Figure 6.25. Inmarsat-B HSD Video and Audio Broadcast Terminals

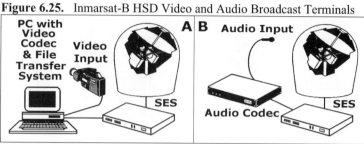

Courtesy of Manual: "Inmarsat Maritime Handbook" by Inmarsat

3. HSD Store-and-Forward Video – With the aid of advanced video codecs, it is possible to digitize, compress and send video material at rate of 56/64 Kb/s achieving almost full motion video, which SES configuration is shown in **Figure 6.25. (A)**. The store-and-forward technique also ensures that the received material is error-free, since the data transfer is achieved using an error detection and repeat transmission solution. These video compression techniques allow news and sports reports to be transferred from yachts, seismic/exploration vessels, war or disaster areas and other remote locations. The shipping and insurance companies can transfer video materials in case of an incident or damage.

4. HSD Audio Broadcast – Several international standards for the coding (digitization and compression) of audio signals provide different degrees of compression to the audio input, music or speech. Broadcasters can use 7.5/15 kHz audio codecs to supply broadcast-quality reports from the field or the high seas directly to the studio. A HSD SES of 64 Kb/s may be used to provide two-way broadcast-quality voice depending on whether simplex or duplex HSD is used, which SES configuration is shown in **Figure 6.25. (B)**.

5. HSD Videoconferencing - This service enables videoconferencing terminals and video phones in mobile units, rural locations or remote offices and can be used for face-to-face conversations with another person or to exchange documents and discuss their contents, which SES scheme is shown in **Figure 6.26. (A)**. Businesses can use Inmarsat-B HSD at speed rates of 56/64 Kb/s for videoconferencing for all remote locations worldwide.

6. HSD PC and LAN Access – It is possible to connect LAN on board ships with other PC LAN systems in the central office of a shipping company or other destinations, which SES configuration is shown in **Figure 6.26. (B)**. These HSD PC interconnections of 56/64 Kb/s will serve operators of LAN on board ships, cruisers and in remote hotels. It is also possible to interconnect different networks, using transparent protocols such as TCP/IP.

Figure 6.26. Inmarsat-B HSD Videoconferencing and LAN Terminals

Courtesy of Manual: "Inmarsat Maritime Handbook" by Inmarsat

Figure 6.27.   Very High-Speed BMCS

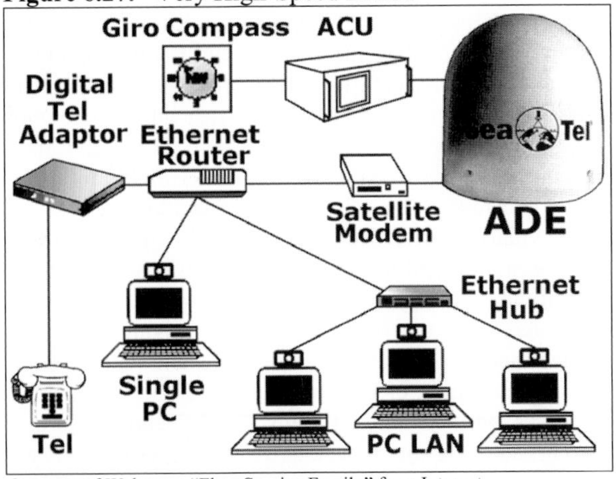

Courtesy of Webpage: "Fleet Service Family" from Internet

6.5.1.5.   Broadband Maritime Communication Service (BMCS)

The new, very high-speed BMCS WaveCall 4003 MSC solution was developed by SeaTel antennas manufacturer of the Globalstar system, to provide high-speed broadband services in small footprint coverage from Northern Europe through the Mediterranean and North to South America on board ships with the equipment configuration shown in **Figure 6.27**.

This BMCS configuration is comprises ADE with radome; three axis stabilized antenna platform; 1 m size backfire or ring focus Ku band antenna, providing unlimited azimuth capability and automatic polarization control and an antenna tracking system controlled by Gyro Compass via the Antenna Control Unit (ACU). The ADE is connected to the BDE unit via a single coaxial cable, serial I/O for remote antenna control using RAM software DVB RCS to Satellite Modem and Ethernet Router. Other features of this system are connections from the Ethernet Router directly to the customer's single PC or to network or PC LAN system via Ethernet Hub and through an RJ-45 Ethernet interface. On the other side is a Tel unit connection RJ-11, which plugs into a Digital Tel Adaptor interfaced to the Ethernet Router. The whole BDE BMCS network, except Tel line, is interconnected through the RJ-45 connection lines to the ADE.

The BMCS WaveCall 4003 configuration can be compatible with Inmarsat Mobile ISDN of Fleet F77 or Inmarsat-B SES because they are interfaced to ISDN LAN routing via the RJ-45 connections and Tel is connected to transceiver via the RJ-11 port. It will need only to provide an adequate antenna using the RF L-band directly to the transceiver and compatible Ethernet devices. This solution can easily be adopted for aeronautical and all military applications and can be easily adapted for rural or fixed sites with roof-mounted simple Tx/Rx satellite antennas without tracking Gyro Compass and Antenna Control Unit, to provide a much cheaper service than the big, Fixed Satellite Hub Infrastructure.

The SeaTel 4003 terminal is Broadband-at-Sea with a superlative antenna solution for home or office at sea. This BMCS terminal provides business-class connectivity for MSC at sea with always-on Internet connectivity at high inbound and outbound speeds. The system provides enough voice lines to support individual stateroom, cabin and office use.

Typically Internet speeds vary, so the BMCS can provide connectivity at 512 Kb/s inbound (satellite-to-ship) and 128 Kb/s outbound (ship-to-satellite) speeds. And if this is not enough bandwidth to transfer data files and images can be upgraded to 1.5 Mb/s inbound and 256 Kb/s outbound. Other systems say they provide global coverage but often at relatively low data speeds or fast inbound speeds but over limited geographical areas. Only the SeaTel 4003 BMCS provides very HSD with the same size as an Inmarsat-B terminal. Otherwise, the broadband services offered with the 4003 BMCS are supplied by WaveCall Communications Inc., a subsidiary of SeaTel. SeaTel manufactures VSAT devices that are designed specifically for marine use. These units are stabilized to withstand tough marine conditions and the ship's motion and turning. This system is suitable for applications such as a commercial and military ocean-going vessels that require always on, sea rig/platforms and offshore explorations, rural and remote industry private corporate networks. The system will offer services such as the fastest connections and data transfer available at sea, virtually state-of-the-art maritime electronics, or surf the web, intranets and LAN extensions and businesses will be able to transfer files and images, VideoConference (VCoIP), digital video broadcast, E-mail, backup databases, VoIP, VPN and more.

6.5.2. Land System Architecture and Operations

On a worldwide scale, millions of trucks, buses, trains and other vehicles can use in-cab communications with their dispatch bases, owners, agents, families and friends, or to deal with emergencies such as damage of cargo, engine breakdowns, collisions and rescue situations on the roads. Transport companies can locate their vehicle fleet and stay in touch with them, no matter where and when they roam. In the same spirit, bus and railway companies can be always in contact with their rolling stock and coach and train personnel and passengers will have possibilities to make phone calls. Otherwise, the voice and messaging service that ensures transport companies and dispatchers can keep permanent contact with their drivers, locomotive operators and vehicle personnel via voice and messaging services. In addition, construction plant-hire companies, large-scale farmers and others, all face a similar problem: how to keep track of their fleets of expensive mobile assets so that they can be used most profitably.

Therefore, irrespective of the location of the fleet vehicle, the mobile terminal can send its geographical position to dispatch. Location and progress of vehicles can be monitored on a PC screen via map-based software. Thus, two-way messaging between dispatch and driver enables status reporting at any time and from anywhere in the world. Distress alert for emergency services and road assistance are value added services of transport security and control through LMSC networks. Besides, a dual mode voice/data service provides cellular facilities when a vehicle is in cellular coverage. In fact, a driver can use cost-effective data messages to support position status reporting and short E-mails via Inmarsat-C standard and in combination with Inmarsat-M or mini-M voice, as illustrated in **Figure 6.28**.

Each vehicle of the fleet can be equipped with a vehicle-borne voice and data satellite terminal with external antenna, enlarged screen for data messages and short E-mail exchanges, Fax facility with small printer, integrated GPS/GLONASS receiver for vehicle location and optional connection points for external systems, such as onboard PC laptop configuration and navigation systems. In this way, long-haul trucks and international tourist buses traveling over all the continents can be equipped with voice satellite equipment to offer telephone service to truck drivers and bus passengers outside cellular coverage.

Figure 6.28. Vehicle Management and En-route Data Processing

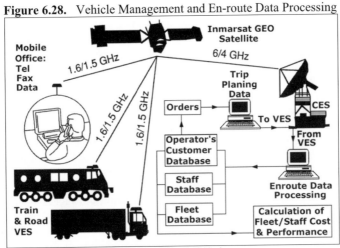

Courtesy of Book: "Never Beyond Reach" by B. Gallagher

6.5.2.1. Vehicle Management and Asset Tracking

In today's long-haul land transport industries, the total fleets' management of commercial vehicles and the active surveillance of cargo from dispatch to delivery is essential for success. Fleet managers increasingly rely on Inmarsat's asset tracking and fleet monitoring solutions, using cab-mounted terminals which, by drawing on sensors all over the vehicle, provide data that keeps the trucks running smoothly, ensuring that shippers are fully informed of the condition of their consignments or helping drivers in an emergency.

With virtually global coverage, Inmarsat provides worldwide monitoring, control and asset tracking system regardless of where customer's fleet or assets are located. Customers can choose the most cost-effective service and technology solutions, such as Inmarsat mini-C portable for vehicles, with the option of solar battery; Pocket-sized Inmarsat-D+ integrated with GPS; Portable laptop-sized GAN terminal and Inmarsat mini-M. All this equipment is suited for VES mounting or as portable terminals and can provide the following services:

Figure 6.29. Regulation and Control of Road traffic

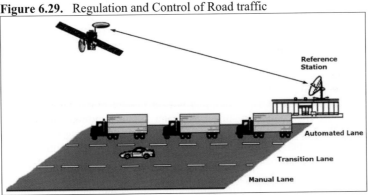

Courtesy of Webpage: "EGNOS Fleet Service Family" from Internet

Position reporting service using the GPS/GLONASS system gives quite accurate, reliable position, speed and bearings; vehicles tracking/assets enables monitoring of mobiles via sensors on vehicle/load conditions by polling/automatically; engine diagnostics; group and two-way messaging; E-mail/Fax and Internet/LAN and corporate data access on the move.

6.5.2.2. Vehicle Traffic Control

The new EGNOS project, similar to the OmniTRACS solution, gives priority to road transport, satellite communications and navigation to provide regulation and minimize traffic jams, see **Figure 6.29**. If all vehicles are fitted with a navigation satellite receiver, such as GPS or GLONASS and a data Tx such as Inmarsat-C or D+, their positions can be relayed automatically to a central station. This information can then be used in a number of ways to control road usage, to charge motorists for using a stretch of road, or to restrict access to congested roads, or to inform drivers of congestion and suggest alternative routes. The next top MSC technology of EGNOS system is a handheld personal receiver, similar to cell phone transceiver, which would use satellite navigation to avoid traffic jams in city centres, find the nearest free parking space, a business building or even the nearest pizza restaurant in an unfamiliar city.

6.5.3. Aeronautical System Architecture and Capabilities

The Inmarsat AMSC system provides two-way voice, Fax and data service for all types of aircraft operating virtually anywhere in the world. It has been developed by Inmarsat and the aviation industry to form a standard for AMSC. It complies with the Standards and the Recommended Practices (SARP) for AMSS developed by the ICAO. It should be noted that the Inmarsat technical definition project provided a significant contribution to the development of ICAO SARP. Therefore, equipment designed and built to these standards may operate worldwide without any restrictions, unless such restrictions are imposed by a national radio licensing authority. Inmarsat AMSS is integrated with two basic elements:
1. Inmarsat Space segment with associated IGN support facilities consisting in a constellation of new Inmarsat-3 birds in 4 Ocean Regions, controlled by NCC and SCC, which network for combined Inmarsat and AMSC/US service is presented in **Figure 6.30**.
2. The ground segment is composed of AES and GES located in four Ocean Regions. The AES is equipped with an aircraft-installed antenna and terminal capable of communicating via satellites with a GES for access to the TTN.

6.5.3.1. Aeronautical Onboard Applications and Services

The Inmarsat Aero systems offer numerous applications that can be supported for three types of users on board aircraft.
1. Cockpit Flight Crew Service – These applications include communications via satellite for ATS, flight operation and other airline purposes, which can be initiated from the air or the ground. The Inmarsat or AMSC solutions can allow the following ATS and operational support services to be performed with much greater efficiency when outside of radio VHF range. Initiated when required by the flight crew or automatically, they include:
a). Direct two-way data link between the pilot and the ATC operator to pass information such as changed clearance, traffic information and requests for changes to flight level.

Figure 6.30. Aeronautical Network Configuration

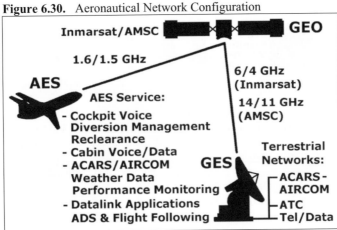

Courtesy of Manual: "T-4000 High Gain Antenna System" by Tecom

b). Position reporting using ADSS over the packet-mode data system. This is an application in which ATC instructs an aircraft to report specified data at regular intervals.

Flight operations that can benefit from the Inmarsat or US-based AMSC systems include:

a). ACARS/AIRCOM data transfer when out of radio VHF range, using packet-mode data.

b). Aircraft system monitoring (engine, airframe, system health), using packet-mode data.

c). En-route flight plans and weather updates via data, Fax or voice.

d). Notams, crew briefings, E-mail and documentation, using circuit or packet data mode.

e). Direct voice contact with company operations or engineering for immediate advice.

The Inmarsat Aero system supports, for cockpit uses only, a special form of air–ground call procedure for access to private networks such as ATC or airline operations centres. Rather than dial a long Tel number, even supposing the device has memory for all the numbers; pilots would use a special short-form number, which would route call to the required centre via the ATC network. Thus, a similar system allows direct voice with ATC facilities, using Inmarsat short-codes which route to the appropriate facility of the public Tel network.

Other airline traffic could include medical advice to crew dealing with unwell passengers, information related to passenger services, such as special meal requirements applied for the next sector of a flight or disabled passenger facilities required on arrival.

2. Cabin Crew Service – Cabin crew applications focus primarily on communications for airline administrative purposes and can be air or ground-initiated such as: crew briefing, flight documentation, medical advice, rostering scheduling, catering management, duty free sales inventory, using the circuits of packet-mode and voice.

3. Passenger Service – Possible cabin or seated passenger applications include telephones employing all of the voice standard signals, dialing tone, ringing signal, etc. by credit and prepaid card activated handsets, which have been developed for airlines. Wall-mounted payphones, similar to urban public phones, can also be installed in passages.

4. Airborne Communication Office – Additional Inmarsat equipment, usually installed in some airborne office or gallery, can offer telephone, PC VDU and Fax machines similar to those found in offices on the ground; connections of PC to ground networks and databases; hotel and rental car reservations, next flight confirmation and/or reservations; Teletext type news; destination information and in-flight shopping.

Figure 6.31. Aero-HSD Swift64 Package

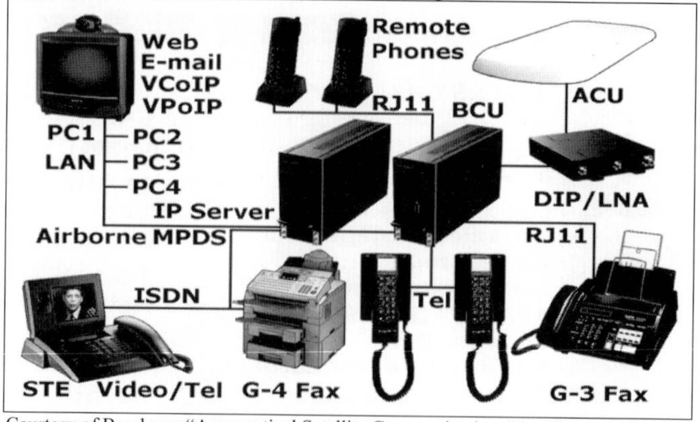

Courtesy of Brochure: "Aeronautical Satellite Communications" by Thrane & Thrane

6.5.3.2. Aero-M4 (Swift64)

The Swift64 HSD transfer is the aeronautical implementation of Inmarsat's GAN service. Following the Maritime Fleet HSD Service, Inmarsat introduced a 64 Kb/s Mobile ISDN and MPDS known as Swift64, which will support the full range of Mobile ISDN compatible communications and TCP-IP Internet connectivity. The Swift64 mobile services have been designed to meet the needs of aircraft passengers, corporate users and the flight deck and are based on technology developed by Inmarsat for land-based services. They are designed to take advantage of existing Inmarsat Aero H/H+/HSD+ installations, making use of the main components already to be found on a large number of corporate jet aircraft. In fact, both services will be delivered through the global and spot beams of Inmarsat 3 satellites. Inmarsat Swift64 provides full global networking coverage for the aeronautical community, offering new Inmarsat services based on Mobile ISDN or the MPDS. Benefits from the extended coverage provided by Swift64 enable craft operating on the deep oceanic flights, to take advantage of reliable AMSC, transforming them into an "office in the air", as shown in the basic Thrane & Thrane's Swift64 package in **Figure 6.31**. This package consists in Aero-HSD transceiver and high gain antenna (TT-5000HSD), an ADE sensor stabilized platform with directional RHCP antenna (TT-3008), a cradle for the handset and handsets. The other illustrated equipment enable: LAN Web/E-mail access, large file transfer and photo transmission, videoconferencing/video phone over IP (VCoIP/VPoIP) using the MPDS network, which charges only for the transferred amount of data sent and received in MG/s; video phone, Secure Telephone Equipment (STE), audio recorder and Group-4 Fax via mobile ISDN and low-speed voice line/cordless Tel, switchboards and Group-3 Fax via RJ11 modular jack.

The performance of Swift64 is based on the link capabilities of the current generation of Inmarsat-3 satellite constellation. Responding to the growing demand for a new Broadband Aeronautical Communication System, Inmarsat is developing a fourth generation that will become operational, currently scheduled for 2004. The Inmarsat I-4 will be able to give global reach to mobile data services parallel to those now being developed for the cellular second-generation General Packet Radio Service (GPRS), with a maximum data rate of 172 Kb/s and the third-generation UMTS, offering better than 384 Kb/s.

1. Mobile Swift64 ISDN Service – The two-way 56/64 Kb/s Mobile ISDN service delivers alternatively an 64 Kb/s Unrestricted Digital Information (UDI) channel; multi-channel avionics; mobile co-operative operation with other Inmarsat aero services provided by Aero H/H+/HSD+ transceiver and antenna; stand-alone installation; operation within the spot beam coverage of Inmarsat 3 satellites and affordable service charges based on per minute usage. The Mobile ISDN service provides full-time use of a high-capacity channel capable of carrying a constant data stream. In this case, ideal applications include the downloading of large files of material such as compressed video or graphics, which occupy all the available bandwidth for significant amounts of time, or any in which speed is paramount, such as satellite newsgathering. This MSC service provides direct and efficient error-free connection with terrestrial ISDN-compatible circuits and systems, allowing airborne LAN to be readily integrated into ground-based private networks. Typical ISDN applications include large file transfer such as audio, graphics, photographs and video clips; voice/G4 Fax; Secure Telephone Equipment (STE), or 33.4 Kb/s PC modem data. In other words, the HSD channel enables real-time image transfer; live videoconferencing; LAN/WAN connection and Internet/WebPages browsing with E-mail messaging facilities.

2. Swift64 MPDS – Service with MPDS 64 KB/s connection delivers full Mobile TCP-IP connectivity to the Internet with per-bit charging and always-on connectivity. The MPDS slices each file into small IP data packets, addresses them and sends them in bursts on a channel being shared by other users. On receipt/delivery, the packets are reassembled to form the original file. These applications include E-mail; secure access to corporate LAN and Intranets; web access; database queries; E-commerce transactions and small and medium-size file transfer. The cabin user pays for the amount of data that passes over the link, so connection can be "always on" without extra charge. The other features can be STU-III and other security add-ons, including "secure tunneling", using the L2TP protocol to create a VPN. In general, the cost-effective extensions of new 64 Kb/s per channel data transfer, both circuit-mode and packet-mode, to aircraft based on the well-established Aero H/H+/HSD+ system, allows access to a range of TTN facilities.

Therefore, ISDN access of Swift64 provides direct and efficient error-free connection to terrestrial ISDN-compatible circuits and systems, allowing the easy integration of corporate and airline airborne platforms into ground-based private networks, while the Swift64 MPDS allows unlimited Internet connectivity and efficient, cost-effective access to company Intranet and global E-commerce solutions.

6.6. Global Business Solutions

Inmarsat's land-based services enable users to create a virtual mobile and semi-fixed office or remote control room anywhere and instantly. It provides a range of voice, video and data applications with a reach that covers almost all of the world's land mass for governments, corporations, construction, agriculture, oil/mining, exploration, expeditions, newsgathering, monitoring/measurements and emergency associations in remote and rural locations.

6.6.1. Inmarsat M4

The Inmarsat M4 service brings the technologies of the world's most advanced MSS to its most remote places. Its portable, lightweight and easy to carry multimedia TES allows users to get HSD and PSTN high-quality voice connectivity from countries where TTN are

Figure 6.32. Inmarsat M4 MES and TES Solutions of Nera WorldCommunicator

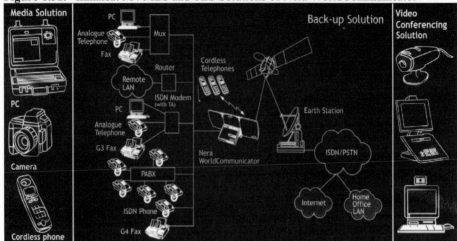

Courtesy of Brochure: "Near WorldCommunicator" by Nera

unavailable or unreliable. With the appropriate software and hardware the Inmarsat M4 service supports all mobile HSD, MPDS and ISDN of up to 64 Kb/s applications including secure encryption (STU III/STE) systems. Inmarsat-M4 service extends a LAN and WAN allowing all companies worldwide accessibility to their remote locations, see **Figure 6.32.** The M4 solution offers access via Inmarsat satellites to all MES and TES, such as Nera's World Communicator (NWC), the same service as in shore-based offices such as LAN, Internet and ISDN/PSTN. This Inmarsat service was the basis for developing Maritime Fleet solutions, Land HSD Truck/Railway Management, Aeronautical HSD and Swift64, Land-based TES HSD/MPDS and fixed and semi-fixed GAN and BGAN office solutions in remote/rural locations. Similar to the Mini-M service, the M4 service will be supported by SIM cards, enabling individualized billing over shared mobile satellite terminals.

6.6.1.1. Global Area Network (GAN)

Inmarsat's GAN integrates the corporate IT infrastructure with a GMSC network and offers HSD 64 Kb/s, Mobile ISDN and the IP-based MPDS, rapidly extending LAN and WAN to where businesses need information. The Inmarsat GAN service offers a combination of three powerful communications tools: voice (including broadcast quality), Fax and 64 Kb/s HSD (ISDN). Hence, together they make possible a range of powerful solutions, including: videoconferencing; video streaming; large file transfer; store-and-forward video; photo transmission; high-quality audio and secure voice and data. There are two basic types of voice service: used via either TES's integral Tel handset, a DECT cordless handset or via equipment connected to the ISDN port, such as an ISDN telephone. In **Figure 6.33. (A)** the NWC TES for use in car or in remote locations is shown. The unit works with PC or Palmtop and is small enough to fit into an A4 document briefcase. A foldout antenna fits neatly alongside the modem unit, which carries the battery, main electronics and connectors. A built-in DECT base station provides on 1.88–1.90 GHz full portability for voice through cordless handsets, which can be used up to 300 metres from the terminal.

Figure 6.33. Nera Inmarsat-TES WorldCommunicator

Courtesy of Brochure: "Near WorldCommunicator" by Nera

Up to 12 handsets may be registered and up to 5 handsets can be used at the same time. The ISDN port on the modem unit allows up to 8 ISDN devices to be connected (such as telephone or Fax and data devices' for example). A powerful on-board lithium-ion battery recharges from a wide range of power sources, including a car cigarette lighter. The modem unit also has a versatile display for satellite signal strength, battery conditions and different interfaces for computer and peripheral devices such as ISDN (RJ45-50), RS-232 (9-pin DSUB), IR port, Universal Serial Bus (USB) port interface and 19 VDC input. The total weight of the unit including antenna and batteries is 3.9 kg. The dimensions of opened antenna are 340 x 774 x 12 mm with a gain of about 17–19 dBi, Rx G/T of −7 dB/K and Tx EIRP of 25 dBW. The antenna is 3 folding panels, manually positioning antenna, with adjustable stand, RF modem transceiver and compass.

For permanent installation, this TES can use the Big dish antenna, illustrated in **Figure 6.33. (B)** or high-gain Quad flat panel antennas that can be mounted on the roof, on a pole or similar fixed structure with an antenna cable up to 200 metres away from the modem unit. Hence, fixed antennas can be installed permanently on the roof of the remote or rural office, as shown in **Figure 6.34**. The unit can be moved from office to terrain in a suitcase or far away by car. Another representative of GAN equipment with almost the same characteristics is the TES unit TT-3080A Capsat Messenger from Thrane & Thrane.

The GAN satellite service is uncompressed, providing broadcast quality speech, suitable for immediate transmission on TV and radio. Fax data can be sent or received at up to 14.4 Kb/s using either a Group-3 Fax machine or a Fax software package through the PC. Group-4 Fax provides a higher speed and quality service using digital scheme and can be sent from a Group-4 fax machine or from an ISDN-equipped PC connected directly to the TES ISDN port. Communication of all forms of data and voice through a single interface is a high quality service at the rate of 64 Kb/s. There are two data services: 64 Kb/s Unrestricted Digital Information (UDI), which is for data using a standard ISDN terminal adapter and 56 Kb/s data, which is also used for data and is mainly for the North American market. The ISDN HSD provides a constantly connected digital channel at up to 64 Kb/s for secure voice, video or data transmission. The subscriber has a dedicated line or channel between their mobile equipment and satellite, which provides a HSD 64 Kb/s bandwidth. This is what is known as a SCPC or Circuit Switched Service (CSS). If two GAN terminals are bonded together at the same time, creating two channels, even faster data transfer speeds of up to 128 Kb/s can be achieved. Mobile ISDN is ideal for voice, video and data functions, especially those that require the highest possible bandwidth and are time critical.

Figure 6.34.   Nera TES for Remote/Rural Indoor Installation

Courtesy of Brochure: "Near WorldCommunicator" by Nera

Inmarsat MPDS is the first global service of its kind in the world, so users can cut their data communications costs and pay for only the amount of data they send and receive, rather than by the time spent online. The addition of MPDS significantly enhances the power and reach of the Inmarsat GAN, adding to and complementing the existing 64 Kb/s Mobile ISDN service. Therefore, a prime application for MPDS is to provide secure access to corporate IP/LAN connectivity, remote LAN access, Internet and Intranet, which involve brief bursts of communication followed by periods of inactivity, turning the MES into a true extension of the company network, wherever the user is located. Users can enjoy fast and reliable database queries and secure packet data, Web browsing, Web mail, E-mail interactive sessions, E-commerce and emerging services include transmission of streaming video and special support for satellite newsgathering teams. In general, MPDS provides two distinct types of access: Public and Private Network Access by setting up a Virtual Private Network (VPN). The idea of the VPN is to give companies the same capabilities at a much lower cost by using the shared public infrastructure rather than a private one. Many firms are looking at using a VPN for both Extranets and wide-area Intranets. In fact, using a VPN involves a technique called secure tunneling, which involves encrypting data before sending it through the public network and decrypting it at the receiving end using the L2TP protocol. Hence, an additional level of security involves encrypting not only the data but also the originating and receiving network address.

Inmarsat MPDS uses the same technology as the Internet. Thus, when someone is browsing the Web, for example, WebPages are downloaded in bursts. There will probably be a lull in transmission while the user examines the content, before clicking on another hyperlink and triggering another burst of data as new pages are brought to the screen. In fact, the link to the Internet remains open all the time. These bursts of information are more usually called packets of data and this is such a common way of transferring information that the whole mechanism has come to be called packet data communications. Since this is the method of

Figure 6.35. TES and Semi-fixed GAN Solutions

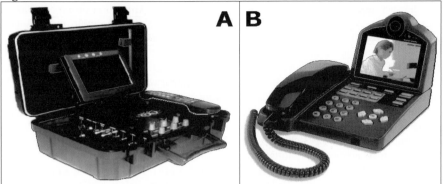

Courtesy of Webpage and Brochure: "GCS 7520 VTC" from Internet & "GAN" by Inmarsat

transferring data on the Internet, it is called the IP, the way in which data is sent from one computer to another over the Internet. The MPDS operates in the same way over the GAN Network's 64 Kb/s satellite channels, in both to-mobile and from-mobile directions and these channels are shared amongst a number of users and are allocated depending on the level of traffic that is being generated. Individual TES send and receive data at 64 Kb/s when the channels are free and can be used by other mobiles. Thus, the important fact with Inmarsat MPDS over GAN is that one is only charged for the amount of data transmitted and not for the length of time spent online.

The first GAN service to launch, Mobile HSD ISDN is available from a large number of Inmarsat partners including Xantic, France Telecom, KDDI, SingTel, Stratos and Telenor. Except for the above mentioned manufacturers Nera and Thrane & Thrane, there are currently two more approved Inmarsat GAN portable (TES) terminals available from EMS Satcom and Glocom.

The Inmarsat GAN service also can be provided by the Inmarsat-B4 HSD rugged portable case terminal (TES) Office and Voyager versions, only 12.5 kg in weight. This global TES provides high-quality access to the international telephone (16 Kb/s), High Speed G3 Fax (9.6 Kb/s), Tlx, Data for File Transfer (9.6 Kb/s), HSD 64 Kb/s Internet /E-mail HTTP and both Videoconference and Real time or store & forward video. This B4 HSD transceiver is easy to operate, with an acoustic signal strength indicator for manual antenna pointing and uses a high-gain transportable quad flat panel or umbrella reflector antenna, which can be fixed indoors or remoted outdoors up to 100 m away, with a flexible cable.

The next example of Inmarsat TES is a product of Global Communications Solutions, Inc., model GCS 7520 VTC-Lite Portable "Talking Head" Videophone. The GCS 7520 is a very small, lightweight, portable terminal designed for video transmission and conferencing from rural or remote operating sites. It is housed in a rugged waterproof case for transport allowing use in harsh environments. Using proven H.320 video technology, the GCS 7520 provides an ideal method of broadcasting or Videoconferencing over ISDN or Inmarsat GAN / M4 satellite terminals at 64 Kb/s or 128 Kb/s (an additional version 2). Features of this TES are: adjustable 6.4" TFT color display and speaker with audio connections for low-Z mic and balanced/unbalanced line audio headphone jack for direct and cue (return) signals for monitoring. It operates from 12VDC or 90-260 VAC and it weighs less than 6 kg. It offers the following applications: government/military surveillances; newsgatherings;

oil/mining, construction/factory sites, maritime, deployable emergency response or disaster relief, distance learning and telemedicine, on and offshore exploration or expeditions and corporate/commercial communications, which modern model of portable videophone is shown in **Figure 6.35. (A)**.

The indoor remote or rural Inmarsat TES solution is a Video telephone set of Motion Media (MM) using ISDN or GAN service of 64 Kb/s HSD via a single dedicated channel, is illustrated in **Figure 6.35. (B)**. Otherwise, an additional version 2 service is also possible by using two Inmarsat GAN terminals bonded together forming two channels, when even faster data transfer speeds of up to 128 Kb/s can be achieved. On the other hand, this indoor video-telephone model can be also attached to the GAN transceiver TES, such as Nera WorldCommunicator or TT-3080A Capsat Messenger from Thrane & Thrane. In addition, these TES can be connected simultaneously in LAN to several hardware systems such as: Voice solutions (ISDN phone, Cordless phones and VoIP); Remote office solutions (Fax, Laptop and Cordless phones); Videoconferencing solutions (Videophone or VPoIP and Web camera) and Multimedia solutions (PC, Camera and Cordless phone). Namely at this point, both TES models are also compatible with the Inmarsat mini-M service.

6.6.1.2. Regional Broadband GAN (BGAN)

Regional BGAN is a revolutionary communications system that enables customers to surf the Web, send E-mail and transfer a host of other data from anywhere within the satellite footprint. This is a go-anywhere wireless packet data service, based on the IP, which offers mobile, high-speed access to the Internet and corporate computer networks via a small, lightweight portable Satellite IP Modem, similar in size to a notebook PC. Because the service is based on IP, it is particularly suited to applications where data is sent in bursts, such as E-mail, browsing the Web and connecting to corporate LAN and Intranets. Other applications include file transfer and FTP for downloading files from the Internet, sharing files with colleagues, as well as E-commerce, for online ordering and procurement and secure end-to-end connectivity over a corporate VPN. In this sense, running as a secure and shared 144 Kb/s channel, Regional BGAN operates at more than twice the speed of current GPRS (General Packet Radio Service) mobile phones. It is based on IP packet technology, so users only pay for the amounts of data they send and receive and not for the amount of time spent online. In such a way, this enables users always to stay connected. A valid SIM card must be inserted for the Regional BGAN Satellite IP Modem to work. This SIM card may be used in other GPRS mobile devices for roaming between networks. Regional BGAN offers a truly exceptional range of business applications, such as IP Consultant, banking, E-fleet, E-commerce, etc. The service supports the commonly used network and Internet applications, providing continuity to professionals' working lives way beyond the boundaries of the office:

1. Instant Remote Access – It is able to access the corporate LAN solution maintains productivity levels from anywhere within the coverage area of Regional BGAN.

2. VPN Connectivity – Connecting to a wide range of corporate VPN ensures end-to-end access to secure information at high speeds.

3. High-Speed Internet Access – The Regional BGAN service enables reliable and fast access to Web content and resources any time, any place inside the coverage area.

4. Store and Forward Video – Fast and convenient, no other mobile wireless service can deliver comparable speed and simplicity.

Figure 6.36.   Regional BGAN Satellite IP Modem

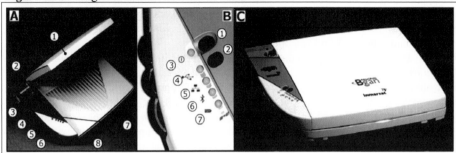

Courtesy of Webpage: "Inmarsat BGAN" from Internet

5. Remote IT Support – Performs software upgrades and runs diagnostic tasks remotely.

6. Digital Image Transfer – With BGAN it is possible to send and receive scanned images of documents and high-quality digital pictures.

7. E-commerce – The new Inmarsat packet data service is ideal for exploiting many other opportunities provided by E-commerce to engage business transactions with total security.

8. Database Queries – Much important information is housed in corporate databases and immediate access to such data stores is crucial to obtain business-critical information and customer records, amongst other data.

9. E-mail – As the backbone of corporate communication, E-mail is crucial to the efficient functioning of business users remaining in touch with other people by sending or receiving information as required.

The Regional BGAN will use a very small, light portable Satellite IP Modem, which is similar in size to a notebook PC, measuring approximately 300 x 204 x 32/42 mm and weighing about 1,594 grams.

In **Figure 6.36. (A)** is illustrated: **(1)** integral antenna, **(2)** compass, **(3)** SIM card, **(4)** battery, **(5)** external power, **(6)** USB and **(7)** indicators.

Figure 6.36. (B) shows control panel items such as: **(1)** Power control button; **(2)** Interface control button; **(3)** Power indicator; **(4)** USB indicator; **(5)** Ethernet indicator; **(6)** Bluetooth indicator and **(7)** Battery indicator and **Figure 6.36. (C)** illustrates the Inmarsat BGAN IP satellite modem ready to "take away and use".

They simply turn on the power, open the lid of the device and point it at the GEO satellite. A high capacity battery and charging system will power the unit, while power can also be used from an 110/220VAC 50/60Hz source. The battery is designed to provide one hour of continuous data communications operation at peak rates and 24 hours in standby mode. The IP Modem must be located so that it has an unobstructed view of the satellite. It is supplied with an integrated GPS antenna and receiver to enable the user to position the Satellite IP Modem correctly. It is connected to a standard PC or hand-held Personal Digital Assistant (PDA). Users access the Internet and other network services in the usual way, via the familiar Microsoft Windows interface or MacIntosh v10.1, where they can run standard applications, such as E-mail packages and Web browsers. However, connection can be by USB with a cable length up to 5 m and by Ethernet, providing connectivity to LAN with cable lengths up to 100 metres, or by Bluetooth, simply offering state-of-the-art wireless connection within a 10-metre range. For USB and Bluetooth access, the computer installation software will create a new Dial-up Network (DUN) access device known as the

Inmarsat Regional BGAN Satellite IP Modem. This device will appear in the list of network access devices presented in the Dial-Up Networking configuration dialog. For Ethernet access, the user will utilize the Ethernet NIC driver software provided by the NIC vendor or built into the Windows operating system.

Regional B-GAN is actually commercially available for mobile and fixed applications in 99 countries, across the northern half of Africa, Western and Eastern Europe, large parts of the Commonwealth of Independent States (CIS), the Indian sub-continent and the Middle East. Once the Inmarsat I-4 AOR satellite enters service, the coverage will extend from North and South America to the Pacific Rim. Thus, Inmarsat will introduce its Full BGAN service, offering voice and data communications at speeds of up to 432 Kb/s, with the entry into operation of the Inmarsat I-4 satellites. Users will be encouraged to migrate to the Full BGAN offering but the Regional BGAN service will still be available until around 2007/8. In additional, Inmarsat has leased capacity on a Thuraya satellite, located at 44° East. The Regional BGAN service will then be transferred to the Inmarsat I-4 satellites, scheduled for launch in 2005.

6.6.2. SCADA (M2M) Satellite Network

Inmarsat-C, mini-C, D/D+, GAN and other standards are very effective ways of remotely collecting basic environmental and industrial data such as for:

1. Leaks of Radioactive Material – Nuclear power stations depend on SCADA to monitor waste storage sites and shipments via satellite and provide an instant alert so that a potential leakage can be averted.

2. Industrial Process Control – Industrial SCADA allows controlling variables such as temperatures, chemical flow rates and emission levels to be remotely monitored 24H a day.

3. Pipeline Monitoring – Oil and gas pipelines (and electricity power lines) can run for a few hundred kilometres, crossing national frontiers as they go and can be vulnerable to acts of nature and human malice. SCADA functions include the monitoring of key operating data, transmission and confirmation of commands by control centres and wellheads and pumping stations in support of oil and gas distribution.

4. Water Resources Requirement – Water resources SCADA installations at remote reservoirs and on pipelines can give early warning of new leaks and allow repair work to be started promptly, at the right place and with the right resources. The result is minimum loss of water, economical repairs and less pressure to increase charges to water users.

5. Automatic Reporting for Ships and Vehicles – The transportation fleet generates a mass of data that ought to be reported and analyzed in the constant search for business advantage such as: positions, mileage and speeds; cargo condition; fuel and water levels; main engine condition and maintenance information, etc.

6. Controlling Lighthouses and Lighting Buoys – It is designated for the control of good and continuous working conditions of lighthouses and lighting buoys.

7. Power Stations Monitoring – This system is designated for remote control of power stations located in far away rural areas. It is necessary to perform voltage controls, breakdowns of systems and other monitoring to keep them in good order.

8. Meteorological Station Reporting – It is designed to conduct monitoring and reporting all meteorological and hydrographic data of remote meteorological station in rural areas.

9. Water Level Control – The satellite SCADA can perform remote water level control of rivers, lakes and water accumulation resources for hydroelectric generators.

Figure 6.37. Nera M2M (SCADA) Satellite Solutions

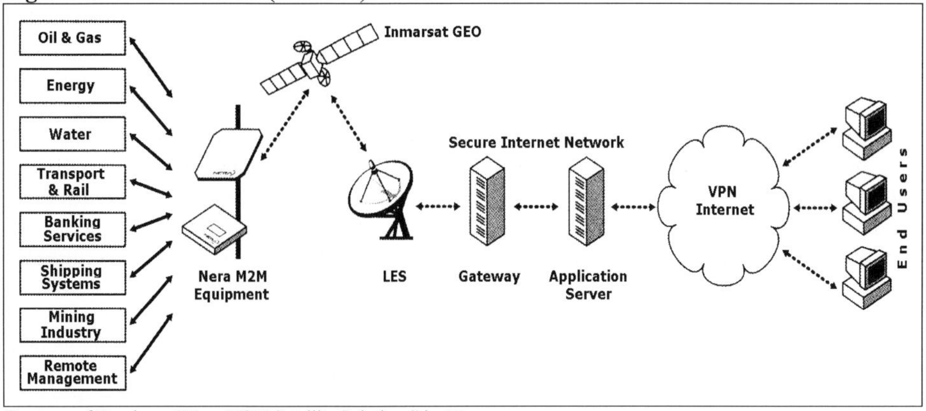

Courtesy of Brochure: "Near M2M Satellite Solutions" by Nera

The modern SCADA solution is Nera M2M GAN equipment, which introduces a new era of efficient machine-to-machine communications efficiency by providing reliable access to whole enterprise, facilitating remote monitoring and control in real-time, as illustrated in **Figure 6.37.** This system is based on the mobile IP and web technology of the Inmarsat MPDS highly cost effective global SCADA service, in that customers only pay for the data transmitted, not the time connected. In effect, Nera M2M (Machine-to-Machine) two-way communication solution via satellite provides constant monitoring, real time data retrieval and input to diagnostics that automatically leads to instant alerting, faster data assimilation and higher levels of decision making capabilities and operating efficiency.

6.7. Inmarsat Emergency and Safety Service

In 1972, IMO, with the assistance of CCIR, commenced a study of new distress and safety systems for maritime communications. After many years of planning and international consultation, the IMO and its member governments developed the new GMDSS (Integrated Radio and Satellite Communications), with the coordination of the CCIR, ITU, WMO, IHO, Inmarsat and Cospas-Sarsat. The GMDSS was incorporated into Chapter 4 of the SOLAS Convention and ships subject to the SOLAS Convention began implementing the GMDSS in 1992 and full implementation took place on 1 February 1999.

All ships use the same safety system but some will carry equipment on a mandatory basis. Namely, the carriage of communication equipment for GMDSS and SAR operations is mandatory for SOLAS Convention vessels (cargo ships of 3,000 GRT and over and passenger ships making international voyages), other ships will fit equipment to the GMDSS standard on a voluntary basis or as required by their national administrations. Most ships, whether SOLAS or not, will find it desirable and convenient to install Inmarsat-type approved ship's equipment which will provide advantages for commercial communications and the added benefit of acceptance for GMDSS operation.

Similarly, following IMO research activities, ICAO proposed the development of an advanced solution for ATM systems based on a digital ATC/CNS infrastructure for ATM known as Future Air Navigation System (FANS). The FANS proposal led to the definition of the ATS data link for ADSS and Control Pilot Data Link Communications (CPDLC).

6.7.1. Global Maritime Safety Satellite Communications (GMSSC)

The GMSSC mission is implementation of two GMSC systems developed by Inmarsat as a commercial and emergency solution and Cospas-Sarsat as emergency mission only. The Inmarsat system, as a part of GMSSC infrastructure, provides the crucial role that MSC plays in emergency and GMDSS communications. Moreover, Inmarsat is ideally placed to provide the GMDSS with very important MSC functions 1, 2, 4, 8 and 9 as specified by the IMO and functions 5 and 7 could also be carried out using a satellite SES. According to this context, all airplanes floating on the sea's surface because of an emergency situation, can be treated as an AES in a maritime distress environment and can use the facilities of GMSSC and GMDSS radio and MSC emergency networks.

Instantaneous communications via satellite to a CES and then directly to an RCC provide the GMDSS function 1 requirements. However, the means of CES–RCC interconnection may vary in each country but can include the use of dedicated lines of the public switched network. Instantaneous access to a satellite is provided by the use of Priority 3 distress, which is automatically included in a distress call made from any SES in the Inmarsat network. Priority 3 establishes a satellite channel, or clears a channel if, in the meantime, all satellite channels are engaged directly to the RCC. The GMDSS function 2 is provided by RCC-to-ship distress and alerting using Group calls to all vessels within a designated sea area. Otherwise, SAR coordination communications can be provided between suitably equipped SES to satisfy functions 4 and 8. Function 9 is ideally covered using the Inmarsat broadcasting of MSI and using the EGC service of SafetyNET/FleetNET transmissions.

As discussed earlier, the new built-in pre-emption and voice prioritization of Fleet F77 can benefit from even greater security and ensure that essential safety and emergency calls can get through instantly and, theretofore, non-essential and low-priority regular MSC traffic can be interrupted immediately. This means that rescue ships or aircraft on the sea's surface and other emergency and rescue services can always have seamless contact with relevant vessels on-scene SAR communications.

In addition, to satisfy most of the IMO requirements within the GMDSS network, Inmarsat is able to provide other maritime safety features, such as automatic ship reporting or polling service, just to enable shore authorities to know which ships are in the area of a causality and automatic transmission (ship-to-shore) of weather observation (WX OBS) to provide a detailed weather forecast for SAR units in the area of causality.

6.7.1.1. Global Distress, Urgency and the Safety Satellite System

The procedure of emergency transmission is the primary means of ship-to-shore distress alerting, via convenient CES in the Inmarsat system. However, the Inmarsat SES equipped vessels can also contact any RCC of their choice by following the calling procedure for routine calls. In this context, the complete international Tel/Tlx number has to be selected. Consequently, the Inmarsat system provides priority alerting for use in distress emergency situations with all standards, except mini-M.

1. Ship-to-Shore Distress Alerting – The Inmarsat Priority 3 (distress) system should only be used for making a distress call by Tlx (SOS) or Tel (MAYDAY) and calling by Tlx or Tel when crewmembers or passengers life is in imminent danger. It is necessary to follow the procedure for certain SES terminals and to select desired CES or RCC. Some of the CES units will automatically route all distress alerts and calls directly to an associated RCC

Figure 6.38. Distress Alert MSC Channels to an RCC

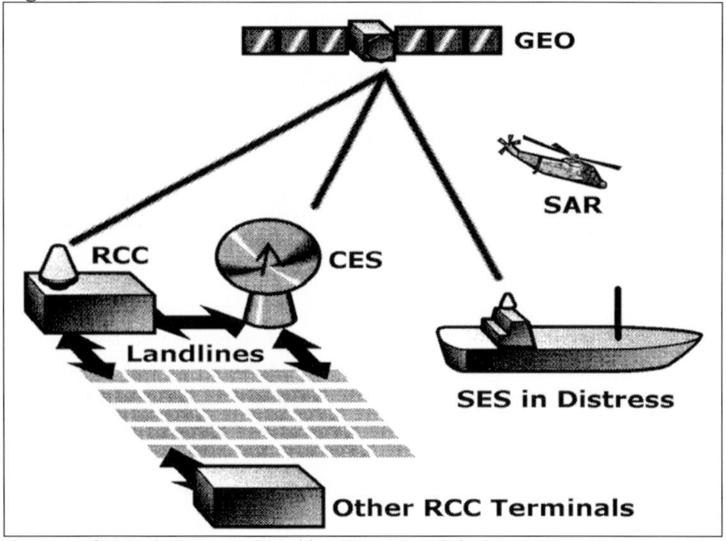

Courtesy of Manual: "Inmarsat-C Maritime Users Manual" by Inmarsat

and if an answer is not received within 12 seconds, the call should be repeated by the ship's operator. Moreover, in some SES, the initiation of a distress priority message is made by using the distress button, usually in red colour. At this point, most SES manufacturers provide instructions for the initiation of distress priority calls, which should be mounted close to the SES operating post. Besides, Inmarsat has also issued technical guidelines to manufactures for a Distress Message Generator (DMG), which consists in SES software to transmit distress messages in a standardized format that provides information on the vessel, its position and the particular emergency, and complies with the requirements of the IMO regulations and recommendations. In this sense, Medical Advice (32), Medical Assistance (38) and Maritime Assistance (39) can be obtained by following the Urgency and Safety procedure and using Priority 0 for these calls. This service can be obtained from some CES by using the two digit codes, which is nominated in brackets.

Inmarsat-C SES uses the signaling channel for distress alerting. Pressing the red distress button enables a short, preformatted alert to be transmitted directly to a CES as a back-up. Distress priority ensures special processing of the LES for rapid transmission to the associated RCC; see distress alert communication channels to an RCC and SAR, illustrated in **Figure 6.38.** Namely, the distress alerting format in an Inmarsat-C SES may be updated manually from the terminal keyboard. However, an automatic position updating may be provided by an integrated GPS or GLONASS receiver or by direct input from the ship's electronic navigation system.

2. Inmarsat-E Distress Alerting – As already explained, the Inmarsat-E EPIRB service is designed to provide ship's distress alerting by broadcasting to an Inmarsat GEO satellite from an L-band EPIRB device. Once activated the EPIRB provides distress alerting within 2 minutes directly to the shore CES and RCC terminals. In such a way, the EPIRB unit transmits 20 alerts within a ten minute time frame and each alert contains the SES ID, position of the distress and identification information to facilitate rescue. Message data may be automatically interfaced with the EPIRB or input manually before release. Both types of EPIRB include a GPS receiver so that the position of the beacons are constantly updated to an accuracy of better than 200 m. This information is transmitted via an Inmarsat satellite to a CES, where it will trigger an alarm, while at the same time it is automatically relayed to an RCC. The Inmarsat-E EPIRB must be mounted on board the ship in such a position that the vessel's nearby large superstructures will not obstruct the L-band signals when operated in situ. It must also float free via a mechanism that operates before reaching a depth of 4 metres. Once triggered, EPIRB terminals will continue transmitting for 48 hours, unless deactivated manually. Moreover, some EPIRB terminals also feature a SAR Radar Transponder (SART) beacon, for easy detection by SAR forces.

3. Inmarsat Fleet Distress Alerting – Inmarsat Fleet F77 offers the most comprehensive MDSS functions of all the Inmarsat family of SES. Fully compliant with IMO Resolution A.888 (21), Fleet F77 offers call prioritization to four levels and real-time, hierarchical call pre-emption in both directions: **(1)** Distress Priority-P3 will pre-empt all other MSC traffic; **(2)** Urgency Priority-P2 will pre-empt both P1 and P0; **(3)** Safety Priority-P1 will pre-empt only P0 and **(4)** Other Priority-P0. Thus, this means that Inmarsat CES must be capable of offering this valuable safety addition. With Fleet F77 SES terminal, the rescue authorities will always get a call through to a ship, even if the voice or data satellite channel is being used continuously.

4. Shore-to-Ship Distress Alerting – This facility can be performed by using the Inmarsat-C EGC SafetyNET service. However, distress alert can be transmitted to a group of ships with the Inmarsat-A or B SES being used as standard, with the exception of the EGC method, using the three following modes: **(1)** all ships calls in the ocean region concerned; **(2)** variable geographical area calls via SES-B to rectangular or circular areas and **(3)** group call to selected ships, which is very useful for alerting SAR units.

5. Shore-to-Ship Alerting through Inmarsat SafetyNET – The EGC receiver is normally an integral part of an SES, though it can be a completely separate unit. It ensures a very high probability of receiving a shore-to-ship distress alert message. When a distress priority message is received, an audible alarm sounds, which can only be reset manually.

6. SAR Co-ordination Communications – For the coordination and control of SAR operations, RCC require communications with the ship in distress as well as with units participating in the rescue operation. The method and modes of communication can be radio and satellite communications according to GMDSS requirements. To increase the speed and reliability of communication between RCC, SAR and on-scene participants it will be recommended to employ the Inmarsat system using SES terminals.

7. On-Scene SAR Communications – On-scene SAR communications are defined as those between the ship in distress and assisting vessels or helicopters and the on-scene Commander or the surface SAR coordinator. These are usually VHF and MF short-range radio communications, however, Inmarsat SES-fitted vessels can, if necessary, use MSC to supplement radio facilities.

8. Promulgation of MSI via Inmarsat Service – In the Inmarsat system, promulgation of MSI for the International SafetyNET service is performed by means of the Inmarsat-C EGC capability. If uninterrupted receipt of MSI is required, or the Inmarsat-C SES is used for above-average amounts of general communications, it is essential for the ship to have a dedicated EGC receiver for taking MSI broadcasts. At this point, an EGC Rx is usually an integral part of an SES-C terminal and may also be fitted as a separate unit together with other types of SES.

Figure 6.39. Navarea/Metarea MSI Broadcasting Areas

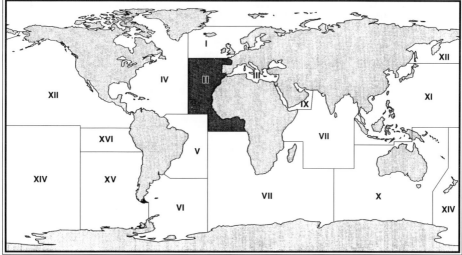

Courtesy of Manual: "Inmarsat-C Maritime Users Manual" by Inmarsat

6.7.1.2. Maritime Safety Information (MSI)

The MSI service is an international system of radio and satellite broadcasting for mariners containing navigation or weather information and warnings. Thus, the Inmarsat-C system has a capability of EGC, which can enable an authorized information provider to broadcast messages through a CES and NCS to selected groups of SES that are equipped with an EGC receiver capability, such as EGC Rx and Inmarsat-C transceivers. Two EGC services are available, such as the EGC SafetyNET for the broadcasting of MSI and EGC FleetNET for broadcasting commercial information to mariner subscribers. All SES fitted with an EGC Rx can receive SafetyNET MSI broadcasts but to receive FleetNET messages, an SES must also include a FleetNET option, already installed or additionally upgraded and be registered as a subscriber with the information provider. In fact, to receive a scheduled broadcast of MSI for a given Navarea/Metarea, EGC SES terminal should be logged-in to the appropriate ocean region at the time of the MSI broadcast, see **Figure 6.39.**

1. EGC SafetyNET Service – The SafetyNET service has been established by Inmarsat to provide a fully automated MSC system capable of addressing messages to individual ships, a predetermined group or all ships in variable geographical areas. However, the EGC alerts may be addressed to a group of ships designated by fleet, flag, or geographical location. A geographical area may be further defined as a Navarea (standard NX forecast areas), METAREA (standard WX forecast areas), a rectangular area defined by latitude and longitude or circular area around a maritime emergency case, as the latter is illustrated in **Figure 6.40.** At any rate, this special service enables information providers who have been authorized by the IMO under the GMDSS mission, to distribute efficient and low-cost means of MSI transmitting from shore to vessels at sea as a safety service for mariners. The SES equipped with EGC capabilities automatically monitors the MSI frequencies, receives and prints out information relevant to the ship safety. The EGC messages can be directed to all ships in a geographic areas or approaching specific regions such as the sea area around a

Figure 6.40. SafetyNET Call to a Circular Area around an Emergency

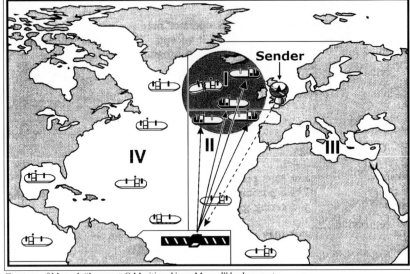

Courtesy of Manual: "Inmarsat-C Maritime Users Manual" by Inmarsat

RCC infrastructure, which service includes: **(1)** RCC stations provide shore-to-ship distress alert and other very urgent information; **(2)** national weather centres provide broadcasting services of Meteorological Warnings (MW) and daily WX forecast; **(3)** hydrographic offices provide Navigational Warnings (NW) and electronic chart correction data and **(3)** international ice patrol provide ice hazard information for the North Atlantic.

2. EGC FleetNET Service – FleetNET allows authorized information providers, such as commercial subscription services, shipping companies and government organizations, which have a registered agreement with CES that supports the FleetNET services, to broadcast messages and commercial information to selected groups of SES terminals or to a virtually unlimited number of predestinated SES terminals simultaneously. In such a manner, each of these terminals has registered with the information provider and been added to a FleetNET EGC closed network, which may belong to a fleet or be registered subscribers to a satellite communication commercial service, as is shown in **Figure 6.41.** Typical FleetNET applications include: **(1)** fleet or company broadcasts to all ships; **(2)** commercial WX service; **(3)** government broadcasts to all vessels on a country's register; **(4)** news broadcasts and **(5)** Market quotations.

Once the SES equipment has been initialized to receive the NCS common signaling carrier for the Ocean Region in which a ship is situated, the operator can select some or all of the information listed by special service codes, which include: All ships call (00), Group call (02), MW to rectangular areas (04), Shore-to-ship distress alert (14) and so on.

3. Enhanced Voice Group Call (EVGC) Network and Service – Using a single voice channel on the Inmarsat-A or B service, Inmarsat have provided an enhanced voice service. The EVGC system provides reception of subscription services including the rebroadcasting of information services and radio programs. In addition, slow-scan Fax transmission can be accommodated within the restricted bandwidth of a voice channel to provide weather Fax charts for any part of the world.

Figure 6.41. FleetNET Call to the Ships of one Company or Fleet

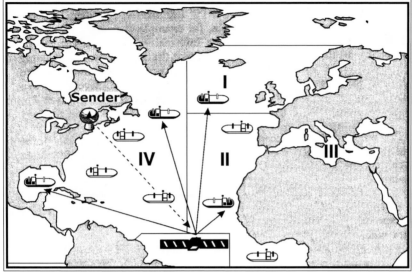

Courtesy of Manual: "Inmarsat-C Maritime Users Manual" by Inmarsat

4. Enhanced Maritime Weather Service – The weather service (WX) for mariners is a very important mission, providing all ships with text and graphic WX forecast, MW and weather routing services for safe and more economical shipping operations. Otherwise, similar enhanced WX service can also be obtained from Shipline Routing Report Centres by daily message reporting via Inmarsat networks to all ships individually subscribed to use this enhanced service, which is illustrated in **Figure 6.42. (A).** In a more general sense, these messages contain the WX situation around related vessel with the best-recommended route to avoid local bad weather conditions and inconvenient wave streams. Moreover, all ships can send on a voluntary basis special OBS messages via CES in Meteorological and WX centres with the purpose of enhancing the WX data base storage to obtain global weather information.

6.7.1.3. Maritime Broadcasting Satellite System (MBCSS)

A modern commercial ship is a multi-million-dollar mobile profit centre, under growing pressure to sail faster, to tighter schedules, with more electronic equipment and with a smaller crew. It also needs to be just another node on the operator's wide-area network, feeding financial and operating data into corporate headquarters, getting instructions and information back, so it can be turned round quickly and managed to optimum efficiency at all times.

Ships fitted with Inmarsat satellite communications solutions can be fully integrated into management systems ashore, by phone, Tlx, Fax, E-mail and the Web browsing, wherever they may roam on the world's oceans to provide MBCSS, which can include: position reporting system, data reporting and polling service, weather data routing, electronic charts, maritime reports and assistance services, planned maintenance, telecontrol and monitoring; telemedicine, teleeducation and so on.

Figure 6.42. Enhanced Maritime WX Service and GPS Augmentation via Inmarsat

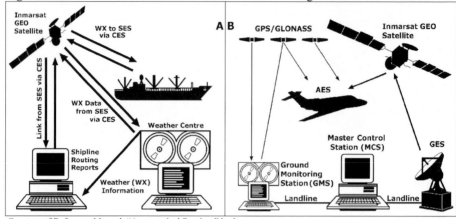

Courtesy of Reference Manual: "Aeronautical Services" by Inmarsat

6.7.2. Global Aeronautical Safety Satellite Communications (GASSC)

Over 4,500 aircraft have been fitted with Inmarsat Aero terminals to improve commercial, corporate, passenger and safety communications. These include more than 1,200 corporate and government aircraft and over 1,700 installations in airline companies. Corporate users favor telephone and Fax services, while the airlines also make use of the data services for commercial and safety AMSC. In addition, more than 1,200 business airplanes, helicopters and military air transports are fitted only with the Aero-C terminals. All these installations will improve commercial and safety Aero communications, enhance current ATC and implement the new Global Satellite Augmentation System (GSAS). Regarding the GSAS concept and other questions relating to GASSC subjects, all readers with a special interest can find three current developed GSAS models in Chapter 13: WAAS, EGNOS and MTSAT, including the navigation payload of the present Inmarsat Civil Navigation Satellite Overlay (CNSO) system.

6.7.2.1. Inmarsat Safety AMSC

The Inmarsat AMSC system can also play a major role in Air Traffic Management (ATM) systems to ensure that a plane gets to its destination both safely and efficiently, to establish GASSC and to enhance alerting and locating services for aircraft involved in accidents. Namely, in respect of the ATM function, ATM must perform three basic tasks, termed as satellite Communications, Navigation and Surveillance (CNS) as follows:

1. Satellite Communications – The AMSC is the exchange of voice and data information such as routine traffic or instructions between the aircraft pilots and air traffic controllers in airports.

2. Satellite Navigation – The GPS, GLONASS or new enhanced Inmarsat GNSS satellite positioning is a process providing pilots with information on the position of the aircraft.

3. Satellite Surveillance – The new aeronautical satellite surveillance solution, known as an Automatic Dependent Surveillance System (ADSS), is the process of detecting the position of the aircraft by ATC.

Until recently, the only means of communication between pilots and ATC was by voice VHF and HF radio equipment. The HF radio is used for long-distance communications in ocean or remote continental airspace areas, while the VHF radio is used for line-of-sight or short distance direct-line communications in regional or domestic airspaces. To overcome the disadvantage associated with HF/VHF radio, the ICAO encouraged the development of AMSC and airborne AES located on board aircraft. A most important application within AMSC is for route communications, which are in relation to safety and regularity of flights primarily along national or international civil air routes.

To negate the effects of unreliable HF/VHF radio communications or unavailability of radar coverage, the ATM system maintained safety by keeping aircraft separated from one another by large distances. Unfortunately, this method of operation was relatively inflexible. Pilots often could not reliably contact ATC to deviate around adverse weather systems or take advantage of any new information on meteorological conditions. The procedure of the current ATM system thereby resulted in aircraft delays, inefficient operation and high fuel costs, all of which were further compounded by a growing air traffic demand along heavily traveled routes.

The new safety/security application, requiring an integrated AMSC system with satellite determination capability, is for aeronautical services like a part of the ICAO's CNS/ATM system. Many AES terminals will support ATC, ADSS and ATM, which allow air traffic controllers to poll the aircraft for positioning, weather, safety and other information.

6.7.2.2. Satellite Air Traffic Control (SATC)

Inmarsat AMSC application plays a major role in the implementation of the ICAO SNC/ATM concept for a new SATC in oceanic and remote airspace. The Inmarsat Aero network will support direct pilot-to-controller voice and data communications and ADSS. Improved routing and enhanced SATC are expected to yield millions of dollars in fuel, safety and other operational cost savings to airline operators, while reduced separations will increase the capacity of oceanic and remote airlines. Moreover, with new, enhanced SATC control of all movement of airplane and vehicle traffic on the airport surface can be improved. The Inmarsat Aero new satellite data link will be used for routine pilot-to-controller kinds of communications and requirements. However, voice communications can be used for non-routine and emergency transmissions. Use of the AMSC data link to integrate aircraft fleets in flight into airline formation systems can yield significant increases in operational and administrative efficiency for the airlines.

There are a variety of AMSC applications that have been developed to support the CNS of ATM system. The applications necessitate high levels of availability, performance and integrity, as defined internationally in the new ICAO Standards and Recommended Practices (SARP) and regionally, by standards such as the Radio Telecommunication Association and Minimum Operational Performance Standards (RTCA-MOPS). The main types of AMSC applications are related either to SATC and airline administrative communications or to passenger services. These are summarized as follows:

1. Air Traffic Control (ATC) – For ATC, AMSC solutions are used by pilots to keep in contact with ground staff in airports and other offices for routine communications such as sending ETA, routine air traffic, request for clearances and advisories and other corporate and safety transmissions. Correspondingly, the ground controllers use AMSC facilities to monitor and direct the position of aircraft, even when outside normal radar range.

According to the ICAO declaration, the data communications will be primary means of pilot-controller information exchange, Notices to Airmen (NOTAM) and voice will be used in emergencies and other non-routine situations.

2. Air Passenger Services (APS) – For APS, AMSC non-corporate communications are used by customers to make phone calls, send data or facsimiles whilst in flight.

6.7.2.3. Inmarsat Satellite Augmentation System (ISAS)

The new ISAS is an integral part of the WAAS Global Satellite Augmentation System (GSAS) aeronautical solution for enhanced satellite Communication, Navigation and Surveillance (CNS) and will provide flight control in the air (domestic and international flying corridors) and control of all movements on the ground (airports), see **Figure 6.42. (B)**. The GSAS configuration requirement will be a combination of existing GEO satellite systems for communications and navigation data; GPS, GLONASS and Inmarsat GNSS, including integrated US-based WAAS, European-based EGNOS and the forthcoming GNSS Galileo. In fact, WAAS and EGNOS employed Inmarsat-3 satellites by contract for their GSAS services. The GMS are receiving GPS/GLONASS signals, performing augmentation and via landline, MCS, GES and GEO sending to AES. Otherwise, the GSAS system will also be convenient for military applications and will offer the following enhanced features:

1) Enhanced Positioning – The GPS/GLONASS old satellite navigation solutions provide position with an accuracy of about 30 m and, therefore, new, more precise navigation and surveillance requirements require enhancement. The enhancement can be performed with an augmentation of these old Satellite Navigation Systems (GPS/GLONASS) with new technology and integration with Inmarsat AMSC and CNS, Radio HF/VHF and Ground Monitoring and Surface Radar Systems. Unlike standard GPS, which can take 30 minutes to notify users of a bad signal, the GSAS system will be capable of notifying users much quicker. By correcting the GPS signal, the GSAS system achieves an accuracy of position between 3 to 7 metres, which is very important for more precise aircraft landings.

2) Reduction of Separate Minima (RSM) – The next important improvement using new GSAS is RSM between aircraft on the air routes almost doubling, for instance the new system will be able to control about 8 aircraft simultaneously in one flight corridor, instead of the 4 controlled by the current system. At this point, the current System has RSM controlled by Radio VHF/HF Radio and Surface Radar systems, which safe flight allows only large distances between several aircraft over the ocean or approaching flight routes. On the other hand, the new Communications, Navigation and Surveillance/Air Traffic Management (CNS/ATM) System is controlling and ranging greater numbers of aircraft for the same air space corridors, which enables minimum secure separations. Significant reduction of separation minima for flying aircraft in air corridors will be available with the introduction of new GSAS for satellite CNS systems.

3) Flexible Flight Profile Planning (FFPP) – The GSAS will provide optimal altitude and route as well as FFPP. The current system is using fixed air routes and flying altitudes, which are controlled only by aircraft navigational aid instruments. Hence, the flying route is composite, variable in flight course and altitude and therefore not the shortest one from departing to arriving at the airport. However, FFPP allows selection of the shortest or optimum routes and flying altitudes between two airports. Thanks to the new GSAS technology of CNS new FFPP will be available for economic and efficient flight operation.

This means that the aircraft engines will use less fuel by selecting the shortest flying route from the new CNS/ATM System than by the preselected fixed route and altitude of current route composition.

4) Surface Movement Guidance and Control (SMGC) – The SMGC infrastructure is a special system that enables a controller in the control tower to guide aircraft on the ground even in very poor visibility conditions at an airport. A controller issues instructions to pilots with reference to a command display in the control tower that gives aircraft position information detected by sensors on the ground. The command monitor also displays reported position information of landing or departing aircraft and all aid vehicles moving on the airport surface. This position is measured by the new Global Navigation Satellite System (GNSS) using data from GPS and/or GLONASS and Inmarsat communication satellites. A controller is able to show the correct taxiway to pilots under poor visibility, by switching the taxiway centreline light and the stop bar light on or off. Development of head-down display and head-up display in the cockpit that gives information on routes and separation to other aircraft is in progress. The major goal of this system is to improve the functions of the control tower and airport infrastructure in each airport. Hence, the control tower as a centre for monitoring the traffic situation on the landing strip on the airport ground environment can also develop monitoring for eventual hijacking. The primary function of this service is to use the GSAS system for the location of each aircraft and vehicle and display them on the command monitor of the control tower. The controller performs ground controlled distance guidance based on this data for all aircraft and vehicles moving on the airport surface. Secondly, to improve Aircraft Cockpit Displays on aircraft position and routes on the windshield (head-up displays) and instrument panel display (head-down display).

Inmarsat provides Satellite-based GPS/GLONASS Regional Augmentation for WAAS and EGNOS solutions serving maritime, land and aeronautical applications. Namely, Inmarsat-3 satellites carry communications payload as well as navigation transponders, which can enable means to enhance GPS/GLONASS performance as follows: integrity accuracy of positioning, accuracy enhancement and availability. In any event, this system requires the monitoring of GPS or GLONASS spacecraft on the ground and unlinking augmentation data signals to the AES terminal via Inmarsat satellite constellation, see **Figure 6.42. (B)**.

6.7.2.4. Automatic Dependent Surveillance System (ADSS)

Whilst Primary and Secondary Surveillance Radar (SSR) have been the core aeronautical systems providing ATM Surveillance Services for over 30 years, the continuous growth in air traffic has led to a need to enhance these surveillance systems to help support increased airspace capacity. Moreover, it has long been recognized that there are parts of airspace where rotating SSR systems are not feasible or are too costly. On the other hand, there is an emerging new technology ADSS that may resolve the above issues. Namely, this solution represents a surveillance technique in which an aircraft transmits on-board data from avionics systems to ground-based and/or airborne receivers. The data may include: aircraft identity, position, altitude, velocity and intent.

The ADSS application is the reporting via Inmarsat Aero network of position and intention information derived from an aircraft's aboard navigation system. Presented on the radar screen like displays at ocean control centres, it will give controllers a real-time knowledge about the air traffic situation, permitting more fuel-efficient routing and reduced separation

Figure 6.43. Aeronautical ADSS and Enhanced Aero WX Service

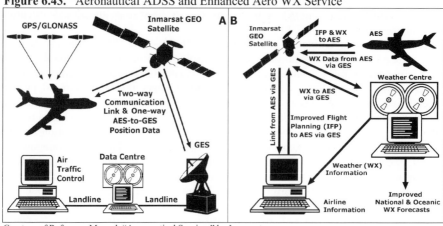

Courtesy of Reference Manual: "Aeronautical Services" by Inmarsat

standards, as illustrated in **Figure 6.43. (A)**. Namely, GPS/GLONASS receiver terminal received signals from GPS or GLONASS satellites, aircraft position and other significant data will be derived on board and transmitted to the ground using AES data link in parallel with other information, without pilot involvement, by air traffic managers' set reporting rates and finally, all received ADSS information in the Data Centre will be forwarded via landline to ATC and displayed to the air traffic manager on radar-like displays. This solution will also be convenient for military applications and it will enable new airborne and ground ATM functions, with the potential to bring extra capacity and increased safety in air corridors. In addition, using satellite ADSS will improve the availability of the real-time flight navigation data from aircraft, support airport ground surveillance, provide surveillance services in areas with none or only locally limited radar coverage and contribute to the compatibility of air and ground systems.

6.7.2.5. Aeronautical Weather Report Services (AWRS)

The ARINC Global Link Value Added Service (VAS) provides a text and graphics and text weather (WX) service receiving weather products from specialized aviation weather service providers, it adds the information to its database and transmits it to requesting aircraft as text or compressed graphic weather images directly to the cockpit via AES terminal. Therefore, weather products can include radar precipitation images, winds aloft, icing, turbulence, precipitation, upper air WX situation (wind speed, direction and temperature), significant WX and lightings. In addition, Meteorological Aviation Reports (METAR) and Terminal Area Forecasts (TAF) can be requested in either standard text format or as enhanced plain language transmissions.

The SITA Gateway service for Digital Automatic Terminal Information Service (D-ATIS) with established data link via Radio VHF or Inmarsat satellite system is another means for Air Traffic Service (ATS) providers to deliver current information on weather and airport conditions to pilots as a NOTAM. In this sense, the ideal means of WX data and other NX transmissions will be on a global basis from one service provider to all airline companies with compliance only to ICAO communication standards and regulations.

The Inmarsat system can provide enhanced WX service from WX centres via satellites to aircraft and OBS messages in the opposite direction, as shown in **Figure 6.43. (B)**. The Weather Centres receive many WX and meteorological data directly from meteorological satellites, Ground Observation Centres, WX data known as OBS messages from ships and aircraft and other information resource centres, processing these WX data and information and afterwards sending via landline telecommunications to an Airline Information Centre, which retransmits these messages in IFP form to AES, via Inmarsat satellites.

The best solution for the transfer of WX messages, charts and NOTAM to aircraft is by using Flight Internet access via the Mobile ISDN Swift64 Inmarsat service. On the other hand, there is an ideal possibility in the future to utilize communication systems similar to the new-developed very high-speed shipborne Broadband Maritime Communication Service (BMCS) as an airborne Broadband Aeronautical Communications Service (BACS). Thus, this Service can be very suitable for transfer of all data and information for corporate, commercial, distress and safety purposes.

6.7.2.6. Future Air Navigation System (FANS)

Global link and Satellite data link communications have been endorsed by the ICAO FANS committee as the primary medium for Oceanic ATC communications. Although the current HF voice-based system has adequately supported all ATC oceanic communications in the past, increasing traffic and message throughput requirements are limited in the current voice-based system. In fact, there are other even cheaper HF radio solutions, such as digital transmission with the possibility of data transfer, but this will be discussed in another forum. In the meantime, HF radio communications for aeronautical application still will be in exploitation for uncovered polar regions by Inmarsat system.

The existing capabilities and coverage of the VHF Aircraft Communications Addressing and Reporting System (ACARS) for Aeronautical Operational Control (AOC) solutions, ATC services, and Airline Administrative Communications (AAC) are expanded with new Global Satellite Supports Airline Operations (GSSAO) and ATC applications. The GSSAO data link includes report on departure/destination locations and movement times, engine monitoring, delays, aircraft position, maintenance reports and winds aloft observations.

On the other hand, complicated ATC instructions, such as oceanic clearances, can now be displayed in written form on aircraft monitors and printers and retrieved at the pilot's convenience. FANS-1/A facilities provide transfer of information between ATC, aircraft and ground operators. Applications include controller pilot data link communications and automatic dependent surveillance.

The SITA system provides pre-FANS and FANS facilities through its new ATS AIRCOM service for ATC and ADSS services. Satellite AIRCOM provides worldwide ACARS air-to-ground communication services that are fully compliant with Airline Electronic Engineering Committee (AEEC) characteristics 618 and 620, using Inmarsat Data-2 capabilities. The AAC between cabin crew and ground control staff can be improved through reliable and timely cabin management, configuration and provisioning. The AOC communications encompass all aircraft flight operations, maintenance and engineering.

The ARINC VAS also provides the FANS facilities mentioned above, including D-ATIS, WX data service and all information for pilots, predeparture and departure clearance, communications between pilots and controllers, ATS Interfacility Data Communications (AIDC), Oceanic clearance delivery, Centralized ADSS, etc.

6.7.2.7. Global Aeronautical Distress and Safety System (GADSS)

In addition to the ATM function, the AMSC provides an important distress alerting and aircraft locating service via the Inmarsat network, GES and RCC terminals, although the AMSC distress and safety system is still not completely developed and implemented. The ICAO has the only "long way" project known as FANS, which does not include enough enhancements regarding distress AMSC solutions. In addition, there is confusion today in global AMSC systems because new satellite applications are mixed with old aeronautical information solutions, instead of generally being defined as two systems as follows:

1) Global Aeronautical Corporate and Commercial System (GACCS) – The GACCS has to enable all Radio and AMSC service between aircraft and airport on one hand and between aircraft and airways companies of offices on the other. This service can include all NOTAM information, ATS Interfacility Data Communications (AIDC), ETA messages, arrival clearance, departure and predeparture clearances, oceanic clearance delivery (OCD), Operation control/maintenance and engineering data, Flight plans and progress, course changes information, position data and reports (0001), controller to pilot data link communications, voice communications between cabin crew and ground staff and all commercial voice/Fax or data messages.

2) Global Aeronautical Distress and Safety System (GADSS) – Without consideration of the main subject of ICAO FANS, the best solution will be the establishment of Future GADDS separately, more effective and similar to a current GMDSS of IMO. At this point, there are necessary predispositions for Future GADSS mission such as: Inmarsat AMSC system and terminals with Inmarsat-E EPIRB; Cospas-Sarsat system with ELT and new DSC (Digital Selective Call) HF/VHF Digital Radio system. Without doubt, the Inmarsat-E floating and other maritime Cospas-Sarsat EPIRB terminals have to be mandatory on board every aircraft in case it has to land at sea, owing to engine or other troubles, or to use as well as ELT with floating possibilities. Following this situation, the SAR procedure for aircraft floating on the sea's surface has to be same as for ships in distress. Besides distress and SAR mission, this system has to provide an additional service including all Safety ATC communications, centralized ADSS position reporting, WX and weather forecast, NX reports, Aeronautical Highlights and Navigation Information Services (AHNIS), hijacking prevention and information, medical service, technical advice, etc. Without doubt, the GADSS network and service will be indispensable and important for the safety of air traffic, similarly to the already successfully developed GMDSS mission by IMO maritime.

In any event, the development of new missions for GADSS has to be led by the ICAO and supported by other communities involved in AMSC services, including Inmarsat and Cospas-Sarsat systems. Without the new GADSS integrated mission, all other solutions and technical implementations will not be complete and successful.

The current function of Inmarsat AMSC is also accomplished by equipping aircraft with special emergency distress satellite beacons, known in the aeronautical industry as an Emergency Locator Transmitter (ELT), which can be detected and located by the special Cospas-Sarsat LEOSAR and GEOSAR satellite system's, LUT and RCC ground terminals. This service has to be integrated with a new solution to the DCS HF/VHF radio system in the new GADSS emergency mission. In this sense, the radio system will be used to provide reliable communications for both Polar Regions uncovered by the Inmarsat System. Otherwise, the readers can find more detailed information about the Cospas-Sarsat LEO and GEO systems in the following chapter.

7

COSPAS-SARSAT LEO AND GEO GMSC SYSTEMS

7.1. Scope of the Cospas-Sarsat System

Vessels sink or become disabled, airplanes crash in disasters, land vehicles and expeditions get lost in the wilderness and other causal emergencies occur that jeopardize many lives and property at sea, in the vast inland regions and air space of the entire world. Continuous advances in technology and rapidly improved use of radio communications has been an historic objective of those charged with solving the Search and Rescue (SAR) problems at sea, on land and in the air. However, early in the history of mobile radio systems, only a hundred years ago radio equipment was installed on ships and aircraft to improve the safety of life and property at sea and in the air. While these improvements generally started well and assisted saving many lives and much property, they were not effective when there was a sudden, catastrophic loss of the supporting platform. For that reason, the most effective improvements occurred to this plan at the end of the preceding century.

Available high technology and requirements for improvement of safety mobile radio were combined in the early 1970s to enable a new system to be developed to counter the really alarming catastrophic loss problem. At that time, small beacons for ships and aircraft needs in the USA, Canada and in Europe were developed in large numbers. These radio beacons are generally small, inexpensive and very lightweight. At that time, the only monitoring platforms were overflying aircraft and in such a manner aeronautical distress frequencies were used on 121.5 MHz and 243 MHz for signal alerting from distress transmitters of radio beacons.

There were immediate benefits and significant problems, such as hasty implementation, cost constraints and operational mistakes combined to produce much unintentional activation (false alarms). Aircraft overflights provided incomplete and sporadic coverage, because high-flying airplanes are able to see electronically large ground areas and localization of the received distress transmission was very difficult. In the meantime, using this system, many lives were saved worldwide and the need for emergency radio beacons grew rapidly. In addition, all SAR authorities continued to develop improved techniques to perform better detection of alerts and location of real distress incidents in such a way as to minimize the problems caused by false alarms.

With regard to previous endeavors, NASA and NOAA in the USA and CNES in France had years of experience in using satellite techniques to locate and collect data from globally deployed weather platforms, high altitude balloons, floating buoys and migrating animals, etc, while Canada had experimented with locating ELT using the Radio Amateur Satellite OSCAR. After that, two separate experiments were undertaken. The first was designed to use two of the above-mentioned VHF aeronautical frequencies and the second to use the UHF 406 MHz frequency band based on the Doppler effect. On the other hand, the former-Soviet Union developed its own Cospas System for SAR missions.

7.1.1. Cospas-Sarsat Organization and Signatories

The Cospas-Sarsat system is a joint international satellite-aided SAR system established and operated by organizations in Canada, France, Russia (ex-USSR) and the United States. This satellite system was initially developed under a Memorandum of Understanding (MU) among Agencies of the former USSR, USA, Canada and France, signed in 1979. Following the successful completion of the demonstration and evaluation phase started in September 1982, a second MU was signed on 5 October 1984 by the Department of National Defense (DND) of Canada, the Centre National d'Etudes Spatiales (CNES) of France, the Ministry of the Merchant Marine (MORFLOT) of the former USSR and the National Oceanic and Atmospheric Administration (NOAA) of the USA. Hence, the system was then declared operational in 1985. On 1 July 1988, the four states providing the space segment signed the International Cospas-Sarsat Program Agreement, which ensures the continuity of the System and its availability to all states on a non-discriminatory basis. In January 1992, the government of Russia assumed responsibility for the obligations of the former USSR. Otherwise, a number of other states, non-parties to the agreement, have also associated themselves with the program and participate in the operation and the management of the system, see the first Cospas-Sarsat emblem in the **Figure 7.1.**

Figure 7.1.   Cospas-Sarsat Emblem

Courtesy of Book: "Satellite Aided SAR" by SNES

7.1.2. The International SAR Program

The Cospas-Sarsat mission is a satellite system designed to provide emergency and distress alert and location data to assist SAR operations, using spacecraft and ground facilities to detect and locate the signals of distress beacons operating on 406 MHz or 121.5 MHz. The position of the distress and other related emergency information is forwarded via LUT LES by the responsible MCC to the appropriate national RCC and SAR authorities. Its objective is to support all organizations in the world with responsibility for SAR operations, whether at sea, on land or in the air.

The detection and location of a ship distress or an aircraft crash is of paramount importance to the SAR teams and to the potential survivors. Hence, studies show that while the initial survivors of an aircraft crash have less than a 10% chance of survival if rescue is delayed beyond two days, the survival rate is over 60% if the rescue can be accomplished within 8 hours. A similar urgency applies in maritime distress situations, particularly where injuries have occurred. In addition, accurate location of the distress can significantly reduce both SAR cost and the exposure of SAR forces to hazardous conditions and clearly improves efficiency. In view of this, Canada, France, Russia and the USA established the Cospas-Sarsat satellite system to reduce the time required to detect and locate SAR events worldwide. The first generation of Cospas-Sarsat system was formed by two compatible, interoperable and international SAR satellite systems:

1) The Cospas system was developed by former Soviet Union experts and in Russian is written КОСПАС (Космическая система поиска аварийных судов и самолетов), which when translated into English, means as: Space system for search of distress vessels and airplanes. In the English acronym it is not indicated that Cospas also includes aircraft.

2) The Sarsat acronym was derived from Search and Rescue Satellite-aided Tracking and the system was designed by Canada, France and the USA.

After extensive technical testing and implementation on two Cospas (Cosmos – in Russian, Космос) and one Sarsat (NOAA) spacecraft, so Cospas-Sarsat system demonstration and evaluation phase began in February 1983 until July 1985.

In the meantime, the successful operational experience of the Cospas-Sarsat system by SAR organizations started on 9 September 1982 with the crash of a light aircraft in Canada, when three Canadian persons were rescued. Namely, the Canadian crashed aircraft was detected by Cospas satellite Cosmos 1383 and it assisted in saving three lives, using SAR satellite aids for the first time in the world. As of May 1986, the system had contributed to saving 576 lives worldwide, of this, 244 maritime, 21 land and 311 air, using the current satellite beacons EPIRB on ships and ELT on airplanes, both operating on the frequency of 121.5 MHz. Since then, the system has been used for hundreds of SAR events and has been responsible for the saving of several thousand lives globally.

The Cospas-Sarsat satellite system has demonstrated that the successful detection, location and SAR of distress signals can be processed by transponders based on LEO in near PEO Cospas and Sarsat spacecraft. This Cospas-Sarsat system was than declared completely operational in 1985. On 1 July 1988, the four mentioned MU state operators providing the space segment signed the International Cospas-Sarsat Program Agreement, which ensures the continuity of the system and its full availability to all other States' administrations on an international and non-discriminatory basis. In addition, a number of states, non-parties to the signed agreement, however, have also associated themselves with the program, such as Norway, Sweden, the United Kingdom, Finland and Bulgaria. Otherwise, these countries are also participating in the joint project to evaluate its effectiveness in their SAR regions. At least a dozen other nations are also actively considering participation at this time.

Following the success of Cospas-Sarsat system, the IMO and the ICAO recommended that ocean-going ships and large aircraft must carry Emergency Position Indicating Radio Beacon (EPIRB) and Emergency Locator Transmitters (ELT), respectively. Therefore, in conformity with previous the attitude, during November 1988, the Conference of Contracting Governments to the International Convention for the Safety of Life at Sea, 1974 (SOLAS Convention) on the GMDSS (1988 Conference) adopted several amendments to the 1974 SOLAS Convention whereby, inter-alia carriage of satellite EPIRB on all convention ships of 300 tons and over became mandatory from 1 August 1993. In addition, various national requirements also exist for the carriage of ELT/EPIRB on different types of craft not otherwise subject to international conventions and some countries have authorized the use of Personal Locator Beacons (PLB), 121.5 MHz and 406 MHz emergency beacons for use on land, namely in remote or rugged areas.

One of the characteristics of a satellite in low PEO of the new LEOSAR (LEO Search and Rescue) system is that it can see the entire globe once every twelve hours, while the contact time with any one coverage area is relatively short. The waiting time between the activation of a distress beacon transmitter and its detection by the overflying satellite is significantly reduced by having more than one satellite available in a certain area. Thus, in a constellation of four satellites the mean waiting time is approximately one hour. Considerable enhancement of system could be done separately either by launching more satellites in PEO or by using additional GEO space segments in integration. These GEO satellites will have the possibility of almost immediate alerts with identity at 406 MHz.

In the meantime, the Cospas-Sarsat Council has considered the development of a GEOSAR (GEO Search and Rescue) satellite system capability as a potential enhancement to the existing LEOSAR (LEO in near-PEO) satellite system. In recent years, the mentioned Cospas-Sarsat participants have been experimenting with 406 MHz beacons via payloads on GEO satellites and several experimental ground terminals. Finally, at the Council's direction, the Cospas-Sarsat joint Committee has developed a GEOSAR Demonstration and Evaluation Plan for the space segment and GEOLUT ground segment providers.

Therefore, during the past 20 years, the Cospas-Sarsat has demonstrated that the detection and location of distress signals from ships, land vehicles and airplanes can be greatly facilitated by global monitoring based on low altitude PEO satellites in LEOSAR global coverage at first and by high altitude GEO satellites in GEOSAR global coverage at the present stage. The system has been used successfully in a large number of SAR operations worldwide for SAR assistance at sea and on land and consequently, has been responsible for the saving of several thousand lives worldwide.

7.2. Cospas-Sarsat Mission and Service

Based on the success of the SAR operations and service to date, the Cospas-Sarsat partners are planning continuation of the system beyond the full exploitation phase of the newly LEOSAR and GEOSAR integration, as well as permanent technical enhancements to improve the operational utility of the data and services provided and to establish new ground LUT, MCC and SAR capacities. All users and participants in the Cospas-Sarsat operation are in agreement that the potential of the system is extremely promising and that it should become a key element of present and future SAR operations for general marine vessels, land vehicles and aviation. Actually, the participants of the Cospas-Sarsat system are the 4 state parties to the Cospas-Sarsat Agreement, 20 ground segment providers, 9 user states, 2 participating organizations and a number of participating countries worldwide. All detailed particulars about Cospas-Sarsat LEOSAR/GEOSAR Satellite Constellation and LEOSAR/GEOSAR ground infrastructure including MCC, RCC and SAR list of service and stations with short a explanation of the IAMSAR and SAR systems, can be found in ALRS "GMDSS", Volume 5 and in Cospas-Sarsat system Data documentation.

The Cospas-Sarsat system is not involved in the production of satellite EPIRB, PLB and ELT beacons but it provides a list of manufacturers. One of the biggest manufacturers of ship's EPIRB satellite beacons is Jotron Electronic A.S., located in Larvik, a small town by the beautiful Oslo fjord, 100 miles south of Oslo. Their major products are GMDSS devices such as portable VHF radio transceivers and SART transmitters and VHF and UHF ground to air devices for airport and offshore applications. The other producers of satellite beacons are ACR, CEIS TM, Sextant, Yaroslavsky Radio EW and so on.

7.2.1. Basic Concept of Cospas-Sarsat System

The Cospas-Sarsat is a satellite-aided SAR system designed only to locate distress beacons transmitting from ships (EPIRB), land (PLB) and aircraft (ELT) on the frequencies of 121.5 and 406 MHz. Certain beacons also can transmit distress signals on 243 MHz but these signals are relayed only by Sarsat satellites and not all LUT stations are equipped with 243 MHz receivers to detect it. At any rate, the 243 MHz RF operates in the same manner as the mentioned 121.5 MHz system. The Cospas-Sarsat system of satellites in LEO is also referred to as the new Cospas-Sarsat LEOSAR system. The PEO satellites used in the LEOSAR system can provide a global but not continuous, coverage for the detection and the positioning of distress beacons using a Doppler shift location technique. However, the non-continuous coverage introduces delays in the alerting process since the user in distress must "wait" for a satellite to pass into the visibility of the distress beacons.

As discussed, since 1996 Cospas-Sarsat participants have been experimenting with 406 MHz payloads on satellites in GEO together with their associated ground stations to detect the transmissions of Cospas-Sarsat 406 MHz beacons. These experiments have shown the possibility of almost immediate alerts at 406 MHz, providing the identity of the beacon transmitting encoded data, such as the beacon position derived from a GPS or GLONASS system. This development is referred to as the UHF 406 MHz GEOSAR system, which Demonstration and Evaluation (D&E) Report highlights the enhancements provided by a GEO complement to the Cospas-Sarsat PEO system in terms of alerting time advantage and the benefits to SAR services of this rapid alerting capability. However, GEOSAR alerts produced by emissions from the first generation 406 MHz beacons do not include any position information, as the Doppler location technique cannot be applied to signals relayed through GEO satellites. A new type of 406 MHz beacon allows for the encoding of positional data in the transmitted 406 MHz message, thus providing for quasi-real-time alerts and position information from the GEOSAR system. The 406 MHz GEOSAR D&E Report also highlights the complementary aspects of GEO and LEO/PEO systems for SAR, particularly on land where obstructions may prevent direct visibility of a GEO satellite.

Unless, as an alternative, every merchant ship can be fitted with an L-band satellite EPIRB operating in sea areas A1, A2 and A3 only, according to GMDSS, the carriage of a float-free satellite EPIRB operating on the 406 MHz in the Cospas-Sarsat system is required on all SOLAS vessels. Also it is mandatory to carry a satellite ELT on every aircraft. Land vehicles or individuals in expedition groups can be equipped with satellite PLB for use out of urban localities. Therefore, the Cospas-Sarsat system is composed of four basic subsystems, as shown in the system block diagram of **Figure 7.2**.

1) The first subsystem is represented by small emergency satellite EPIRB, PLB and ELT transmitter beacons designed to radiate alert and distress signals in the VHF 121.5/243 and UHF 406 MHz bands for both LEOSAR and GEOSAR spacecraft systems.

2) The second subsystem is the LEOSAR and GEOSAR satellite configurations, which are capable of receiving these distress messages and retransmitting them at corresponding RF to the LEOLUT or GEOLUT stations for processing. The 406 MHz LEOSAR data are also processed and stored on board the spacecraft for direct or later stored data transmission.

3) The third subsystem is LUT (LEOLUT and GEOLUT) that receive the relayed distress signal from beacons via LEOSAR/GEOSAR satellites. These signals are then processed within the LUT stations to provide position and location of the distress signals transmitted by LEOSAR or GEOSAR satellite beacons and these data are then transferred to the MSS.

Figure 7.2. Cospas-Sarsat LEOSAR and GEOSAR System Block Diagram

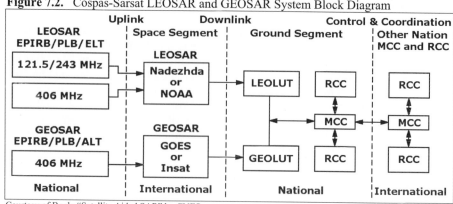

4) The fourth subsystem is the Control and Coordination system infrastructure, which is fully accomplished by the MCC base in each of the participating countries also by every communications arrangement between these various MCC offices. Thus, their duties are to exchange distress incident data, or any other relevant information may be arranged between all participating MCC, as required. The final destination MCC office is then responsible for providing the data to the appropriate national RCC for SAR action.

The Cospas-Sarsat satellite configuration ensures full mutual interoperability between two LEOSAR and two GEOSAR operational subsystems, respectively, because a set of compatible parameters at each subsystem interface were agreed, documented and established. In this sense, certain key parameters, particularly on the space LEOSAR and GEOSAR hardware were verified separately before launching.

The Cospas-Sarsat LEO and GEO subsystems shown in **Figure 7.3.** are used to implement two data systems and two satellite coverage modes for the detection and location of EPIRB, PLB and ELT emergency satellite beacons operating in the three mentioned international RF bands. The LEOSAR subsystem has the following two data systems:

Figure 7.3. Cospas-Sarsat LEOSAR and GEOSAR Integrated System

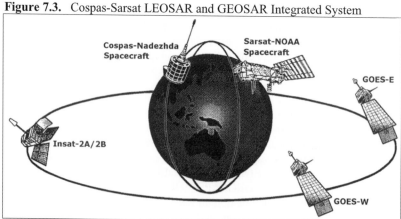

Figure 7.4. Basic Concept of Cospas-Sarsat LEOSAR and GEOSAR System

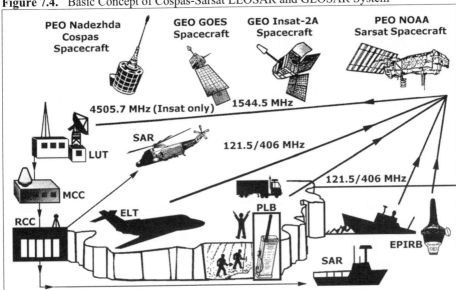

Courtesy of Manual: "GMDSS Volume V" by ARLS

1. Repeater Data System – In the Sarsat (NOAA) system a repeater data solution on board the spacecraft relays the 121.5/243/406 MHz signals directly to a LUT station. At the LUT special processing is used to extract the weak VHF signals from the noise and recover the Doppler information for determining the beacon position. Moreover, in the Cospas system a repeater on board the spacecraft relays only the VHF 121.5 MHz band.

2. Processes Data System – A 406 MHz data processor is included on both LEOSAR types of spacecraft (Cospas and Sarsat), which receives and detects all three UHF applications of satellite beacons. Thus, the Doppler shift is measured, which will be explained in the next section and the satellite beacons identification and status data are recovered. This information is time tagged, formatted as digital data and later on transferred to the repeater downlink for real-time transmission to any LUT station in view. Simultaneously, the data is stored on both types of spacecraft for later transmission to the NOAA ground stations in the case of a Sarsat and to any LUT in the case of a Cospas Nadezhda spacecraft. On the other hand, the new generation of Cospas-Sarsat GEO system does not need a process data system because GEOLUT stations are always in direct view of all GEOSAR 406 MHz beacons within the corresponding satellite coverage.

7.2.2. LEOSAR and GEOSAR Satellite System

The basic concept of the Cospas-Sarsat satellite-aided SAR mission is illustrated in **Figure 7.4.** The signals radiated by either an EPIRB, a PLB or an ELT on 121.5/243 and 406 MHz are received by a first and/or second generation low PEO and GEO spacecraft respectively, equipped with suitable transponders and antenna systems. These distress signals detected by satellite receivers are then relayed by Nadezhda, NOAA and GOES on 1544.5 MHz and by Insat satellite transmitter on 4505.7 MHz to a visible or any corresponding LUT station, where the signals are processed to determine the location of the satellite beacons and/or a mobile in distress. An alert is then relayed, together with location data of the beacon and other information via MCC, either to a national RCC, another MCC or to the appropriate SAR authority and SAR Point of Contact (SPOC) to initiate SAR operation and activities.

Doppler positioning using the relative motion between the spacecraft and the LEOSAR beacon has been chosen as the only practical means of locating distress signals. The carrier RF radiated by the beacon is reasonably stable during the period of mutual beacon and satellite visibility. As was discussed, the beacons frequencies are currently in use on VHF band of 121.5 MHz (international aeronautical emergency frequency) and on a UHF band of 406.0 to 406.1 MHz exclusively reserved for distress beacons operating with satellite systems. Although the UHF beacons are more sophisticated than the VHF beacons, through the addition of identification and situation codes in the messages, complexity is still kept to a minimum by the retention of the Doppler location concept. Another VHF band on 243 MHz is usually only in service for military and Sarsat distress communications.

To optimize Doppler location, a LEO near polar orbital plane is used. The Doppler effect is an event causing frequency changes of the received radio signals for a time during the relative motion of the beacon transmitters and/or satellite receivers:

$$fD = f'V'/c = 2(f'V \cos\alpha/c) = V \cos\alpha/ \lambda \quad [Hz] \tag{7.1.}$$

because $1/ \lambda = f'/c$, where f' is the frequency of EMW, α is the angle between real relative speed (V) and radial relative speed (V') shared by Tx and Rx. The Doppler frequency (fD) is directly proportional to the relative speed of Tx and Rx and inversely proportional to the wavelength. Hence, when one of the Cospas-Sarsat satellites is approaching the position of beacon its Rx is detecting higher nominal RF rates radiated, when the satellite is just over the position of the beacon, both RF are identical and when the satellite is distant from the beacon, the received frequency becomes lower than the radiated one. This RF divergence is Doppler shift, which can be shown by the letter S curve form of the RF over time. The form and steep slope of the Doppler curve together with the position of the satellite on every point of the curve is used for determining the position of the beacon transmitter.

The Doppler location concept provides two positions for each beacon: the true position and its mirror image relative to the satellite ground track. This ambiguity is obviously resolved by calculations that take into account the Earth's rotation. If the beacon stability is good enough, as with 406 MHz beacons, which are designed for this purpose, the true solution is determined over a single satellite pass. In the case of 121.5 MHz beacons, the ambiguity is resolved by the result of the second satellite pass if the first attempt is unsuccessful. Location accuracy is also significantly better with 406 MHz beacons, see **Table 7.1**.

Table 7.1. Comparison of VHF and UHF Satellite Beacons

Characteristics	VHF 121.5 MHz	UHF 406 MHz
Detection probability	(not applicable)	0.88
Location probability	0.9	0.9
Ambiguity resolution	17.2 km	90% within 5 km
Capacity	10	90

The altitude of the Cospas spacecraft orbit is approximately 1,000 km, while that of Sarsat spacecraft is 850 km. These low orbit altitudes result in a low uplink power requirement, a pronounced Doppler shift and short intervals between successive passes of spacecraft.

Figure 7.5. LEO Satellites in Near-PEO Constellation

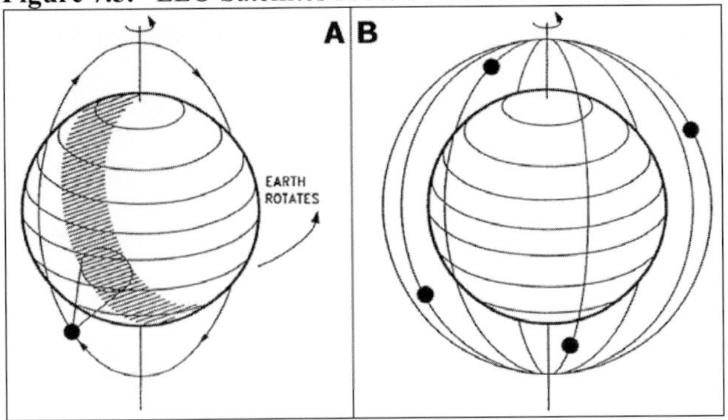

Courtesy of Manual: "Instruction to the Cospas-Sarsat System" by Cospas-Sarsat

The near PEO could provide full global coverage but 121.5 MHz alerts are produced only if the relayed signals are received by a LUT. Namely, this constraint of the 121.5 MHz system limits the useful coverage to the geographic area about 3,000 km radius around each LUT where the satellite can be simultaneously in the visibility of the transmitting beacon and the receiving LUT station. The improved performance of the 406 MHz maritime EPIRB is the reason these small devices were selected by IMO for the GMDSS and included in the 1988 amendments to the 1974 SOLAS Convention. Soon after the ICAO, in a similar way, recommended beacons (ELT) for avionics utilization.

As illustrated in **Figure 7.5. (A)** a single PEO satellite circling the Earth around the poles, eventually views the entire Earth's surface. The "orbital plane", or path of the satellite, remains fixed, while the Earth rotates underneath. At most, it takes only one half rotation of the Earth (i.e., 12 hours) for any location to pass under the orbital plane. With a second satellite, having an orbital plane at right angles to the first, only one quarter of a rotation is required, or 6 hours maximum. Similarly, as more satellites orbit the Earth in different planes, the waiting time is further reduced. Thus, the COSPAS-SARSAT system design constellation is four satellites, which provide a typical waiting time of less than one hour at mid-altitudes, see **Figure 7.5. (B).**

7.3. Overall Cospas-Sarsat System Configuration

Within the Cospas-Sarsat overall system configuration it is important to note that two distinct LEOSAR spacecraft have been considered which are substantially different in their implementation and one GEOSAR satellite system with two types of GEO satellites.

7.3.1. Cospas-Sarsat VHF 121.5/243 MHz System

The basic concept of the Cospas-Sarsat VHF 121.5 MHz system is illustrated by **Figure 7.6.** and is composed of the following modules: 121.5 MHz distress beacons; PEO satellites of the Cospas-Sarsat LEOSAR system and the associated ground receiving stations, LEOLUT, of the LEOSAR system.

The LEOSAR uses the 121.5 MHz aeronautical emergency frequency. As discussed, this frequency served ELT for installation on aircraft. However, they can also be used on board

Figure 7.6. Basic Concept of Cospas-Sarsat System

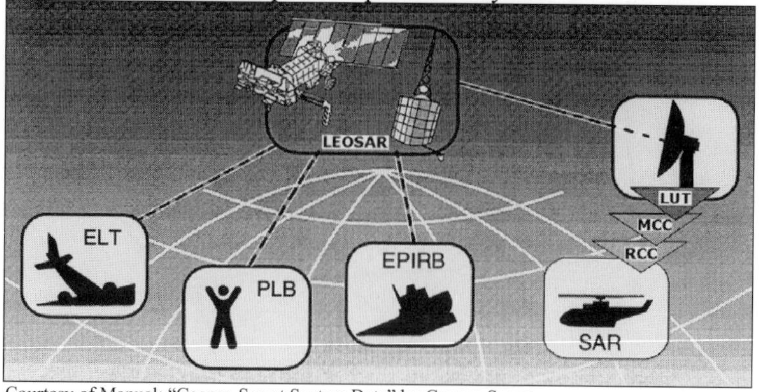

Courtesy of Manual: "Cospas-Sarsat System Data" by Cospas-Sarsat

ships as EPIRB and for vehicles as PLB or personal utilizations. These beacons transmit distress signals that are relayed by LEOSAR satellites to LEOLUT, which process them to determine the beacon or mobile location. The alert, which consists in the computed position of the transmitter, is then relayed, via an MCC, to the appropriate SAR SPOC or RCC. To optimize Doppler performances, satellites in a low-altitude near-PEO are used, however the location accuracy of 121.5 MHz beacons is not as good as the accuracy achieved with 406 MHz beacons due to insufficient frequency stability of the 121.5 MHz emissions. The low altitude results in a low uplink power requirement, a renounced Doppler shift and short intervals between successive satellite passes. The PEO could provide full global coverage but 121.5 MHz alerts are produced only if the relayed signals are received by a LUT. This constraint of the 121.5 MHz system limits the useful coverage to a geographic area of about 3,000 km radius around each LUT, where the satellite can be simultaneously in the visibility of the transmitting beacon and the receiving LUT.

The Cospas-Sarsat LEOSAR system design constellation is four satellites that provide a typical waiting time of less than one hour at mid-latitudes. However, users of the 121.5 MHz system have to wait for a satellite pass that provides for a minimum of 4 minutes simultaneous visibility of the radio beacon and a LUT. This additional constraint may increase the waiting time to several hours if the transmitting beacon is at the edge of the LUT coverage area. As discussed, the Doppler location provides two positions for each beacon: the true position and its mirror image relative to the satellite ground track. In the case of 121.5 MHz beacons, normally, a second pass is required to resolve the ambiguity.

The local coverage mode at 121.5 MHz has proven very effective in providing quick alert and location data for the large number of existing 121.5 MHz beacons. The transmission characteristics of these beacons do not permit satellite on-board processing of the signal, which would be required for global Earth coverage. However, the system cannot generally distinguish between a 121.5 MHz distress beacon transmission and any other 121.5 MHz signal. This limitation results in a large number of processing anomalies, which could cause false alerts in the form of Doppler locations, which do not correspond to distress beacons.

Sarsat satellites are equipped also with 243 MHz repeaters, which allow the detection and location of 243 MHz distress beacons. This system's operation is identical to the 121.5 MHz system operation, except for the smaller number of satellites available. Namely, only Sarsat LEOLUT generally processes both 121.5 MHz and 243 MHz repeater signals.

7.3.2. Cospas-Sarsat UHF 406 MHz System

The Cospas-Sarsat UHF 406 MHz system is composed of 406 MHz radio beacons carried aboard ships EPIRB, personal PLB or used as aircraft ELT; PEO satellites in LEO from the LEOSAR system and GEO satellites from the GEOSAR system and the associated LUT stations for the satellite systems, referred to as LEOLUT or GEOLUT, respectively.
Frequencies in the 406.0–406.1 MHz band have been exclusively reserved only for radio beacons operating with satellite systems. The 406 MHz beacons have been specifically designed to provide improved performance in comparison with the older VHF 121.5 MHz beacons. They are more sophisticated because of the specific requirements of the stability of the transmitted frequency and the inclusion of a digital message, which allows the transmission of encoded data, such as unique beacon identification. A second generation of 406 MHz beacons has been introduced since 1997, which allow for the transmission in the 406 MHz of encoded positional data acquired by the beacon from GPS systems, using internal or external navigation receivers. This feature is of particular interest for GEOSAR alerts, which otherwise would not be able to provide any positional information.

7.3.2.1. UHF 406 MHz LEOSAR System

The 406 MHz LEOSAR system uses the same PEO satellites as the 121.5 MHz system and operates with the same basic constraints, which result from the non-continuous coverage provided by LEOSAR satellites, although with significantly improved performance. The use of LEO satellites provides for a strong Doppler effect in the uplink signal, which allows the use of Doppler positioning techniques. The 406 MHz LEOSAR system operates in two coverage modes for the detection and location of beacons: the local and the global coverage modes. In the local mode, a LUT tracking a satellite receives and processes signals from transmitting beacons in the field of view of the satellite. In the global mode, LUT receives and processes data from the 406 MHz beacons transmitting from anywhere in the world.
1. 406 MHz Local Mode – When the satellite receives 406 MHz beacon signals, the on-board SARP recovers the digital data from the beacon signal, measures the Doppler shift and time-tags the information. The result of this processing is formatted as digital data and transferred to the repeater downlink for transmission to any LUT in view. The data are also simultaneously stored on the spacecraft for later transmission and ground processing in the global coverage mode. In addition to the 406 MHz local mode provided by the 406 MHz SARP instruments, a 406 MHz repeater, on Sarsat satellites only, can also provide a 406 MHz local coverage mode of operation, similar to the 121.5 MHz local mode.
2. 406 MHz Global Mode – The 406 MHz SARP system provides near global coverage by storing data derived from the on-board processing of beacon signals, in the spacecraft memory unit. The content of the memory is continuously broadcast on the satellite downlink. Thus, each beacon can be located by the LUT, which tracks the satellite. This provides 406 MHz global coverage and introduces ground segment processing redundancy. The 406 MHz global mode also offers an additional advantage over the local mode in respect of alerting time when the beacon is in a LUT coverage area. As the beacon message is recorded in the satellite memory at the first satellite pass in the visibility of the beacon, the waiting time is not dependent upon achieving simultaneous visibility with the LUT. The total processing time can be considerably reduced through the broadcast of the satellite beacon message to the first available LUT.

7.3.2.2. UHF 406 MHz GEOSAR System

The basic 406 MHz GEOSAR System in integration with the LEOSAR system is shown by
Figure 7.4. It consists in 406 MHz repeaters carried on board various GEO satellites and
the associated ground facilities called GEOLUT. The GEOLUT stations have the capability
to detect the transmissions from Cospas-Sarsat type approved 406 MHz beacons relayed by
the GEO. A single GEO satellite provides GEOSAR uplink coverage of about one-third of
the globe, except for the polar regions. Three GEO satellites equally spaced in longitude
can provide continuous coverage of all areas of the globe between approximately 70° N and
70° S. As a GEOSAR satellite remains fixed relative to the Earth, there is no Doppler effect
on the Rx RF and this positioning technique cannot be used to locate the distress beacon.
To provide rescuers with beacon position information, such information must be either
acquired by the beacon through an internal or an external navigation receiver and encoded
in the beacon message, or derived, with possible delays, from the LEOSAR system.

7.3.3. Complementarity of the 406 MHz LEOSAR and GEOSAR Systems

The use of satellites in LEO does not permit continuous coverage. This results in possible
delays in the reception of the alert. The waiting time for detection by the LEOSAR system
is greater in equatorial regions than at higher latitudes. The GEO satellites provide
continuous coverage, hence they have an immediate alerting capability, but access to the
GEO satellite can be masked due to ground relief or obstructions, particularly on land, at
high latitudes. The GEOSAR satellites do not provide coverage of the polar regions. The
LEOSAR satellites will eventually come into the visibility of any beacon on the surface of
the Earth, whatever the terrain and the obstructions, which may mask the distress
transmission. In terms of coverage, the specific characteristics of LEOSAR and GEOSAR
systems are clearly complementary. The rapid alerting capability of the GEOSAR system
can be used by SAR forces, even when no position information is provided in the beacon
message. Such information can be used effectively to resolve a false alarm without
expending SAR resources, or to initiate SAR operation on the basis of information obtained
through the beacon registration data. The 406 MHz satellite beacon signals from LEOSAR
and GEOSAR systems can also be combined to produce Doppler locations, or to improve
location accuracy. These are two examples of combined LEOSAR/GEOSAR operations.
Other aspects to the complementarity of the GEOSAR and the LEOSAR system are
highlighted in the report of the UHF 406 MHz GEOSAR Demonstration and Evaluation
(D&E) performed by Cospas-Sarsat Participants from 1996 to 1998.

7.3.4. Distribution of Alert and Location Data

The alert and location data generated by LEOLUT or GEOLUT stations are forwarded to
appropriate SPOC via the Cospas-Sarsat MCC network. Since a single distress incident is
usually processed by several LUT stations, in particular in the 406 MHz global mode of the
LEOSAR system, the alert and location data are sorted by MCC to avoid unnecessary
transmission of identical data. The principle of continuous downlink transmission to all
LEOLUTs in visibility of a satellite results in simpler downlink transmission procedures
and a high level of redundancy in the ground processing system, both at 121.5 MHz within
overlappings LUT coverage areas and at 406 MHz worldwide. A 406 MHz and 121.5 MHz

Figure 7.7. Cospas-Sarsat Space Segment

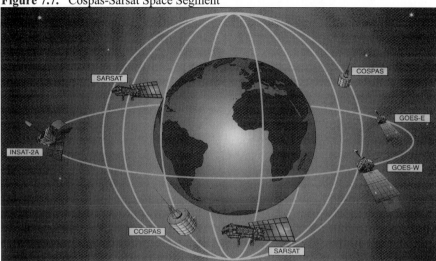

Courtesy of Manual: "GMDSS Volume V" by ARLS

beacon alerts are always distributed to the RCC or SPOC that has responsibility for the area where the distress is located. On the other hand, the same principle applies to alerts generated by the 406 MHz GEOSAR system, since several GEOLUT stations can detect the same 406 MHz beacon transmission. When a 406 MHz GEOSAR alert is received with encoded position information, that alert is forwarded to the RCC or SPOC that has responsibility for the area where the beacon is located. When no location is available in the 406 MHz alert (if insufficient data does not allow a LEOLUT to compute a Doppler position or when no position data is encoded in the 406 MHz beacon message received by a GEOLUT), the alert is forwarded to the SAR authorities of the country where the beacon has been registered.

7.4. Cospas-Sarsat Space Segment

As discussed earlier, the Cospas-Sarsat space segment has been designed and implemented to meet the requirements of the distress and safety mission as the result of a corporate effort from the various partners involved in this Program. At first, only the Cospas and Sarsat PEO satellite configuration was developed, currently known as the LEOSAR space segment and recently an additional constellation infrastructure of the Cospas Sarsat system was developed, named the GEOSAR space segment. Therefore, both segments are today integrated elements of the Cospas-Sarsat worldwide enhanced emergency mission. The nominal LEOSAR system configuration comprises four satellites, two Cospas and two Sarsat, while the GEOSAR global constellation is configured by one Insat and two GOES satellites, as is illustrated in **Figure 7.7.**

The Cospas-Sarsat LEOSAR and GEOSAR space segment configuration and relation with Ground segment is shown in **Figure 7.8.** The distress alerts sent by LEOSAR or GEOSAR beacons are relayed through corresponding LEOSAR or GEOSAR satellites to the ground network of LEOLUT or GEOLUT to the MSS, RCC/SPOC infrastructures and SAR forces.

Figure 7.8. Cospas-Sarsat System Block Diagram

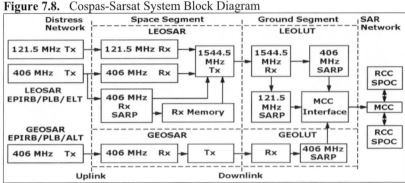

Courtesy of Manual: "Introduction to the Cospas-Sarsat System" by Cospas-Sarsat

7.4.1. LEOSAR Cospas Payload

The LEOSAR Cospas-Sarsat constellation includes two types of PEO satellites, of which the major particulars, with names of satellite/payloads and RF, are presented in **Table 7.2**. Russia supplies two LEO Cospas satellites placed in near-polar orbits at 1,000 km altitude and equipped with SAR instrumentation at 121.5 and 406 MHz. The USA supplies the two NOAA meteorological satellites of the Sarsat system placed in sun-synchronous, near-polar orbits at about 850 km altitude and equipped with SAR instrumentation at 121.5/243 and 406 MHz, supplied by Canada and France. Each PEO satellite makes a complete orbit of the Earth around the poles in about 100 min, traveling at a velocity of 7 km/s. The satellite views a "swath" of the Earth's surface over 4,000 km wide as it circles the globe, giving an instantaneous "field of view" about the size of a continent. When viewed from the Earth the PEO satellite crosses the sky in about 15 minutes, depending on the maximum elevation angle of the particular pass, see the footprints of the LEOSAR satellites and the 40 LEOLUT stations marked in **Figure 7.9**. The LEOSAR satellite has three basic units:

a). A platform moving in PEO as a mounting for the other payload units (this platform is not dedicated to the SAR mission and generally carries other payloads).

b). A 121.5 MHz SAR Repeater (SARR) unit on Cospas satellites and 121.5; 243 and 406 MHz repeater units on Sarsat satellites designed for retransmission of distress signals in the local coverage mode.

Table 7.2. LEOSAR Space Segment Instrument Status

Spacecraft	Payload	406 MHz SARP		406 MHz SARR	121.5 MHz SARR	243 MHz SARR
		Global	Local			
Nadezhda-1 (1)	Cospas-4	O	O	NA	O	NA
Nadezhda-3 (2)	Cosaps-6	O	O	NA	O	NA
Nadezhda-5	Cosaps-8	O	O	NA	O	NA
Nadezhda-6	Cosaps-9	O	O	NA	O	NA
Nadezhda-7	Cosaps-10	IO	IO	NA	IO	NA
NOAA-11	Sarsat-4	O	O	O	O (3)	NO
NOAA-14	Sarsat-6	NO	NO	O	O	O
NOAA-15	Sarsat-7	O	O	O	O	LO
NOAA-L	Sarsat-8	O	O	O	O	NO
NOOA-M	Sarsat-9	O	O	O	O	O

Legend: O = Operational; NO = Not Operational; NA = Not Applicable; IO = Initial Operation; LO=Limited Operations; SARP= SAR Processor; SARR = SAR Repeater; (1) = Limited Operation in Southern Hemisphere; (2) = Reduced 406 MHz Message throughput and (3) = Some Electromagnetic Interference Degradation

Figure 7.9. LEOSAR Spacecraft Footprint

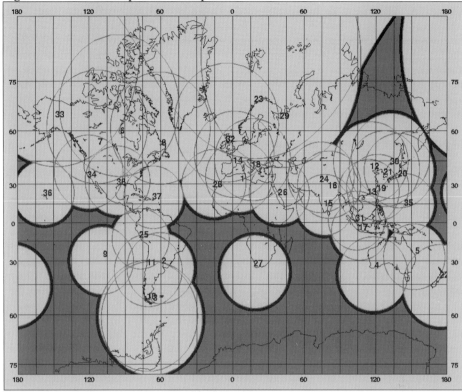

Courtesy of Manual: "Cospas-Sarsat System Data" by Cospas-Sarsat

c). A SAR Receiver-processor and memory (SARP) unit on Cospas and Sarsat satellites is designed to receive, process and store signals received on 406 MHz for retransmission in the local and the global coverage mode.

In general, both LEOSAR Cospas and Sarsat payload consists in an on-board transponder, which includes 121.5; 243 (this RF is only for Sarsat), 406 MHz receivers and a 1544.5 MHz transmitter with Rx and Tx antenna systems, respectively.

1. SARR Unit - The SARR, as a repeater unit of Cospas and Sarsat system receives either 121.5 or 243 MHz and 406 MHz signals transmitted by activated distress satellite beacons. After amplification and frequency conversion, the signals are retransmitted on the 1544.5 MHz downlink, as depicted in **Figure 7.8**. This unit provides Automatic Level Control (ALC) to maintain a constant output level. Thus, the 1544.5 MHz Tx on the repeater unit performs the following tasks: accepts input from the uplink receivers; adjusts the relative power level in accordance with ground command; phase modulates a low frequency carrier with the composite signal; multiplies the RF to produce 1544.5 MHz; amplifies the power level and transmits the composite baseband signal via the spacecraft downlink antenna.

2. SARP Unit – The main functions of the SARP Rx-processor in the satellite transponder are: demodulating the digital messages received from satellite beacons; measuring the received frequencies and time-tagging the measurements. All these data are included in the output signal frame for transmission to LUT on the 1544.5 MHz downlink illustrated in **Figure 7.8.** The frame is transmitted at 2400 b/s in the processed data mode and subsequently stored in memory. The data from the satellite SARP on board memory are transmitted on the downlink in the same format and at the same bit rate as local mode data. The LUT stations thus receive the stored beacon messages acquired during previous orbits. If a beacon signal is received during the stored memory dump, the dump is interrupted so that the signal can be processed and the resultant message interleaved with the stored data. Appropriate flag bits indicate whether the data are real-time or stored and the time at which full playback of the stored data was accomplished.

7.4.1.1. Cospas Repeater Functional Description

The Cospas navigation satellites are equipped with systems for the accurate determination of orbital parameters with the precise RF standards. The former USSR initiated Cospas program in 1977 and first generation navigation satellite Cosmos 1000 was proposed as basic apparatus for emergency radio signals from the crashed ships and aircraft locations. The design was determined by navigation equipment consisting in formatter of space borne real time scale; ephemeral data formation, correction and storage unit and navigation message Tx. This satellite emitted navigation messages at 150 and 400 MHz coherent RF and is transmitted to the users time marks and orbital data. Nothing more is required to ship receiving equipment to obtain the ship's position with portable digital computers. In this way, space borne navigation real-time scale formatter and required ephemeris control show the expediency of repeater and processor installation on this spacecraft. To also perform the main Cospas-Sarsat tasks, this type of satellite has to carry out the transponder payload for reception and retransmission of distress signal from emergency satellite beacons to the LUT ground stations. At the same time, some problems appeared to be resolved as follows:
a). Satellite SAR repeater and processor installation on an already designed satellite.
b). Compatibility of navigation and SAR satellite antenna and feed assembly.
c). Coordination of SAR equipment and other satellite systems.
d). Ensuring power supply, attitude control and temperature conditions for SAR equipment.
Resolving all these problems to concrete technical standards was determined by the SAR program, owing to specific features of the navigation spacecraft, coordination of all SAR performances and the international Cospas-Sarsat satellite system. The designers had to change the satellite's arrangement, to make all the necessary calculations on strength, heat and power data, to modify some systems and assemblies so that the new Cospas satellite meets all the requirements, as well as to check up all experimentally accepted technical concepts before the flight test. After successful resolution of all enumerated problems, many satellites of the old Cosmos and new Nadezhda series were launched for the Cospas missions, which orbital parameters and spacecraft characteristics are shown in **Table 7.3.**
Thus, the Cospas navigation satellite also includes LEOSAR transponder as an additional communication payload to the compulsory GNSS equipment and a set of ancillary systems required for the good operation of the platforms. All the LEOSAR equipment is redundant and commutable by remote control. The device operating modes, activation and deactivation, as well as the modulation index adjustment are also remotely-controlled.
The Cospas LEOSAR payload equipment consists in two receiving/processing modes: the VHF real-time data retransmission and UHF processed and stored data retransmission and there are five basic components, shown in **Figure 7.10.**

Table 7.3. Orbital Parameters of Cospas Nadezhda Spacecraft

Background	Launch vehicle: Kosmos 11K65M
Owner/Operator: Cospas-Sarsat	Typical users: Cospas-Sarsat SAR, Civilian and
Present status: Operational	Maritime Navigation
Altitude: 712 km (Perigee); 712 km (Apogee)	Cost/Lease information: Nil
Type of orbit: PEO (LEO)	Prime contractors: Morflot (Cospas) - Russia
Duration of orbit: About 105 min	Other contractors: Yuzhnoye
Inclination angle: 98,1°	Type of satellite: Navigation
Eccentricity: Less then 0.2	Stabilization: Gravity gradient (accuracy ±8°)
Number of satellites: 2 operational & 3 spare	Design lifetime: Over 2 years
Coverage: Global with minimum 4 satellites	Mass in orbit: 825 kg
Additional information: 14 PEO positioned in plane	Dimensions deployed: 2 x 3.7 x 20 m,
14 of constellation	**Communications Payload**
Spacecraft	Frequency bands:
Name of satellite: Надежда-6 (Nadezhda-6)	Service uplink: VHF 121.5 & UHF 406 MHz
Launch date: 28 June, 2000	Feeder downlink: L-band 1544.5 MHz

1. VHF 121.5 MHz Band Receiver – This unit performs VHF 121.5 MHz band real-time reception of distress signals for direct retransmission to LUT sites. Because of the limited power of the EPIRB/PLB/ELT Tx and taking account of the quite low level of the signals received from emergency satellite beacons and of the high level of interference, the satellite Rx has been designed so as to have good linearity and frequency forming characteristics. The Rx comprises three intermediate frequencies. At this point, the 25 KHz frequency band is fixed in amplifier No 3. Preliminary filtering is achieved in amplifier No1 by means of crystal filters. In fact, a summary of the technical characteristics is as follows: nominal frequency of the Rx is 121.5 MHz; pass band signal at 1 dB is 25 kHz; input signal dynamics are –110 dBW/–170 dBW; noise temperature is 450°K/1.000°K; RF stability is 3.10^{-10} in 10 mn; output RF is about 47 kHz; survival level is 30 dBW and return to nominal performance after receiving a level of –60 dBW is approximately 2 ms.

2. UHF 406.025 MHz Band Receiver-Processor – This unit serves the real-time receiving and processing of the UHF 406.025 MHz signals transmitted by emergency radio beacons,

Figure 7.10. Cospas LEOSAR Satellite Transponder

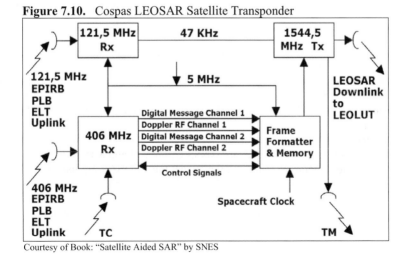

Courtesy of Book: "Satellite Aided SAR" by SNES

and direct retransmission to LEOLUT terminals. It includes two portions, one being linear and the other digital. The linear portion brings the input frequency from 406.025 MHz to 311 kHz through a dual frequency change. Amplification and filtering are also achieved in this portion. The digital portion performs all the other signal processing and identification operations. The 311 kHz output signal of the Rx is digitalized. The data recovery unit analyzes the spectrum and provides the information necessary to the control unit, which manages the two processing channels. These two channels are capable of simultaneously processing the transmissions from at least two beacons: Doppler measurement, then time tagging measurement and demodulation of the beacon's digital message. A summary of the technical characteristics is as follows: central RF is 406.025 MHz; pass band at 1 dB is 23 kHz; input signal dynamics is –108 dBm/–128 dBm; noise temperature is about 600°K; Doppler measurement accuracy is 0.35 Hz r.m.s and bit error rate is 2.10^{-5}.

3. Format Encoder – This unit is actually the frame formatter and memory, whose main role is to structure the downlink digital stream of UHF 406.025 MHz stored data and simultaneous real-time retransmission of the processed data. Its principal components are processor, mainframe and memory, while among the rest of its functions are: to perform time-tagging of the Doppler RF measurements; to store all data related to the satellite beacons (RF received, time-tagging, digital messages) and to obtain formatting of the downlink. On the Cospas satellite, memory reading is continuous, every new reading being initialized as soon as the previous one is completed. When data are received from beacons, reading is interrupted in order to provide real-time transmission of these data. Reading is resumed immediately after the transmission of the signals. A summary of the technical characteristics is: Doppler measurement timing accuracy is about 1 ms; output message is 8 and 24 b/words; memory capacity is 2.048 messages and output data rate is 2,400 b/s.

4. UHF 1544.5 MHz Band Transmitter – The role of the Cospas LEOSAR Tx is to shape a 1544.5 MHz signal that is phase modulated by the 2,400 b/s digital stream and by the analog output of the 121.5 MHz Rx. It employs a temperature-controlled crystal oscillator. The linear modulator operates at a frequency of 386.125 MHz. After modulation, the output RF is multiplied by 4 and the final amplification takes place on the 1544.5 MHz RF. Namely, before entering the linear phase modulator; modulation signals are amplified by a wideband linear amplifier. There is a two-level limiter in this amplifier, which prevents the instantaneous value of the summed modulating signal exceeding a certain level. The modulation index adjustment is achieved by means of changes of signal modulating voltage, which are subsequently passed to the input of the wideband linear amplifier. Modulation index control is achieved by remote control. Index adjustment is separate on each channel. Suppression of each channel is also possible by remote control. A summary of the technical characteristics is: output RF is 1544.5 MHz; output stability is 10^{-6}; output power is 4 W; off band signals are ≤60 dB; modulation index is: output of the 121.5 MHz receiver (47 kHz) from 0.21 ±0.01 to 1.3 ±0.03 rads by 15 steps of 0.5–0.8 rads and 2,400 bits/sec 5-step digital channel and Modulation linearity is: ±0.5% (For θmax = ±2.4 rads).

5. Cospas Antenna System – This system provides three antennas: two Rx and one Tx on the spacecraft in support of the Cospas payload. Cospas receive antennas (SRA for 121.5 MHz and SPA for 406 MHz) have the following characteristics: Polarization is LHCP for 121.5 MHz, RHCP for 406 MHz; Gain is about maximum value of +6 dBi for 121.5 MHz Rx antenna and Frequency are 121.5 MHz ±20 kHz and 406.05 MHz ±50 kHz. Cospas Tx antennas have the following characteristics: polarization is LHCP; gain is about maximum value of –2 dBi and frequency is 1544.5 MHz ±500 kHz.

7.4.1.2. Sarsat Repeater Functional Description

The NOAA LEOSAR payload contains uplink and downlink frequencies that are the same as the Cospas payload, the only difference being that the Sarsat transponder also relays VHF 243 MHz band transmissions. The Sarsat program was realized by Canada, France and the USA, using meteorological NOAA satellites to install the SAR payload for the detection and relaying the distress signals from vessels and airplanes alert locations. The design of the Sarsat payload was determined by meteorological equipment and so, all the problems that appeared were resolved on a similar basis to that of the Cospas designers.

Immediately following the successful launch of US NOAA-8 meteorological spacecraft together with Sarsat payload in March 1983 and the turning on of the Sarsat instruments, a series of activation and evaluation tests was conducted in order to verify the satisfactory operation of the Canadian and French built Sarsat integrated devices. After that, an extensive series of system performance tests was carried out at VHF 121.5/243 MHz band in order to measure the following performance parameters: multiple access capacity, detection probability and threshold, location accuracy and ambiguity resolution. Finally, planning the post launch technical test and evaluation of the Sarsat payload was controlled by the Post Launch System Checkout Plan. This plan divided the testing into two phases:

1) The activation and evaluation tests were conducted to verify the satisfactory operation of the Sarsat repeater and to provide an initial estimate of the system performance. These tests were performed primarily by Canada and the USA, while France was carrying out similar tests using the 406 MHz on-board SAR processor. The tests included measurements of the following repeater performance parameters: downlink spectral occupancy, EIRP counter and carrier frequency, spurious repeater outputs, receiver bandwidth and dynamic range, transmitter modulation index and calibrated S/No measurements.

2) The Sarsat system performance tests were intended to provide a detailed assessment of the overall performance. This section presents a brief summary of the particular tests as follows: Canada generally used controlled test signals from a single site and conducted ELT detection threshold test, the US approach was to perform a field test program for position location and, in this sense, France made use of both controlled test signals as well as the influence of satellite emergency beacon frequency stability on location accuracy and ELT activation. The Sarsat LEOSAR payload consists in two receiving/processing modes: the VHF/UHF real-time data retransmission and UHF processed and stored data retransmission and five further basic components, shown in **Figure 7.11**.

1. Search and Rescue Repeater (SARR) – The SARR Sarsat module is supplied by DND Canada and was developed by the SPAR Aerospace Company, see the lower part of the equipment in **Figure 7.11**. The electronic module contains an antenna divided into two separate parts for detecting both 121.5/243 MHz VHF and 406 MHz UHF beacon signals. The VHF signal then comes into a diplexer, where it performs delimitation of 121.5 MHz and 243 MHz towards two separate filters and finally these filtered signals lead away into two separate receivers. Meanwhile, the UHF signal goes into the filter and then into the 406 MHz receiver. These three real-time processed signals from SARR are forwarded via the 1544.5 MHz Tx on to the antenna and towards visible LEOLUT stations.

2. Search and Rescue Processor (SARP) – The SARP module is supplied by CNES from France and developed by the Electronique Serge Dassault Company, see the upper part of **Figure 7.11.** The electronic module contains a UDA antenna, which detects UHF signals and thereafter forwards them via diplexer and filter towards the UHF 406.025 MHz Rx.

Figure 7.11. Sarsat LEOSAR Satellite Transponder

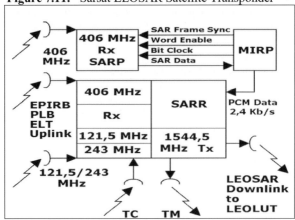

Courtesy of Book: "Satellite Aided SAR" by SNES

These UHF signals are processed in the MIRP (Manipulated Information Rate Processor) unit and then memorized in the Global Coverage SAR Data Storage for later relaying when the LUT comes into view or direct forwarding, like PCM data via SARR and 1544.5 MHz Tx on to antenna at any visible LUT station. It should be noted that the global coverage data ground storage and transmission system are not shown. At this point, these data are sent, together with the other meteorological information to the Suitland processing centre, which draws them from the telemetry data and then sends them to the MCC. Moreover, as from the fourth payload, this solution will be replaced by a memory integrated to the SARP module to provide direct distribution of the stored data to all of the ground LUT stations by means of the SARR 1544.5 MHz downlink signals.

3. SARR Transponder – This module is fully redundant and self-contained with its own power supply and TT&C functions. **Figure 7.11.** presents one of the transponder channels with the processing of signals in the Rx and Tx modules. The 121.5/243 and 406 MHz Rx provides global coverage for LEOSAR satellite beacons, which weak signals are received by the antenna and are processed in the Rx's filters, RF amplifier, mixers and IF amplifier. The IF signals are combined and phase-modulate a 1544.5 MHz carrier, amplifying and driving via the antenna for transmission to the LEOLUT stations. In order to achieve this relatively simple function, various contradictory requirements must be reconciled. The location accuracy depends on the preservation of the signal received through the repeater. The RF and power of the signal vary according to the Doppler effect and to the distance. As discussed, when the satellite comes closer to the beacon, the received RF signal rate is higher than is nominally radiated and the power becomes higher, when the beacon is just at the subsatellite point, both frequencies are identical and the power of the signal has a maximal rate. When the satellite goes far away from the beacon, the received frequency becomes lower than the radiated one and the signal always becomes lower. These requirements result in a comparative broadband, in well-controlled gain and modulation characteristics and in a high sensitivity. As interference from the ground is considerable, good linearity and correct rejection of the off-band signals can also be essential. To meet the above-mentioned requirements, during the construction of the project, antennas are used with suitable values of gain/EIRP, Rx with acceptable selectivity, careful selection of the frequency plan, highly selective low-loss HF filters, low-noise HF amplifiers with good linearity, FET mixers, local oscillators and linear modulators. Very stable frequency sources are used so that the absolute RF of the signal is accurately determined. A 50 dB dynamics Automatic Gain Control (AGC) is also employed. The value of its control voltage is remotely measured for ground processing to perform any eventual corrections during next stage of exploitation. A summary of the SARR technical characteristics is as follows: input RF are 121.5 MHz ±12.5 kHz, 243.0 MHz ±23.0 kHz, 406.05 MHz ± 40.0 kHz; output RF is 1.544,50 MHz ±400.0 kHz; Noise factor is 5.0 dB for 121.5 MHz and 3.5 dB for 243/406 MHz; survival level is –50 dBW continuously without degradation; modulation index is 0.63 rads r.m.s for a nominal 3 rads peak input; output level of the Rx is a remotely controlled attenuator –15 x 1 dB; Tx RF stability is 2.60^{-6} for transmission duration; Rx mass is 15 kg; output power 7 W and the power consumed is 45 W.

4. The SARP Receiver Processor – This unit processes the message signals transmitted by the 406.025 MHz distress satellite beacons. The transmissions are distributed in a random manner in terms of time and Doppler RF, which allows the problem of multiple access to be solved. When a distress signal is received, the processor measures the Doppler RF and the time of arrival. It demodulates the signal and integrates all of these data into a digital signal. They are then transmitted to the ground in real time via the SARR's 1544.5 MHz Tx and in deferred time via the channel provided by the ATN satellite. The signal is first sent to the Rx, which uses a dual RF change. The Rx is of the constant gain type, controlled by an AGC with internal reference. Its dynamics are 23 dB. The data recovery unit located at the receiver output explores the 24 kHz band in 90 ms. When a distress signal is detected, the control unit then assigns the signal to one of the two processing units by means of an algorithm, which has been designed so as to optimize the random access performance. This algorithm is based on the input signal frequency and its level also on the operational status of the processing units. At this point, when a signal is attributed to a processing unit, a control phase loop is locked onto this signal. Moreover, after demodulation, a bit and frame synchronizer allows the message to be processed. Simultaneously, the Doppler RF is measured by counting through the oscillator of the loop. The Doppler measuring time is determined by using a time base derived from the ultra stable oscillator of the equipment. The SARP's format encoder delivers an output message comprising eight 24-bit words (short message), or nine 24-bit words (long message). The length of the message is indicated by one of the bits in the satellite beacon's distress message. A buffer memory allows the SARP's unit output rate to match that of the format encoder sampling the MIRP of the ATN platform. Thus, this encoder inserts the processed distress data in a new format used by the 1544.5 MHz transmitter, where the SARP data are included in sequence of 1,200 bits/0.5 sec., namely a bit rate of 2,400 bits/sec. A summary of the SARP's technical characteristics is as follows: input RF is 406 MHz ±12 kHz; Rx input level is –108 dBm to –131 dBm; accuracy of the Doppler measurement rate is ≤1 ms; Doppler measurement accuracy is ≤0.35 Hz r.m.s input signal $≤1.10^{-10}$ over 100 ms; processing capacity is 90 beacons evenly distributed within the circle of visibility with a probability of 95%; SARP weight is approximately 15 kg and power consumed is approx. 15 W.

5. Sarsat Antenna System – This System provides four antennas: three Rx and one Tx on the spacecraft with necessary diplexers and filters in support of the Sarsat payload. Namely, the SARR Rx Antenna (SRA) consists in two coaxial quadrifilar antennas. The two outer quadrifilar are used for the 121.5 and 243 MHz Rx, while the inner one is used for the 406.05 MHz Rx. Hence, the SARP Rx antenna signal comes from the quadrifilar UHF Data

Figure 7.12. GEOSAR Spacecraft Footprint

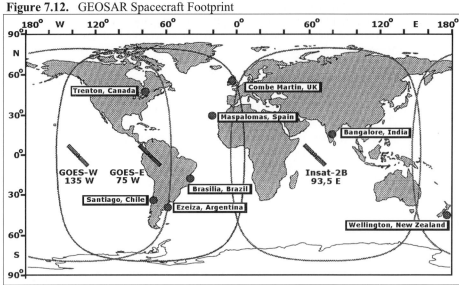

collection system Antenna (UDA). Sarsat Rx antennas have the following characteristics: Polarization is RHCP and frequencies are for SARR 121.5 MHz ±20 kHz, 243.0 MHz ±30 kHz and 406.05 MHz ±50 kHz, while for SARP is 406.05 MHz. The SARR L-band Tx Antenna (SLA) is a quadrifilar antenna that has been optimized to produce a hemispherical pattern. Sarsat transmit antenna has the following characteristics: polarization is LHCP and frequency is 1544.5 MHz ±500 kHz.

7.4.2. GEOSAR Space Segment

The GEOSAR space segment is composed of three operational GEO multipurpose satellites with the capability to relay only the transmissions of the Cospas-Sarsat 406 MHz beacons. These three satellites carry various payloads in addition to the Cospas-Sarsat payload of the 406 MHz SAR mission. The GEOSAR payload consists in the 406 MHz antenna, a Rx and the downlink Tx. In this sense, the frequency of the downlink Tx may vary with the GEO platform carrying the 406 MHz payload. In **Table 7.4.** are shown all the characteristics of GEOSAR operational and planned satellites, including working frequencies.

The Cospas-Sarsat GEOSAR Space Segment consists in SAR instruments on board three operational satellites in GEO constellation. The SAR instruments are radio repeaters that receive distress beacon signals in the 406 – 406.1 MHz band and relay these signals to special ground GEOLUT stations for processing, beacon identification and associated data. The GEOSAR system transponders and other equipment are carried by the satellites shown in **Figure 7.2**.

The operational GEOSAR payloads available on board GEO satellites are provided by the USA and India. Russia and EUMETSAT (European Meteorological Satellite Organization) also have plans for equipping GEO satellites with 406 MHz SAR payloads. The coverage provided by GEOSAR satellites with GEOLUT stations is illustrated in **Figure 7.12**.

Table 7.4. The GEOSAR Operational and Planned Spacecraft

Spacecraft	Launch Date	Position	Status	Rx RF MHz	Tx RF MHz	Country/Firm
GOES-E	04/1994	75° W	O	406	1544.5	USA
GOES-W	04/1997	135° W	O	406	1544.5	USA
Insat-2B	1993	93.5° E	O	406	4505.7	India
Insat-3A	Projected	93.5° E	P	406	4505.7	India
Insat-3D	Projected	83° E	P	406	4507	India
Luch-M E	Projected	95° E	P	406	11381.05	Russia
Luch-M W	Projected	16° W	P	406	11381.05	Russia
MSG	Projected	0°	P	406	1544.5	Eumetsat
Legend: O = Operational and P = Planned						

A functional diagram of SAR instruments on GEOSAR spacecraft is shown in **Figure 7.13.** The GEOSAR instruments were independently developed and integrated into spacecraft that have different mission requirements. This has resulted in differences in Tx repeater designs that affect the output signal depending on the system of GEO satellite. Hence, these differences must be considered in developing a GEOLUT. Namely, the GEOSAR repeater receives 406 MHz beacon signals within the field of view of the 406 MHz receive antenna beam. The beacon signals are processed by the repeater and transmitted via the antenna on the downlink signal for reception by a GEOLUT. The downlink centre frequency and antenna pattern characteristics vary among the different repeater implementations as described in subsequent sections. In such a way, transmitters of the US-based Geostationary Operational Environmental Satellite (GOES) and the Meteosat Second Generation (MSG) of Eumetsat use 1544.5 MHz; the Indian National Satellite System (INSAT) repeater works on 4505.7 MHz Tx RF and the Russian-based Luch-M Tx uses 11381.05 MHz RF band. The GEOSAR ground stations transmit telecommands (TC) to spacecraft and vice versa, the ground stations receive telemetry data from spacecraft.

7.4.2.1. GOES Repeater Functional Description

The US GEOSAR repeater is redundantly configured and consists in the following units:
1) Two 406 MHz low noise amplifiers (shared with another satellite subsystem);
2) Two dual-conversion 406 MHz receivers;
3) Two 3 watt phase modulated L-Band transmitters;
4) One 406 MHz receive antenna and one 1544.5 MHz transmit antenna; and
5) Command and telemetry points in a transponder interfaced with the spacecraft telemetry and command subsystem.

Figure 7.13. GEOSAR Satellite Transponder

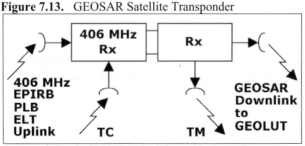

Courtesy of Manual: "GEOSAR Payloads System" by Cospas-Sarsat

The signals coming from the beacon are received on the UHF antenna, fed through the antenna diplexer and then switched to an LNA device in one of the redundant pair of Data Collection Platform Repeaters (DCPR). The DCPR LNA is used as a part of the SAR implementation to accommodate circuit efficiency on the spacecraft. The LNA outputs are connected to the redundant pair of SAR Rx. The signal applied to the selected Rx is down-converted for bandpass filtering in accordance with one of two commandable bandwidth modes; a narrow band mode of 20 kHz or a wide band mode of 80 kHz. The filtered output signal is further down-converted to near baseband and fed via amplifiers to the SAR Tx. The overall gain of the SAR Rx can be command selected into a fixed gain or ALC mode.

The outputs of the Rx are provided to the redundant pair of SAR Tx devices. The selected SAR Tx phase modulates the signal, multiplies the signal to 1544.5 MHz and amplifies the modulated carrier to 3 W. The phase modulated signal has the nominal modulation index set such that the carrier suppression is 3 dB with the receiver in the ALC mode or with the receiver in the fixed gain mode operating with two nominal beacon signals plus the noise. A baseband limiter restricts the modulation index from exceeding 2 radians. The Tx output is applied through a 4 MHz bandwidth filter to the helical antenna and radiated with an EIRP of +15.0 dBW.

7.4.2.2. Insat Repeater Functional Description

The SAR instruments are fitted on the Insat 2A/2B, 3A and 3D satellites. On each of these satellites the SAR payloads share some common circuitry with the Data Collection System (DCS) meteorological instruments. All Insat SAR payloads are of similar design; however, some technical parameters for the Insat 3D payload differ slightly from the others. The 406 MHz signals from distress beacons within the coverage area of the SAR Rx antenna are fed to a pre-selected helical band pass filter, which helps to suppress out-of-band interference. The filtered signal is passed to an LNA to achieve the required input noise figure and is down converted twice, such that an uplink signal at 406.05 MHz would appear at 100 kHz

Table 7.5. Orbital Parameters of GEO Insat 2B Spacecraft

Background	Type of satellite: GEO
Owner/Operator: Cospas-Sarsat	Stabilization: 3 Axis
Present status: Operational	Design lifetime: 10 years
Orbital location: 93.5° East	Mass in orbit: 905 kg
Altitude: About 36.000 km	Launch weight: 1,900 kg
Type of orbit: Inclined GEO	Dimensions deployed: 1.93 x 1.7 x 1.63 m
Inclination angle: 0.03°	Electric power: 1.024 kW @ EOL
Number of satellites: 1 operational & 1 spare	TWTA Power: C-band 4 W; S-band 50 W
Coverage: IOR	Telemetry beacon: 4191/4197.5/4187.5 MHz
Additional information: This spacecraft was	Command beacon: 6415/6419.5/6423.5 MHz
substituted Insat 2A satellite in position 74° E	**Communications Payload**
Spacecraft	Frequency bands: a) GEOSAR bands:
Name of satellite: Insat 2B	Service uplink: VHF 121.5 & UHF 406 MHz
Launch date: 22 July 1993	Feeder downlink: L-band 4505.7 MHz
Launch vehicle: Ariane V58	a) Communications: S & C-Band
Typical users: Cospas-Sarsat SAR, TV, Telecom and	Number of transponders: 2 S & 18 C-band
Meteorology	Channel bandwidth: 36 MHz
Cost/Lease information: Nil	Channel polarization: C-Orthogonal Linear
Prime contractors: ISRO - India	& S-LHCP

and an uplink signal at 406.025 MHz would appear at 75 kHz. The resulting signal is passed to a transistorized limiter circuit and to a low pass filter. The filtered signal is phase modulated and multiplied to achieve a modulation index of '1 radian at 71.7 MHz.

The main functions of the phase modulator are to: reduce noise in the down link signal and provide a continuous down link carrier for LUT tracking receivers. The output signal from the modulator is filtered, combined with the DCS IF signal (70.2 MHz/70.05 for Insat 3D), up-converted to 4505.7 MHz (4507.0 MHz for Insat 3D) and applied to a Solid State Power Amplifier (SSPA). Namely, for the Insat 3A and 3D satellites, the composite DCS SAR signal is combined with the output from the VHRR and CCD payloads and passed to an extended C-band Multiplexer (MUX). Finally, the signal is routed to an extended C-band antenna which provides an EIRP of 4,0 dBW for Insat 3A (end of life at the edge of Indian coverage) and 3.8 dBW for Insat 2B/Insat 3D. As a typical example of GEOSAR spacecraft is presented the technical characteristics of Insat 2B in **Table 7.5**.

7.4.2.3. Luch-M Repeater Functional Description

The Russian Luch-M SAR payload is redundantly configured and contains:
1. The 406 MHz receive antenna, LNA and Rx;
2. The 11381 MHz transmitter antenna and Tx.
The 406 MHz signals from Cospas-Sarsat distress beacons are received by the Luch-M UHF antenna. The signal is down converted twice to an Intermediate Frequency (IF) of 18.55 MHz after which it is filtered. The 3dB beam width of this filter is 600 kHz. This IF signal [s1(t)] is unconverted then combined with signals from other instruments on-board the satellite [s2(t)]. This composite signal is amplified, unconverted to 11381.05 MHz and amplified to a power of 3.75 W. The composite amplified signal is then filtered before being transmitted via the satellite 0.6 m parabolic antenna.

7.4.2.4. MSG Repeater Functional Description

The European MSG SAR transponder comprises the following:
1) An UHF receive antenna, which is made up of an array of 16 crossed-dipoles located close to the periphery of the main satellite drum. The dipoles of this array are electronically switched in order to form an electronically de-spun beam that fully covers the Earth.
2) An input filter together with a redundant UHF Rx provides LNA amplification in the SAR channel.
3) A non-redundant SAR transponder, which provides channel filtering, amplification and up-conversion for the SAR channel. The SAR channel has fixed gain and bandwidth.
4) A wave-guide Output Multiplexer (OMUX) in which the SAR signals are multiplexed with the other L-band downlink signals.
5) An L-band Tx antenna comprising an array of dipoles arranged in 32 columns, each with 4 dipoles connected in parallel. The columns of this array are also electronically switched to make a de-spun antenna beam that fully covers the Earth.

7.5. Cospas-Sarsat Ground Segment

The Cospas-Sarsat ground segment contains three subsystems: Cospas-Sarsat emergency beacons, LUT ground stations and MSS associated to RCC and SAR infrastructures.

Figure 7.14.   Types of Cospas-Sarsat EPIRB, PLB and ELT Satellite Beacons

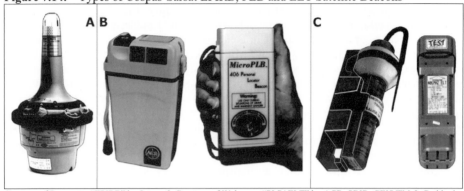

Courtesy of Prospects: "EPIRB" by Jotron & Courtesy of Webpage: "PLP/ ELT" by ACR, SBIR, CEIS/TM & Goddard

7.5.1. Cospas-Sarsat Emergency Satellite Beacons

The Cospas-Sarsat satellite beacons are designed for use for maritime (EPIRB), personal or land (PLB) and aeronautical (ELT) applications. There are 2 types of beacons: the first type transmits an analog signal on 121.5 MHz and the second transmits a digital identification code on 406 MHz band and a low-power "homing" signal on 121.5 MHz. A list of all manufacturers of satellite beacons is presented in the Cospas Sarsat System Data brochure.

7.5.1.1. Emergency Position Radio Beacons (EPIRB) for Maritime Application

The EPIRB maritime satellite beacons are designated for use on board ships and oil/gas sea platforms. The UHF 406 MHz EPIRB beacons are divided into two main categories:
Category I EPIRB beacons are activated either manually or automatically. The automatic activation is triggered when the EPIRB is released from its bracket. Hence, these units are housed in a special bracket equipped with a hydrostatic release. This mechanism releases the EPIRB at a water depth of 1 – 3 m. The buoyant EPIRB then floats to the surface and begins transmitting the distress signals. This type of beacon has to be mounted outside the vessel's cabin, on the deck where it will be able to "float free" of the sinking vessel. Both categories of EPIRB beacons can be detected by both LEOSAR and GEOSAR systems.
Category II EPIRB beacons are manual activation only units and in this sense, they should be stored in the most accessible location on board, where it can be quickly accessed in an emergency situation. At all events, before eventual use and testing of EPIRB beacons it will be necessary to follow the manufacturer's recommendations and guidelines for general beacon testing and inspection procedures. In addition, it will be also very important to register EPIRB beacons, which will help rescue forces to find it faster in an emergency and allow it to make an important contribution to the safety of others by not needlessly occupying SAR resources that may be needed in an actual emergency.
All VHF 121.5 MHz EPIRB beacons are often referred to as Category B/mini-B with manual activation only. Although these units do work with the LEOSAR they cannot be detected by the GEOSAR system that provides instantaneous alerting for 85% of the globe. In addition, 121.5 MHz beacons are a large source of wasted effort by SAR forces. Most of UHF 406 MHz false alerts can easily be resolved with a phone call. In contrast, every 121.5

MHz false alert must be tracked to the source using direction finding equipment. These reasons (and more) have led the International Cospas-Sarsat Program to phase out the 121.5 MHz satellite alerting as of 1st February 2009.

There are many manufacturers of EPIRB units but the most eminent is Jotron's 406 MHz Tron 40 GPS and 40S shown in **Figure 7.14. (A)**. Both beacon designs are specially based on the GMDSS Regulations as float free to operate with the Cospas-Sarsat system. The Tron 40 GPS is designated further to enhance the life-saving capabilities of conventional beacons such as Tron 40S which can be effectively used with the LEOSAR system only. The integrated 12 channel GPS in Tron 40 EPIRB accept continuous positional information from the standard GPS and provides vital SAR operations in a shorter time. Whenever a distress message transmitted by Tron 40 GPS is detected by LEOSAR, the delayed alert remains the same as for Non-GPS EPIRB, such as Tron 40S, within about 90 minutes but the position accuracy is improved considerably from a radius of 5 km to an amazing 100 m. Upon detection of a distress message sent by Tron 40 GPS by GEOSAR, the improvement of integrated GPS proves its benefits as the alert is immediate, within a maximum of 5 minutes, still providing an accurate position of within 100 m. Both Tron 40S and 40 GPS are designed to be installed on board ships in the special FB-4/FBH 4 bracket, which is a float free and automatic release bracket, while a manual MB-4 bracket may also be used.

The next Jotron's interesting product is the Tron 1E MKII personal emergency handheld floating beacon, shown in **Figure 4.19. (A)**. This unit uses aircraft and regional mode VHF 121.5 MHz RF to transmit distress signals via LEOSAR satellites. This unit is made for manual operation for many applications such as personal and aeronautical but in general is a floating EPIRB for maritime utilization. The user must keep the unit horizontal, with the antenna always above the water level. If the antenna is kept under the water level, the searching craft will not be able to receive the emergency signal. The combined VHF 121.5 and UHF 406 MHz are Jotron's products Tron 45S and Tron 45SX, see **Figure 4.19. (C)**.

The Jotron Tron VDR beacon is a float-free Voyage Data Recorder (VDR) memory unit with EPIRB. This unit has built-in GMDSS EPIRB 406 MHz/5 W Cospas-Sarsat GEOSAR Tx and 121.5 MHz/100 mW homing Tx. It duplicates and stores data from the VDR system on board ships via standard Ethernet interface and keeps the stored VDR information on the surface even if the vessel goes down. Tron VDR is a necessity even as an optional device according to the IMO resolution A.861 (20), due to the possibility of avoiding costly under- water search. Storage capacity is up to 8 GB in three memory capacities of VDR video, audio and data including black box.

7.5.1.2. Personal Locator Beacons (PLB) for Land Applications

The PLB beacons are portable units that operate in much the same way as EPIRB units. These beacons are designed to be carried by an individual walking or moving in a vehicle. Unlike ELT and some EPIRB, they can only be activated manually and operate exclusively on 406 MHz. Like EPIRB and ELT, all PLB units also have a built-in, low-power homing beacon that transmits on 121.5 MHz. This allows rescue forces to home in on a beacon once the 406 MHz satellite system has obtained its position. Some newer PLB also allow GPS units to be integrated into the distress signal via the GEOSAR satellite system. This GPS-encoded position dramatically improves the location accuracy down to 100 m. In the USA for instance, PLB beacons are now authorized for nationwide use, granted by the FCC at the beginning of July 2003.

The ACR GyPSI 406 PLB beacon transmits with a digitally coded distress signal on UHF 406 MHz also uses homing signals on 121.5 MHz, shown in **Figure 7.14. (B left)**. It is small, lightweight and can easily be carried in a pack or a large pocket. In addition, this unit's recessed, waterproof GPS and programming interface (NMEA 0183) transmits GPS data for an even faster response.

A state-of-the-art miniature PLB is being developed through Small Business Innovative Research (SBIR) contracts funded by NASA and DOD. This small, lightweight and easily carried unit is made by Microwave Monolithics, Inc., in Simi Valley, California and meets the complete Cospas-Sarsat beacon specifications and certification, which is presented in **Figure 7.14. (B right)**.

7.5.1.3. Emergency Locator Transmitters (ELT) for Aeronautical Applications

The ELT terminals have been standard aircraft equipment for many decades and pre-date satellite tracking. Early versions of ELT beacons started to operate at a frequency of VHF 121.5 MHz and later VHF 243 MHz. When much ELT electronic equipment started operating at the same frequency, there was a large increase in the level of false alerts. As a result, a new ELT technology emerged, using digital signals operating at a frequency band of UHF 406 MHz, which improved this defect with quicker tracing and determination. The 406 MHz system was either made mandatory or recommended by regulatory bodies of ICAO and domestic civil aviation organizations for commercial aircraft as of 1st January 2002. As discussed, due to the obvious advantages of 406 MHz beacons and the significant disadvantages of the older 121.5 MHz beacons, the International Cospas-Sarsat Program have made a decision to phase out the 121.5 MHz band satellite alerting on 1st February 2009. In any event, all pilots are highly encouraged both by Cospas-Sarsat and their participants to consider making the switch to 406 MHz ELT.

The CEIS/TM line of aeronautical ELT units have been extensively tested and approved in the USA and many European countries. In the meantime, fitted on most airlines is the 406 model AO6T ELT illustrated in **Figure 7.14. (C left)**. This unit transmits on three frequencies: 406 MHz, 121.5 MHz and 243.5 MHz. With a maximum weight of 2.6 kg it is portable from aircraft to aircraft. It can be activated automatically, has a built-in, special "G" switch per FAA requirements and includes a bracket to affix it to an aircraft. The AO6T can be remotely controlled via a control panel and if required, it may be recoded in the field. Otherwise, the prototype UHF 406 MHZ GPSELT satellite beacon of NASA Goddard Company, in the testing phase, is presented in **Figure 7.14. (C right)**.

7.5.2. Local User Terminals (LUT)

The configuration and specific capabilities of each of the LUT ground Earth stations in the Cospas-Sarsat participating nations varies somewhat to meet national requirements. The Cospas and Sarsat satellite downlink signal formats were agreed to ensure interoperability between the various spacecraft and LUT stations. Further details on these signal formats are available from any LUT station. In **Figure 7.16.** is shown a typical EMS LEOLUT or GEOLUT 600 station with antenna **(A)** and Rx/processor **(B)**. This configuration is similar and interoperable with products of EMS for both LEOLUT and GEOLUT terminals.

The EMS LEOLUT small and lightweight antenna is an innovative design, which allows very easy installation without structural modifications, in virtually any location including a

Figure 7.16. LEOLUT or GEOLUT Configuration

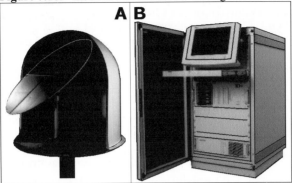

Courtesy of Webpage: "LEOLUT and GEOLUT Products" by EMS

rooftop. It is extremely resistant to high wind loading. Controlled by small stepper motors, its low weight allows the motor to smoothly drive the elevation over the azimuth positioner through the range of motion required to track LEOSAR Cospas and Sarsat satellites. The antenna positioner operates effectively in a temperature range of –40°C to +50°C, allowing for installation anywhere in the world without expensive and often unreliable heating or air conditioning equipment. The LEOLUT antenna is enclosed in a radome to protect it from the environment and to ensure continuous operation in all weathers. The reinforced radome and aluminum base are corrosion-resistant for years of maintenance-free use. The antenna assembly is a type of phased array of 1.43 cm in diameter x 1.0 cm thick with feed coplanar micro strip feed; polarization is LHCP; RF is 1544.5 MHz; antenna gain is 25.3 dBi; noise figure is 0.7 dB maximum at 25°C; antenna beam width is 8°; positioner is 37.5cm x 23.5cm x 90.5cm high; elevation is 0 to 180° and azimuth (max.) is +/–270°.

Detection and location determination of Cospas-Sarsat alerts are the primary functions of the EMS LEOLUT and GEOLUT 600 station, based on over 20 years of EMS experience in the design and evolution of LUT technology. Hence, this station uses industry standard components to provide mission critical reliability and scalability to ensure many years of continued good service. A fully automated high performance Earth station, the LEOLUT 600, monitors, receives and processes distress signals relayed from Cospas-Sarsat satellites. After each satellite pass, it processes the signals received from beacons to update location information. The LEOLUT 600 monitors time and frequency data to detect and locate the active beacon and then sends location coordinates and other pertinent information to the MCC for analysis and distribution to SAR resources. A key function of the EMS LEOLUT 600 is its ability to internally maintain satellite orbit data. This equipment is the only LUT in the world today that can perform combined LEO/GEO beacon processing as defined in the Cospas-Sarsat specifications. The LEO/GEO processing can improve the determined beacon location accuracy by using data received from a GEOLUT. In some instances, this combined processing can provide beacon location solutions when LEOLUT processing alone cannot. The EMS LEOLUT 600 and GEOLUT 600 systems work together to provide a comprehensive and effective SAR system. With their advanced and combined LEO/GEO processing functionality, these EMS systems provide SAR users with response capabilities unmatched by any other system provider. The integrated EMS LEOLUT and GEOLUT 600 is a fully-automated high-performance satellite ground station that receives and processes distress alerts from both LEOSAR and GEOSAR Cospas-Sarsat satellites, determines the beacon location and forwards data to SAR forces.

The LEOLUT/GEOLUT service operators are expected to provide the SAR community with reliable alert and accurate location data, without restriction on its use and distribution. The Cospas-Sarsat Parties provide and operate the space segment to supply LEOLUT and GEOLUT operators with system data required to operate their LUT. To ensure that data provided by an LUT are reliable and can be used by the SAR community on an operational basis, the Cospas-Sarsat has developed LEOLUT/GEOLUT performance specifications and LEOLUT/GEOLUT commissioning procedures. Hence, the LEOLUT/GEOLUT operators provide regular reports on their LUT operation for review during Cospas-Sarsat meetings.

7.5.2.1. LEO Local User Terminal (LEOLUT)

The system configuration and capabilities of each LEOLUT station may vary to meet the specific requirements of the participating countries but the Cospas and Sarsat spacecraft downlink signal formats ensure interoperability between the various spacecraft and all LUT meeting Cospas-Sarsat specifications. Consequently, all Cospas-Sarsat LEOLUT stations must, as a minimum, process the 2.4 Kb/s Processed Data Stream (PDS) of the UHF 406 MHz SARP, which provides the 406 MHz global coverage. Processing of 406 MHz SARP data (i.e., those generated from 406 MHz transmissions processed by the satellite SARP) is relatively straightforward, since the Doppler RF is measured and time-tagged on-board the spacecraft. All 406 MHz data received from the satellite memory on each pass can be processed within a few minutes of pass completion. There are about 40 LEOLUT stations worldwide in service of the LEOSAR Cospas-Sarsat system.

For the VHF 121.5 MHz signals, each transmission is detected and the Doppler information calculated, while a beacon position is then determined using these data. A similar type of processing is also applied to the UHF 406 MHz signals in the 406 MHz SARR band, which can be processed by LEOLUT either separately or combined with the 406 MHz SARP data. Namely, the LEOLUT can improve their Doppler processing of 406 MHz SARP or SARR data by combining it with GEOSAR data. The combined LEO/GEO processing allows the Cospas-Sarsat System to produce Doppler locations in some cases where the data from one LEOLUT is insufficient to produce a location. The combined LEO/GEO processing can also lead to an improvement in Doppler location accuracy. In order to maintain location accuracy, a correction of the satellite ephemeris is produced each time the LUT receives a satellite signal. The downlink carrier can be monitored to provide a Doppler signal using the LUT location as a reference, alternatively, highly stable 406 MHz calibration beacons at accurately known locations can be used to update the ephemeris data.

The antenna and receiving system of Sarsat LEOLUT down converts the signal to an IF RF and a linear demodulator produces the composite baseband spectrum, which is filtered and separated into the various bands of interest. As the signal is received, the processing of each of the several bands is accomplished depending upon the specific capabilities of the LEOLUT stations. Analog type recording of the signal provides a back-up mode in case of processor failure. The VHF 121.5/243 MHz band data is partially processed during the satellite pass and then post-pass processing is accomplished to provide all position locations within thirty minutes after the satellite pass. The 2400 b/s UHF 406 MHz band preprocessed data can be computed in one operation and the time for processing is a function of the capacity of the particular system design. Thus, to improve position location accuracy, an orbit correction update is produced for each satellite received by the LUT. Two methods are used to update the orbit. In one method, the downlink carrier is tracked to provide a Doppler signal using the LUT location as a reference. In the other, local calibration platforms operating a 406 MHz, with accurately known locations, are used.

The antenna and the receiving system of the Cospas LEOLUT detect the spacecraft signal and convert it to about 3 MHz band. This low frequency signal is then A/D converted and after demodulation and processing, is implemented digitally. The UHF 406 MHz band preprocessed data is computed in much the same way as in the Sarsat system.

7.5.2.2. GEO Local User Terminals (GEOLUT)

The GEOLUT receive and process distress alerts from UHF 406 MHz beacons relayed by the GEO satellites of the GEOSAR system and provide permanent monitoring of the RF band. The GEOLUT consists in the following components: antenna and RF subsystem (antenna electronics/receiver); processor; time reference subsystem and MCC interface.

Almost as soon as a beacon is activated in the GEOSAR satellite coverage area, it can be detected by the LUT. In **Table 7.6.** are presented all current GEOLUT stations worldwide serving in the corresponding GEOSAR satellite coverage. As there is no relative movement between a transmitting beacon and the satellite, it is not possible to use the Doppler effect to calculate the beacon position. However, when location information provided by external or internal navigation devices is included in the digital message of a 406 MHz beacon, this position data can be sent with the alert message to the MCC for retransmission to the appropriate MCC, RCC or SPOC.

Table 7.6. Current GEOLUT Infrastructures Associated to Particular GEO Spacecraft

GEO Spacecraft	GEOLUT Infrastructures Associated to GEO Satellites							
	Trenton Canada	Ezeiza Argentina	Brasilia Brazil	Santiago Chile	Bangalore India	Maspalomas Spain	Combe Martin /UK	Wellington New Zealand
GOES-E	Yes	Yes	Yes	Yes		Yes	Yes	
GOES-W	Yes							Yes
Insat-2B					Yes			

7.5.3. Mission Control Centres (MCC)

MCC missions have been set up in most of those countries operating at least one LUT. Their main MCC functions are to: collect, store and sort the data from all LUT stations and other MCC; provide data exchange within the entire Cospas-Sarsat system and distribute alert and location data to associated RCC or SPOC.

Most of the data fall into two general categories: alert data and system information. Alert data is the generic term for Cospas-Sarsat 406 and 121.5 MHz data derived from distress beacons. The 121.5 MHz alerts only include the LEOSAR Doppler position. For 406 MHz beacons, alert data always include the beacon identification and may comprise Doppler and/or encoded location data and other coded information. System information is used primarily to keep the Cospas-Sarsat System operating at peak effectiveness and to provide all users with accurate and timely alert data. It consists in satellite ephemeris and time calibration data used to determine beacon locations, the current status of the space and ground segments and coordination messages required to operate the Cospas-Sarsat System. All MCC in the system are interconnected through appropriate networks for the distribution of system information and alert data. To ensure data distribution reliability and integrity, Cospas-Sarsat has developed MCC performance specifications and MCC commissioning procedures. There are many MCC offices worldwide, whose regular reports on operations are provided during Cospas-Sarsat meetings. Exercises are performed from time to time to check the operational status and performance of all LUT/MCC and data exchange.

Each MCC is responsible for distributing all alert data for distresses located in its Service Area (SA). An MCC SA includes aeronautical and maritime SAR regions in which the national authorities facilitate or provide SAR services and include such regions of other countries that have appropriate agreements for the provision of Cospas-Sarsat alert data. The MCC SA coordinates among Cospas-Sarsat Ground Segment Providers (i.e., LUT-MCC Operators), through the Cospas-Sarsat Joint Committee, taking into account: geographical location and common SAR regions boundaries; communication capabilities and existing national SAR arrangements and existing bilateral operational agreements. When a Cospas-Sarsat beacon transmission is located outside the SA of the MCC, which receives the alert, the alert message is either forwarded to the MCC serving the area where the distress has been located, or is filtered out if the alert data has already been received through another LUT/MCC. To further improve the distribution of operational information amongst a growing number of Cospas-Sarsat MCC, MCC SA regions have been regrouped into a small number of Data Distribution Regions (DDR), with one MCC in each region, acting as a node in the communication network, being requested to take responsibility for the exchange of data between the DDR.

Each MCC distributes Cospas-Sarsat alert data to its national RCC and to designated SPOC in the countries included in its SA. A SPOC is generally a national RCC that can accept or assume responsibility for the transfer of all Cospas-Sarsat alert data on distresses located within its national area of responsibility for SAR, defined as its Search and Rescue Region. Cospas-Sarsat alert messages are exchanged among MCC stations according to standardized message formats, while messages between MCC and their national LUT and RCC are formatted to satisfy national requirements. The detailed structure of messages for distribution of alert data and system information are described in document C/S A.002 "Cospas-Sarsat Mission Control Centres Standard Interface Description (SID)".

The Cospas-Sarsat SID formats have been defined to allow for the transmission of alert messages on any communication system as agreed between MCC stations and between a Cospas-Sarsat MCC and its associated RCC or SPOC. Alert messages in SID formats can be routed and processed by MCC either automatically or manually. In addition to telephone (voice and/or facsimile) communications, all Cospas-Sarsat MCC stations are required to have access to at least two international networks for transmitting alert messages in order to provide maximum availability and adequate flexibility for the exchange of alert data. The MCC stations transmit alert data to other MCC, SPOC or RCC using the following systems: international telex; Automatic Fixed Telecommunication Network (AFTN) of Civil Aviation and packet data networks (X.25). The AFTN addresses, telex numbers, telephone or facsimile numbers are given in the Cospas-Sarsat Data Distribution Plan (C/S A.001). Prior coordination may be necessary between the MCC operator and SPOC in its Service Area to agree on communication systems and interfaces to be used. In order to ensure that Cospas-Sarsat alert and location data are efficiently distributed, IMO and ICAO have invited each of their member governments to designate a national SPOC and to provide the required information on available international communication links (Tel, Tlx, AFTN or X.25), either to IMO, ICAO or the Cospas-Sarsat Secretariat.

8

NON-GEO GMSC SYSTEMS

Non-GEO GMSC systems are new space solutions for GMPSC such as Big LEO and Little LEO, including many other combinations of satellite orbits and systems as follows:

1) MEO MSS: Odyssey (abandoned) and ICO (Inmarsat-P) still in the development phase.
2) Big LEO MSS: Globalstar and Iridium (both will be introduced).
3) Little LEO and Asset Tracking MSS: E-Sat, Falsat, GEMnet, IMC SatPhone, Leo One, Orbcomm, SES Americom and Signal. Leo One and Orbcomm only will be introduced.
4) Broadcasting and HEO MSS: Archimedes and T-Sat.
5) Hybrid MSS Constellations: Celestri, Geostelesat, Ellipso (with Borealis and Concordia Subsystems) and Marathon (with Arkos and Mayak Subsystems).
6) Broadband MSS: M2SAT, Virtual GEO, VITAsat, S2COM and SWANcom.
7) Broadcast MSS: MEASAT, Sirius, XM Radio and WorldSpace.

All the above-mentioned MSS and forthcoming systems will be introduced in a new book with the title: "Global Mobile Personal Satellite Communications (GMPSC)", including regional GEO MSS such as: ACeS (Garuda), AMSC (ACS/Motient), Artemis (LLM/ESA), Emsat (EMS/Italsat), Insat (ISRO), MSAT (TMI/MSV), N-Star, Optus (former MobileSat), Solidaridad and Thuraya (EMARSAT). Accordingly, in this new book project will be also included regional navigation and tracking MSS OmniTRACS/USA and EutelTRACS including all GSAS, GNSS, GSPS and the new European Galileo system.

8.1. Big LEO GMSC Systems

The handheld portable satellite phones and semi-fixed site satellite telephones are a very new communications tools available for business people, all professionals in transportation and fixed environments and everyone who wants to have satellite telephone access using big LEO GMSC systems at sea, on land and in the air. In addition, compared to the little LEO systems, big LEO satellite systems are expected to be bigger and to have more power and bandwidth to provide different services to their subscribers. The bigger size of the satellites in these systems enables them to have more complex data processing in the satellite than the simple store-and-forward feature of satellites in the little LEO systems. The big LEO systems provide a wide variety of services, such as voice, data and Fax, SMS and paging, search and rescue, disaster services, environmental and industrial monitoring, cargo tracking and location, position determination, see **Figure 8.1.**

More than ten years in the process, several GMSC providers in the USA and Europe began as a plan to develop new MSS multipurpose applications, handheld manufacturing, ground access technology and voice transmission protocols in order to enhance the commercial and military telecommunications industry in the new millennium. In September 1991, Inmarsat announced its strategy for the future development of GMSC under the heading Project-21. In fact, the culmination of this strategy was the introduction of a handheld telephone prototype before the entire world under the service name Inmarsat-P. It was recognized that in order to implement this service, new space segment architecture would be required.

Figure 8.1. Big LEO System Concepts

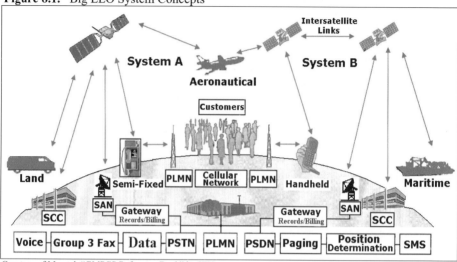

Courtesy of Manual: "GMPCS Reference Book" by ITU

At that time, Inmarsat evaluated a number of possibilities for the Inmarsat-P space segment including enhanced GEO, new LEO and MEO satellite configurations. In any event, these investigations subsequently led to the identification of a MEO satellite constellation as the optimum solution and the eventual establishment of ICO company in order to finance the development of the new MSS. The first GMPSC system, ICO Global Communications, formerly known as the project Inmarsat-P Affiliate Company, was established in January 1995 as a commercial spin-off of the Inmarsat organization.

In addition, other GMSC providers Globalstar, Iridium, Ellipso, Odyssey, Aries and AMSC proposed to exploit the big LEO satellite constellation. However, on 31 January 1995 only Globalstar, Iridium and Odyssey were awarded licenses by the FCC to operate in the USA. The American organization TRW also proposed to exploit the MEO satellite solution using a configuration of satellites named Odyssey. The Odyssey constellation was to consist in 12 satellites, equally divided into three orbital planes, inclined at 55° to the Equator. The satellites were to be placed 10,600 km above the Earth. The FCC awarded TRW a license to establish its MEO satellite system in 1995, with the caveat that building of the first two spacecraft should commence by November 1997. Odyssey was predicted to start service in 1999, at an estimated cost of 3.2 billion US$. Unable to find another major investor willing to support the project, Odyssey was abandoned in December 1997. Also during this time, ICO and TRW had been involved in a patent dispute over the rights to exploit the MEO spacecraft architecture. With the abandonment of the Odyssey project, all legal actions were dropped and TRW took a 7% stake in ICO.

In August 1999, 2 weeks after Iridium filed for Chapter 11 bankruptcy protection in the USA, ICO Global Communications followed suit. Namely, this was after failing to secure sufficient funding to implement the next stage of system development for the upgrade of its terrestrial network to allow the provision of high-speed Internet services via its satellites. The investment of 1.2 billion US$ in ICO by Teledesic was announced in November 1999. Teledesic, although primarily aiming to provide named Internet in the Sky services to fixed users via its constellation of non-GEO satellites, also foresees the possibility of offering services to the maritime and aeronautical sectors. On the 17 May 2000, New ICO, formerly ICO Global Communications, successfully emerged from bankruptcy protection. In the meantime, some of these GMSC operators/providers had failed or changed orbital system configuration. Industry experts expected that only two or three of these providers could survive past the year 2000 because of limited finance budgets, which range between 800 million to 3.37 billion US$ for each proposed system. Otherwise, today only Iridium and Globalstar are completely operational, while the ICO system intends to launch its satellite services late in 2003, after an extensive period of customer testing in 2002.

8.1.1. Globalstar

 Loral Space & Communications, with Qualcomm developed the concept of Globalstar at a similar time to Iridium. Globalstar gained an operating license from the USA FCC in November 1996. Then, the first launch of four Globalstar satellites occurred in May 1998 by Delta rocket from Cape Canaveral and completed the deployment of 48 satellites plus four spares, using Delta and Soyuz-Ikar rockets.

Globalstar is a LEO satellite-based digital telecommunications system that offers wireless telephone and other telecommunications services worldwide, starting from the end of the last century. The communications system is designed to provide worldwide digitally crisp voice, data and facsimile services to portable, mobile and fixed user terminals. To the user, operation of a Globalstar phone is similar to that of a cellular phone but with one main advantage: while a cellular phone works only with its compatible system in its coverage areas, the Globalstar system will offer worldwide coverage and interoperability with current and future public switched telephone and land mobile networks.

The system uses CDMA and FDMA methods with an efficient power control technique, multiple beam active phased array antennas for multiple access, frequency reuse, variable rate voice encoding, multiple path diversity and soft handoff beams to provide high quality satellite service to users anywhere in the world, even when affected by propagation interference and environmental conditions. Globalstar CDMA is a modified version of the IS-95, which was originally developed by Qualcomm Incorporated. The Globalstar system consists in three major segments such as: the Space, Ground and User segments including a Terrestrial Network, as shown in **Figure 8.2**.

8.1.1.1. Space Segment

The Globalstar system has a constellation of 48 satellites in 8 planes with 6 satellites per plane inclined at $52°$ to the Equator at an altitude of 1,414 km LEO and 4 in-orbit spares parked at a lower altitude. The low orbits permit low-power user phones, similar to cellular. The constellation is a 48/8/1 Walker Delta pattern with $52°$ inclination, designed to provide global Earth coverage between $70°$ N and S latitudes. The satellite payload is a "bent-pipe" transponder which includes: two antenna arrays which form two sets of 16 spot beams on the Earth's surface for service uplink (user-to-satellite) and downlink (satellite-to-user); horn antennas for feeder uplink (GES-to-satellite) and downlink (satellite-to-GES); Tx and Rx antennas and circuitry for TT&C. Globalstar satellites function as a relay between the user segment and the ground segment; therefore, they merely transmit the signals received

Figure 8.2. Globalstar GMSC Network

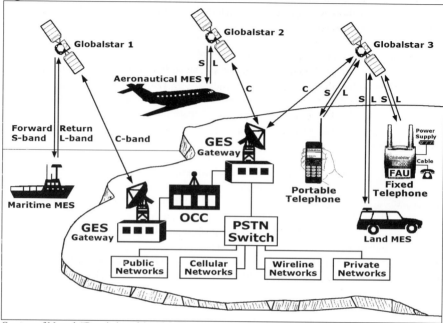

Courtesy of Manual: "Description of the Globalstar System" by Globalstar

Figure 8.3. Globalstar Satellites Coverage

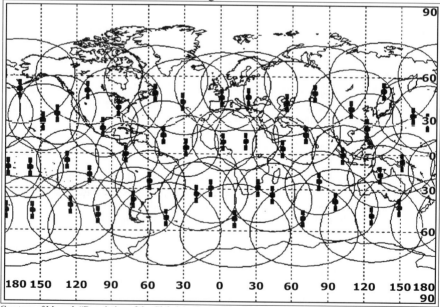

Courtesy of Manual: "Description of the Globalstar System" by Globalstar

from user terminals to the Gateways and vice versa. Gateways are similar to base stations in cellular systems; however, the primary difference compared to terrestrial cellular is that the user signals are relayed through satellites to the Gateways.

The Globalstar GMSC infrastructure provides coverage from any point on the Earth's surface to any other point worldwide with multiple overlapping satellite beams, exclusive of both polar regions as shown in **Figure 8.3**. The contours shown indicate that a User Terminal (UT) within the area can communicate with the satellite at an elevation angle above $10°$. Constraining the UT to operate with satellites that have higher elevation angles with reference to the UT will reduce the overlapping coverage but would provide an advantage in that it would reduce the power demands placed on the UT to close the link. This would result in longer battery life for the UT. In this sense, the satellite orbits are optimized to provide highest link availability in the area between two Poles. Service is feasible in higher latitudes with decreased link availability.

There are two problem areas in the coverage: the first is over the Equator where the beam is narrow and there has been some study using additional satellites to cover the equatorial area. The second area that is not covered well is the polar area and some study on supplementing coverage with Molniya orbit satellites has been performed.

The Globalstar communication satellite is a simple, low-cost satellite designed to minimize both satellite costs and launch costs. The satellite and spacecraft orbital planes are shown in **Figure 8.4. (A)** and **(B)**, respectively. The orbital parameters of Globalstar spacecraft are presented in **Table 8.1**.

Table 8.1. Orbital Parameters of Globalstar Spacecraft

Background	Prime contractors: Space System/Loral
Owner/Operator: Globalstar, USA	Other contractors: Alenia Spazio
Present status: Operational	Type of satellite: Big LEO
Altitude: 1,414 km	Stabilization: 3-Axes
Orbital period: 114 min	Design lifetime: 7.5 years
Type of orbit: LEO	Launch weight: 450 kg
Inclination angle: $52°$ of orbital planes	Mass in orbit: 125 kg
Number of orbital planes: 8	Batteries: 64 A/h
Number of satellites/planes: 6	Electric power: 1100 W (EOL)
Number of satellites: 48 LEO	SSPA power: 100 to 400 hundred per satellite at
Coverage: Global between $70°$ N and S	less than 1 W will be built into phase array antenna
Additional information: The Globalstar space	**Communications Payload**
segment has 12 spare satellites; Each satellite is	Frequency bands:
fitted with bent-pipe transponder; The system	User uplink: 1.610-1.621 GHz
requires GES to be about 65	User downlink: 2.483-2.500 GHz and
Spacecraft	Feeder link: 5.091-7.055 GHz
Name of satellite: Globalstar	Multiple access: CDMA
Launch date: Different	Number of transponders: 16 spot beams resulting
Launch vehicle: Delta 2 & Soyuz	in 2400 circuits
Typical users: All Mobile applications	Channel polarization: LHCP
Cost/Lease information: Construction and launch	EIRP: 26.8 – 36.3 dBW
estimated at 3.26 billion US$ 600 million US $	G/T: – 11.5 dB/K

The Globalstar satellite transponder is transparent, thus, unlike the Iridium system, without cross or intersatellite-links and on-board traffic processing, all traffic switching service happens on the ground and traffic routing is through the existing fixed PSTN with associated networks. A satellite phased array antenna produces 16 elliptical spot beams that

Figure 8.4. Globalstar Satellite and Space Constellation

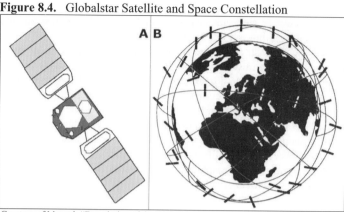

Courtesy of Manual: "Description of the Globalstar System" by Globalstar

provide continuous multiple satellite global coverage, path diversity and position locations. Conversely, lowering the angle to the satellite will increase the overlapping coverage. Thus, small changes can dramatically increase the coverage area, which is particularly apparent in the polar regions. If operated at low elevation angles, polar areas that otherwise could not be covered can receive service. In both polar areas, overlapping coverage would be increased and power demands may be increased because the look angle to the satellite is limited. High gain directional antennas become practical for fixed and even portable installations. The payback is that Globalstar could now serve areas that otherwise might be unserviceable. To a degree, some of these same considerations discussed for polar areas apply to equatorial areas, where the overlapping coverage is less than 100%.

8.1.1.2. Ground Segment

The Globalstar Ground network segment consists in Gateways, the Satellite Control Centre (SCC), Operations Control Centre (OCC) and the Globalstar Business Office (GBO) that are interconnected via a Globalstar Data Network (GDN).
Functions of the Gateways as a GES are to provide the interface between the satellites and the PSTN/PLMN, to provide TT&C and control link between the SCC and the satellites and to allocate satellite resources on a call-by-call basis. The Globalstar system has many Gateways distributed all around the world, which receive and transmit feeder link signals from and to the satellites and provide interconnection to the PSTN. Gateways are designed for unmanned operation; each consists in up to four 5.50 metre antennas and electronics equipment installed in a building or shelter. The Gateway connects the Globalstar space segment to terrestrial switching equipment, receives telephone calls from the terrestrial switching equipment and generates CDMA carriers to transmit through the satellite. The satellite then retransmits the signal to UT. This UT equipment may be either handheld, fixed or mobile and located anywhere within the satellite antenna footprint. In the return direction, the UT transmits to the satellite(s) and the satellite(s) retransmit the signal to the Gateway. The Gateway connects the call to terrestrial switching equipment, which can then connect to any subscriber using the standard telephone system. Connections can also be made to terrestrial cellular subscribers or to other Globalstar UT stations.

The Globalstar system also includes two OCC to manage and control system planning and execution. Each is completely capable of operating the network and managing the satellite constellation. There are two to circumvent the possibility of earthquake, power grid failure or other disaster. One is located in San Jose, California and one is located near Sacramento. Each includes: OCC, SCC and GBO. The OCC manages the satellites, controls the orbits and provides T&C services for the satellite constellation. In order to accomplish this function on a worldwide basis, the OCC communicates with T&C units collocated at selected Gateways. The T&C units share the RF links with the Gateway communications equipment to relay commands and to receive telemetry. The SCC manages all satellite telemetry, track, command and control functions and launches operations.

To support the GBO, the Globalstar Accounting & Billing System (GABS) is collocated with the GOCC and the SOCC. The GABS is responsible for all financial activities associated with Globalstar.

8.1.1.3. User Segment

The user segment includes three different kinds of UT equipment, such as handheld unit, mobile-mounted unit and fixed units. User terminals with omni-type antennas are designed to support data rates up to 9.6 Kb/s. A variable rate vocoder is used that varies its rate each framecording to voice activity. This automatically reduces transmitter power for lower vocoder rates, which means, on average, less interference to other users and higher system capacity. Satellite diversity is utilized in the system; if a call is transmitted through multiple satellites, the user terminal and the gateway receive at least 2 and usually more signals and coherently combine them, which brings diversity gain, reduces the required link margin on each individual link and increases the capacity. Diversity overcomes the adverse effects of propagation such as blocking, shadowing and fading. With the constellation, double satellite coverage is available nearly 100% of the time; therefore, UT devices can provide diversity as required. Both forward and reverse link power control is used to adjust the Gateway and UT powers to the minimum required to maintain high performance. The power is increased only as needed, which means less interference to other users and increased capacity. At this point, the UT units in a particular location on the surface of the Earth are illuminated by a 16-beam satellite antenna as it passes overhead. Hence, UT units can be served by a satellite 10 to 15 minutes out of each orbit. A smooth transfer process between beams within one satellite and between many satellites provides unbroken communications for users. Coverage is maximized in the temperate areas with at least two satellites in view, providing path diversity over most of the area. There is some small sacrifice in multiple satellite coverage at the equator and at latitudes above 60°.

1. Handheld Terminals – the Globalstar handheld terminal looks like a standard cellular telephone. Accordingly, there are multiple mode handset sets that operate with the local cellular system or Globalstar, such as:

a). Tri-mode UT offers a global roaming solution for USA-based AMPS/IS-95 (Advanced Mobile Phone System) North American analog system for cellular users; or the IS-95 CDMA digital coverage area; or the Globalstar MSC service, the presented Qualcomm tri-mode satellite phone for AMPS/CDMA/Globalstar services in **Figure 8.5. (Qualcomm)**.

b). Dual-mode UT offers global service for GSM cellular class 4 phone or Globalstar MSC service users, as shown in the Ericsson and Telit dual-mode GSM/Globalstar satellite phones in **Figure 8.5. (Ericsson and Telit)**.

Figure 8.5. Globalstar Handheld User Terminals

Qualcomm
Tri-mode
AMPS
CDMA
Globalstar

Ericsson
Dual-mode
GSM
Globalstar

Telit
Dual-Mode
GSM
Globalstar

Courtesy of Manual: "Description of the Globalstar System" by Globalstar

2. Mobile Terminals – Globalstar MSS offers three types of MSC terminals similar to the Inmarsat system:

a). Maritime MES terminals, portable and fixed, with ADE and BDE equipment. **In Figure 8.6. (A)** is shown the ICS550 maritime radiotelephone, a product of the ICS and Telit companies. This equipment is designed for sea-going vessels of all sizes to operate anywhere within the Globalstar satellite coverage area, with the possibility to switch if required to GSM cellular telephone networks when close to the coast. Otherwise, the Telital SAT550 hand terminal may be removed as a part of MES for use ashore.

Figure 8.6. Globalstar Maritime, Land and Aeronautical MES Terminals

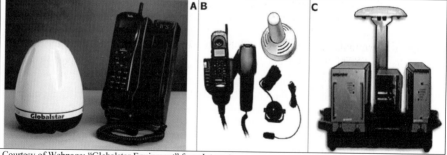

Courtesy of Webpage: "Globalstar Equipment" from Internet

Figure 8.7. Globalstar Payphone and Fixed Satellite Terminal

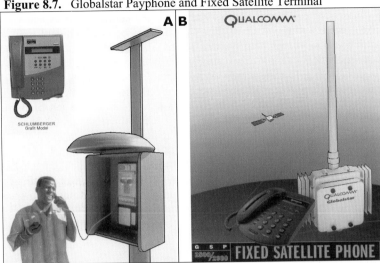

Courtesy of Prospect: "Globalstar System" by Globalstar

b). The land VES terminal of MSS Qualcomm's Globalstar GCK-1410 hands-free car kit equipment, with wired handset telephone and external antenna is shown in **Figure 8.6. (B).** This car dual-mode kit will complement existing fixed and cellular telephone networks by switching from terrestrial cellular telephony to satellite telephony as required.

c). The Aeronautical AES terminal Medium Data Rate Satcom System (MDSS), developed by Qualcomm for use aboard aircraft offers high-speed up to 128 Kb/s, high-quality digital voice and data communications for airborne applications, see **Figure 8.6. (C).** Globalstar's two-way MSC capability allows access to and from airborne platforms equipped with MDSS. Using the Globalstar satellite system, the MDSS supports any aviation application including high-speed access to Internet, E-mail, or private networks. In this sense, other applications include, real-time video and audio monitoring of aircraft cabins and cockpits; an ATC service to alert aviation authorities of emergency situations, allowing ground support teams to rapidly assess and respond to crisis situations; remote control of on-board aircraft cameras; transmission of real-time aircraft flight data to the ground; on-the-ground access to and possibly automated real-time monitoring of flight data and cockpit voice recorders; dedicated voice for Air Marshals to the cockpit and ground; in-flight emergency safety and medical services and back-up transponders with aircraft identification, altitude, speed and location data. This AES also has an extension to 600 Kb/s under study.

3. Fixed terminals – Globalstar in general offers two types of fixed satellite terminals:

a). Payphone terminals are single line Globalstar fixed units used to connect a payphone service into the PSTN for rural and remote areas out of cellular coverage, as shown in **Figure 8.7. (A).** Access to the Globalstar network is via an antenna mounted outside the booth with a clear view of the sky. This antenna can be connected to the CDMA radio unit, also in use in the USA.

b). Fixed single line device offers communication service in remote office environments, which indoor telephone kit and outdoor antenna are illustrated in **Figure 8.7. (B).** The antenna can be mounted in a convenient position on the roof, wall or mast with a clear view of the sky and connected to the subscriber's equipment. The system is compatible with all RJ11 type subscriber equipment such as wall, desk and cordless phones and value added devices like Fax/answering machines. The Globalstar also offers standard trunk interface for compatibility to local switching systems, such as PABX. The fixed UT equipment has a performance equivalent to the MES except that the antenna gain and transmitter power may be even higher. In fact, fixed terminals do not require path diversity to combat fading and blockage and must support seamless beam-to-beam and satellite-to-satellite hand off.

Therefore, since there is no hand off between the local cellular system and Globalstar, if the user crosses a service boundary between the local cellular system and Globalstar, the call could be dropped and must be placed again. The indicators tell the operator that the mode has changed. The system will not clash in a boundary area, thus all users/MES can select the preferred mode. If cellular is preferred and coverage is not available, the UT will drop the call. The call can be placed in Globalstar mode and the call will continue until the phone is in an idle state. The Globalstar system in general offers voice, data speeds from 9.6 Kb/s to 200 Kb/s, circuit switched data similar to dial-up Internet services, packet-switched data, wireless Internet, SCADA and integration with GPS for satellite navigation.

8.1.2. Iridium

The concept for the Iridium MSC system was proposed in late 1989 by Motorola engineers and after the research phase, Iridium LLC was founded in 1991, with an investment of about 7 billion US$. Maintaining its lead, Iridium LLC became operational MSC system on 1st November 1998. After a period of bankruptcy, the Iridium service was relaunched on March 28, 2001. This system was backed by 19 strategic investors from around the world and 17 investor partners also participated in the operation and maintenance of 12 ground station "Gateways" that link the Iridium satellites to terrestrial wireless and landline public telephone networks. The 12 Gateway operators also served as regional distributors of Iridium services in their designated commercial territories.

The Iridium system is a satellite-based network designed to provide truly global personal and mobile service of voice, facsimile, paging and data solutions, while the GPS capability is under development. With complete coverage of the Earth, including polar regions, the Iridium MSS delivers essential access to and from remote or rural areas, where no other form of communication is available. The Company office is situated in Leesburg, Virginia where the Satellite Network Operations Centre is located and Gateway facilities in Tempe, Arizona and Oahu, Hawaii. Through its own Gateway in Hawaii, the U.S. Department of Defense relies on Iridium for global communications capabilities. Iridium is a member of GSM-MoU Association with arguments to provide complementary and value-added global roaming to augment cellular offerings. The Iridium system comprises three principal components: the satellite network, the ground network and the Iridium subscriber products including phones and pagers, as illustrated in **Figure 8.8.**

8.1.2.1. Satellite Network

The satellites are in a near-polar orbit at an altitude of 780 km. They circle the Earth once every 100 minutes traveling at a rate of 26,856 km/h. Each satellite is cross-linked to four other satellites; two satellites in the same orbital plane and two in an adjacent plane. Thus,

Figure 8.8. Iridium GMSC Network

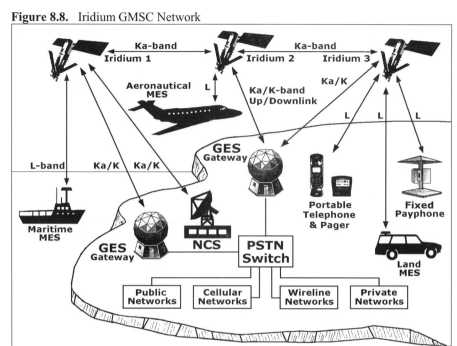

Courtesy of Prospect: "Technical Background" by Iridium

the Iridium constellation consists in 66 operational satellites and 14 spares orbiting in a constellation of six polar planes. Each plane has 11 mission satellites performing as nodes in the telephony network. The 14 additional satellites orbit as spares ready to replace any unserviceable satellite. This constellation ensures that every region on the globe is covered by at least one satellite at all times. Each Iridium satellite provides real global coverage and roaming over the entire globe with 48 spot beams and the diameter of each spot is about 600 km, see **Figure 8.9. (A)**. The 66 satellites enable 3,168 cells, of which only 2,150 need to be active to cover the whole surface of the Earth. At this point, each cell covers about 15

Figure 8.9. Iridium Spot Coverage, Iridium Satellite and Space Constellation

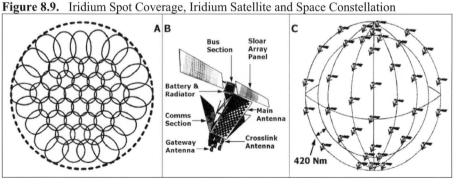

Courtesy of Prospect: "Technical Background" by Iridium

million km^2 and each satellite simultaneously serves an average of 80 and a maximum of 240 cells. The global throughput varies between nominally 171 and 500 thousand simultaneous calls. As the spacecraft moves with great speed, the user encounters adjacent beams about once a minute. The Iridium spacecraft with main components is shown in **Figure 8.9. (B)** and the satellite constellation of 66 satellites in **Figure 8.9. (C)**.

Each satellite is fitted with 3 antennas for communication with other spacecraft, Gateways and mobile terminals. Each antenna communicating with ground terminals will use 16 spot beams, making a cellular type honeycomb of 48 cells from each satellite. Thus, this allows considerable reuse of the same frequencies within different non-adjacent beams, something which is essential if frequency and system congestion is to be avoided. The outer beams will be turned off as satellites approach the poles, thus avoiding overlapping coverage and conserving power. The allocated frequency is divided into 12 sub bands and each sub band is reused four times on a single satellite. Since at high northern or southern latitudes some outer beams are not used, 2,150 spot beams are actually active to cover the globe, so the frequency reuse factor is 2,150/12 = 180. The system is designed for each spot beam to support 80 channels; hence the channel capacity worldwide is 2,150 x 80 = 172,000 channels. Uplink and downlink frequencies are identically allocated in the range around 1.6 GHz. At this point, using 50 Kb/s TDMA bursts in uplink and downlink, 4.8 Kb/s voice or 2.4 Kb/s data full duplex communication service is available. The Iridium system uses only one-way links at a time, which is known as time-duplexing and the user can rapidly switch modes between receive and transmit. The use of one set of frequencies for up and downlinks simplifies the user's hardware. Like the GSM system the user will be handed-off between beams in the same satellite and when required from one satellite to the next.

Each Iridium satellite will have communication links to the bird immediately ahead and behind on the same plane and up to four links with satellites on adjacent planes for cross or intersatellite hand-off. Ka-band intersatellite links with four cross-links on each satellite: front, back and two in adjacent orbits, provide reliable, high-speed communication between neighboring satellites and connect a subscriber to a GES via various possible paths. This flexibility improves call delivery efficiency and system reliability. The orbital parameters of the Iridium satellites are presented in **Table 8.2**.

8.1.2.2. Ground Network

The ground network is comprised of the System Control Segment (SCS) as well as telephony Gateways (GES) and is used to connect into the terrestrial telephone lines. The SCS is the central management component for the Iridium system. It provides global operational support and control services for the satellite constellation, delivers satellite tracking data to the Gateways and performs the termination control function of messaging services. The SCS consists in three main integrated components: four TT&C sites, the Operational Support Network (OSN) and the Satellite Network Operation Centre or Network Control Station (NCS). The primary linkage between the SCS, the satellites and the Gateways is via K-Band feeder links and cross-links throughout the satellites.

Gateways are the land stations that enable connection to and from the PSTN via high-gain K-band parabolic antennas, to track Iridium satellites for services and network operations. They support the interconnection of mobile subscribers via the Iridium network to the terrestrial PSTN and provide network management functions of the entire infrastructure. Each GES is connected with up to 4 satellites and today there are 12 GES worldwide.

Table 8.2. Orbital Parameters of Iridium Spacecraft

Background	Design lifetime: 8 years per satellite
Owner/Operator: Iridium LLC	Mass in orbit: 550 kg
Present status: Operational	Launch weight: 670 kg
Orbital period: 100 min and 28 sec	Dimensions deployed: 4.3 m high, 7.3 m solar
Altitude: 780 km	array tip to tip
Type of orbit: LEO	Electric power: 1200 W (EOL)
Inclination angle: 86.4° of orbital planes	SSPA power: Phased Array Main Mission
Number of orbital planes: 6	Antenna (48 beams per satellite)
Number of satellites/planes: 11	Telemetry beacons: Downlinks 19.4-19.6 and
Number of satellites: 66 Big LEO	Uplinks 29.1-29.3 GHz
Number of spot beams: 48, each of 600 km diameter	**Communications Payload**
Coverage: Global coverage including both Poles	Frequency bands:
Additional information: Satellites have intersatellite	User uplink/downlink: 1621.35-1626.5 MHz
links, on-board processing, link margin is 16 dB	Feeder links:
Spacecraft	Uplink: 29.129.3 GHz (Ka-band)
Name of satellite: Iridium	Downlink: 19.4-19.6 GHz (K-band)
Launch date: Started in November 1998	Cross-link: 23.1823.38 GHz (Ka-band)
Launch vehicle: Proton, Delta and Long March LMC	Modulation: QPSK
were used	Multiple access: FDMA/TDMA
Typical users: Satellite-based Mobile Voice, Paging	Number of transponders: 1 Processing
and Data Services	Channel bit rate (uplink): 2.4 to 4.8 Kb/s
Cost/Lease information: About 5 billion US$	(downlink): 50 Kb/s
Prime contractors: Motorola	Channel capacity: 236/1100 channels
Other Contractors: Locked Martin (Bus), Raytheon	Channel bandwidth: Waveform F/TDMA; Service
(Main Mission Antenna), COM DEV (Feeder &	link 10.5 MHz, feeder Link 107 MHz
Cross Link Antennas)	Channel polarization: Service link Circular
Type of satellite: LM 700	EIRP: 8.5 dBW
Stabilization: 3-axes	G/T: – 23 dB/K

The design of the Iridium network allows voice and data to be routed virtually anywhere in the world. They are relayed from one satellite to another until they reach the satellite above the Subscriber Unit (handset) and the signal is relayed back to Earth. As shown in **Figure 8.10.** and **8.11.** a handset can be handheld satellite telephones and pocket-size pagers to receive short messages and public payphones or fixed and portable units and three mobile applications for ships, vehicles and airplanes. Therefore, each satellite provides 1100 voice channels with high QoS. The Iridium call-processing architecture is based on the GSM digital cellular standard, although the GES needs unique management due to satellite constellation. Thus, the relatively short distance of Iridium satellites reduces the delay and enhances the quality of the telephone conversation between MES and ground subscribers. The phone call is transferred from cell to cell and from satellite to satellite as the spacecraft rise and set during their orbital motion, approximately one revolution around the Earth per hour. Thus finally, Iridium expects to serve 2,7 million subscribers by the year 2005.

8.1.2.3. User Network

The user network configuration consists in three different kinds of Mobile Terminal Units (MTU) such as handheld units, mobile-mounted units and fixed units. User terminals with omnidirectional antennas are designed to support data and facsimile at 2.4 Kb/s as well as numeric and alphanumeric paging. With the Iridium system all communications services voice and paging are delivered regardless of the user location or the availability of PSTN.

Figure 8.10. Iridium Handheld and Portable Satellite Phones and Pagers

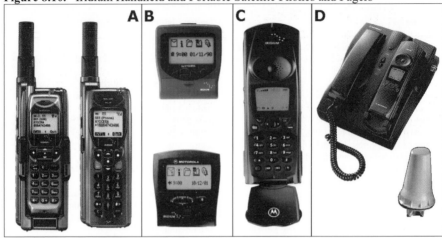

Courtesy of Prospect: "Personal Communication Products" by Kyocera and Motorola

Hence, a variety of subscriber units is available to communicate with the Iridium network, including dual-mode handsets; numeric and alphanumeric pocket pagers; portable, fixed and payphone terminals and specialized maritime, land and aeronautical equipment.

1. Handheld Terminals – An Iridium Portable Telephone is quite similar to a conventional handheld cellular unit because dimensions, weight, battery lifetime and so on, are similar to cellular phones. Thus, it can operate in dual-mode: cellular or Iridium mode, so the Iridium terminal is also a cellular terminal and could be used where cellular networks are available.

a). Dual-mode handheld terminals provide a global roaming for cellular and Iridium users or mono-mode for Iridium users only, as shown **in Figure 8.10. (A)** on the left and right, respectively. These two SS-66K models are products of the Japanese producer Kyocera and the unit on the left-hand side is dual-mode integrated adapter placed inside a standard cellular handset terminal along with all of its memories and functions, in such a way that it is possible to roam both cellular GSM and Iridium networks. In **Figure 8.10. (C)** is shown the Motorola mono-mode satellite series 9500 handheld telephone, for Iridium users only.

b). An Iridium Pager system offers the first true global satellite roaming capability in small, belt-worn, personal message receivers. In **Figure 8.10.** is illustrated the Kyocera model SP-66K Iridium pager **(B-top picture)** and Motorola 9501 Iridium pager **(B-lower picture)**.

2. Fixed terminals – Iridium provides two types of fixed satellite terminals:

a). Payphone fixed terminals are single line Iridium fixed sets used to interface a payphone service into the PSTN for rural and remote areas out of cellular coverage. The portable or semi fixed, redeployable and free-standing payphone booths are designed to provide public access to the Iridium service. These units have their own satellite antenna and transceiver equipment and are able to operate on standard or solar power, reducing costs for the development of other expensive communication services.

b). Fixed or portable single line equipment offers communications services in remote/rural offices and can be a transportable unit in briefcase/car with external antenna, shown in the Motorola 9570 Portable Dock in **Figure 8.10. (D)**. This device is designed for integrated operation with the Motorola 9500 handheld portable telephone sets, to provide charging for

Figure 8.11. Iridium Maritime, Land and Aeronautical MSC Equipment

Courtesy of Prospect: "Mobile Communication Products" by S.P. Radio and Motorola

it and two additional batteries when connected to an AC/DC power source. An external antenna can be connected by special cable from temporary or fixed locations mounted on mast, roof or building walls. The terminal has a speakerphone for voice conferencing and a lightweight cord (passive) handset for private communications and enables data/Fax port access to 2.4 Kb/s asynchronous data service via SIM card as usual for cellular service.

3. Mobile Terminals – Iridium offers three types of MSC terminals similar to Inmarsat and Globalstar systems as follows:

a). Maritime MES terminals portable and fixed with ADE and BDE equipment. **In Figure 8.11. (A)** is shown the maritime Sailor Iridium single channel fixed terminal SC4000 designed by S.P Radio A/S from Denmark. This unit provides one-channel voice/data at a rate 2.4 Kb/s of O-QPSK modulation, can be mounted on board ships, with one external helical omnidirectional antenna and interfaced to Tel handset, Tel/PBX, data RS232 and position information NMEA183. The same model has the possibility of multichannel service, providing 4 channels at a rate of 2.4 Kb/s of O-QPSK modulation with 4 separate helical omnidirectional antenna and the same interface solutions. The Japanese company Kyocera offers an Iridium model IM-S100 Maritime Phone with the possibility to use it on board different types of ships. This is a single channel transceiver unit with external helical omnidirectional antenna, a capacity of 2 handsets connection and with possibility to use and charge a single-channel Iridium handheld Kyocera SS-66K model.

b). The land VES terminal of Motorola's Iridium model 9520 is a permanently installed mobile telephone, as shown in **Figure 8.11. (B)**. It is designed for in-vehicle operation with hands-free functionality and antenna mount options: magnetic, permanent and fixed mast. Its transceiver meets Military Standard 810 (MILSPEC 810). Italian company Telit offers a similar model for car-mounting the SAT550 and Japanese Kyocera offers the HF-S100.

c). The aeronautical AES terminal was developed by Honeywell-Allied Signal Leat s.r.l for use aboard aircraft and offers voice and data communications for airborne applications, shown in **Figure 8.11. (C)**. The system is composed of an ITU 100 Iridium Transceiver Unit, a WH100 Handset Control Device, a JB-100 Junction Box and a SATCOM ANT 100 Blade external mount antenna. This terminal provides the following requisites: the system function does not interfere with the other systems on board; if the system should be subject to any type of failure, this will not affect the normal function of other equipment on board, or create any danger for the remaining flight and the installation of all devices, including the antenna, is designed so that they will not have any effect or compromise the aircraft structure or dynamics, especially in the case of emergency landings.

8.2. Global Little LEO GMSC Systems

Little LEO satellites are so-called because they are very small (measuring around one cubic metre and weighing in at around 100 kg) and because they occupy what is known as a Low Earth Orbit. This means they occupy orbital slots between 700 km and 1,500 km from the Earth, which is low, relative to other satellite systems. The LEO system is a non-GEO solution offering fast and inexpensive services and getting a foothold in the market well ahead of their big brothers. The majority of LEO ventures intend to use the satellites as either bent-pipe or store-and-forward systems. The bent-pipe system in real time relays all messages directly between users; while the store-and-forward approach means that a satellite receives information from a ground station, stores it in on-board memory, continues on its orbit and releases the information to the next appropriate ground station, or user. Thus, users will be able to access the new Little LEO systems using small handheld messaging units incorporating a low-power omnidirectional antenna and weighing less than half a kilogram. The principle difference between the proposed offerings of Little LEO operators and other MSS is that they concentrate on providing data services, rather than handling real-time voice traffic. The kinds of services users can expect from a Little LEO provider are messaging, including E-mail and two-way paging, limited Internet access and Fax. Important markets for Little LEO will include remote data transfer, digital tracking (for the transportation management market), environmental monitoring and SCADA.

8.2.1. Leo One

 The Leo One MSS communication network configuration is designed to use a constellation of 48 Little LEO satellites and to provide very low cost and high quality wireless data communications for business, industry, transport, government and all subscribers worldwide. The system was projected and authorized by the FCC in February 1998 and commercial service was expected to begin in 2002 or 2003. Consequently, operation of the Leo One satellite system will provide low data rate transmission using store-and-forward packet communications with coverage of all points between the Arctic and Antarctic Circles.

The Leo One satellite system will serve the market for short data messaging and digital data, including tracking and fleet mobile management, monitoring and remote control, emergency services and transaction processing. At any rate, the Leo One system is not designed to transfer large files or support voice communications. The prolific expansion of the Internet, the declining cost of GPS, the rapid growth in the number of communications devices and systems with embedded intelligence and the emergence of wireless data as an important future service on terrestrial wireless networks, are all forces converging to make the potential market for the Little LEO satellite mobile configuration more attractive. In this sense, automotive telematics, logistics transportation management services, wireless alarms and transaction services are likely to be important markets for the Leo One network.

8.2.1.1. Space Segment

Leo One Worldwide Inc., a privately-held company based in St. Louis, Missouri, USA, has been licensed by the US FCC to construct, launch and operate a Non-Voice Non-GEO MSS (NVNG MSS), commonly known as a Little LEO. The Leo One GMSC constellation consists in 48 operational satellites plus 2 on-orbit and on-the-ground spares. Each Leo One satellite will have two UHF and one VHF quadrifilar helical antennas for Tx and Rx signals. The Leo One network has user terminal to satellite links at 149 MHz and satellite to user terminal links at 137 MHz. Each satellite supports 15 service uplink channels and only one service downlink channel. Additionally, only one Gateway uplink feeder channel and one Gateway are available on each satellite. The orbital parameters of the Leo One constellation are given in **Table 8.3**. The 50° orbital plane inclination allows coverage from the Arctic Circle to the Antarctic Circle using Leo One spacecraft, see **Figure 8.12**.

The Eurockot Launch Services GmbH company has been selected by Leo One to launch its LEO MSS constellation. Both companies are working toward the same goal, i.e. to provide competitive and reliable services for the benefit of customers. Namely, reliability is a top priority of the Leo One project by selecting the highly reliable Rockot launch vehicle, which will minimize the risk associated with the launch of the Leo One satellite constellation. Thus, launches for Leo One will be carried out from Baikonur Cosmodrome in Kazakhstan, where Eurockot will operate dedicated launch and payload preparation facilities. Eurockot is the joint venture of Daimler Chrysler Aerospace AG (DASA) and the Khrunichev State Research & Production Space Centre to market the Russian Rockot launching vehicle as a readily available and proven launch system to the LEO market.

Otherwise, Leo One selected other business units, Dornier Satellitensysteme (DSS) GmbH (a corporate unit of DASA) and Lockheed Martin Space Electronics & Communications, to participate together with Eurockot in the construction of its satellite network. In addition to the existing contract, these agreements complete the new arrangements necessary for the construction and launch of all 48 satellite constellations, including 2 spares.

Table 8.3. Orbital Parameters of Leo One Spacecraft

Background	Other contractors: Multiple
Owner/Operator: Leo One Worldwide, USA	Type of satellite: Little LEO
Present status: Planned (Operational in 2003)	Stabilization: 3-Axes
Altitude: 950 km	Design lifetime: More than 5 years, with 7 years of
Type of orbit: LEO	consumables
Inclination angle: 50° of orbital planes	Launch weight: Each satellite should weigh 192 kg
Number of orbital planes: 8	Mass in orbit: 125 kg
Number of satellites/planes: 6	Dimensions stowed: 1.1 x 1.1 x 0.54 m
Number of satellites: 48 LEO	Electric power: 560 W (EOL)
Coverage: Global between 73° N and S	SSPA power: 34 W
Additional information: Right ascension of	**Communications Payload**
ascending node on 0, 45, 90, 135, 180, 225, 270 and	Frequency bands:
315°. The system has 1 on-orbit and 1	User uplink: 148-150.05 MHz
on-the-ground spare satellites	User downlink: 137-137.025 MHz and
Spacecraft	Feeder link: 400.15-401 MHz
Name of satellite: Leo One	Data rate (uplink): 2.4 to 9.6 Kb/s
Launch date: To be launched since in 2002	Data rate (downlink): 24 Kb/s
Launch vehicle: Rockot (9 launches)	Multiple access: FDMA
Typical users: All Mobile applications of	Number of transponders: 15 channels per satellite
two-way messaging (vehicle tracking, fleet	Channel capacity: U/L 960 b/s,
management, pager alerts and monitoring)	D/L 24 kb/s
Cost/Lease information: About 500/600 million US $	Channel polarization: LHCP
Prime contractors: Eurockot Launch Services GmbH,	EIRP: 13.3 dBW
Dornier Satellitensysteme GmbH and Lockheed	G/T: –30.6 dB/K
Martin Space Electronics & communications	Saturation flux density: –125.0 dBW/m^2

Figure 8.12. Leo One Space and Ground Configuration

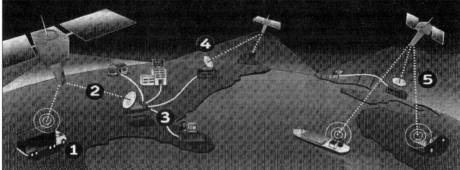

Courtesy of Handbook: "MSS" by ITU

The Leo One USA Corporation has successfully developed and verified its new proprietary Leo One satellite antenna. The antennas are now in the engineering development phase and were designed to optimize links with transceivers on the Earth within the satellite footprint at any visible point. Radiation patterns of the prototype satellite antennas based on the Leo One system design are successful. Satellites of the Leo One constellation need to be able to communicate with transceivers and in order to transmit the same power density to any point on Earth under the satellite, the antennas on the Leo One satellite must radiate in an "isoflux" pattern. The quadrifilar helical antennas were chosen because they are particularly suited to produce "isoflux" coverage as well as to satisfy other system requirements. The proprietary satellite antenna design is expected to provide many significant benefits in the quality and reliability of service to be provided to all subscribers. Finally, the three antennas were initially designed on a computer using specialized antenna analysis software and prototypes were fabricated and tested on the deployment mechanism. Measurements of the antenna radiation patterns were taken and compared against the theoretical predictions. The patterns were in excellent agreement, experimentally verifying this design; therefore, satellite antennas are now in the engineering development phase.

8.2.1.2. Ground Segment

The Leo One system is designed to support higher data rates and thus to achieve greater spectral efficiency than other Little LEO systems. Furthermore, Leo One is the only Little LEO system designed to support near-real-time data messaging delivery. Leo One's use of VHF and UHF frequencies will outperform and be more reliable than the higher frequency signal from a Big LEO and assures high quality signals in areas of dense foliage, heavy rain and different man-made obstructions, in addition to facilitating the use of very low cost and low power terminals.

Leo One has developed a new Integrated Communications Facility (ICF) that can serve as a Gateway Ground Station (GGS) or relay station in the NVNG MSS Leo One network, commonly known as a Little LEO system. The system of 48 LEO satellites will contain approximately 20 GGS or ICF operating in the VHF and UHF bands, capable of providing near-real-time store-and-forward messaging services to users around the world. The GGS contains no moving parts, does not utilize radomes and requires minimum acreage. The siting and construction costs, hardware and operating expenses are anticipated to be less for the GGS terminal than for a conventional Gateway site. At this point, other benefits include no downtime during repairs.

The UHF segment of the GGS (ICF) consists in a four-sided array with a 32-element array per face. Each array has the ability to automatically track 3 satellites simultaneously and the four arrays in combination provide full 360° ground coverage. The four-sided UHF array sits atop a similarly constructed VHF array. Approximately 20 such facilities will be located worldwide to support the Leo One constellation. The first full face of the UHF array segment of the Leo One GGS has been completed. The 32-element array is also designed to provide a gain of 16 dBi over a scan range of ±45° in azimuth and 10° to 90° in elevation. The array face is tilted 30° from the vertical to achieve the elevation scan range. The assembly of four arrays sits atop a similarly constructed VHF four-sided array, which operates over the frequency range of 136 to 151 MHz. The first VHF 32-element array face is under construction. The VHF array has the same electrical performance as the UHF array and the four-sided assemblage also provides coverage over a full hemisphere beginning at 10° above the horizon. In addition to serving as a Gateway or relay station, the ICF may be used for administrative offices and maintenance shops. The interior of the ICF provides space for housing various administrative functions, such as operations, billing and accounting, required for the operation and support of a satellite communication system.

The Leo One network infrastructure presented in **Figure 8.12**. operates in conjunction with and as an extension of terrestrial data networks through the use of GGS that provide the interconnection to the existing terrestrial infrastructures. The Mobile User Terminal (MUT) or MES provides packet data connection with the satellites and has the ability to send and receive digital data. On the uplink, the users access several communications channels and transmit data to the satellites at data rates from 2.4/9.6 Kb/s. Downlinks from the satellites are distributed for the particular user's terminals. Messages can be sent between users, like persons or mobiles shown on the same figure, where one or both are equipped with a Leo One terminal for MSS. The MES terminal mounted in a truck **(1)**, sends a message to the nearest in-view satellite **(2)** and then the satellite forwards the message to a GGS for validation and optimal routing **(3)**. The GGS delivers the message to its recipient via the best route to the Internet, PDN or PSTN interfaces. In some special cases, the receiving GGS will route the message to another GGS and then through the next visible satellite communications path for delivery **(4)**. In fact, messages can also be initiated by subscribers connected to landline networks, routed to a GGS and delivered via Leo One satellite to a MUT installed on board a ship's ES service terminal or any other vehicle **(5)**.

The Leo One satellites operate in a store and forward mode to receive, store and transmit the data as required. The overlapping beam coverage on the surface of the Earth by the Leo One satellite footprints is designed to provide near-real-time communication links for the users. The satellites, in conjunction with the GGS, control the user terminal access to the Network. The GGS, operating with the Network Management Centre (NMC), controls the Network operation including packet routing, satellite tracking and ephemeris, billing and subscribers database.

The Leo One MES used by the subscribers have their technical parameters listed in **Table 8.4**. The user terminals are approximately 160 cm^3 in size, have 7 W output power and are battery powered where necessary.

The Leo One GGS terminals used as interconnecting Gateways between MES and certain numbers of satellites, containing the technical parameters listed in **Table 8.5**. Each GGS is located in corresponding positions from where Leo One satellites are in better view.

Table 8.4. Leo One MES Characteristics

Subscriber Uplink		Subscriber Downlink	
Technical Parameters	Value	Technical Parameters	Value
Band (MHz)	148-150.05	Band (MHz)	137-138/400.15-401
Tx Power (W)	7	Tx Power (W)	17.5
Tx EIRP (dBW)	8.5	Tx EIRP (dBW)	18.1
Max Tx antenna gain (dBi)	0	Max Tx antenna gain (dBi)	5.7
Channel bandwidth (kHz)	15	Channel bandwidth (kHz)	25 and 35
Rate (Kb/s)	9.6/OQPSK	Rate (Kb/s)	24/OQPSK - 9.6/FSK
Polarization/Tx wave	Linear	Polarization/Tx wave	RHC
Sat Rx G/T (dB/K)	−22.9	Subscriber Rx G/T (dB/K)	−30.8
Max Rx antenna gain (dBi)	5.7	Max Rx antenna gain (dBi)	0
Rx antenna pattern	Isoflux		
C/(I+N) (dB)	5.5	C/(I+N) (dB)	5.1

Table 8.5. Leo One GGS Characteristics

Gateway Uplink		Gateway Downlink	
Technical Parameters	Value	Technical Parameters	Value
Band (MHz)	148-150.05	Band (MHz)	400.15-401
Tx Power	1.2 W	Tx Power	15 W
Tx EIRP	17.8 dBW	Tx EIRP	17.5 dBW
Max Tx antenna gain	18 dBi	Max Tx antenna gain	5.7 dBi
Channel bandwidth	50	Channel bandwidth	60
Rate	50/OQPSK Kb/s	Rate	50/OQPSK Kb/s
Polarization/Tx wave	RHC	Polarization/Tx wave	RHC
Sat Rx G/T	−22.9 dB/K	Subscriber Rx G/T	−9.9 dB/K
Max Rx antenna gain	5.7 dBi	Max Rx antenna gain	17 dBi
C/(I+N)	8.5 dB	C/(I+N)	8.5 dB

8.2.1.3. Network Operation and Applications

The user MES terminals provide data links to the satellites for sending and receiving data messages as shown in **Figure 8.12.** The sender's message goes to the nearest in-view satellite where it is linked to the local Gateway for validation and optimal routing to the recipient. About 20 GGS are interconnected via a Leo One terrestrial backbone network.

In the case of the mobile recipient, the message is then returned to the satellite and stored briefly, until the intended receiver is in view, before delivery to the transceiver unit of recipient. If necessary, GGS can relay messages between satellites for faster delivery. If the intended recipient is a fixed site, as would be the case for a company tracking remote devices attached to a mobile fleet, the message is delivered to the dispatch recipient via interconnection with the Leo One terrestrial backbone network. In the latter example the final link from the Leo One backbone to the customer premises is via Internet; dial-up line or by a dedicated connection.

The number of satellites and the orbit inclination ensure that there is always at least one satellite in view for latitudes up to $64°$. Beyond that, there may be a short wait. Each satellite has a beam footprint of about 12 million square kilomtres. The system can provide near- real-time data communications for users at data rates from 2.4 to 9.6 Kb/s. The Leo One network is designed to be highly reliable because every message will be acknowledged to guarantee delivery. Otherwise, the multiple classes of service ensure high priority handling of emergency and all alarm-type messages. Redundancy will be built into the terrestrial backbone infrastructures and the system will be able to tolerate multiple satellites failures with no customer-perceptible degradation in service. In this way, the Leo One system further minimizes risk by maintaining spare satellites on-orbit and by using highly reliable launch services and space-qualified satellite components and subsystems. The overlapping satellite footprints and high satellite sparing result in a very high level of redundancy in the constellation, so customers can safely rely on Leo One for mission-critical applications. The Leo One network will also utilize multiple levels of encryption along with sophisticated error-checking to ensure all customers can rely on the integrity and security of their data communications. This is an important feature for commercial applications related to recording of revenues, remote control of assets or security.

8.2.1.4. System Estimated Costs and Features

The overall installed system cost of the Leo One network is between 500 and 600 million US$. Transceivers will cost 100 to 500 US$ but after several years of system operation, user terminal costs are estimated to be under 100 US$. Service costs will be competitive with terrestrial-based data communications systems, however, monthly service costs for use of the system and transmission of data would range from 1 to 50 US$ per MES, depending on the amount of data transmitted to and from users.

Leo One will initially focus on serving those applications with a requirement or a preference for near-real-time message delivery in 2003 with about 71 million terminals and 7.8 billion US$ in potential revenue. With its design for a revolutionary new mobile satellite system, as mentioned, the major market strategy is that the Leo One system will provide a new level of inexpensive low-cost, economic and reliable near-real-time data mobile and fixed satellite communications service for industrial, business and personal data communications.

The number of 48 satellites and the highly inclined orbits of 50° provide near-real-time communications to the most populated regions of the Earth between the polar circles. The low satellite altitude of 950 km produces short transmitter path lengths to the satellites resulting in low transmitter power requirements for the user terminals and the satellites. This reduces the overall system cost and the consequent cost of service. Data rates ranging from 2.4 to 9.6 Kb/s allow for short duration packet data bursts of less than 500 ms.

Adoption of the Little LEO services will just be starting to enter its real growth phase. These include the ability to deliver reliable, near-real-time service, where messages are delivered in less than 5 minutes and often less then 1 minute in a targeted geographic region. This Network will maintain virtually 100% service availability in its target markets, even in a situation of multiple, simultaneous satellite failures.

Together with DBS Industries (E-Sat) and Final Analysis (FAISAT), Leo One is preparing to challenge Orbcomm Global for a share of the LEO MSS market. Leo One specifically wants to expand into the vehicle monitoring and asset tracking markets. Namely, fixed and mobile applications can be served offering vehicle tracking, status monitoring, emergency alerting, messaging, paging and positioning.

There are significant differences between Leo One and existing Little LEO systems with regard to the percentage of time satellites in view, capacity and measures taken to ensure data integrity and security. These differences allow Leo One to address applications that will remain unserved until Leo One is in service and they will put it on a strong competitive footing to gain its proportionate share of the markets that can be served by other proposed Little LEO systems.

8.2.2. Orbcomm

 The Orbcomm system is a wide area packet switched and two-way data transfer network providing satellite communication, tracking and monitoring services between mobile, remote, semi fixed or fixed Subscriber Communication Units (SCU) and Gateway Earth Stations (GES) or Gateway Control Centres (GCC) accomplished via the constellation of LEO satellites and Network Control Centres (NCC). Namely, Orbcomm is a mobile satellite-based system that offers affordable global wireless data and messaging communications services via LEO satellites to provide worldwide coverage. The system is capable of sending and receiving two-way alphanumeric packet messages, similar to the well-known two-way paging, SMS or E-mail. The Orbcomm network enables two-way monitoring, tracking and messaging services through the world's first commercial LEO satellite-based data communications system, which applications include tracking mobile assets such as trailers and containers, locomotives, rail cars, heavy equipment, fishing vessels and barges; monitoring and controlling fixed sites such as electric utility metres, water levels, oil and gas storage tanks, wells, pipelines and environmental projects and a two-way messaging service for consumers, commercial and government entities.

Orbcomm Global, L.P., from Dulles, Virginia, USA equally owned by Teleglobe and the Orbital Sciences Corporation, provides global services via the world's first LEO satellite-based data communications system. The FCC granted Orbcomm a commercial license in October 1994 and the Commercial service began in 1998. Orbital Sciences was the prime contractor for the design project of Orbcomm satellites. The Company owns and operates a network consisting in 36 LEO satellites and four terrestrial Gateways deployed around the world. Small, low-power and commercially proven SCU can connect to private and public networks, including the Internet, via these satellites and Gateways. Through this network, Orbcomm delivers information to and from virtually anywhere in the world on a nearly real-time basis. The Orbcomm satellites have a subscriber Tx that provides a continuous 4.8 Kb/s stream of downlink packet data, which is capable of transfer even at 9.6 Kb/s.

Vital messages generated by a variety of applications are collected and transmitted by an appropriate mobile or fixed SCU to a satellite in the Orbcomm constellation. The satellite receives and relays these messages down to one of four US GES. The GES then relays the message via satellite link or dedicated terrestrial line to the NCC. The NCC routes the message to the final addressee, through the Internet via E-mail to a personal computer, through terrestrial networks to a subscriber communicator or pager and to dedicated telephone line or facsimile, as illustrated in **Figure 8.13**. Messages originating outside the USA are routed through international GCC in the same way to its final destination. In reverse mode, messages and data sent to a remote SCU can be initiated from any computer using common E-mail systems, including the Internet, mail and X.400. The GCC or NCC then transmits the information using Orbcomm's global telecommunications network.

Orbcomm serves customers through Value Added Resellers (VAR) that provide expertise in specific industries. These Orbcomm VAR provide whole product solutions and customer support to end-users. Customers from around the world currently rely on Orbcomm for a wide range of mobile and fixed site data applications including:

1) Monitoring and controlling assets at remote or rural sites for oil/gas extraction, pipeline operations, storage, custody transfer and electric power generation and distribution;

2) Messaging for truck fleets, owner operators and remote workers;

Figure 8.13. Orbcomm System Overview

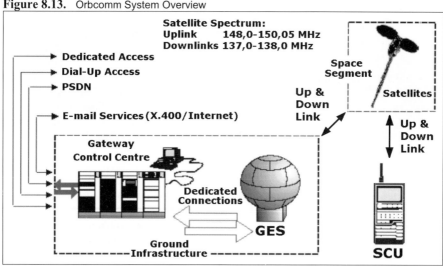

Courtesy of Manual: "Orbcomm System Overview" by Orbcomm

3) Tracking and managing construction equipment, locomotives, rail cars, trucks, trailers, containers, vessels, aircraft and locating and recovering stolen vehicles and cargo and
4) Weather data for general aviation.

The Orbcomm system allows users to track, monitor and manage remote assets. Through a network of LEO satellites and regional GES, users can communicate with their mobile or fixed assets anywhere in the world. This system is operational, robust and provides service to customers worldwide today. Orbcomm is in a position to offer low-cost and high-quality service to each customer. Therefore, Orbcomm committed staff are dedicated to fulfilling the specific needs of all potential users.

8.2.2.1. Space Segment

Orbcomm communication network consists in 36 operational satellites in LEO orbit at about 825 km above the Earth's surface, as shown in **Figure 8.14. (A).** The main function of Orbcomm's satellites is to complete the link between the SCU and the switching capability at the NCC in the USA, or a licensee's GCC in other countries. The satellites are "orbiting packet routers" ideally suited to "grab" small data packets from mobile sensors in vehicles, containers, vessels or remote fixed sites and relay them through a tracking Earth station and then to a GCC. The Orbcomm satellites constantly move, so large obstructions do not prohibit communications and coverage is available in remote rural areas. In comparison, cellular coverage depends on tower location, usually centered around major highways and cities and cannot reach remote areas. Moreover, the GEO satellite system requires large, costly and power-intensive hardware. Large data files (such as graphics) or emergency response latencies are, however, not appropriate applications for Orbcomm.

As mentioned, the last Orbcomm constellation consisted in 36 satellites in orbit:
1) Planes A, B and C are inclined at 45° to the equator and each contains eight satellites in a circular orbit at an altitude of approximately 815 km.

Figure 8.14. Orbcomm Satellite Constellation and Parts of Deployed Satellite

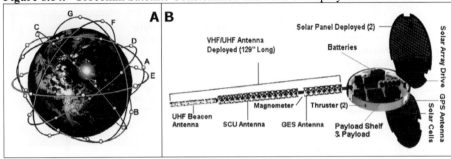

Courtesy of WebPages: "Orbcomm Satellite" by Orbcomm

2) Plane D is also at 45° containing seven birds in a circular orbit at an altitude of 815 km.
3) Plane F is inclined at 70° and contains two birds in a near-PEO at an altitude of 740 km.
4) Plane G is inclined at 108° and contains two satellites in a near-polar elliptical orbit at an altitude varying between 785 km and 875 km. Plane E is in circular equatorial orbit.

The Orbcomm network depends on the number of satellites and Gateways in operation and the user's location. As the satellites move with the Earth, so does the approximately 5.100 km diameter geometric footprint of each satellite. This system provides redundancy at the system level, due to the number of satellites in the constellation. Thus, in the event of a lost satellite, Orbcomm will optimize the remaining constellation to minimize the time gaps in satellite coverage. Consequently, the Orbcomm constellation is tolerant of degradations in the performance of individual satellites.

To date, 36 Orbcomm satellites have been launched, using Pegasus XL and Taurus launch vehicles. Each of the satellites is based on the Orbital Microstar satellite bus. Undeployed, the Orbcomm satellite resembles a circular disk and the spacecraft weighs circa 43 kg, measuring approximately 1 m in diameter and 16 cm in depth. Circular panels hinge from each side after launch to expose solar cells. These panels articulate on the 1-axis to track the Sun and provide 160 W. The satellite's electrical power system is designed to deliver circa 100 W on an orbit-average basis, near its expected EOL in a worst-case orbit. The satellite solar panels and antennas fold up into the disk (also called the "payload shelf") with the remainder of the payload during launch and deployment. Once fully deployed, the spacecraft length measures about 3.6 m from end to end with 2.3 m span across the solar panels disks. Long boom is a 2.6 m VHF/UHF gateway antenna. **Figure 8.14. (B)** shows the main parts of a fully deployed satellite. Each spacecraft carries 17 data processors and seven antennas, designed to handle 50,000 messages per hour. The Orbcomm satellite transponder receives by 2400 b/s at 148 to 149,9 MHz and transmits by 4800 b/s at 137 to 138 MHz and 400.05 to 400.15 MHz. The system uses X.400 (CCITT 1988) addressing and message size is typically 6 to 250 bytes (no maximum). The most important orbital parameters of the Orbcomm constellation are presented in **Table 8.6**. The communication subsystem is the principal payload flown on the satellite, consisting in five major parts:

a). Subscriber Communications Section as the main payload part consists in one subscriber Tx, seven identical receivers and the associated receive and transmit filters and antennas. Six of the receivers are used as subscriber receivers and the seventh is used as the DCAAS Rx. The subscriber Tx is designed to transmit an operational output power of up to circa 40 W, although the output is less during normal operations. So, the power of each Tx can vary

Table 8.6. Orbital Parameters of Orbcomm Spacecraft

Background	Prime contractors: Orbital Science Corporation
Owner/Operator: Orbcomm Global LP, USA	Type of satellite: Microstar (Little LEO Project)
Present status: Operational	Stabilization: Magnetic with gravity gradient assist
Altitude: 775/739 km	Design lifetime: 4 years
Type of orbit: LEO	Mass in orbit: 1,385 kg
Inclination angle: 45°/70°	Dimensions stowed: 1.83 x 12.50 m circular Electric
Number of orbital planes: 4/2	power: 135 W (EOL)
Number of satellites/planes: 8/2	SSPA power: 10 W
Number of satellites: 32/4 LEO	**Communications Payload**
Coverage: Worldwide	Frequency bands:
Additional information: system offers data and	Service/Feeder uplink 148.0-150.05
asset tracking messaging with 14 GES all over	Service/Feeder downlink 137.0-138.0 MHz
the world	Multiple access: FDMA/TDMA
Spacecraft	Number of transponders: 6 Uplink Rx; 2 Downlink
Name of satellite: Orbcomm	Tx; Ka-band operation
Launch date: Started in November 1998	Channel capacity: 15 Gb/s total data rate
Launch vehicle: Pegasus XL & Taurus	Channel polarization: Circular
Typical users: Global Mobile Messaging Service	EIRP: Varies over coverage area
Cost/Lease information: Approximately 900	G/T: Varies over coverage area
million US $	Saturation flux density: High

over a 5 dB range, in 1 dB steps, to compensate for aging and other lifetime degradations. The SDPSK modulation is used on the subscriber downlink at a data rate of 4800 b/s. (It is capable of transmitting at 9600 b/s.). The satellite uplink modulation is SDPSK; with a data rate of 2400 b/s. Raised cosine filtering is used to limit spectral occupancy.

b). The Orbcomm Gateway Communication Section contains both the Gateway satellite Tx and Rx. Separate RHCP antennas are used for the transmit and receive functions. In fact, the Orbcomm Gateway Tx is designed to transmit 5 W of RF power. The 57.6 Kb/s downlink signal to the GES is transmitted using an OQPSK modulation in a TDMA format. The Gateway Rx is designed to demodulate a 57.6 Kb/s TDMA signal with an OQPSK modulation. The received packets are routed to the onboard satellite network computer.

c). The Satellite Network Computer receives the unlinked data packets from the subscriber and the Orbcomm Gateway receivers and distributes them to the appropriate Tx. The computer also identifies clear uplink channels via the DCAAS Rx and algorithm and interfaces with the GPS receiver to extract information pertinent to the communications system. Several microprocessors in a distributed computer system aboard the satellite perform the satellite network computer functions.

d). The UHF TX is a specially constructed 1 W Tx to emit a highly stable signal at 400.1 MHz. The Tx is coupled to a UHF antenna designed to have a peak gain of circa 2 dB.

e). The Satellite Subscriber Antenna Subsystem consists in a deployable boom containing three separate circularly polarized quadrifilar antenna elements.

The Attitude Control System (ACS) is designed to maintain both nadir and solar pointing. The satellite must maintain nadir pointing to keep the antenna subsystem oriented toward the Earth. Solar pointing maximizes the amount of power collected by the solar cells. The ACS subsystem employs a three-axis magnetic control system that operates with a combination of sensors. The satellite also obtains knowledge of its position through its on-board GPS receiver. Satellite planes A/B/C are designed such that the satellites maintain a separation of 45° (± 5°) from other satellites in the same orbital plane. Planes D/E are designed for a 51.4° spacing. The supplementary, highly inclined satellite planes (F/G) are designed such that the satellites are spaced 180° apart (± 5°). The springs used to release the satellites from the launch vehicle give them their initial separation velocity. A pressurized gas system will be used to perform braking maneuvers when the required relative in-orbit satellite spacing is achieved. An Orbital Sciences Corporation formation-keeping technique will maintain the specified satellite intra-plane spacing. Therefore, one of the benefits of this technique is that, unlike GEO satellites, it does not affect the satellite's life expectancy in fuel usage.

8.2.2.2. Ground Segment

The Orbcomm ground segment, which has most of the intelligence of the Orbcomm system, comprises Gateway Earth Stations (GES), Control centres and both mobile and fixed SCU user terminals. Otherwise, the space segment is controlled by one Satellite Control Centre (SCC).

Gateways, which include the GES, GCC and the NCC, are located at Orbcomm headquarters in Dulles. Within the USA, there are four GES located in Arizona, Georgia, New York State and Washington State. The NCC also serves as North America's GCC and manages the overall system worldwide. Orbcomm Gateways are connected to dial-up circuits, private dedicated lines, or the Internet. The SCU hand-held devices for personal messaging are fixed and mobile units for remote monitoring and tracking applications.

1. Gateway Earth Station (GES) – Orbcomm is committed to continuing the deployment of additional regional GES to provide near-real-time service for all major areas of the world, as well as developing and launching a new generation of satellites that will enhance and expand the current system's capabilities. All Orbcomm's GES terminals link the ground segment with the space segment and will be in multiple locations worldwide. The GES provide the following functions: acquire and track satellites based on orbital information from the GCC; link ground and space segments from multiple worldwide locations; Transmit and receive transmissions from the satellites; transmit and receive transmissions from the GCC or NCC; monitor status of local GES hardware and software and monitor the satellite system level performance "connected" to the GCC or NCC.

The GES terminal is redundant and has two steerable high-gain VHF antennas that track the satellites as they cross the sky. The GES transmits to a satellite at a frequency centered at 149.61 MHz at 56.7 Kb/s with a nominal power of 200 W. The GES receives 3 W transmissions from the satellite at 137 to 138 MHz range. These up and downlink channels have a 50 KHz bandwidth. The mission of the GES is to provide an RF communications link between the ground and the satellite constellation. It consists in medium gain tracking antennas, RF and modem equipment and communications hardware and software for sending and receiving data packets. An Orbcomm licensee requires a Gateway to connect to Orbcomm satellites in view of its service area. Namely, the Gateway consists in a GCC and one or more GES sites, as well as the network components that provide interfacility communications.

2. Gateway Control Centre (GCC) – The GCC terminals are located in a territory that is licensed to use the Orbcomm system and provide the following functions: locate wherever Orbcomm is licensed; link remote SCU with terrestrial-based systems; communicate via X.400, X.25, leased line, dial-up modem, public and private data networks and E-mail networks including the Internet; efficiently integrate the Orbcomm infrastructure with new or existing customer MIS systems, etc.

3. Network Control Centre (NCC) – The NCC is responsible for managing the Orbcomm communications network elements and the USA Gateways through telemetry monitoring, commanding and mission system analysis. It provides network management of Orbcomm's satellite constellation and is staffed 24 hours a day by Orbcomm-certified controllers and has the following main functions: monitoring real-time and back-orbit telemetry from the Orbcomm satellites; sending real-time and stored commands to the satellites; providing the tools and information to assist engineering with resolution of satellite structure and ground anomalies; archiving all satellite and ground telemetry data for analysis; monitoring the performance of the USAGES terminals and so on. The NCC manages the entire Orbcomm satellite constellation and its processes and analyzes all satellite telemetry. The NCC is responsible for managing the Orbcomm system worldwide. Through OrbNet, the NCC monitors message traffic for the entire Orbcomm system and manages all message traffic that passes through the US Gateway. The NCC is staffed 24 hours a day, 365 days a year and is located in Dulles, Virginia. A backup NCC system was established in 2000, which permits the recovery of critical NCC functions in the event of an NCC site failure.

4. Satellite Control Centre (SCC) – The SCC serves in a territory that is licensed to use the Orbcomm system and provides control of the Orbcomm satellite constellation.

5. Satellite Communication Unit (SCU) – The SCU equipment are both mobile and fixed terminals used for connection to the Orbcomm satellite network through Gateway stations. The SCU terminal is a wireless VHF modem that transmits messages from a user to the Orbcomm system for delivery to an addressed recipient and receives messages from the Orbcomm system intended for a specific user. Manufacturers have different proprietary designs and each model must be approved by Orbcomm and adhere to the Orbcomm Air Interface Specification, Subscriber Communicator Specifications and Orbcomm Serial Interface Specification (if an RS-232 port is available). Different versions of SCU terminals are currently available, which include "black-box" industrial units that have RS-232C ports for data uploading and downloading. Current options on a number of SCU include internal GPS receivers and/or additional digital and analog input and output ports.

a). Magellan GSC 100 is the world's first handheld satellite terminal that allows sending and receiving text/E-mail messages to and from anywhere in the coverage area, see **Figure 8.15. (1)**. This unit offers communication and navigation using the Orbcomm network and GPS system. Integrated GPS receiver capabilities allow one to identify position, plot and track course, store waypoints and send this information to anyone, anywhere in the world. Unlike traditional landlines, cellular/paging systems, the GSC 100 and Orbcomm network operate from isolated parts of the world, where TTN systems do not reach. Messaging features allow worldwide messaging via Orbcomm MSC service, send and receive brief, global E-mail messages called GlobalGrams to any Internet E-mail address, easy-to-use menu-driven interface, storing up to 100 messages and 150 addresses, sending and receiving messages at pre-selected time intervals and automatic wake-up. The GPS features provide navigation and pointing location worldwide, displays position, speed, distance, time-to-go, continuously points to the destination and keeps on a true course, displays the trip's progress with a track plotter, stores up to 200 user-defined waypoints, relays present location by inserting GPS position into GlobalGram message. This unit is equipped with telescopic whip antenna, rechargeable NiCad battery package and universal AC converter, software update, data and power extension cables and instruction manuals. Optionally, it is possible to supply external GPS antennas, fixed Site VHF Antennas, Combined GPS/VHF Magnetic Mount Antennas and Combined GPS/VHF Roof or Trunk Top-Mount Antennas.

Figure 8.15.   Different Type of SCU Terminals

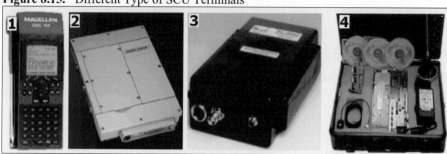

Courtesy of WebPages: "Orbcomm Satellite" by Orbcomm

b). Panasonic Trailer Tracking KX-G7121 PANATRACKER has been created to help meet the demand for a rugged, low-cost and reliable trailer-tracking product. Utilizing a KX-G7101 model as its core it is shown in **Figure 8.15. (2)**. This device is designed to be mounted in a covert manner (within the bulkhead's two plywood shells) as typically found in most trailers today. Its built-in battery allows the device to communicate for up to two months should the trailer be disconnected from the tractor. When connected to the tractor, a built-in charger maintains the battery's charge. Programming is similar to the KX-G7100 series, with the addition of several powerful, easy to use "macro" type commands, designed specifically for this application. The KX-G7100 is currently a second-generation product and has become the most popular of Orbcomm SCU. It is easily programmed with ready-to-use commands that are accessible using a simple terminal emulator or PC. Additional functionality and customization can be had by purchasing the Software Developer's Kit (SDK), thereby tapping into and utilizing the dedicated User Application Programming area provided in the SCU. Otherwise, this terminal has two programming SDK modes for maximum operational flexibility. One mode utilizes the resident Panasonic "KX" command set, accessible using standard terminal emulator software such as Windows HyperTerminal. A second mode allows a C language program to be embedded and executed on the SCU for additional functionality. The Panasonic SDK contains all the tools needed to create, debug and implement a custom application. Use of this powerful tool eliminates the need for external microcontrollers, substantially reducing application hardware costs. Furthermore, the SDK gives to users complete control of enhancements and revisions by allowing the development of applications in-house. Minimum Hardware Requirements are PC-AT or compatible, CPU 80386 or higher and hard disk minimum 3MB for installation. The new Panasonic SCU KX-G7200 will replace the highly successful KX-G7100 communicators, thousands of which are currently in operation throughout the world.

c). The StelComm EL-2000 (Stellar Communicator Manager) model is illustrated in **Figure 8.15. (3)**. The Windows applications Development Tool SelComm enables customers to program, configure, monitor and debug their Stellar Communicator.

d). Quake Q1500 is a modem with GPS and Magnetic Mount VHF Antenna using a 12 VDC Power Supply and Windows-based Programming Software, shown in **Figure 8.15. (4)**. This briefcase kit contains everything necessary to program custom applications and exercise all basic hardware functions.

9

GLOBAL SATELLITE AUGMENTATION SYSTEMS (GSAS)

The GNSS applications are represented by old fundamental systems for Position, Velocity and Time (PVT) military determination systems such as GPS and GLONASS for US or former-USSR requirements, respectively. The GPS or GLONASS are old GNSS giving positions to about 30 metres and they therefore suffer from certain weaknesses, which make them impossible to use as the sole means of navigation for ships, particularly for civil aviation. In this way technically, GPS and GLONASS systems used autonomously are incapable of meeting civil aviation's very high requirements for integrity and position availability and precision in particular are insufficient for certain critical flight stages. Because these two systems are developed for military utilization, many countries and international organizations would never be dependent on or entrust people's safety to GNSS systems controlled by one country. Augmented GNSS solutions of GSAS were recently developed to improve the mentioned deficiencies of current military systems.

9.1. Development of Global Navigation Satellite System (GNSS)

GNSS augmentations are available to enhance standalone GPS or GLONASS satellite PVT performances for maritime, land and aeronautical applications. Moreover, user devices can be configured to make use of internal sensors for added robustness in the presence of jamming, or to aid in vehicle navigation when the satellite signals are blocked in the "urban canyons" of tall city buildings. Some special applications, such as maritime and especially aeronautical, require far more accuracy than standalone GPS or GLONASS.

Accuracy can be improved by removing the correlated errors between two or more satellite receivers performing range measurements to the same satellites. One receiver is calling the reference Rx and is surveyed in, namely its geographical location is precisely-known. One method of achieving common error removal is to take the difference between the reference Rx's surveyed position and its electronically derived position at a discrete time point. This difference represents the error at the measurement time and is denoted as the differential correction, which may be broadcast via data link to the user receiving equipment such that the user Rx can remove the error from its received solution. Alternatively, in non-real-time applications, the differential corrections can be stored along with the user's positional data and applied after the data collection period. This non-real-time technique is typically used in surveying applications.

If the reference station (RS) is within line-of-sight of the user, the mode is usually referred to as local area differential. As the distance increases between the user and the RS, some ranging errors become decorrelated. This problem can be overcome by installing a network of RS throughout a large geographic area, such as a country or continent and broadcasting the differential corrections via GEO satellites. The RS relay their collected data to one or more Central Processing Stations (CPS), where differential corrections are performed and satellite signal integrity is checked. Afterwards, the CPS sends the corrections and integrity

Figure 9.1. GSAS Worldwide Architecture

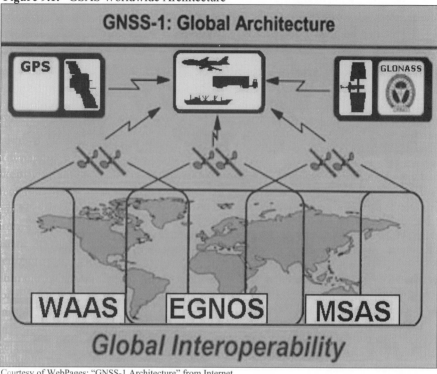

Courtesy of WebPages: "GNSS-1 Architecture" from Internet

data to a GES for uplink to the GEO satellite. This differential technique is referred to as the wide area differential system, which is implemented by system known as Wide Augmentation Area (WAA), while, the another system, Local Augmentation Area (LAA) is an implementation of a local area differential.

The LAA is an implementation of a local area differential system for local CNS airport or maritime ports and are approaching utilization. The WAA is an implementation of a wide area differential system for wide area CNS maritime, land and aeronautical applications, such as Inmarsat CNSO and the newly developed Wide Area Augmentation System (WAAS) in the USA, the European Geostationary Navigation Overlay System (EGNOS) and Japanese MTSAT Satellite-based Augmentation System (MSAS). These three systems are part of the worldwide GSAS network and integration segments of interoperable Global GNSS-1 architecture, as shown in **Figure 9.1**. Actually, CNSO is part of GNSS offering this service via Inmarsat satellite constellation, while forthcoming GSAS projects in Russia (CIS), Australia and China including eventual development of other regions: Africa and South America, etc., as shown in **Figure 9.2**. will complete CNS system globally.

These three augmentation satellite systems are inter-compatible and each constituted of a network of GPS and GLONASS constellation observation stations and own or leased GEO satellites. The Inmarsat CNSO system offers GNSS satellite payload, while the European system EGNOS, which will provide precision to within about 5 metres, will be operational in 2004. It also constitutes the first steps towards forthcoming Galileo, the future European

Figure 9.2. Integration of GSAS Regions Worldwide

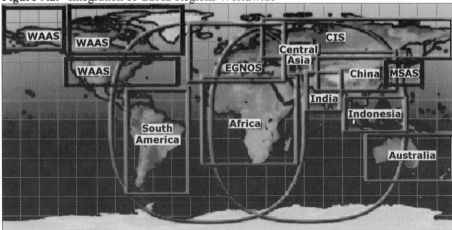

Courtesy of WebPages: "GNSS-1 Architecture" from Internet

system for global navigation by satellite. EGNOS uses leased Inmarsat AOR-E and IOR satellites and its own ARTEMIS satellite. The WAAS is using Inmarsat satellites and MSAS is using its own satellite constellation, both will be operational in 2003 and 2005, respectively. Although the positioning accuracy associated with the overlay is a function of numerous technical factors, including the ground network architecture, the expected accuracy for the US FAA WAAS will be in the order of 7.6 m (2 drms, 95%) in the horizontal plane and 7.6 m (95%) in the vertical plane.

9.2. Global Determination Satellite System (GDSS)

Position determination by satellite system has long been used to help meet the navigation requirements especially of ships and aircraft. Current developments of LMSS have given rise to a demand for similar services from terrestrial users. The service provider who can offer global, low-cost position locating facilities serving all of those markets stands to profit from a large and expanding customer base.

9.2.1. Passive GDSS

In a passive system, signals are transmitted as a continuous stream of data and picked up by mobile receivers. The users than calculate own position from the received data. This system requires the users to have nothing more than a receiver and is the best solution when it is the mobile user, rather than a static observer, who needs to know where he is and at frequent intervals. Ships, aircraft and even land vehicles come into this category.

Passive systems, as shown in **Figure 9.3.**, have been in operation for a long time, are well established and generally have the following features:

1) For a two-dimensional fix, three satellites must be visible to the users. Four satellites are required for a three dimensional-fix.

2) The user can determine his position independently and without alerting others to his presence. The number of users and the frequencies with which they can obtain updates are

Figure 9.3. Passive Satellite Determination

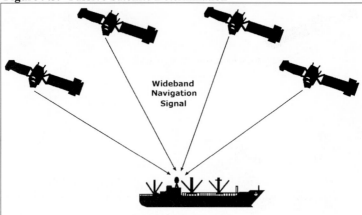

Wideband
Navigation
Signal

Courtesy of Book: "Never Beyond Reach" by B. Gallagher

not limited by system power or bandwidth. As a result, however, passive systems tend to be economical of such resources, particularly when the number of users is large or when frequent position updates are required. Thus, system costs are independent of the number of users and the amount of usage

3) Equipment costs are low because the user is not required to have a transmitter, though he may carry communications equipment for other purposes. This equipment can be readily used in conjunction with other systems such as GPS, GLONASS, Loran-C and/or even in combinations.

4) The Space and especially Ground Segments are simple, partly because users share the responsibility for position calculations.

9.2.2. Active GDSS

In an active system the mobile user transmits a signal to the satellite transponder and the position is calculated by the ground operator's central computing facility. Positioning data can then be relayed back to the mobile unit or to any other location, see **Figure 9.4**.

The mobile user's position can be determined only when he transmits the necessary signal. Typically, the signal travels via satellite to a central facility, where the position is then calculated. Since the position of the user is known at the central facility, the active system can be used for surveillance, especially for aeronautical flight and ground air traffic control. Surveillance is a basic need of air traffic controllers, who have to know the position of aircraft accurately and quickly. More recently, some sectors of the land mobile community have also identified a requirement for surveillance services. Managers of tracking fleets and railway rolling stock, for example, would be able to keep track of special cargoes such as hazardous or perishable goods. The marine sector is calling for similar facilities for vessel tracking. Active systems have the following characteristics:

1) Two-dimensional active mobile position fixing requires two visible GNSS satellites, while three-dimensional requires three.

2) The mobile user must be able to communicate with the central facility. Namely, he does not need computing capacity, since position calculations are handled centrally.

Figure 9.4. Active Satellite Determination

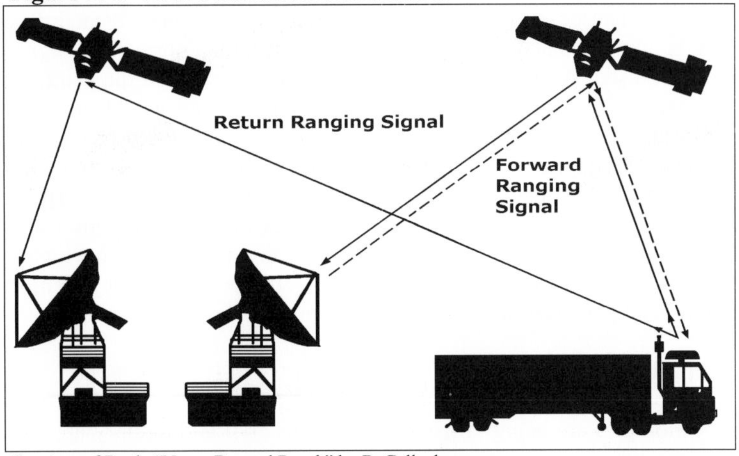

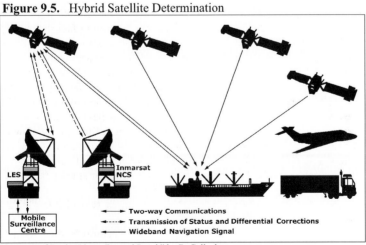

Courtesy of Book: "Never Beyond Reach" by B. Gallagher

3) Active radio determination systems can provide information on the position of one user in relation to another, which is important for some land and aeronautical applications.
4) The central facility can improve positioning accuracy through the use of data from reference locations.

9.2.3. Hybrid GDSS

In a hybrid system, the mobile user receives ranging signals from a number of satellites and position is calculated by measuring the time or phase differences of the signals. Therefore, ranging information can be supplied where two or more satellites are visible to the mobile unit, see **Figure 9.5.** This system is important for developing aeronautical CNS solutions.

Figure 9.5. Hybrid Satellite Determination

Courtesy of Book: "Never Beyond Reach" by B. Gallagher

The measurements are the same as those, which take place with the Loran-C, GPS or GLONASS system. These are then transmitted via satellite to a central facility, where the user's position is computed and displayed on the screen, which normally designates which satellites are to be used. A hybrid system allows both the user and the central facility to calculate the position. In an aeronautical determination, the user would probably perform navigation calculations on board the aircraft, since one usually needs navigation data as well as surveillance. With most land mobile applications the central facility would perform the position determination calculations and use the results for surveillance purposes.

Therefore, hybrid systems have most of the advantages of active systems while offering compatibility with GPS, GLONASS and any potential civilian satellite navigation system. Characteristics of hybrid systems include spectrum efficiency, accuracy, uniformity and integrity, commercial potential, simplified user equipment, flexible control, etc.

9.3. GNSS Applications

The GNSS technologies are being utilized in numerous civil and military applications that range from leisure hiking to spacecraft launching guidance. In any event, the old and new applications affect all sectors of transportation such as maritime, land and aeronautical.

9.3.1. Maritime Navigation Satellite System (MNSS)

The GNSS system has been embarked on board, at first of military and after on merchant and recreation maritime communities. Navigation is enhanced on all bodies of waters, from oceanic travel to rivers, especially in inclement weather. Several nations are developing local area Differential GPS networks to increase system accuracy for harbor, anchorages, harbor approach and river usage. The Commonwealth of Independent States with Russia is considering the implementation of a local area differential GLONASS network. Wide area differential GPS has been utilized by the offshore oil exploration community for several years. One area in which, therefore, differential GNSS will play a large role is in Vessels Traffic Services (VTS). The combination of a datalink and differential GNSS Rx permits the broadcast of the vessel's position to a control centre. The VTS system is used for collision avoidance of ships in navigation and to expedite the flow of traffic during periods of restricted visibility and ice cover. Thus, it can be used in conjunction with the Electronic Chart Display Information System (ECDIS), which displays a vessel's position in relation to charted objects, navigation aids, land, as well as unseen hazards.

9.3.2. Land Navigation Satellite System (LNSS)

The surveying community has relied on differential GPS to achieve measurement accuracy in the millimetre range. Similar techniques are in use within the railroad communities to obtain train location with respect to an adjacent set of tracks. Thus, the GPS system is a key component in Intelligent Transportation Systems (ITS). In terms of vehicle applications, the GNSS system will be used for route guidance, tracking and emergency short messaging. Integrating a GNSS Rx with a street database, digital moving map display and processor will allow the vehicle driver to obtain directions and/or the shortest most efficient route. Combining a cellular phone or MSC will enable vehicle tracking from ADSS and/or from emergency messaging. In such a way, a vehicle position can be automatically reported to a control centre for fleet management. The activation of a "panic" button by the vehicle driver broadcasts an emergency message, vehicle characteristics and vehicle location to law enforcement authorities for assistance, similar to Cospas-Sarsat PLB devices.

9.3.3. Aeronautical Navigation Satellite System (ANSS)

The aviation community and ICAO have propelled the utilization of a GNSS and various augmentation systems to provide better guidance for the en-route precision approach phases of flight. The ICAO requirement defines a system that contains at least one or more satellite navigation systems as a GNSS. Thus, the continuous global coverage capability of GNSS permits aircraft to fly directly from one location to another, provided factors such as obstacle clearance and required procedures are adhered to. Incorporation of a data link with a GNSS Rx enables the transmission of aircraft locations to other aircraft and/or to ATC. This function, called ADSS, is in use in some Pacific Ocean Regions as an outgrowth of ICAO FANS working group activities. Key benefits are ATC monitoring for collision avoidance and optimized routing to reduce travel time and, consequently, fuel consumption. The new satellite ADSS solution is also being applied to airport surface surveillance of both aircraft and ground support vehicles.

9.4. GSAS EGNOS

 Satellite navigation is currently based on the US GPS and Russian GLONASS military satellite navigation systems. An evolution of these systems is known as the GNSS, which provides an overlay function and supplementary services. Thus, the major contribution to GNSS-1 is the European Geostationary Navigation Overlay Service (EGNOS), an important initiative by ESA, the European Commission (EC) and Eurocontrol. The EGNOS uses existing GEO communications birds, such as Inmarsat-3 and the Artemis, to transmit overlay signals almost identical to those of GPS and GLONASS to provide three main civil navigation functions as follows: improvement of accuracy and compensation for coverage gaps; continuously updated and independently-monitored data and information on the system's integrity will be broadcast to identify faulty satellites and wide-area differential corrections will be relayed from ground stations. The EGNOS system as an augmentation to GPS/GLONAS, will be fully interoperable within the USA (WAAS) and Japan (MSAS). In fact, the built-in expansion capability of EGNOS will extend its service areas within the Broadcast Area of GEO satellites to cover Africa, Eastern European countries and Russia.

The ESA has overall responsibility for the design and development of the EGNOS system for the safety of navigation over the European continent. It has placed a contract, with a consortium led by Alcatel Space Industries of France, to develop the EGNOS system. When EGNOS is up and running, an operator will be selected to take responsibility for daily operations. The EC is responsible for international cooperation and coordination and for making sure that the views of all modes of transport feed into EGNOS design and implementation. Eurocontrol is defining the needs of civil aviation and playing a major role in system testing. Several leading civil aviation operators and other firms are supporting EGNOS development. The EGNOS development and validation will cost approximately €300 million. The ESA is contributing €200 million together with member states and European national civil aviation organizations, while the EC is contributing €100 million.

9.4.1. Development of the EGNOS System

The EGNOS system is implementing the European contribution to the GNSS-1, which will provide and guarantee navigation signals for maritime, land and aeronautical, modular and mobile trans-European network applications. In particular for civil aviation, EGNOS AOC (Advanced Operational Capability) will meet primary means navigation requirements for all phases of flight from en-route to Non-Precision Approach (NPA) as well as Precision Approach (PA), with a decision height capability down to Category 1 Landing within the European Civil Aviation Conference (ECAC) area. For maritime applications, positioning accuracies in the range of 4–8 m will be provided in European coastal waters and better than 30 m in the open ocean waters of the European Maritime Core Area (EMCA). Land road and railway applications in continental Europe will benefit from the same range of accuracies provided for aviation over continental ECAC and it will be 5 – 10 m. In particular, civil aviation requirements for PA and NPA phases of flight cannot be met by GPS or GLONASS only. Maritime users may also require some sort of augmentation for improving GPS and GLONASS performances for more precise positioning when ships are approaching ports or are going to the berth in waters with heavy traffic. Land users have a need for the improvement of determination, monitoring and tracking of their fleets.

The first generation Global Navigation Satellite System, GNSS-1, as defined by the experts of the ICAO/GNSS Panel, includes the basic GPS and GLONASS constellations and any system augmentation needed to achieve the level of performance suitable for civil aviation applications. Thus, the EGNOS system as a regional Satellite-Based Augmentation System (SBAS) will be equivalent to the WAAS or MSAS and first European implementation of GNSS. Namely, development of EGNOS is part of the European Satellite Navigation Program (ESNP) involving GNSS-1 activities including the Local Area Augmentation System (LAAS) of GNSS-2 activities and the recently approved Galileo GNSS definition studies. In the context of GNSS-1, ESA is responsible for the AOC EGNOS system design, development and qualification of the EGNOS system.

The EGNOS implementation program was launched in 1996 through the impetus of the CNES (Centre National D'Etudies Spatiales) and STNA. The program is managed by ETG (European Tripartite Group) composed of the EEC, ESA and Eurocontrol. The program is developed by ESA and Alcatel Space was made responsible as the project management entity. EGNOS, Europe's first venture into satellite navigation, will augment the two military satellite navigation systems now operating, the US GPS and Russian GLONASS systems and make them suitable for safety critical applications such as flying aircraft or navigating ships through narrow channels. Among the GNSS elements standardized by ICAO, the FAA launched the development of WAAS for US territories. As not to have to depend on the USA for navigation service to be used over its territory, Europe launched its own equivalent GSAS European system EGNOS, similar to US-based WAAS, with three main solutions such as: **(1)** Satellite Base Augmentation System (SBAS); **(2)** Ground Base Augmentation System (GBAS) and **(3)** Wide Area Augmentation System (WAAS).

The EGNOS program was organized in accordance with ESA methods. The industrial development phase, after the definition phases, was officially launched in June 1999, during the Le Bourget Airshow and the work effectively began a few months later. It is planned that ESA will deliver EGNOS to the future operator in late 2003. There will then be a few months to wait for validation of the signal before the EGNOS system is declared to be operational. This means that EGNOS should be in use by the end of 2004.

The development of GNSS will be carried out in two main stages:
1. GNSS-1 System – The GNSS-1 System will be the first-generation integrated system, based on signals received from the existing American GPS and/or Russian GLONASS constellations and civil augmentation systems using space-based, ground-based and mobile autonomous-based techniques. The US WAAS, Japanese MSAS and EGNOS are currently in the development phase and thus, there are several projects in the initial phases of developing a similar system, such as in Russia, Australia and China.
2. GNSS-2 System – The GNSS-2 System, as a second-generation, will provide services to civil users and will be under civil operation and control by 2010.
The EGNOS system development program comprises two different phases: **(1)** Initial phase and **(2)** AOC Implementation phase. The EGNOS Initial Phase was successfully concluded in November 1998, with the Preliminary Design Review (PDR). In addition, the EGNOS AOC Implementation Phase started in December 1998 and thus, it is planned to be completed in mid-2003, with the Operational Readiness Review (ORR), which will encompass the verification of the overall system performance and its operations. Key milestones include the Critical Design Review (CDR) in late 2000 and the Factory Qualification Review (FQR) in mid-2002.

9.4.2. Network Architecture and System Description

In general, the EGNOS system is a joint project of the EC for the safety of air navigation. It is Europe's contribution to the first stage of the GNSS and is a precursor to Galileo, the full GNSS planned for development in Europe. Consisting in three GEO satellites, a network of ground stations and user terminals, EGNOS will achieve its aim by transmitting a signal containing information on the reliability and accuracy of the positioning signals sent out by GPS and GLONASS. It will allow mobile users in Europe and beyond to determine their position to within 5 m, compared with about 20 m at present and will quickly alert users of any satellite failures. The EGNOS AOC architecture, as usual, comprises three main infrastructures: Space, Ground and User segments.

9.4.2.1. Space Segment

The EGNOS AOC Space Segment is composed of three payloads with global Earth coverage, based on board two Inmarsat and one ESA ARTEMIS navigation spacecraft. Otherwise, the EGNOS AOC will require additional transponders to guarantee availability during at least 15 years mission duration. Namely, EGNOS GNSS relies on the availability of GEO satellites equipped with navigation transponders to broadcast a GPS look-alike signal containing integrity and wide-area differential corrections to users. In this sense, the EGNOS system uses signals from the GPS and the GLONASS constellation.
The operational system uses three GEO satellites to disseminate this data: Inmarsat-3 AORE at 15.5°W; Inmarsat-3 IOR at 64°E and ESA ARTEMIS at 21.5°E. The navigation payloads on all these satellites are essentially bent-pipe transponders so that a message uploaded to a satellite is broadcast to all users in the GEO broadcast area of the satellite.
The coverage area serviced by EGNOS will be the ECAC service area, comprising the Flight Information Regions (FIR) under the responsibility of ECAC member states (most of the European countries, Turkey, the North Sea and the Eastern part of the Atlantic Ocean), as shown in **Figure 9.6**.

Figure 9.6. EGNOS Space Segment

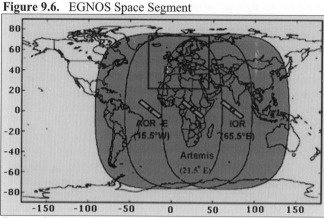

Courtesy of WebPages: "EGNOS Spacecraft Architecture" from Internet

The EGNOS performance objectives will therefore be such that EGNOS AOC will have the technical capability to provide a primary means service of navigation for en-route oceanic and continental, NPA and CAT-I PA within the ECAC area. The EGNOS network has potentially the capability to offer en-route and NPA services over the full GEO broadcast area and discussions are being pursued with international partners to provide this capability, in order to offer to users a full, seamless service. The final scope of the system is to provide coverage of subpolar areas ranging from the East coast of the United States to Japan. According to **Figure 9.6.**, there is the possibility to use the space facilities of the EGNOS system and develop ground infrastructure for regional GSAS systems for Central Asia, South America, Africa and India, although India has its own Insat GEO system.

Figure 9.7. EGNOS Network Architecture

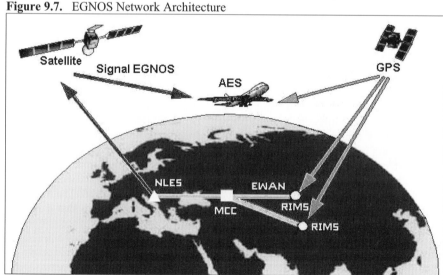

Courtesy of WebPages: "EGNOS Spacecraft Architecture" from Internet

9.4.2.2. Ground Processing and Support Segment

The EGNOS ground infrastructure consists in GNSS (GPS, GLONASS and GEO) Ranging and Integrity Monitoring Stations (RIMS), which are connected to a set of redundant control and processing facilities called Master (Mission) Control Centre (MCC), Navigation Land Earth Station (NLES) and EGNOS Wide Area Network (EWAN), as is illustrated in **Figure 9.7**.

1. Remote Integrity and Monitoring Station (RIMS) – It is planned to deploy 34 RIMS, which are data collection stations spread over the EGNOS service area. Some of these are outside the EGNOS area so as to dispose of certain information by anticipation. Thus, for security reasons, each RIMS infrastructure is composed of two independent, dissimilar reception channels, RIMS A and RIMS B. With the aim of detecting certain GPS satellite dysfunctions, i.e., Evil wave forms, some stations will be equipped with a third collection channel, RIMS C. The RIMS centre measures the positions of each EGNOS satellite and compares accurate measurements of the positions of each GPS and GLONASS satellite with measurements obtained from the spacecraft's signals. The RIMS then send this data to the MCC, via a purpose-built communications network. In France will be located three RIMS in Paris, Aussaguel and Kourou.

2. Master Control Centre (MCC) – The MCC are composed of a Central Process Facility (CPF) in charge of calculation, validation and distribution of messages to be transmitted and a Central Control Facility (CCF). This receives all the data from the various elements of EGNOS network and is in charge of managing all the interfaces between these elements and of configuring EGNOS according to the status of these elements. It prepares the data that will be archived in order to study EGNOS performances off line. At this point, the EGNOS system is composed of 4 MCC infrastructures located in Gatwick (UK), Langen (Germany), Torrejon (Spain) and Fucino (Italy). This number of ground facilities ensures that EGNOS is always available. At any given time, the four CPF stations are in continuous operation, when one single CPF, as a master CPF, supervises EGNOS. In other words, this master CPF is monitored by a second CPS, the hot backup, which can take over instantaneously from the master CPF in case of failure. A third CPF, the cold backup, then takes the role of hot backup. The fourth CPF is not in operational status. It is used for validation of new versions and for tests. In order to maintain the competence of each site and its operational capacity, a rotation of CPF roles among the sites will be organized. The MCC determines the integrity, Pseudo Range differential corrections for each monitored satellite, ionospheric delays and generates GEO satellite ephemeris. Thus, this information is sent in a message to the Navigation Land Earth Station (NLES), to be unlinked along with the GEO Ranging Signal to GEO satellites. These GEO satellites downlink this data on the GPS Link 1 (L1) frequency with a modulation and coding scheme similar to the GPS one. The four MCC missions determine the accuracy of GPS/GLONASS signals received at each station and determines position inaccuracies due to disturbances in the ionosphere. All the deviation data is then incorporated into a signal and sent via the secure communications link to the NLES or so-called Up Link Stations (ULS), which are widely spread across Europe. The six NLES send the signal to the three EGNOS satellites, which then transmit it for reception by GPS and GLONASS users with an EGNOS Rx. Considerable redundancy is built into EGNOS so that the service can be guaranteed at practically all times. There is also redundancy in the NLES. Only three are needed to operate EGNOS, one for each satellite and the other three are in reserve in case of failure.

3. Navigation Land Earth Station (NLES) – The NLES is in charge of transmitting the signal to GEO satellites and maintaining perfect synchronization of the message with GPS time. For each satellite, there are two NLES, geographically separated and controlled by the MCC. Thus, the planned sites are Aussaguel (France), Fucino (Italy), Sintra (Portugal), Goonhilly (UK), Torrejon (Spain) and Raisting (Germany). The first stage EGNOS AOC architecture foresees only seven NLES, five of which will uplink EGNOS messages to the Inmarsat III AORE and IOR birds for test and validation purposes and two of which (one primary and one back-up) will uplink EGNOS messages to the Artemis navigation transponders. The main functions of the NLES will be to generate a GPS-like signal and transmit this to a GEO transponder; synchronize this signal to EGNOS Time (ENT) at the output of the GEO L1-band antenna; control the code/carrier coherency and transmit the GIC and WAD messages to satellites in GEO orbit.

4. EGNOS Wide Area Network (EWAN) – The EWAN configuration is in charge of inter-site communications between the various ground elements of EGNOS. Actually, this network is composed mainly of low-speed links connecting the RIMS stations to an MCC centre and high-speed links connecting the MCC centres between themselves and with the NLES stations.

5. Performance Assessment and Check out Facility (PACF) – The PACF will provide technical support for EGNOS in the areas of operational use and maintenance. Hence, it is a platform that operates off line and is not involved in the real-time loop of the operational system. In fact, this infrastructure tasks the following areas: analysis of EGNOS performance and simulation; analysis of faults and failures; specification of evolutions in the system; definition and validation of operational procedures; staff training; maintenance support; system configuration control and archiving of system data. The PACF is based on a set of workstations connected by a local network, designed around data servers. It receives all the system data off line and, for the purposes of analysis and testing; it receives data from the EGNOS network stations in real time. Furthermore, it has a data server, used to receive information from the various world centres diffusing data connected with satellites of both the GPS/GLONASS and INMARSAT constellations and the environment. The PACF will be located in Toulouse and will be operated by STNA and CNES.

6. Application Specific Qualification Facility (ASQF) – The ASQF solution provides the technical means required for the validation and certification of the EGNOS network applications. In particular, for civil aviation utility, the ASQF solution will be in charge of validating the EGNOS performances concerning Radio Navigation Performances (RNP) and the analysis of the failure modes for the EGNOS certification. It will be used by EUROCONTROL for approval of its use and validation of the operational procedures. The ASQF will be located in Torrejon (Spain) and operated by AENA.

9.4.2.3. User Service Terminals (UST)

The multiple EGNOS System Test Bed (ESTB) user receivers are available on the market to test the system and perform demonstrations. In the context of the EGNOS Contract, a specific Test Bed User Equipment (TBUE) has been developed by Thales Avionics and is based on the existing TSO C129A family of products. The 15-channel receiver allocates two channels to GEO satellites and 13 to GPS or GLONASS capability that will be added at a later stage. In addition, this segment also comprises a data recorder and a computer for navigation processing.

Figure 9.8. EGNOS UST ESA 1 and Personal-Nav 400

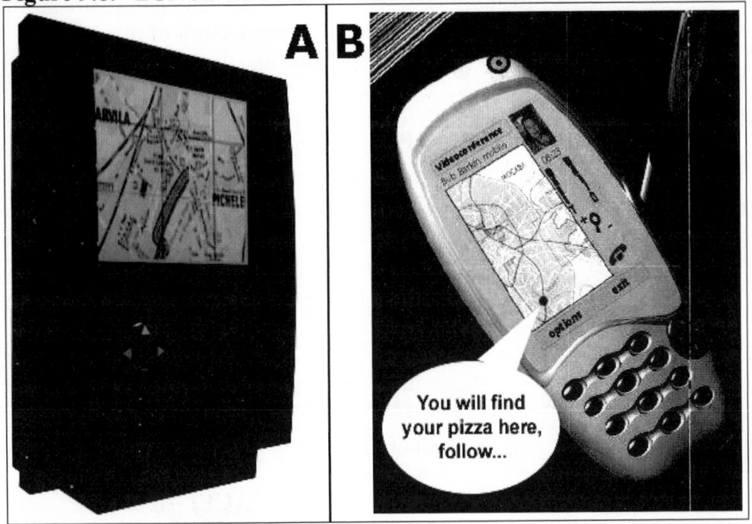

Courtesy of WebPages: "EGNOS Test Bed User Equipment" from Internet

The EGNOS users will have at their disposal a few models of multimodal prototype Rx units. In fact, as work on EGNOS proceeds, these will be further developed and utilized. The multimodal prototypes will enable users to carry out few tests on the EGNOS system: static and/or dynamic platform testing; user EGNOS Rx validation and system performance demonstration comparison with reference position: geodetic marks (static), trajectography data (dynamic), such as the model ESA 1 prototype shown in **Figure 9.8. (A)**.

The EGNOS Standard receiver was also developed to verify the Signal-In-Space (SIS) performance. In the meantime a set of prototype user equipment has been manufactured for civil aviation, land and maritime applications. That prototype equipment will be used to validate and eventually certify EGNOS for the different applications being considered. In such a way, a handheld personal receiver (like a cell phone) would use satellite navigation to avoid traffic jams in city centres, find the nearest free parking space, or even the nearest pizza restaurant in an unfamiliar city, as shown in the Personal-Nav 400 in **Figure 9.8. (B)**. Precise position anytime via the Internet and EGNOS system is possible since the end of 2002, thanks to the Signal in Space through Internet (SISNeT) technology developed by the European Space Agency. This technology combines the powerful capabilities of satellite navigation and the Internet. As a result, the highly accurate navigation information that comes from the EGNOS SIS is now available in real time over the Internet.

9.4.3. ESTB Experimental Program

The EGNOS System Test Bed (ESTB) is the EGNOS trial prototype developed to provide a broadcasting Signal in Space (SIS) since February 2000. It is used to support and test the development of the EGNOS system, to demonstrate EGNOS to potential users, to prepare for the introduction of EGNOS and to test the possibility of expanding this system outside Europe. The ESTB provides users with a GPS-augmentation signal that enables them to calculate their position to an accuracy of within a few metres.

Figure 9.9. EGNOS ESTB Architecture

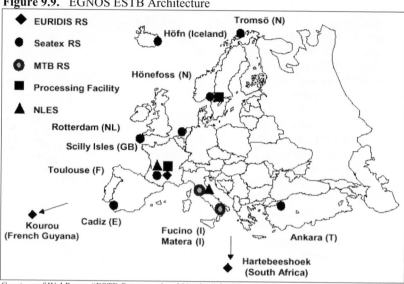

Courtesy of WebPages: "ESTB Pre-operational Version" from Internet

The ESTB is a pre-operational version of EGNOS, which will allow Europe to support the demonstration of the operational benefits of GNSS to user communities. The test bed will be used for all modes of transport (air, land and maritime) that require positioning services to accuracies of a few metres and more particularly, safety-critical services. For aviation users, for instance, EGNOS will provide en-route navigation as well as non-precision approach (NPA) and precision approach (PA) phases of flight. The ESTB system presented in **Figure 9.9.** contains the following infrastructures:

a). Some RS gathering data for corrections and integrity message purposes deployed over Europe in host sites;

b). Central Processing Facilities (CPF) are computing on line differential corrections and integrity data, based at Hönefoss in Norway (NMA premises);

c). Geostationary Ranging Station (GRS) implementing a wide triangular observation base for ranging purposes with the stations located in Aussaguel (France), Kourou (French Guiana) and Hartebeeshoeck (South Africa),

d). Mission (Master) Control Centre (MCC) located in Toulouse (France, CNES premises) computes ranging messages and prepares navigation messages to AORE payload;

e). Test Master Station (TMS) located at Fucino (Italy) prepares all navigation messages to IOR payload,

f). Two NLES based on existing stations of the Inmarsat network, one located in Aussaguel (France), implementing the broadcast link to the AORE Navigation payload, the other located at Fucino (Italy), implementing the broadcast link to the IOR Navigation Payload. Each NLES is compliant with requirements from DA3 (SDM for Inmarsat Geostationary Navigation Overlay Service)

g). Ground network composed of few sub-networks to transmit data from RS to CPF, from GRS to MCC, from CPF to MCC or TMS and MCC or TMS to the appropriate NLES.

There are three different types of reference stations:

1) Stations for GEO ranging function (using a signal similar to GPS), which are part of the Euridis ranging system. There are three stations: one in Toulouse, France, one in Kourou, French Guiana and one in Hartebeeshoek, South Africa.

2) ESTB-specific reference stations. There are ten stations that serve as data collection points for the GIC/WAD function and which use the existing reference station facilities of Racal Survey's SkyFix and NMA's SatRev systems. At this point, all the ESTB-specific reference stations are equipped with a GPS/GEO receiver, a GPS/GLONASS receiver, an atomic clock, computers with processing software, routers, dual-frequency antenna and an archiving device to store data. In addition these stations also transmit real-time data.

3) Reference stations from the Mediterranean Test Bed (MTB) create GIC and WAD.

The ESTB experimental system is the EGNOS system prototype, which has been used to broadcast Signal in Space (SIS) since February 2000. It is employed to support and test the development of the new EGNOS system, to demonstrate it to potential users, prepare it for introduction and to test the possibility of expanding this system outside Europe. The ESTB provides users with a GPS-augmentation signal that enables them to calculate their position to an accuracy of within a few metres. Since the first EGNOS signal was switched on, around 10 demonstrations and one workshop for potential users have been held. In this sense, the trial has commenced successfully and will last until the EGNOS system starts with full and official operations.

Therefore, before EGNOS becomes fully operational in 2004, ships captains, drivers of land vehicles, airline pilots and others equipped with an EGNOS receiver can tune into a test signal broadcast by one of the EGNOS satellites. Initially, this test signal broadcasting began only via Inmarsat AORE satellite, while the area over which it became available was extended when Inmarsat IOR also started to broadcast on a continuous basis.

The redundancy that will be built into the fully-fledged service is not yet available as the ESTB elements are only a fraction of those planned for EGNOS. At present, ESTB consists in 10 RIMS dotted about Europe, though this number may increase once the MTB system is connected to ESTB together with one Processing Centre which collects data for modulation onto the EGNOS signal. The RIMS send the data they have collected on the GPS and GLONASS signals to a Single Processing Centre (SPC) in Norway, which computes the corrections needed for each satellite and incorporates them into a signal. The SPC then sends the signal via terrestrial networks to a station in Toulouse, France for up-linking to the AORE satellite. It will shortly also send the signal to a ground station in Fucino, Italy for uplinking to the IOR satellite. The EGNOS ESTB users will receive the signal from one of these two satellites on specially designed EGNOS receivers. Since the first EGNOS signal was switched on, numerous successful demonstrations covering all possible modes of transport and GNSS applications, as well as a workshop for potential users, have been held. The next workshop was scheduled for November 2001.

9.4.4. EGNOS Processing Centres and Functions

The ESTB experimental system has two Processing Centres, both responsible for: ranging creation, ESTB data collection, archiving and post processing; GIC (Ground Integrity Channel) and WAD (Wide Area Differential) messaging; ESTB system monitoring and control GEO and GPS-related data are processed in real time, while in this first stage the GLONASS performance assessment is limited to off-line analysis.

The EGNOS system will provide the following functions:

1. GEO Ranging (R-GEO) – Transmission of GPS-like signals from 3 GEO satellites for the AOC phase (for the FOC additional GEO satellites will be provided). Thus, this will augment the number of navigation satellites available to users and, in turn, the availability of satellite navigation using RAIM. An additional distance measurement will be performed to supply further pseudo-distance signal from a GEO satellite.

2. GNSS Integrity Channel (GIC) – Broadcasting of integrity information on the status of the satellites observed by the ground segment. This function will increase the availability of GPS/GLONASS/EGNOS safe aero navigation service up to the level required for NPA.

3. Wide Area Differential (WAD) – Broadcasting of differential corrections. This system will increase the GPS/GLONASS and EGNOS navigation service performance, mainly its accuracy, up to the level required for precision approaches down to CAT-I landing. At this point, the basic differential corrections is the diffusion of corrections concerning calendars and clocks relating to satellites observed by the ground segment and to be applied to the pseudo-distances coming from the satellites observed by the ground segment. The next function is precise differential corrections: diffusion of ionospheric corrections relating to satellites observed by the ground segment to be applied to the pseudo-distances coming from the satellites observed by the ground segment.

The first processing centre, located on the premises of the French Space Agency, Centre National D'Etudies Spatiales (CNES) in Toulouse, France, is responsible for the ESTB GEO ranging function, which is part of EURIDIS. This centre receives the pseudo-range measurements from the three EURIDIS Reference Stations (RS) taken from the AORE and GPS satellites, determines AORE orbit, generates the standardized ranging message in real time and sends this information to the nearest NLES centre. The ESTB data archiving and post-processing is also undertaken at this centre.

The second processing centre is hosted by the Norwegian Mapping Authority (NMA) at its SATREF centre in Hönefoss, Norway. At present, data are received continuously from the ten RS and it is planned to receive data on an almost continuous basis from the MTB reference stations as well. The GIC/WAD messages are generated online using EGNOS prototype software developed by GMV, which is also a subcontractor for EGNOS central processing algorithms.

The GIC/WAD data are sent to the Inmarsat-32 AOR-E NLES via the Toulouse processing facility and/or to the Inmarsat-3 IOR NLES, via the NLES in Fucino, Italy. The Hönefoss processing facility also has a data management unit for data collection and archiving. Two NLES are used in the ESTB: one located in Aussaguel, close to Toulouse (France), is part of the EURIDIS Ranging system and transmits via the Inmarsat-3 AORE satellite, while the other NLES, situated in Fucino (Italy), provides access to the Inmarsat IOR satellite. Thus, a real-time communication network based on frame-relay links, allows the transfer of the reference station data to the processing centres and the navigation messages from Hönefoss to the NLES. There are three stations for GEO ranging functions (a signal similar to GPS) which are part of the Euridis ranging system: one in Toulouse, France; one in Kourou, French Guiana and one in Hartebeeshoek, South Africa.

There are 10 stations that serve as data collection points for the GIC/WAD function and which use the existing reference station facilities of Racal Survey's SkyFix and NMA's SatRev systems. All stations are equipped with a GPS/GEO receiver, a GPS/GLONASS receiver, an atomic clock, computers with processing software, routers, dual-frequency antenna and an archiving device to store data. In addition, all these ground stations transmit real-time data. Finally, the EGNOS support facilities include the Development Verification

Platform (DVP), the Application Specific Qualification Facility (ASQF), both located in Torrejon (Spain) and the Performance Assessment and System Checkout Facility (PACF) located in Toulouse (France).

Therefore, the main objectives of the ESTB are to validate the ICAO SARPS, as well as the design and algorithms to be used and finally, to participate in promoting the EGNOS system. The ESTB mission uses the EURIDIS elements and integrates other elements from SATREF, which was developed by the Norwegians and provides integrity data and certain differential corrections. Namely, by integrating the SATREF data into the EURIDIS signal, a prototype is obtained that transmits the whole of a complementary message. The first version of the ESTB started transmitting at the end of January 2000 from Toulouse.

The Euridis satellite navigation system developed by Alcatel Space, as Prime Contractor, provides France's initial contribution to the EGNOS program, the European component in the first-generation worldwide navigation satellite system (GNSS-1). In the context of the EGNOS program, GNSS Euridis provides users with a permanent navigation signal in addition to those supplied by the GPS and GLONASS constellations. In any event, this signal significantly increases the availability and integrity of the mobile user's position. Euridis navigation service will be available in the Inmarsat-3 AOR-E satellite footprint area. Otherwise, the GNSS EURIDIS program was run by CNES (National Space Study Centre) and DNA (Air Navigation Directorate). It was an in-kind delivery for the EGNOS program. The aim of this project was to validate the feasibility of the "EGNOS ranging" function. Hence, this was developed by T4S and delivered in 1999 to ESA.

9.4.5. Benefits and Evaluation of the EGNOS System

The purpose and benefits of the EGNOS system is to implement a system that fulfils a range of user service requirements by means of an overlay augmentation to GPS and GLONASS based on the broadcasting through GEO satellites of GPS-like navigation signals containing integrity and differential correction information applicable to the navigation signals of the GPS and GLONASS satellites, the EGNOS own GEO overlay satellites and the signals of other GEO Overlay systems (provided they can be received by a GNSS-1 user located inside the defined EGNOS service area). EGNOS will address the needs of all modes of transport, including mobile maritime, land and aeronautical users.

Space technology is recognized as having a key role to play in maximizing safety in the transport of passengers and goods. The EGNOS system is based on a signal that is suitable for use by aircraft, ships, trucks, trains and other forms of transport. Other SBAS systems, such as the US WAAS and the Japanese MSAS, will be dedicated exclusively to air navigation. According to some estimates, the worldwide market for satellite navigation could be worth about €50.000 million by the year 2005. The EGNOS GNSS program is thus an opportunity for Europe to foster the development of a substantial market with good potential for creating new businesses and jobs in a wide range of industries.

1. Maritime Applications – The performance objectives for maritime application utilities are generally broken down into open sea, coastal and harbor navigation. Thus, the related determination accuracy requirements considered today are: Ocean and open sea navigation about 1–2 Nm, Coastal navigation is 0.25 Nm and Approaching and harbor navigation is 8–20 m. Even without EGNOS or other GSAS, GPS/GLONASS can easily meet sea and coastal navigation precision requirements. However, for harbor approach and the berthing of ships, differential techniques have to be applied as a DGPS. A European forum has been

been set up to identify the possible maritime applications for the GNSS network, which includes: navigation, seaport operations, traffic management, casualty analysis, offshore exploration and fisheries. Once in operation, EGNOS will be able to meet most of these requirements and will complement the services already provided by marine radio beacons.

2. Land Applications – In general, land vehicles do not need radio navigation as such but rather radio positioning and tracking. The two main land applications under development worldwide making use of GPS receivers are route optimization and fleet management. Thus, depending on the application, the accuracy required for the various systems ranges from a few to a hundred metres or more. In many cases, they then require the use of differential corrections. EGNOS is one of the keys to managing land transport in Europe, whether it is by road, rail or inland waterways. It will increase both the capacity and the safety of land transport. Not only airlines but also companies that operate transport services need to know where their vehicles are at all times, as do other public services such as the police, the military, ambulance and taxi services.

As well as improving safety, EGNOS will be an invaluable aid to managing transport operations. Managers will be able to know exactly when a consignment has been held up and its exact location. This will also improve customer services, as clients can be notified of delays and the reason for them and when necessary, breakdown crews can be sent out immediately. EGNOS has many other potential uses. It can help farmers in aerial crop spraying, fishermen to locate their catch and the police to detect fraud. EGNOS can also be used for leisure activities such as hiking, sailing and climbing. Every day more and more potential uses are being found for EGNOS. Computer and telecommunication networks around the world need an extremely accurate clock reference, a kind of "world speaking clock". At this point, the EGNOS system will be able to broadcast a reliable time standard with unprecedented accuracy to fixed and mobile users.

Satellite navigation will help to regulate road use and minimize traffic jams. If all vehicles are fitted with a navigation satellite receiver and a data transmitter, their position can be relayed automatically every few seconds to a central station. This information can then be used in a number of ways to control road usage. Furthermore, it could, for example, be used to charge motorists for using a stretch of road, to restrict access to congested roads, or to inform drivers of congestion and suggest alternative, quieter routes.

3. Aeronautical Applications – The performance objectives for aeronautical applications are usually characterized by four main in-flight parameters: accuracy, integrity, availability and continuity of service. Thus, the values for these parameters are highly dependent on the phases of flight. The typical aircraft operations signal-in-space performance requirements are determined for Accuracy Lateral (AL)/Accuracy Vertical (AV), respectively as follows:
a). En-route is 2 Nm/N/A;
b). En-route Terminal is 0.4 Nm/NA;
c). Initial Approach (IA), Non-Precision Approach (NPA) is 220 m/NA;
d). Instrumental Approach with Vertical Guidance (IAWVG) is 220 m/9.1 m and
e). Category I Precision Approach (PA) is 16 m/7.7-4.4 m.

Neither GPS nor GLONASS can meet the above listed typical phases of flight integrity, availability and continuity of service objectives without an augmentation system, although their performance in terms of accuracy alone could meet the requirements of only en-route, terminal area navigation and non-precision approaches. These requirements are currently being finalized by the ICAO GNSS-Panel under the form of Standards and Recommended Practices (SARPS). The world's commercial airways fleet is expected to double in the next

20 years. This will result in crowded routes leading to fuel wastage and delays, which could cost millions of dollars annually. The potential benefits will assist air traffic control to cope with increased traffic as well as improving safety and reducing the infrastructure needed on the ground.

But not only airlines will benefit from satellite navigation systems. Companies operating transport services at sea, by road or rail need to know where their mobiles are at all times, including police, ambulance, rent-a-car and taxi services. At this point, some European car manufacturers are already featuring satellite navigation systems in their top-of-the-range vehicles and cheap, handheld receivers are becoming widely used by recreational sailors, climbers and hikers. At all events, as well as improving safety, a European contribution to a GNSS will greatly contribute to improving regional economic prosperity, industrial returns, employment and quality of life in Europe.

Of the three transport communities, civil aviation navigation requirements are the more sophisticated, most mature and the most stringent related to the speed of moving and in terms of integrity and continuity and hence, the EGNOS performance objectives are mostly tailored to fulfill the needs of civil aviation, covering also the needs of land and maritime mobile users. In addition, the spacecraft needs a more precise determination because of very high speed and sophisticated approaches to the airports especially during very bad weather conditions. Moreover, ships can use very slow speed during approach to the berth and with the assistance of tugs and radar navigation, even under bad weather conditions.

9.5. GSAS MTSAT (MSAS/JMA)

MTSAT Japan Meteorological Agency (JMA) is producing a new integrated, on-board payload system of the Multi-functional Transport Satellite (MTSAT) as a successor to the first generation spacecraft GMS-5, in close cooperation with the Civil Aviation Bureau (CAB) and the Japanese Ministry of Transport (MOT). This system will be a satellite-based ATC for Japan and nearby regions.

The MTSAT project will have two main missions: a Meteorological for the JMA and an Aeronautical mission for the CAB. This system provides the AMSS and SBAS capabilities for ATS providers and aircraft operators in the Asia/Pacific region, to implement the ICAO CNS/ATM for the environment, operated by the CAB. The SBAS integrated with MTSAT forms the Japanese regional MTSAT Satellite-based Augmentation System (MSAS), which is designed to meet the ICAO SARP and be interoperable with the existing satellite system. The first MTSAT spacecraft was planed to be operational in 2003 and to cover the airspace throughout most of the Asia/Pacific area in 2004. This project will offer an opportunity for ATS providers and world aircraft operators to have highly reliable CNS facilities.

In February 1995, the MOT placed a contract for a MTSAT project with the SS/L Toshiba and Alcatel Consortium. The MTSAT Satellite-based ATC will optimize use of the world's air routes and enable aircraft to take off and land automatically. This advanced bifunctional system includes an optical instrument for weather observation and an aeronautical payload for the ATM system. It will dramatically improve air traffic safety and efficiency over the Asia and North Pacific regions. Later, in Gaithersburg, Maryland, on March 24, 1997 Lockheed Martin Federal Systems was selected as a member of a team led by NEC Corp., Tokyo, to design and develop the Japanese government's MTSAT GNSS, similar to the European EGNOS and US-based WAAS, as a part of GSAS. The award has a potential value of about 36 million US$ to the NEC team.

As a member of the NEC team, Lockheed Martin will have responsibility for three major subsystems in the Japanese development program, the first outside of the US to employ wide area satellite navigation technology for en-route and precision approach for civil aviation. Lockheed Martin will design and develop MSAS subsystems that monitor and control the all-important ground-based operation, which automates safe and correct system operation. These systems will also provide operators with the tools to manage the real-time flight-safety critical MSAS system. Then, Lockheed Martin will also design the ground communications network that will link all MSAS sites blanketing the Japanese home islands, extending to Australia and Hawaii for ranging sites. The MSAS system will be used to aid civil air navigation in the Japanese Flight Information Region (FIR) and will be fully compatible with the regional WAAS owned by the US FAA and European EGNOS.

9.5.1. Aeronautical Mission

The navigation transponder of MTSAT payload is a key part of the entire system. Namely, it sends GNSS signals to aircraft in the same way as other GPS satellites and improves the accuracy and availability of the positioning system. Thanks to the large number of small Earth stations, the GNSS signal is able to incorporate information on GPS spacecraft status and correction factors, thus greatly improving the reliability and accuracy of the present GPS system. Otherwise, the final GPS accuracy will be just a few metres, allowing air traffic to be controlled solely by satellite, without any ground radar facilities.

To complement the navigation channels, communication channels allow bidirectional (two-way) transmission between an aircraft and GES. The aircraft sends its position and flight data to the ATC authorities and to the relevant airline. This enables aircraft movements to be managed to enhance air safety and to improve operating efficiency. The satellite will forward flexible and safe routing information to aircraft, as determined by the ATC centre, improving fuel consumption, reducing flight times and enhancing the safety systems. The CNS service of the MSAS is divided into Aeronautical Communications, Navigation and Surveillance. As usual, the MSAS system consists in space and ground infrastructures.

9.5.2. Space Segment

The GMS-4 (Geosynchronous Meteorological Satellite) was launched in September 1989 for JMA and is widely used around the Pacific rim. After successful service, this spacecraft was finally moved to a graveyard orbit, 600 km above GEOs altitude, on 24 February 2000. The next GMS-5 weather satellite was launched in the spring of 1995 and it became operational in June 1995. This satellite reached its 5-year design lifetime serving for JMA in the Spring of 2000. The GMS-5 spacecraft operations are to be extended until December 2003 because MTSAT-1 failed to reach orbit.

In the late 1990s, MTSAT-1 was constructed by Space Systems Loral, integrated with an Imager by ITT Fort Wayne and finally shipped to Japan in March 1999. The launch, from the Japanese Tanegashima Space Centre in Kagoshima in 1999 on an H-2 rocket of the NASDA mission, failed to put the spacecraft into the GEO at 140° E. The MTSAT bird, shown in **Figure 9.10.**, is designed to be used in three different orbital GEO positions: 135°, 140° and 145° E and was scheduled to replace GMS-5 JMA spacecraft. This satellite carries meteo and aero payloads, including a precise pointing mechanism to optimize performance of optimal transponder transmission signals, whatever its orbital position.

Figure 9.10. MTSAT Satellite Configuration

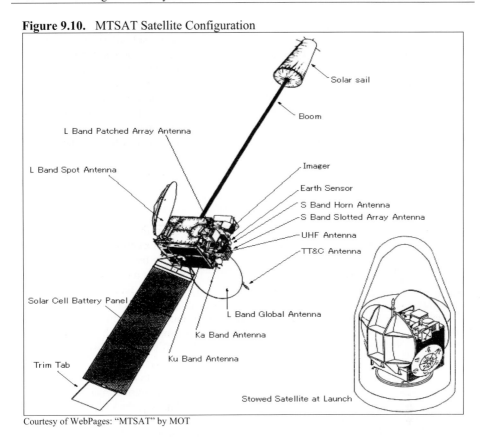

Courtesy of WebPages: "MTSAT" by MOT

The MTSAT-1 replacement spacecraft, MTSAT-1R, was scheduled for launch in March 2003 as a Single MTSAT of Phase I. Then in March 2000, a contract for MTSAT-1R was again awarded to SS Loral. In November 2001, the MTSAT-1R launch was rescheduled to mid-Summer 2003 and to be operational by December 2003, due to delays created by US technology transfer restrictions. Loral Space & Communications has won a contract to build new MTSAT-1R and deliver it in 2002 to provide digital voice/data communications, navigational services for aircraft and gather weather data for users throughout the Asia and Pacific regions. Accordingly, the satellite will be positioned at 140°E, using the L, Ku and Ka-bands. The MTSAT-1R spacecraft project will be a version of Space System & Loral's three-axis, body-stabilized F-1300 bus. It will provide 10 years of service in its aeronautical and five years in its meteorological mission. In this sense, the Japan's CAB will use MTSAT-1R to increase the efficiency of aircraft flight routes, provide flexible flight profile planning, enhance air travel safety and improve the quality of aeronautical communications. This multipurpose spacecraft will be the successor to the GMS-5 and previous Himawari series. The JMA will use MTSAT-1R to deliver observation data to a processing station and provide cloud imagery and continuous weather data from the Asia and Pacific region, such as cloud and vapor distributions, cloud-motion wind vector, sea surface temperature, information on typhoons, low pressure and frontal activity.

The next MTSAT-2 multipurpose spacecraft is scheduled for launch in June 2004 and is expected to be operated as a hot stand-by, using the dual MTSAT of Phase II. The MTSAT-2 spacecraft contract was awarded late in 2000 to MELCO, known as Mitsubishi Electric Space Systems, with MTSAT-2's Imager to be supplied by ITT Fort Wayne. The MTSAT-2 satellite will be the first to employ Mitsubishi's new communications bus. The major improvement in the satellite Imager requirements is the noise performance of the infrared channels. The MTSAT-2 image navigation and calibration are included in the spacecraft contract and to use the latest computers and algorithms. The orbital parameters of MTSAT-1, which are similar to MTSAT-1R and 2 spacecraft, are shown in **Table 9.1**.

Table 9.1. Orbital Parameters of MTSAT Spacecraft

Background	Launch weight: 2,900 kg
Owner/Operator: JMA, Japan	Dimensions stowed: 2.4 x 2.6 x 2.6 m
Present status: Launch failed	(box-like structure)
Orbital location: 140° E (long) & 0.1o NS/EW (lat)	Dimensions in orbit: 33.1 x 10.7 x 4 m;
Altitude: 36,000 km above the equator	2.7 KW Ga As Solar arrays with 3 panels each 2.4
Type of orbit: GEO	x 2.6 m; 3.3m Solar sail on a 5.1m boom
Coverage: Global beam of MTSAT covers most of	Batteries: Ni-Cd
Asia/Pacific airspace. Spot beams cover heavy traffic	Telemetry beacons: Ku and S-band
areas to meet the demand of increasing air traffic	Command beacons: Unified S-band
Additional information: MTSAT-1R is to be	**Communications Payload**
launched in 2003 & MTSAT-2 in June 2004	Frequency bands:
Spacecraft	1) Aeronautical mission:
Name of satellite: MSAT-1	Feeder link: Ku-band Tx/Rx 14/11 GHz
Launch date: 15.11.1999.	(4 spot beams); Ka-band Tx/Rx 30/20 GHz (3 spot
Launch vehicle: H-2S	beams)
Typical users: Aeronautical and Meteorological Data	Service link: L-band Tx/Rx 1.6/1.5 GHz
Cost/Lease information: 95 US $	(1 global beam and 6 spot beams)
Prime contractors: SSL and JMA	2) Meteorological mission: UHF band and S-band
Other Contractors: Civil Aviation Bureau	Number of transponders: 2
Type of satellite: Attitude Controlled GEO	EIRP: Feeder link Ku/Ka-band 27 dBW
Stabilization: 3-axes	Service link Global/Spot 40/43 dBW
Design lifetime: More than 10 years (Aeronautical	G/T: Feeder link Ku/Ka-band -1 dB/K
Mass in orbit: 1,250 kg	Service link Global/Spot –9/–243 dB/K

The MTSAT spacecraft also has an innovative aircraft communication purpose payload, AMSS service, which will be similar to the Inmarsat system. The heart of the payload is an IF processor that separates all the incoming channels and forwards them to the appropriate beam in both directions: forward (ground-to-aircraft) and return (aircraft-to-ground). A separate 24 dB gain control function is available on each of two direction paths to provide maximum operational flexibility. The MTSAT coverage areas are shown in **Figure 9.11.** for the global beam and **Figure 9.12.** for the spot beams **(A)** and Feeder links **(B)** and both are available in L-band. Global beam of MTSAT covers most of the Asia/Pacific airspace and includes all parts of the Earth that can be seen from the satellite. An RF power of 150 W is available for communications using four SSPA in parallel and 30 W RF power is available for the GNSS signal. Spot coverage consists in 6 spot beams over Japan and the North Pacific, including heavy traffic areas, to meet the demands of increasing air traffic. At this point, an 80 W multiport amplifier ensures maximum system flexibility by allowing the total RF power to be varied dynamically between the spots. Two large reflectors of 3.2 m diameter and one flat array are used to generate the coverage areas.

Figure 9.11. Global Beam MTSAT Coverage

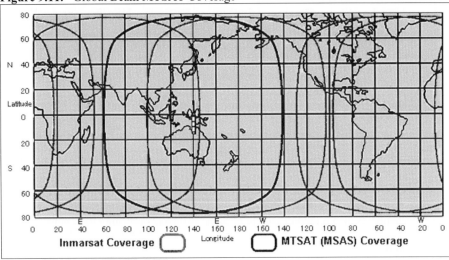

Courtesy of WebPages: "MTSAT" by MOT

The MTSAT signal characteristics are generally based on the ICAO Annex 10 (SARP) and Inmarsat SDM and comply with the Radio Regulations and ITU-R Recommendations. The MTSAT signal characteristics are summarized in **Table 9.2**. The MTSAT spacecraft has two types of satellite links related to the Ground Earth Stations (GES):

1. Forward GES to Satellite Direction – The MTSAT GES are located throughout the region. Their signals are received by either a Ku-band or a Ka-band antenna system. Thanks to the high frequency used, the reflector size of the antennas is quite small (500 m for Ku-band and 450 mm for Ka-band). The reflector is movable via focusing motors, so that it can work with the satellite in any of the three possible GEO positions.

Incoming signals are amplified, converted to IF, filtered and routed within the IF processor where they are then up-converted and transmitted in L-band to the AES.

2. Return Satellite to GES Direction – The L-band signal received from all approaching AES are processed in the same way and retransmitted to GES via a Ku-band and Ka-band antenna. Output power of the Ku-band and Ka-band AES transmitters is just 2 W thanks to

Figure 9.12. Spot Beam and Feeder Links MTSAT Coverage

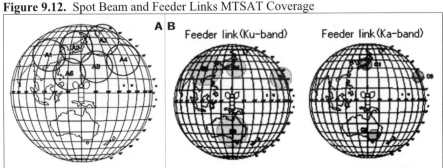

Courtesy of WebPages: "MTSAT" by MOT

Table 9.2. MTSAT Signal Characteristics

Channel		Direction	Usage	Bearer rate	Modulation	Bandwidth
P	Packet mode/TDM	Forward-link	Signalling and message continuously transmitted from GES	600 bit/s	1/2 FEC A-BPSK	5 kHz
				10.5 kbit/s	1/2 FEC A-QPSK	10 kHz
R	Random access (Slotted ALOHA)	Return-link	Signalling and message	600 bit/s	1/2 FEC A-BPSK	2.5 kHz
				10.5 kbit/s	1/2 FEC A-QPSK	10 kHz
T	Reservation TDMA	Return-link	Long message	600 bit/s	1/2 FEC A-BPSK	2.5 kHz
				10.5 kbit/s	1/2 FEC A-QPSK	10 kHz
C	Circuit mode SCPC	Forward-link	Voice message	21 kbit/s	1/2 FEC A-QPSK	17.5 kHz
		Return-link		8.4 kbit/s	2/3 FEC A-QPSK	7.5 kHz

the HG antenna. It is also possible to provide station-to-station channels in either the Ku or Ka-band to enable stations working with different spots to communicate with one another. The GNSS channel is also routed to ground on same two bands for calibration purposes.

9.5.3. Ground Segment

The MTSAT Ground Segment infrastructure consists in several Ground Earth and System Control stations located at present in almost corresponding positions in Japan. The final system will include a large number of GES spread throughout the region. An important feature of these stations is that they have been built to withstand earthquakes, which also required a special antenna design.

1. Ground Earth Stations (GES) – In order to provide service continuously, even during natural disasters, two aeronautical satellite centres have been implemented at two different locations in Japan. The first is GES Kobe, approximately 500 km West of Tokyo and the second is GES Hitachi-Ota, approximately 100 km Northeast of Tokyo. The Aeronautical Satellite Service (ASS) provided by these two GES are in charge of all communication functions through the satellite. For instance, Kobe GES, with a 13 m diameter antenna, transmits and receives satellite signals in the Ku, Ka and L-band. A very high EIRP of 85 dBW and a high G/T ratio of 40 dB/K are achieved in the Ku and Ka-band, respectively and ensure very high availability of the feeder link. The L-band terminal, which is similar to the aircraft terminal, is used for system testing and monitoring. Approximately 300 circuits are available simultaneously in both the transmit and receive directions. This station also includes dedicated equipment for testing the satellite performance after launch and for permanent monitoring of the traffic system. In addition, top-level management software is provided to configure the overall system and check the system's status.

2. Aircraft Earth Stations (AES) – Special part of the MTSAT Ground Segment are AES terminals approaching to the region including GNSS service. This terminal is similar to Inmarsat aero standards, which contain: ACU as an antenna and BCU as a transceiver with peripheral equipment. The BCU voice or data terminals can be used for flight crew, cabin crew and passenger applications. The AES terminal is an aircraft-mounted satellite radio capable of communications through spacecraft in the MTSAT system, providing Tel, Fax and data two-way service anywhere inside the satellite footprint. The characteristics of the MTSAT AES terminal are based on the ICAO Annex 10 (SARP), Inmarsat SDM Module and UTU Radio Regulation, the major parameters of which are summarized in **Table 9.3**.

Table 9.3. MTSAT Spacecraft Characteristics

Type	Low gain antenna	High gain antenna
G/T	better than −26 dB/K	better than −13 dB/K
Maximum carrier e.i.r.p. (P_{max})	more than 13.5 dBW	more than 25.5 dBW
Power control range	1 dB steps from P_{max} to P_{max} − 15 dB	1 dB steps from P_{max} to P_{max} − 15 dB
Antenna characteristics	Non-directional	Steerable BW = 45°

3. Satellite Control Stations (SCS) – The SCS terminal is located in the same building as the GES and utilizes an antenna with the same diameter. This station has to control the satellite throughout its operational life. Two frequency bands are used: S-band in normal operation and Unified S-band (USB) while the satellite is being transferred to its final orbit, or in the event of an emergency, should the satellite lose its altitude. In S-band the EIRP is 84 dBW and for security reasons, the EIRP in USB is as high as 104 dBW. An SCS displays the satellite's status and prepares telecommands to the satellite. Furthermore, the satellite position is measured very accurately (within 10 m) using a trilateral ranging system instead of measuring one signal, which is sent to the satellite then returned to the Earth. The Station sends out two additional signals, which are retransmitted by the satellite to two dedicated ranging stations, which return the same signals to the SCS via satellite. This technique allows the satellite's position to be measured in three dimensions. A dynamic spacecraft simulator is also provided to check telecommands.

4. GNSS System – The GNSS system is known as the MTSAT MSAS and consists in a large number of Ground Monitoring Stations (GMS) and Master Control Stations (MCS). The GMS terminals are small, autonomous Earth stations housed in a shelter. Each Monitoring Station computes its location using GPS and MTSAT signals. Any differences between the calculated and real locations are used by the system to correct the satellite data. Data is sent to the MCS via the public network. The MCS collects all the information from each GMS. Complex software is able to calculate accurately the position and internal times of all GPS spacecraft and MTSAT satellites. Finally, the GNSS signal, incorporating the status of the GPS spacecraft and corrections, is calculated and sent to the traffic station for transmission to MTSAT satellites.

9.5.4. MTSAT Network Architecture

As was mentioned, the MTSAT (MSAS) system has a dual satellite mission, for the JMA, performing meteorological functions and for the Japanese MOT, implementing ATS/CNS communications, navigation air traffic control and surveillance functions.

Though the sky is assumed to be unlimited, all aircraft fly over a pre-determined route and as traffic jams occur with the traffic volume on the roads, so the number of airliners on a route has limitations. According to the increase of air traffic volume in recent years, the new air navigation system that will replace the current one was initiated and approved by the ICAO in 1991.

This new solution is called the CNS/ATM system, which is composed of three CNS; Communications, Navigation and Surveillance components and comprises ATM in the global Aeronautical Navigation System (ANS). Accordingly, this system utilizes Global Satellite Positioning and Communications, Automatic Surveillance and the latest spacecraft technology, which improves safety, capacity and the economics of flight operations.

Figure 9.13. The MTSAT System Configuration

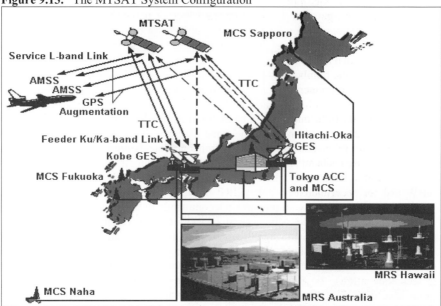

Courtesy of WebPages: "MTSAT" by MOT

The MTSAT will perform the key role in this new system as aeronautical communication satellite, especially on the North Pacific route, which is the main, and most expanding air route between Japan and the USA. This system will also contribute to improve both safety and capacity in Asia Pacific regions; the system configuration is shown in **Figure 9.13**. In addition, Space and Ground Segments include an Aeronautical Control Centre (ACC) in Tokyo and other control and monitoring stations of the ground network.

9.5.5. Aeronautical Mobile Satellite Service (AMSS)

The AMSS functions of new MTSAT system includes the provision of all the aeronautical communications defined by the ICAO, such as ATS, Aeronautical Operational Control (AOC), Aeronautical Administrative Communications (AAC) and Aeronautical Passenger Communications (APC). These MSC services could be available for ATS providers and aircraft operators in the Asia Pacific region through data link service providers. Direct access to the MTSAT network could also be possible through the implementation of dedicated GES in other states covered by MTSAT spacecraft. The AMSS provided by the MTSAT system for the AES is interoperable with that of the Inmarsat network.

The MTSAT system will provide direct cockpit controller-pilot communication in voice, facsimile and data, GPS augmentation information and Automatic Dependent Surveillance capabilities. The MTSAT/MSAS system will not only be capable of handling oceanic ATS communications within the Japanese FIR, connecting Asia and the US but will also be offered to the Civil Aviation Community (CAC) in the Asia/North Pacific Region as an aviation infrastructure, which could facilitate the development and implementation of the ICAO CNS/ATM systems.

The MTSAT/MSAS service provides all aircraft with GPS augmentation information to improve safety and navigational performance requirements, namely to find out the response to the demands of integrity, continuity and availability, which are essential to the use of GPS for aircraft operation as the sole means of navigation. In order to provide aircraft with sufficient GPS augmentation information, a certain number and location of GMS will be required. Therefore, the number and location of GMS required for each state in the region will depend on the requirements for the level of navigation services and reception of GPS signals. The Japanese CAB has implemented for the MTSAT two Monitoring and Ranging Stations (MRS) in Australia and Hawaii, four GMS and two MCS in Japan. For this reason, the Asia/Pacific States could implement MSAS with a lower number of GMS than other SBAS. Since most of the Asia/Pacific region will be covered by two MSAT spacecraft, in this way the integrity and availability are higher than another SBAS within this region.

Figure 9.14. Current and New CNS/ATM System

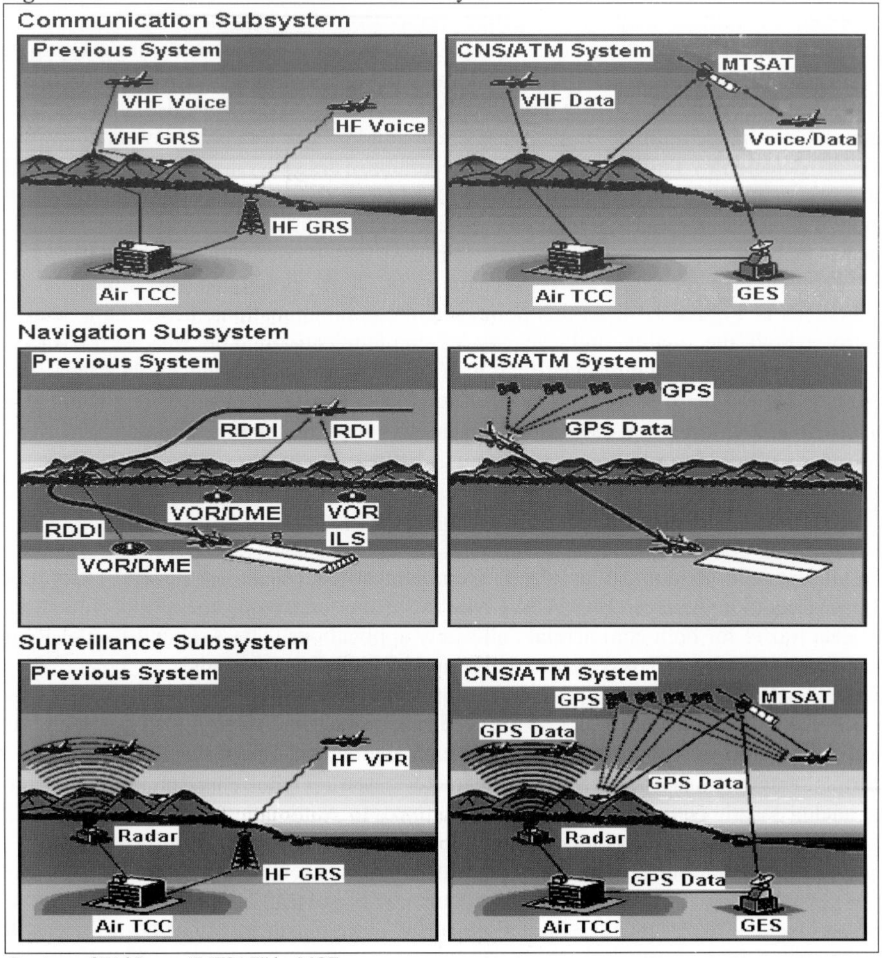

Courtesy of WebPages: "MTSAT" by MOT

9.5.6. Current and New CNS/ATM Aeronautical Mission

The aeronautical mission of the MSAS is divided into Aeronautical Communications, Navigation and Surveillance subsystems (CNS). It will provide three major aeronautical services: Aeronautical Mobile Satellite Service (AMSS), Global Navigation Satellite System (GNSS) and Automatic Dependent Surveillance System (ADSS).

Current communication facilities between aircraft and ATC are executed by VHF and HF voice (radiotelephone system), see Communication Subsystem in **Figure 9.14**. The VHF voice link between aircraft on one the hand and Ground Radio Station (GRS) and Traffic Control Centre (TCC) on the other, may have the possibility to be interfered with by high mountainous terrain. Moreover, the HF link may not be established due to lack of available frequencies, intermediation, unstable wave conditions and to heavy rain or thunderstorms.

Current navigation possibilities for recording and processing Radio Direction Information (RDI) and Radio Direction Distance Information (RDDI) between aircraft and ATC are performed by ground landing navigation equipment, such as the Instrument Landing System (ILS), VHF Omnidirectional Ranging (VOR) and Distance Measuring Equipment (DME), see the Navigation Subsystem in **Figure 9.14**. This subsystem needs more time for ranging and secure landing, using an indirect way of flying in a semicircle.

Current surveillance utilities for receiving Radar and HF Voice signals between aircraft and TCC are detected by Radar and Ground HF Stations. This subsystem may have similar HF voice communications problems and when aircraft are flying behind high mountains they cannot be detected by Radar, see the Surveillance Subsystem in **Figure 9.14**.

On the contrary, the new CNS/ATM System utilizes the communications satellite and it will eliminate the possibility of interference by high mountains, see all three CNS Subsystems in **Figure 9.14**. At this point, satellite communications, including a data link, improves both the quality and capacity of communications. The weather (WX) data, NOTAM and flight planning data may also be directly input to the Flight Management System (FMS). The GPS Navigation Subsystem Data provide almost a direct landing line, so surveillance information cannot be interfered with by mountainous terrain. The display on the screen will eliminate misunderstandings between controllers and aircraft pilots.

9.5.6.1. Flight Radio and AMSS Communication System

The present mobile Aeronautical Radio Communications (ARC) for general international purposes has to be replaced by MTSAT/MSAS system to enhance cockpit-to-ground voice and data traffic for both commercial and safety applications. The Asia and Pacific Region use a current service of Tokyo Radio on HF and VHF frequency bands.

Above the Pacific, the main type of voice communications is HF (non-DCPC). Over land, VHF voice communications (DCPC) are used. The VHF data link for D-ATIS/AEIS is also used. This data link uses the Aircraft Communication and Reporting System (ACARS), with a character transmission system (character-based). Thus, terrestrial communications use analog telephones and the Aeronautical Fixed Telecommunication Network (AFTN). This communication system has, like ACARS, been operational for over 20 years.

Japan is responsible for providing the air traffic services for the trunk routes in the North Pacific, connecting Asian and American continents. In order to fulfill this important role, it is most essential to conquer the limitations of the current system. It has been forecast that the amount of the air traffic over the North Pacific in 1992 will have trebled by 2010.

Meanwhile, in order to respond to the increase in the volume of communications data that has accompanied the large increase in flight traffic, periodic communications have moved to the data link and data transmission has become the core type of flight communications. The media needs to be divided to reflect this change in communications content, which has seen voice communication used mainly for irregular emergency situations. A transmission system based on basic new digital technology (bit-based) needs to be introduced to make wholesale improvements in communications ability and to replace ACARS for security.

Gradually, AMSS voice and data links have come into use and may replace HF and VHF radio. Namely, AES can be used for communications with GES via MTSAT or Inmarsat satellites for aeronautical commercial, safety and social purposes.

9.5.6.2. MSAS GNSS

The GPS satellite-based navigation system can be used worldwide to control the positions of aircraft and to manage air traffic. It supports aircraft navigation well in all phases, including take-off and landing utilities. In fact, GPS has some performance limitations and it cannot consistently provide the highly precise and safe information in the stable manner required for wide-area navigation services. To assure safe and efficient air traffic navigation of civil aircraft, GPS performance needs to be augmented with another system that provides the four essential elements of air navigation as follows: integrity, precision, availability and continuation of services. As is known, the MSAS system is the wide-area augmentation system for GPS, similar to the US WAAS and the European EGNOS. Once in operation, this state-of-the-art system will assure full navigation services for aircraft in all flight phases within the Japanese FIR through MTSAT coverage.

The L1/L2 frequency band is nominated for the transmission of signals from GPS spacecraft in ground and air directions. These signals can be detected by the GMS, MRS and GPS receivers of flying aircraft. The MSAS satellite transponder uses the L1 frequency band to broadcast GPS augmentation signals in the direction of GES and AES. The Ku-band is used for unlinking GPS augmentation signals to the MTSAT spacecraft. The whole ground infrastructure and Network Communication System is controlled by MCS.

The components of the MSAS navigation system are shown in **Figure 9.15**. To provide GPS augmentation information, all ground stations, which always monitor GPS signals, are necessary in addition to MTSAT. This special navigation infrastructure, which is composed of MTSAT, GPS wide-area augmentation system and these ground stations, is called the MSAS network.

9.5.6.3. Wide Area Navigation System

The Wide Area Navigation (WANAV) system is a way of calculating own position using the Flight Safety Satellite Equipment (FSSE) facilities and installed air navigation devices to navigate the desired course. Until now, the airways have made mutual use of the FSSE, which often led to broken line routes. However, in the case of WANAV (RNAV – an original version) routes it has been possible to connect in an almost straight line to any desired point within the area covered by the satellite equipment and service.

In any event, setting the WANAV routes has made it possible to ease congestion on the main air routes and has created double tracks. This system enables more secure and economical air navigation routes.

Figure 9.15. Future MSAS Navigation System

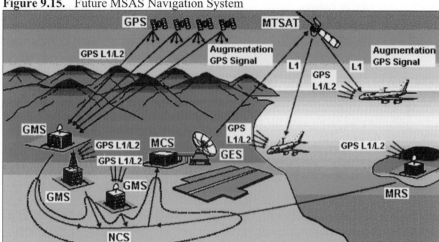

9.5.6.4. MTSAT Automatic Dependent Surveillance System (ADSS)

The current surveillance system in Japan is mainly supported by VHF GRS. This system enables display of real-time positions of the nearby approaching flying aircraft using radar and VHF voice radio equipment. Due to its limitations, the VHF service being used for domestic airspace cannot be provided over the ocean. Meanwhile, out of radar and VHF coverage and range on the oceanic routes, the aircraft position is regularly reported by HF radio voice or via data terminals to the HF GRS.

Figure 9.16. Future Surveillance System

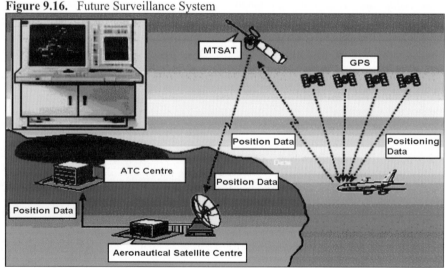

Figure 9.17. SELA Subsystem

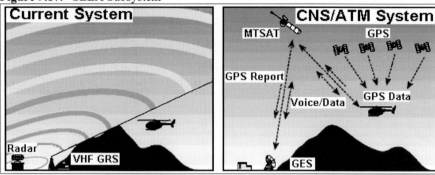

Courtesy of WebPages: "MTSAT" by MOT

Consequently, the advanced CNS/ATM MSAS system utilizes the ADSS data function, which automatically reports all current aircraft positions measured by GPS to ATC, as illustrated in **Figure 9.16**. In such a manner, the approaching aircraft receives positioning data from GPS spacecraft and then sends its current position for recording and processing to the ATC Centre via GES.

The display looks just like a pseudo-radar coverage picture. The coming ADSS system will increase air safety and reduce aircraft separation, improve functions and selection of the optimum route with more economical altitudes. In addition, the system will also increase the accuracy of each aircraft position and reduce the workload of both controller and pilot, which will improve safety. In this sense, aircraft can be operated in a more efficient manner and furthermore, since the areas where VHF radio does not reach due to mountainous terrain will disappear, small aircraft, including helicopters, will be able to obtain meteorological data on a regular basis. These functions are mandatory to expand the traffic capacity of the entire air region and for the optimum air route selection under limited space and time restraints.

9.5.7. Special Effects of the MTSAT System

Special effects of the MTSAT system used for secure communications, navigation, ranging and control of air and surface traffic are Safety Enhancements in Low and High Altitudes, Reduction of Separation Minima, Flexible Flight Profile Planning and Surface Movement Guidance and Control.

These effects are very important to improve aircraft communication facilities in any phase of flight, enable better control of aircraft, provide flexible and economic flight with optimum routes, enhance surface guidance and control and to improve security.

9.5.7.1. Safety Enhancements at Low and High Altitudes

A very important effect of the new MTSAT CNS/ATM Satellite system is to provide Safety Enhancement at Low Altitudes (SELA) via GES, shown in **Figure 9.17**.

Current system for short distances between aircraft and GRS is provided by VHF voice equipment, so the pilot will have problems establishing voice cockpit radio communications when the aircraft's flying position is in the shadow of high mountains.

Figure 9.18. SEHA Subsystem

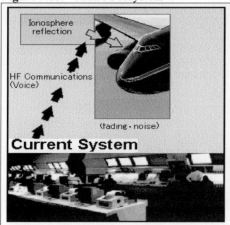 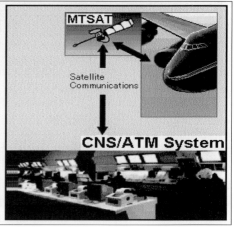

Courtesy of WebPages: "MTSAT" by MOT

Meanwhile, all aircraft in flight can receive satellite navigation and communications even at low altitude and where there is no navigation and communications coverage due to mountainous terrain. This is very important for secure flying during bad weather conditions and reduced visibility.

The MTSAT system is able to also provide Safety Enhancement at High Altitudes (SEHA) shown in **Figure 9.18.**, by using faded radio or the noise-free satellite system.

9.5.7.2.   Reduction of Separation Minima (RSM)

One of the very important MTSAT navigation effects is the Reduction of Separation Minima (RSM) between aircraft on the air routes by almost half, as shown in **Figure 9.19**. The current system has an RSM controlled by conventional VHF and HF Radio and Radar systems, which allows only large distances between aircraft. Besides, the new CNS/ATM system controls and ranges greater numbers of aircraft for the same air space corridors, which enables minimum secure separations, with a doubled capacity for aircraft. Hence, a significant reduction of the separation minima for flying aircraft will be available with the widespread introduction of the new augmentation satellite technologies on the CNS system.

Figure 9.19. RSM Subsystem

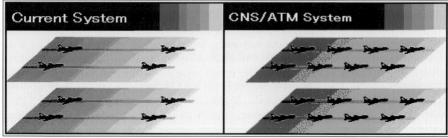

Courtesy of WebPages: "MTSAT" by MOT

Figure 9.20. FFPP Subsystem

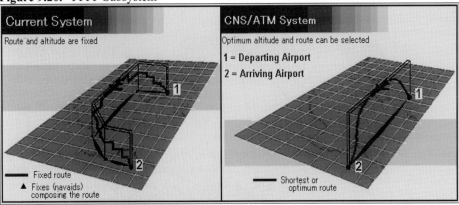

Courtesy of WebPages: "MTSAT" by MOT

9.5.7.3. Flexible Flight Profile Planning (FFPP)

The next positive effect of MTSAT spacecraft system is Flexible Flight Profile Planning (FFPP) of optimal altitude and route, illustrated in **Figure 9.20**. The current system uses fixed air routes and flying altitudes. At this point, the fixed route is controlled by the aircraft on-board navigation instruments only, which is a composite and not the shortest route from departure to arrival at the airport. Moreover, FFPP allows the selection of the shortest or optimum route and flying altitude between two airports. In the other words, with thanks to new wide augmentation satellite technologies on CNS system FFPP will be available for more economic and efficient flight operation. This means that the aircraft's engines will use less fuel by selecting the shortest flying route of New CNS/ATM system than by selected the fixed route and altitude of current route composition.

9.5.7.4. Surface Movement Guidance and Control (SMGC)

The Surface Movement Guidance and Control (SMGC) infrastructure is a special system that enables a controller to guide aircraft on the ground, even in poor visibility conditions at an airport. In this sense, the controller issues instructions to pilots with reference to a command display in a control tower that gives aircraft position information detected by sensors on ground, shown in **Figure 9.21**.

The command monitor also displays reported position information of landing or departing aircraft and all auxiliary vehicles moving onto the airport's surface. This position is measured by GNSS, using data from GPS and MTSAT/MSAS satellites. A controller is able to show the correct taxiway to pilots under poor visibility, by switching the taxiway centreline light and the stop bar light on or off. The development of head-down display and head-up display in the cockpit that gives information on routes and separation to other aircraft is in progress. The following segments of SMGC are shown in **Figure 9.21.**:

1) GPS Satellite measures the aircraft or airport vehicle's exact position.

2) MTSAT is integrated with the GPS satellite positioning data network. In addition to complementing the GPS satellite, it also has the feature of communicating data between the aircraft and the ground facilities, pinpointing the aircraft's exact position.

Figure 9.21. SMGC Subsystem

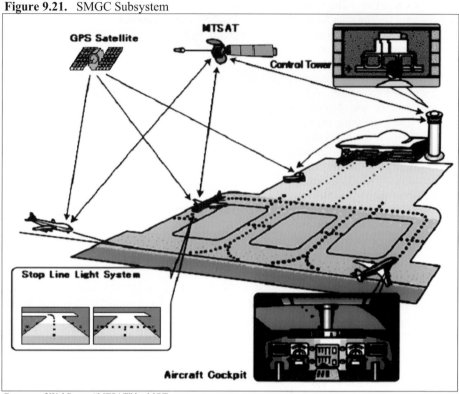

3) Control Tower is the centre for monitoring the traffic situation on the landing strip around the airport's ground environment. The location of each aircraft and vehicle is displayed on the command monitor of the control tower. The controller performs ground-controlled distance guidance for the aircraft and vehicles based on this data.

4) Stop Line Light System is managed by the controller, who gives guidance on whether the aircraft should proceed to the runway by turning on and off the central guidance line lights and stop line lights as a signal, indicating whether the aircraft should proceed or not.

5) Aircraft Cockpit displays the aircraft position and routes on the headwind protective glass (head-up displays) and instrument panel display (head-down display).

9.5.8. Ground Air Traffic Control Automation System (ATCAS)

The ground ATCAS is a supporting system, which conducts smooth ATC services as well as managing aviation safety and on time scheduled aircraft operations as follows:

1. ATC Data Processing System (ATCDPS) – Provides the following systems: Flight Data Processing (FDP); Radar Data Processing (RDP); Automated Radar Terminal System (ARTS); Terminal Radar Alphanumerical Display (TRAD); Oceanic Air Traffic Control Data Processing (ODP); Air Traffic Flow Management (ATFM); Oceanic Track Generator (OTG) and Spot Management and Planning (SMAP).

2. CADIN System – The Common Aeronautical Data Interchange Network (CADIN) is a general term for the data terminal stations installed in all airports and related facilities, the Information-Communication Network comprises these stations and three subsystems as follows: Aeronautical Fixed Telecommunication Automatic Exchange (AFTAX), Domestic Telecommunication Automatic Exchange (DTAX) and Input Data Processing System (IDP), which are main communication centres of the network. In addition, these systems are connected to AFTN, ATC Data processing system, Meteorological Agency, Self-Defense Agency and Airlines system, etc. Therefore, the CADIN system is the general term the data processing exchange systems (AFTAX, DTAX and IDP), which are the main data communication networks. In the other words, this communication system conducts to delivers, exchanges, handles and processes many kinds of information, such as Flight Plan, NOTAM, airport information (using runway, approach procedures, WX, etc.) and pilot reports in flight to allow airmen and the relevant aviation facilities the efficient use of the information needed for safe aircraft operation. At any rate, the main function of CADIN is to keep the above-mentioned database systematically arranged and stored and to relay and exchange various messages necessary for SAR operations by connecting to AFTN, ATC data processing system, Meteorological Agency, the Defense Agency as well as airline companies, etc.

a). AFTAX is composed of two main computers for processing all incoming messages and data received from aircraft and other resources via HF/VHF Radio or MTSAT AMSS equipment. The first is the Message Switching Main Computer (MSMC) and the second is the Data Base Main Computer (DBMC). In fact, all four Central Processor Units (CPUs) are ACOS 3500 computers.

b). DTAX performs the following major functions: Message Exchange; ATS Message Data Processing; NOTAM Handling; Weather Information Handling; Pre-flight Information Bulletins and Displaying the Graphical Data.

c). IDP is located only in Tokyo and is very important to process the input database of ATS messages only.

3. IDEC System – The Integrated Development and Evaluation Centre (IDEC) System has 4 test functions and only 1 ATC simulator. The system is designed for the development and evaluation of programs for ATC data processing systems to have ATC services, etc and to operate smoothly. In addition, this system plays an important role in contingency management as a back-up system for operating systems installed at the ACC and so on. The main functions of IDEC are to develop and evaluate the programs of FDP, RDP, ARTS and CADIN, to evaluate the programs by connecting more than one system and perform total systems simulation and the function of contingency management with regard to FDP, RDP and CADIN.

9.5.9. Special Ground Services

A Special Ground Service serves all airports in the region and includes all additional activities in connection with FIR and the Control Area System for Japan and the Pacific Region, RCC Control System, Automatic and en-route information services, HF/VHF Radio and MTSAT Concept diagram for Oceanic Data Link and ODP system, Message Flow of FDP System and Future ATM System.

1. Flight Information Region (FIR) and Control Areas – The sky of Japan is divided into two FIR, Tokyo and Naha, which are controlled by ATFM Centres located over Japan.

2. RCC Control System – This system performs Position Control and the whole of nearby SAR operations for this region; Information Position Management 1 and 2 of the SAR zone in accordance with ICAO procedures; Position Coordination, using office telephones; the RCC Main Computer host manages the whole RCC system and communications services of Tokyo DTAX and uses a Projector (70") to display basic SAR information and maps.

3. Automatic Terminal Information Service (ATIS) – The ATIS data system is input from terminals placed in 19 main airports around Japan. This information input from AEIS terminals is registered in the Data Link Processing (DLP) system database at 4 air traffic control sections along with NOTAM, PIREP and volcano information input from CADIN. The data is regularly updated and service is provided based on the latest information.

4. Aeronautical En-route Information Service (AEIS) – The AEIS system has the role of offering up-to-date information quickly and effectively to all aircraft in flight (other than those arriving at and departing from airports) by HF for long distances and on the VHF radiotelephone for short distances. The new MTSAT AMSS will improve the ATIS, AIES and the safety of flying in the region for Japan, Eastern Asia and the North Pacific Ocean.

5. Oceanic Data Link (ODL) – The ODL system enables the transfer of data and information between Air and Ground ADS and CPDLC services by HF/VHF via Tokyo Radio and by AMSS via MTSAT satellites, through corresponding ground infrastructures.

6. Message Flow of FDP System – A flight plan submitted to the flight operations office by a pilot before departure is sent via DTAX and AFTAX to IDP operations at Tokyo ACC, thence to FDP system operated at the same ACC. The processed data are then sent to the ATC facilities concerned, such as RDP, control towers of major airports, FADP, ODP and ARTS. The ACC terminals other than Tokyo ACC receive necessary flight plan data via their respective Distribution and Concentration Communications Processor (DCCP).

9.6. GSAS WAAS

The basic GPS service fails to meet the accuracy, availability and integrity that is needed by many mobile users, therefore some types of differential GPS corrections are developed to improve the accuracy of the GPS signal. Another system being developed in the USA to provide DGPS correction is the Wide Area Augmentation System (WAAS). In order to meet the requirements for better accuracy and integrity of GPS, the US FAA has designed the new WAAS system. It is what the name implies, a geographically expansive augmentation to the basic GPS service. The WAAS service improves the accuracy, integrity and availability of the basic GPS signals and allows GPS to be used as a primary means of navigation for en-route travel and NPA in the USA. The coverage area for this system includes the entire US and some outlying air areas such as Canada and Mexico.

The Flight Standards Service (FSS) is the sponsor of the FAA-based Satellite Operational Implementation Team (SOIT). The SOIT is responsible for providing the leadership role in coordinating the operational implementation of existing and emerging satellite navigation system technologies into the National Airspace System (NAS). Since its formation in August 1991, SOIT has been instrumental in the development of the criteria, standards and procedures for the use of unaugmented GPS as well as GPS augmented by the WAAS and Local Area Augmentation System (LAAS). Therefore, to increase the accuracy, integrity and availability of the GPS signals, the FAA is developing the satellite-based WAAS and the ground-based LAAS.

9.6.1. Comparison of WAAS and DGPS Service

On 24 August 2000, the FAA announced that their space-based, L-band WAAS, as a part of GNSS had become available for use by "some aviation and all non-aviation" users. The FAA announcement has prompted numerous inquiries to the Coast Guard regarding the maritime use of WAAS for ships and the status of the Coast Guard DGPS system. Its goal was to recommend the optimum integrated system to meet aviation, maritime and land navigation needs. In this sense, the study concluded that a combination of two systems: the FAA's Wide/Local Area Augmentation Systems (WAAS/LAAS) and the USCG's DGPS system was the optimum mix. This integration, consisting in the L-band line-of-sight WAAS for aviation users and the terrain-following medium frequency DGPS for both maritime and land users, meets the vast majority of the nation's precise navigation and positioning needs. In fact, the WAAS mission is not yet fully operational and has currently a testing status, while undergoing further development. It is not certified for use as a safety of life navigation system in the maritime navigation environment, although it may be used, with caution, to improve overall situational awareness but it should not be relied upon for safety-critical maritime navigation. Thus, the Maritime DGPS Service is fully operational and meets all the standards for the harbor entrance and approach phases of navigation.

The WAAS is not optimized for surface (maritime and land) use; rather, it was designed primarily for aviation. It is eventually intended to support aeronautical en-route through precision approach air navigation. The current WAAS test signals are transmitted by two GEO satellites on a line-of-sight, L-band frequency. This means that if anything obstructs the view of the portion of the sky where the satellite is, the WAAS signal will be blocked. Since GEO satellites are positioned over the equator, the further north users are, the lower the GEO satellites are in the sky, increasing the likelihood of an obstruction. In contrast, the medium frequency (285 to 325 kHz) radio beacon-based Maritime DGPS Service is optimized for surface (maritime and terrestrial) applications because its ground wave signals "hug the Earth" and wrap around objects. This means that the Coast Guard DGPS system is well suited for the marine environment (in the "ground clutter") where a GEO satellite can be blocked by terrain, harbor equipment and other man-made and natural objects. However, the Coast Guard's system was designed with the surface (maritime and land) user in mind. It was neither designed nor intended to meet aviation requirements.

Although aviation users could potentially obtain some modest benefit from the DGPS for applications such as surface traffic management at all airports or for general aviation, it could not attain the type and level of an aeronautical service for which WAAS and LAAS are designed without significant re-engineering. The DGPS has already been adopted globally as an international maritime standard established by the 1994 ITU document R-M.823. At this point, the DGPS solution meets IMO Resolution A-815 (19) standards for navigation in harbor entrances and approaches. Afterwards, over 40 nations fully embraced this robust technology and are implementing DGPS services identical to WAAS.

On average, the WAAS and DGPS accuracy are virtually the same, although the DGPS accuracy is better when the user is near a DGPS transmitting site. The WAAS architecture is designed to provide uniform 7 m accuracy (95%) regardless of the location of the mobile receiver, within the WAAS service area. The DGPS is designed to provide better than 10 m navigation service (95%) but typically provides better than 1 metre horizontal positioning accuracy (95%) when the user is less than 100 nautical miles from the DGPS transmitting sites. Accuracy then degrades at a rate of approximately 1 metre per hundred nautical miles as the user moves away from the transmitting site. A total of 56 maritime DGPS sites provide coastal coverage of the continental US, the Great Lakes, Puerto Rico, portions of Alaska and Hawaii and portions of the Mississippi River Basin.

Once WAAS becomes fully operational, the combination of Coast Guard and FAA systems is expected to provide a robust, complementary service to all modes of transportation. It is looking forward to the day that the industry provides the public with a fully integrated receiver, one that uses all available radio navigation systems to provide unprecedented accuracy, integrity and availability. Despite the differences between DGPS and WAAS, it should always be kept in mind that both services ultimately rely upon a single navigation system, not augmented GPS, which is vulnerable to interruption at any time. This lends additional credence to the recommended practice of using all available means of navigation and not relying upon any single system.

9.6.2. Overview of WAAS Project

Among the GNSS elements that will be standardized by the ICAO board, the complements to GPS SBAS and Ground-Based Augmentation System (GBAS) are as follows:

1. Regional SBAS – The complementary information is diffused by satellite by means of a pseudo-GPS signal and covers a wide geographical area, such as Europe and the USA.

2. Local GBAS – The complementary information is valid over a limited area (airport or town) and is diffused by VHF or can be used as well as MSS.

On this basis, the FAA launched the development of regional WAAS, whose objective was ultimately to provide one single navigation system over the whole of the US territories. The WAAS program is a combination of ground and space-based equipment to augment the standard positioning service of the GPS. It is being designed as a milestone of the next generation civil aviation CNS service. The fundamental mission of WAAS is to provide a primary means of navigation for en-route, terminal, non-precision and precision approach phases of flight in the NAS. At this point, the functions being provided by WAAS are: differential GPS corrections (to improve accuracy), integrity monitoring (to ensure that errors are within tolerable limits with a very high probability and thus ensure safety) and ranging (to improve availability). Therefore, separate differential corrections are broadcast by WAAS to correct GPS satellite clock errors, ephemeris errors and ionospheric errors. Ionospheric corrections are broadcast for selected Ionospheric Grid Points (IGP), which are lattice points of a virtual grid of lines of constant latitude and longitude at the height of the ionosphere. The WAAS system integrates the Space Segment of leased Inmarsat-3 GNSS satellite payloads and the Ground Segment, which consists in a primary and secondary GES or Navigation Earth Station (NES), Wide-Area Reference Stations (WRS), Wide-Area Master Stations (WMS), Terrestrial Communications Subsystem (TCS) and GEO satellites to attain improved availability, accuracy and integrity beyond the standard GPS system.

The Mitre Corporation's Centre for Advanced Aviation System Development (CAASD) at the request of the FAA Satellite Navigation Program Office performed WAAS feasibility studies including participating in the early flight tests of an experimental WAAS in the early to mid 1990's. Working with the FAA Technical Centre, CAASD helped to analyze the accuracy of the experimental system when it was used to provide guidance to aircraft performing approaches on the East and West coast of the US. Feasibility studies included a few performance tests of alternative ionospheric correction and integrity algorithms (error boundings). Besides, CAASD helped to establish performance demands for the WAAS and provided technical data to the FAA team that evaluated contractor responses to the WAAS Request for Proposal (RFP). After the contract award, CAASD assisted in the transfer of technology to the prime contractor. The CAASD has also provided technical advice to the FAA on the WAAS design in the areas of performance and safety since the contract award. The CAASD has also been involved in the modeling and simulation of WAAS availability performance. The CAASD's SBAS Worldwide Availability Tool (SWAT) has been used in sensitivity analyses to help determine the optimal mix and location of land resources (such as WRS and GEO) and to determine the impacts of design changes that alter equipment performance or location. Furthermore, it has also been used as a tool to demonstrate to air traffic planners the behavior of a space-based navigation system (i.e., orbiting sensors) and to help to determine operational strategies for dealing with low performance areas.

In the period between 1993 and 1994, the FAA established the National Satellite Test Bed (NSTB) that included the FAA tech centre in Atlantic City, 18 NTSB reference stations in the USA and 3 in eastern Canada and a ground uplink centre in Southbury. The first trials were a joint effort between the FAA and Canada. Since 1996, successful airborne flight tests were conducted in eastern Canada by NAV Canada aircraft, while in 1997 the NTSB began broadcasting an WAAS-like signal into space. In 2000, the FAA organized a WAAS Integrity Performance Panel (WIPP) to assess the feasibility of achieving GNSS Landing System (GLS) performance from WAAS. The CAASD was involved in performing numerous analyses and trade studies in support of the WIPP GLS efforts and helped determine the feasibility of and to define an implementation roadmap to GLS.

The FAA initiated a 3-phase developmental approach to complete the WAAS network:

1. Phase 1 (1998–2002) – Started with initial WAAS commissioned with 25 WRS, 2 WMS uplinks and 2 GEO satellites. The WAAS enables NAS-wide en-route navigation, terminal navigation and NPA. It also supports Category I PA within a limited coverage area.

2. Phase 2 (2003–2007) – Will initiate with full WAAS, expected with additional WRS, WMS and GEO birds. Redundant coverage of the entire NAS-Initial WAAS operational restrictions will be removed. The national LAAS ground structures will be deployed at approximately 150 airports. Precisely surveyed ground stations with multiple GPS receivers and processors will be established, including one or more pseudolites and VHF data link-supports Category II/III precision approach at all runway ends of an airport. An additional approach lighting system will be deployed, the Cat I precision approach, where outside WAAS coverage or where higher availability is required than WAAS can provide. The reduction of ground-based NAVAIDS (VOR and NBD) will be initiated and the added 2nd and 3rd civil frequencies will improve GPS accuracy, availability and robustness.

3. Phase 3 (2008–2015) – To continue reducing ground-based NAVAIDS, when VOR and DME will support only operations along principal air routes and non-precision approach at many airports. Thus, the ILS will support precision approach at high-activity airports. Full constellations of GPS satellites with 2nd and 3rd civil frequencies available for WAAS and LAAS have to be modified accordingly to: Dual-frequency avionics to mitigate unintentional jamming and complete phase-out of all on-airport NAVAIDS (VOR, NDB).

The WAAS, as a GPS satellite-based system that provides differential GPS corrections and integrity warnings throughout the entire continental US and other regions, will provide better accuracy, integrity and availability for en-route, terminal and Category I precision approach guidance through this wide area. In this sense, the Litton PRC team has awarded a contract to combine a cost benefit analysis for the FAA WAAS program, which will be used to justify the FAA's investments in ground and space equipment and will include cost

Figure 9.22. WAAS Network Configuration

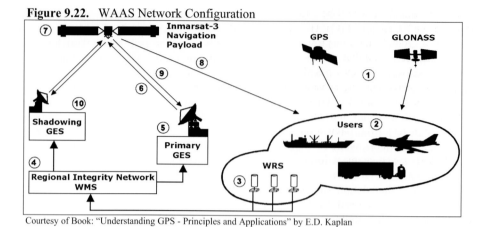

estimates of avionics devices and the benefits to all NAS users. Rannoch will lead this analysis for Litton PRC, working with a few team members. The study builds on previous Litton PRC team research to determine the costs and benefits for the Category II/III LAAS.

9.6.3. WAAS System Configuration

The WAAS was designed as the future primary means of satellite navigation for en-route aviation travel, NPA and selected PA and was intended to provide the following services:
1) The transmission of integrity and health information on each GPS or GLONASS satellite in real time to ensure users do not use faulty satellites for navigation. This feature has been called the GNSS Integrity Channel (GIC).
2) The transmission of ranging signals in addition to the GIC service, to supplement GPS, thereby increasing GPS signal availability. Increased signal availability also translates into an increase in RAIM availability, known as Ranging GIC (RGIC).
3) The transmission of GPS or GLONASS wide area differential corrections, in addition to the GIC and RGIC services, to increase the accuracy of civil GPS and GLONASS signals. This service is called Wide Area Differential GNSS (WADGNSS).
The combination of the overlay services is referred to as the WAAS, shown in **Figure 9.22.** As observed in this figure, mobile users **(2)** receive navigation signals **(1)** transmitted from GPS/GLONASS satellites. These signals are also received by WRS of integrity monitoring networks **(3)** operated by governmental agencies. The monitored data are sent to a regional Integrity Network Central Processing Facility of WMS **(4)**, where the data is processed to form the integrity and WADGNSS correction messages, which are then forwarded to the GES **(5)**. At the GES, the spread spectrum navigation signal is precisely synchronized to a reference time and modulated with the GIC message data and WADGNSS corrections. This composite signal is transmitted to a satellite on the C-band uplink **(6)**. On board the Inmarsat bird **(7)**, the navigation signal is frequency-translated to the mobile user on L1 **(8)** and to the C-band **(9)**, which is used for maintaining the navigation signal timing loop. The timing of the signal is done in a very precise manner in order that the signal will appear as though it was generated on board the satellite as a GPS ranging signal. At all events, the shadowing GES **(10)** functions as a hot standby in the event of failure at the primary GES.

Therefore, the WAAS will be used as a primary means of navigation during all phases of flight through Category I (Cat I) PA, with greater efficiency with respect to minimizing the number of ground facilities providing intercontinental coverage. For PA, it makes use of a differential technique that provides separate corrections for satellite clock error, ephemeris error and ionospheric delay based on observations from a minimally-distributed network of WRS. The corrections are broadcast via special transponders on GEO satellites in the form of digital data contained in GPS-like signals. WAAS will prove enormously beneficial to both the FAA provider and users, enabling the FAA to eventually decommission a vast inventory of terrestrial navigation aids, (VOR, DME) at significant savings in operations and maintenance cost, while serving as a critical enabling technology for free flight.

The WAAS space segment formally consists in the 24 operational GPS satellites and of two Inmarsat GEO satellites. The GEO satellites downlink the data to the users on the GPS L1 RF with a modulation similar to that used by GPS. Information in the navigational message, when processed by a WAAS Rx, allows the GEO satellites to be used as additional GPS-like satellites, thus increasing the availability of the satellite constellation. In this sense, the WAAS signal resembles a GPS signal originating from the Gold Code family of 1023 possible codes (19 signals from PRN 120-138). The WAAS signal is transmitted on RF L1 or 1575,42 MHz and is modulated using a spread-spectrum code thus providing a WAAS pseudo range measurement. These ranging signals improve Dilution of Precision (DOP) and Receiver Autonomous Integrity Monitoring (RAIM). In addition, the WAAS signal is modulated with a 250 b/s data message containing GPS satellite health information, vector position corrections and ionospheric mapping terms. This data message separates WAAS from normal GPS by increasing integrity (satellite health), improving the accuracy (vector corrections) and improving the availability (additional pseudo range) of the system. Therefore, WAAS improves accuracy in two ways by reducing the range measurement error by sending differential corrections for each satellite and by adding new ranging signals thereby improving the geometry.

The vector corrections include fast corrections containing the satellite clock error; long-term corrections containing more slowly varying errors such as the satellite location and ionospheric corrections (Van Dierendonck). The fast corrections are sent every 10 to 12 seconds and only one correction per satellite is sufficient for the entire WAAS coverage area. Long-term and ionospheric corrections are sent much less frequently (about every 2 minutes) as they do not vary much over the entire WAAS network. The GEO satellites reduce the need to update these corrections as rapidly as normal GPS satellites. The CAASD SWAT system, as a base for the WAAS project, is a Markov-process model that calculates a variety of statistical measures of required navigation service at locations and times of interest. Thus, the SWAT inputs include a model that accounts for the relative geometry between satellites, GES and aircraft; error models for known inaccuracies in DGPS user position solutions (based on environment and equipment) and the failure and restoration statistics for the space-based resources. SWAT can also model deterministic scenarios in which the subsystem failures are predetermined. As a resource analysis tool, the SWAT can help to determine rough-order-magnitude costs for fielding WAAS as part of GNSS, to perform sensitivity analyses for optimal mixes and locations of augmentation resources and to determine impacts of mid-lifecycle design changes that alter equipment performance or location. As an operational tool, the SWAT can be used by air traffic planners to determine operational strategies and procedures for dealing with lower performance areas as an alternative to adding space or ground resources.

Early in 1997, the SWAT was chartered by the FAA to help with all operational strategies and procedures dealing with the critical nature of satellite navigation, due to the inclined orbits of GPS satellites. One of SWAT's earlier uses was to validate the cost feasibility of wide-area DGPS in the USA. Initial results confirmed the utility of WAAS and sensitivity analyses over the last several years have been used by the FAA to optimize WAAS resource location, to validate the WAAS contractor system design and to assess trade-offs between GPS, WAAS and LAAS (Local Area DGNS).

The trial of signal availability was performed at the end of August 2000. After a successful 21-day stability test of the WAAS signal in space, the US DOT's FAA declared that it is available for some aviation and all non-aviation uses. The test demonstrated that the system can operate without interruption, providing a stable and reliable signal to augment GPS. For non-aviation users, the signal supports a variety of applications in recreation, boating, agriculture and surveying. Using this chain, the WAAS navigational message improves the GPS signal accuracy from 100 metres to approximately 7 metres. The system provides 1–2 metres horizontal accuracy and 2–3 metres vertical accuracy throughout the contiguous US. Raytheon will operate the system for the FAA on a continuous basis, interrupting it only as necessary to upgrade or test the system. Unfortunately, to receive a WAAS signal, a GPS receiver must be capable of receiving and decoding WAAS. In some cases, the GPS-based receiver may be upgraded, using a special WAAS software modification.

9.6.3.1. WAAS Space Segment GEOSAT

The current WAAS GEO communications satellite GEOSAT constellation consists in two Inmarsat-3 satellites, leased from Inmarsat by contract in January 1997, through September 2006. The non-optimal coverage area provided by these satellites results in a potential single thread of failure for the entire WAAS system in the eastern four-fifths of the CONUS. The two Inmarsat-3 POR and AOR-W satellites are equivalent to one optimally placed satellite between the orbital slots of $119°$ and $129°$ West. The FAA is pursuing the Independent Review Board (IRB) recommendations regarding an Inmarsat satellite used by the WAAS ground network. The studies performed on the best satellite configuration for the FAA were presented to the IRB. They found that the greatest near-term risk to system availability is the current dependence on only two Inmarsat satellites (called bent-pipes) for the transmission of integrity and ranging corrections. These Inmarsat-3 satellites have poor ranging accuracy and vulnerable uplinks. In the case of one of them failing, about half of the US will lose coverage until service is restored or a spare satellite is employed.

They also found that the optimal long-term configuration is four satellites with autonomous navigation payloads, broadcasting all three civil GPS signals. The IRB recommends that an additional, "bent-pipe" satellite be procured as soon as possible as an interim measure. At the same time, planning and initial actions will be undertaken to deploy the final 4-satellite WAAS constellation, with autonomous payloads that broadcast all three civil signals. With regard to the IRB recommendation to procure an additional bent-pipe satellite as soon as possible as an interim measure, the FAA is actively pursuing the acquisition of a near-term additional GEOSAT service for on-orbit capability in early FY04. The benefits of the 3rd GEO include additional availability and coverage, incorporation of L5 signal capability for increased performance and interference mitigation and decreased failure of existing Inmarsat satellites. Potential near-term GEO-hosted opportunities are limited for on-orbit operations in FY04 and include the following satellites: NIMIQ-II, Inmarsat I-4 and ANIK

Figure 9.23. WAAS Ground Infrastructure

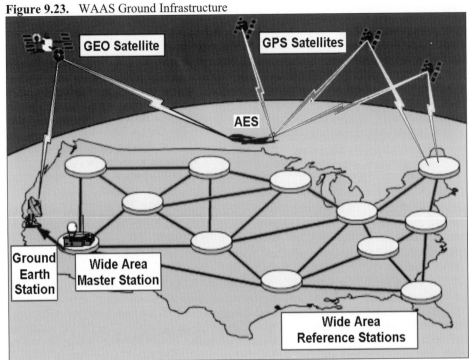

GEO Satellite

GPS Satellites

AES

Ground Earth Station

Wide Area Master Station

Wide Area Reference Stations

Courtesy of WebPages: "WAAS" from Internet

F3. Due to the long lead times required for satellite payload development, this effort required expenditure of funds in October 2001. "Lost" GEO opportunities include: NASA TDRS H I/J, Navy UHF-11, NOAA GOES M/N/O/P and ANIK F2. The FAA WAAS models show increased coverage and availability in Maine, Washington, Southeast Alaska, Southern Florida and Southwest California with the addition of the optimally-placed GEO. Therefore, the status of the long-term solution, with regard to the IRB recommendation to deploy a final four-satellite WAAS constellation, is that the FAA is pursuing the acquisition of GEOSAT service capability to replace the Inmarsat-3 constellation at the end of its lease period. These additional satellites will supplement the near-term GEO acquisition and, integrated within the WAAS architecture, will provide enhanced performance availability and large coverage.

9.6.3.2. WAAS Ground Segment

The WAAS service corrects GPS signals from the 24 orbiting GPS satellites, which can be in error because of satellite orbit and clock drift or signal delays caused by the atmosphere and ionosphere, or can also be disrupted by jamming. The WAAS network in the USA is based on approximately 25 WRS, covering a very large service area and monitors GPS satellite data, see **Figure 9.23.** Signals from GPS satellites are received and processed at 25 WRS, which are distributed throughout the USA territory and linked to form the WAAS network. Thus, each of these precisely surveyed reference station receives GPS signals and

Figure 9.24. WAAS Ground Infrastructure

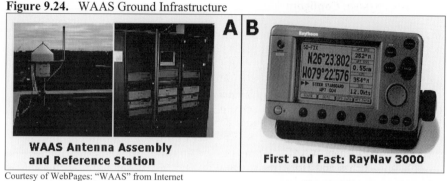

| WAAS Antenna Assembly and Reference Station | First and Fast: RayNav 3000 |

Courtesy of WebPages: "WAAS" from Internet

determines if any errors exist, while two master stations collect data from these reference stations, assesses signal validity, compute corrections and create the WAAS correction message. Data from the WRS are forwarded to the WMS, which process the data to determine the differential corrections and bounds on the residual errors for each monitored satellite and for each IGP. The bounds on the residual errors are used to establish the integrity of the ranging signals. The corrections and integrity information from the WMS are then sent to each GES and unlinked along with the GPS navigation message to the GEO. The GEO downlinks this data to the users via the GPS L1 frequency with GPS type modulation. Therefore, the message is broadcast on the same frequency as GPS to receivers that are within the broadcast coverage area of the entire WAAS network. These two GEO communications satellites also act as an additional navigation constellation for the aircraft, thus providing supplemental navigational signals for position determination. Otherwise, each satellite covers a part of the hemisphere, except for both polar regions. The user receiver, installed aboard a boat, ships, land vehicles or aircraft, combines the GPS signals with the WAAS message to arrive at a more accurate position. Otherwise, each WAAS ground-based station or subsystem configuration communicates via the terrestrial landline TCS infrastructure.

The WRS is a special ground reference station with antenna and adequate equipment located at a precisely surveyed position, as shown in **Figure 9.24. (A)**. As already discussed, the current WAAS system uses only 25 WRS spread over the entire territory of the Continental US and eastern Canada. The WRS infrastructure continuously receives and collects GPS data for various satellites and then sends the data to their WMS, which interpret the data from each WRS and calculate the errors and health of each satellite. Two WMS calculate observed satellite errors for the entire WAAS system and then forward these corrections to a primary GES. The GES receives GPS corrections and transmits them to GEO satellites using a ground uplink system on the GPS L1 frequency. The GEO WAAS satellite receives corrections and forwards them to users, who are equipped with special Rx equipment, as shown in the Raytheon GPS/WAAS Rx 2 in **Figure 9.24. (B)**. The WAAS satellite signal type is compatible with GPS, so new WAAS-enhanced GPS receivers will not be much more expensive than GPS receivers (possibly 50 US$ or more). Some type of GPS receivers or chart plotters can be upgraded with WAAS software without additional cost, by contacting the manufacturer to be converted for WAAS signal utilization. This promises to be the best electronic navigation system for recreational users or small ships, with GPS a close second.

9.6.4. LAAS System Configuration

The Local Area Augmentation System (LAAS) is intended to complement the WAAS service using a single differential correction that accounts for all expected common errors between a local reference and users. The LAAS will broadcast navigation information in a localized service volume of approximately 30 Nm via VHF radio system, with the potential to use the satellite service as well. This service volume would typically encompass a specific airport or airports within close proximity. Although the service volume of LAAS is much smaller than WAAS, it provides greater accuracy than WAAS and should enable PA service beyond the capability of WAAS. This service includes all Cat I PA requirements (higher availability than WAAS), Category II instrument approaches and Category III instrument approaches and landings. In addition, the quality and accuracy of the LAAS signal should provide new service, such as airport surface navigation and sensors for ADSS in low visibility and its architecture will permit the inclusion of other GNSS elements, such as the GLONASS, when deemed to be operationally acceptable. Thus, the LAAS will operate independently of WAAS while being fully compatible with it and enabling satellite-based backup to the WAAS service within the US.

The LAAS Ground Facility (LGF) will be modular in design to accommodate CAT I through CAT III requirements and will include 4 Reference Receivers (RR), RR antenna pairs, redundant VHF Data Broadcast (VDB) feeding a single VDB antenna and equipment racks. Thus, these sets of equipment can be installed on airport property, where LAAS is intended to provide service. The LGF receives, then decodes and monitors GPS satellites information and produces correction messages. To compute corrections, the ground facility calculates position based on GPS and then compares this position to their known location. The LAAS architecture includes a ground station with multiple receivers, pseudolites and a VHF radio navigation data link. The location of the ground station is carefully selected to ensure optimum performance. The LAAS ground station compares the position calculated using its GPS receivers to the actual surveyed position of the station. The difference between these positions is then placed in a correction message, which is broadcast to all users within a 20 to 30 mile radius. The output of the LAAS Rx will provide precise positioning indications to a cockpit display subsystem and/or to a surveillance subsystem. The Airport Pseudolites (APL), which are ground-based Tx of GPS-like signals, will be selectively used at airports to provide additional signals to meet availability requirements.

9.6.5. Benefits and Evaluation of WAAS System

The current radios are based on 1960s technology. In fact, there is no radar coverage over the ocean areas, so pilots must report their positions verbally or have them automatically sent through a relay station. For the controller, surveillance equipment, primarily radar, detects the position of the many moving aircraft. The radar monitoring the surface movement of aircraft and other vehicles spins much faster than those radars covering en-route and terminal airspace. New tools, like satellite surveillance, have been developed as part of WAAS combined with surface radars, to help the controllers move more aircraft safely through the system. This additional navigational accuracy now available in the cockpit will be used for other system enhancements and for surface control. This is the Automatic Dependent Surveillance System (ADSS), currently being evaluated by the FAA and airlines, which is taking advantage of this improved accuracy of flight control.

Another cornerstone of the future for the FAA is improved navigational information available in the cockpit. The use of GPS will become more widely accepted. The WAAS will supplement GPS and provide pilots with the accuracy they need for most flights. This improved accuracy helps the pilots to know their positions, which increases the safety of flights. WAAS also enables improvements in efficiency, by providing access to more runways in poor weather, due to the precise navigation service it provides. The LAAS will provide localized services for final approaches in poor weather conditions at major airports in the USA. Airports that require LAAS will be most of the top 100 airports in the US and a few selected other locations that need the local signal due to other technical reasons.

Currently, WAAS coverage is only available in North America and access to the signal is free. The user may not be able to receive the signals if trees or mountains obstruct the line of sight to the satellites over the equator. The WAAS signal reception is ideal for open land and marine applications. The WAAS provides extended coverage both inland and offshore compared to the maritime and land-based DGPS. WAAS does not require additional receiving equipment, while DGPS does. Since 2001 WAAS infrastructure is available for recreational boating or small ships and as supplementary navigation for aircraft. The system is not expected to be approved for precision landings for aircraft until 2003. It is designed for aircraft landing and has positional accuracy as good as 3–7 metres, several times more accurate than available GPS navigation systems and better than Cat I PA standards for aircraft. The system was developed for the safety of precision aircraft landing, where a 10 metre position error would miss the runway.

Prior to the commissioning of the WAAS for aviation for the Instrument Flight Rule (IFR), the FAA will conduct a series of activities including developmental testing and evaluation and operational testing and evaluation of the system. When in operation, WAAS will provide pilots with en-route navigation and vertical guidance for PA to runways over a limited portion of the continental US. In the meantime, a signal capable of supporting non-safety applications, such as an aid to Visual Flight Rule (VFR) flight, is currently available. The FAA is moving directly to a Lateral Navigation/Vertical Navigation (LNAV/VNAV) capability using WAAS. This capability will facilitate improved instrument approaches to include vertical (glide path) guidance to an expanded number of airports. Additionally, procedures can be put in place that can be used now, to reap early benefits. Thus, expected achievement of LNAV/VNAV capability is in 2003. Concurrently, the FAA will evaluate the approach to achieve GNSS Landing System (GLS) capability in later years. To guide this activity, the FAA has enlisted the support of experts in the field of satellite navigation. This group, known as the WAAS Integrity and Performance Panel (WIPP), will advise the FAA on the most cost-effective and expedient solutions as the FAA progresses through the more challenging aspects of WAAS implementation.

Planned expansion of the US-based WAAS network will include Canada, Iceland, Mexico and Panama and has the potential to expand to other countries as well. At the same time, Japan and Europe are building similar systems, known as MTSAT/MSAS and EGNOS, respectively, which are planned to be interoperable with the US WAAS. The merging of these systems will create an integrated GNSS/GSAS CNS worldwide seamless navigation capability similar to GPS but with greater accuracy, availability and integrity.

Additionally, the FAA is involved in the ICAO's GNSS Panel (GNSS-P), which supports the development of standards and procedures for satellite navigation for civil aviation applications worldwide.

10

GLOBAL STRATOSPHERIC
PLATFORM SYSTEMS (GSPS)

Global Stratospheric Platform Systems (GSPS) are the newest space technique with top technologies for fixed and mobile applications, including military solutions. These systems are using unmanned or manned aircraft and on solar or fuel energy airships and carrying payloads with transponders and antenna systems. With a few very cheap remote controlled and solar powered airships as a better solution, a territory can be covered of some region or country including urban, suburban and rural areas, farms and other environments with a low density of population. There are four general telecommunications architectures, which can be used to deliver broadband wireless local loop service to consumers. Two of these architectures are space-based GEO and Non-GEO satellite systems and the other two are terrestrial rooftop cellular-like millimetre wave repeaters and stratospheric relay platforms. The GSPS network offers better solutions than all cellular systems, with greater speed of transmission than even optical modes, roaming will be better, without shadowing problems and disturbances inside buildings and service will cost less. The GSPS mission can be integrated with current satellite and cellular systems; the system is more autonomous and discrete and will be the best solution for military and all mobile applications. For instance, the Halo Broadband GSPS Millimetre Wavelength (MMW) Network of the Angel Company provides data densities nearly one thousand times higher than proposed satellites, see **Table 10.1.**, while having round trip time delays appropriate for interactive broadband services. Whereas, the delays through satellite network nodes, even through LEO satellite nodes, are too long for many interactive applications, delays are 25 or 1,000 times longer for LEO or GEO then for Halo Networks, respectively. In fact, the Halo parameters are similar to a variety of metropolitan environment spectrum bands of the Local Multipoint Distribution Service (LMDS) band near 28 GHz.

Table 10.1. Comparison of Data Density and Signal Delays

Node Type	Node Data Density		Round Trip Delay	
	Min (Mb/s/km^2)	Max (Mb/s/km^2)	Min (millisec)	Max (milieu)
LMDS	3	30	0.003	0.060
Halo	2	20	0.10	0.35
LEO (Broadband)	0.002	0.02	2.50	7.50
GEO	0.0005	0.02	200	240

10.1. Aircraft GSPS

The new aircraft projects offer cost-effective systems for GSPS by using special unmanned and non-fuelled solar-cell powered planes with an estimated endurance of several months and piloted aircraft with fuel engine propulsion for operating on a daily basis. This system will be more effective and reliable will confirm future practical use of systems such as: General Atomic AVCS Network; Halostar (Halo) Network; Heliplat (HeliNet) Hale Network; SkyTower Global Network and other, forthcoming solutions.

10.1.1. General Atomic AVCS Network

 Building upon its worldwide leadership position in the new design, manufacture and deployment of Unmanned Aerial Vehicles (UAV), General Atomics AVCS is developing the next generation of Wireless Stratospheric Communications delivery technology. Thus, by using a stabilized aerial platform as the electronic interface between mobile, car-mounted, or home/office-based wireless local loop subscribers and the necessary cellular infrastructure, it is possible to minimize the requirement to build any cell sites, towers and other related, very expensive components of typical cellular or Personal Communications Service (PCS) systems. Eliminating these costs and increasing geographic coverage by using this new system can make serving many low-density and rural areas cost-effective and can be deployed in a fraction of the time it takes to build PCS.

10.1.1.1. Aerial Vehicle Communications System (AVCS)

The General Atomics Company is working on a special project of GSPS named an Aerial Vehicle Communications System (AVCS), see **Figure 10.1**. This system consists of the space segment of UAV as a high-altitude aircraft carrying payload and the ground segment. The payload is equipped with corresponding transponder and antenna systems, while the main parts of the ground segment are user terminals and AVCS ground stations. The AVCS user terminals can be as usual fixed, mobile and handheld equipment roaming within the radiation of the aircraft spot beam antenna system. The users calls are routed through the aircraft's transponder and AVCS ground station to terrestrial subscribers and vice versa. The ground station is actually Base station equipped with suitable transceivers for receiving and sending signals from and to aircraft and with facilities to control the aircraft's position and its payload equipment.

Figure 10.1. General Atomic AVCS Network

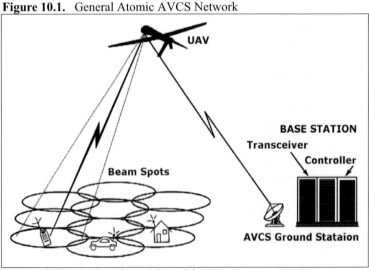

Courtesy of Webpage: "General Atomic System" from Internet

Figure 10.2. Prototype of High Altitude Aircraft Altus

Courtesy of Webpage: "General Atomic System" from Internet

Therefore, everyone in a position under the CDMA coverage umbrella can communicate using the latest CDMA air interface from subscribers unlinking to the UAV and down linking to the ground station, which delivers the signals to the normal cellular or other system signal-processing platform. At the same time, the base station equipment controls the flight of UAV plane to ensure that it stays in the proper relationship to the coverage areas on the ground, allowing the on-board antennas to do their work. The UAV carries sophisticated guidance control equipment that utilizes the GPS system to maintain its flight integrity. Traffic from the PSTN simply uses the reverse path to be connected.

10.1.1.2. The Altus High-Altitude Aircraft

The research team of the General Atomic Company together with the NASA, the US Navy and the Department of Energy (DOE) developed a prototype high-altitude aircraft, Altus, for the future AVCS Network project, shown in **Figure 10.2**. This stratospheric aircraft will be ideal for Telecommunications Relay, Cellular Relay and Commercial Applications. Operational with NASA and the DOE, the Altus was at first deployed in support of atmospheric research for the DOE, with future plans to use the high altitude capabilities to further atmospheric research, understand the genesis of and predict hurricane paths and damage potential, as well as many other advanced scientific applications.

Using the proven technology of previous GNAT and Predator aircraft there were two aircraft prototypes developed: Altus I for operations at about 13.70 km and Altus II for operations at 19.80 km. Both aircraft are using fuel as a motive power, which is a disadvantage in comparison with aircraft using solar energy. The wingspan of the aircraft is 16.76 m, the length is about 6.71 m and they can be equipped with a special large payload capacity of about 148.5 kg to carry transponders and antenna systems.

10.1.1.3. Scope of the AVCS System

Putting the array of antennas in stationary orbit high enough to avoid commercial traffic and almost all adverse weather conditions can dramatically increase the amount of ground coverage for many wireless systems. Thus, a satellite-based system obviously can cover the largest territory because they are hundreds of miles above the Earth. They are also very expensive and today require unique subscriber terminals and very sophisticated electronics to deal with the signal propagation delay caused by their distance from Earth. In this sense, the AVCS project will eliminate in the near future some of the mentioned and other technical problems caused by terrestrial or satellite communication systems.

In addition, launching a communications satellite is an expensive and lengthy process. The AVCS can accomplish broader coverage, increased capacity and deeper penetration into low-density areas at a fraction of the cost of satellite-based services and without the requirements for cell sites, towers, backhaul facilities or microwave. Thus, if somebody would like to find out how big an area can be covered by a complement of two aircraft, which can also provide backup for each other, it is necessary to imagine a sample coverage overlay for the San Francisco Bay area. This area is limited by Bodega Bay to the north and Monterey Bay to the south, covering a great part of the open sea and continental territory up to the boundary line of Brentwood, which is parallel with the coastline. This is pretty impressive, especially since covering the same area with individual cell sites would require dozens of real estate parcels, obligatory permits, power systems, land lines, or microwave for backhaul or intercellular connectivity, towers and maintenance for all of the above.

10.1.2. SkyTower (Helios) Global Network

 AeroVironment officially formed SkyTower, Inc. in October 2000 to pursue commercial communication opportunities enabled by its own proprietary unmanned solar-electric plane. On the other hand, the SkyTower team, together with the leading telecommunications system providers. started to validate the technical and economic viability of the SkyTower stratospheric platform for multiple applications and begin the development of commercial systems.

SkyTower conducted successful symposiums in Tokyo, Japan in February 2001 and Taipei, Taiwan in March 2001, introducing the SkyTower technology and business opportunities to industry and government representatives. The company executes agreements with multiple telecommunications service providers to jointly assess and pursue opportunities in their respective regional markets. Thus, SkyTower held its first Board of Advisors meeting in April 2001. Advisory board members include current and former executives from leading telecommunications and aerospace companies such as Cisco, Global Crossing and Loral.

10.1.2.1. Development of the SkyTower Aircraft

The first unmanned, high altitude solar-electric aircraft designed under the NASA project was Pathfinder, developed in 1995 with a wingspan of about 29.87 m which flew to 15.39 and 21.79 km in 1995 and 1997, respectively, see **Figure 10.3. (A)**. A second modified Pathfinder program, known as Pathfinder-Plus, with a bigger wingspan of 36.88 m, flew to 24.44 km, higher than any other propeller-driven aircraft. This record flight was the 39th consecutive successful flight test of the Pathfinder platform. After climbing to this altitude above the Hawaiian island of Kauai, the unoccupied Pathfinder-Plus transmitted several hours of third-generation mobile voice, data and video service to the ground, where it was received on a standard NTT DoCoMo 3G-video handset, shown in **Figure 10.3. (B)**. At all events, building on Pathfinder's successes, so AeroVironment introduced a next-generation

Figure 10.3. SkyTower Aircraft Pathfinder and NTT DoCoMo 3G-handset

Courtesy of Webpage: "SkyTower System" from Internet

aircraft with a wingspan of 62.78 m, called the Centurion, which was test flown in 1998 at Edwards Air Force Base in the US. The wingspan was then further extended to 75.28 m and the previous model of aircraft was renamed the Helios prototype.

The Helios prototype, which successfully completed initial low altitude flight-testing at Edwards Air Force Base in 1999, was then equipped with high efficiency solar cells and underwent high-altitude flight-testing in the summer of 2001 in Hawaii at the U.S. Navy Pacific Missile Range Facility in Kauai. On 13 August 2001, on its second high altitude flight, Helios flew to 29.52 km, shattering the world altitude record for both propeller and jet-powered aircraft (the SR-71 spy plane was the previous record holder, having flown to 25.93 km in July, 1976). Its 18-hour flight duration on that day is a record for solar-powered aircraft, thus setting the stage for continuous multi-day/month flight upon integration of the fuel cell energy storage system already under development. Namely, AeroVironment and NASA officially started to work on light fuel cell technology that will allow the aircraft to retain enough power from the day's sunlight to remain aloft at night. Until this time, the aircraft takes off after dawn and glides back to Earth in the evening after the sunlight is gone. In February 2002 was NASA's milestone bench test of complete day/night cycle operation of the regenerative fuel cell energy storage system. At this point, in a major milestone meeting NASA performance requirements, engineers from NASA and AeroVironment successfully completed functional tests of a prototype regenerative fuel cell energy storage system for the Helios aircraft in April of 2002.

In June and July of 2002, SkyTower/AeroVironment, in collaboration with the Japanese Ministry of Post and Telecommunications (CRL/TAO) and NASA, successfully completed several wireless tests in Kauai, the world's first commercial applications transmitted from over 18 km in the stratosphere. The two applications tested, HDTV and 3-G mobile, further validate the viability of the SkyTower's unmanned High Altitude Platform Station (HAPS) for use by wireless service providers for a broad range of telecommunications applications. The final test, a telecommunications demonstration scheduled for mid-July, provided an opportunity for participants to see this breakthrough technology first-hand.

The commercial version of Helios, which is the ultimate evolution of the Pathfinder and Helios Prototypes, will incorporate a fuel cell energy storage system to provide power for flying throughout the night. It will be capable of continuous flight for months at a time at altitudes of 15–22 km. A full-sized fuel cell and electrolyzed energy storage system for the

Figure 10.4. SkyTower Space and Ground Segment

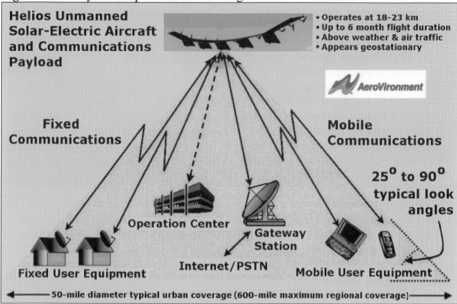

Courtesy of Webpage: "SkyTower System" from Internet

Helios Prototype is now operating in AeroVironment's test facility. As part of the NASA Environmental Research Aircraft and Sensor Technology (ERAST) program, a reduced weight version of this system will be integrated into the Helios stratospheric aircraft to enable continuous multipurpose communications service.

The next step will be in 2003 when the world's first multi-day flight in the stratosphere has to be conducted by Helios aircraft integrated with fuel cell energy system. Therefore, in the course of this experimental flight. Fixed broadband demonstrations from the stratosphere with user speeds in excess of 50 Mb/s will be performed. The last event is in 2004, the launching of the first fixed broadband commercial service.

10.1.2.2. Network System Configuration

For decades, the orbiting Earth satellites have been considered telecommunications' final frontier. In the meantime, a pair of new communications ventures is aiming their sights a lot lower the GSPS orbit solutions, namely AeroVironment is exploring the concept of a flying telecommunications platform, a high altitude aircraft that would blanket a metropolitan area with an array of telecommunications services. Behind the project is the fact that fixed-wing planes are much cheaper than satellites to build, launch and operate.

The SkyTower Space segment is based on AeroVironment's solar electric unmanned airplane technology, developed in conjunction with NASA. SkyTower's business opportunity is to build and operate solar/electric airplane-based telecommunications networks initially for fixed broadband access ("the last mile") service providers only. Thus, other commercial telecommunications applications like 3-G mobile voice and video service, direct broadcast and narrow band voice, will follow.

In **Figure 10.4.** are illustrated SkyTower Space and Ground configurations consisting of Helios unmanned solar-electrical powered aircraft and communications payload operating at a minimum of 18 km in the stratosphere above all traffic and weather influence, with up to 6 months flight duration; and with access links for fixed, semi-fixed, all mobile and ground communications facilities and equipment. The ground segment is covered by the Helios communications payload in diameters of 50 miles for urban areas and of 600 miles for regional roaming, including rural environments. The fixed and mobile communications equipment can be linked via Helios aircraft and Gateway stations to the Internet and PSTN networks. In the proper manner, the communications service in each region is controlled by a dedicated Operations Centre. The summer's payloads on board the aircraft as a part of its "next-generation telecommunications demonstration", will include digital TV broadcast equipment and a kind of cellular telephone technology known as an International Mobile Tele-communications-2000 (IMT-2000), which can carry voice and data, Internet and video images over cellular phones, then voice (VoIP), videoconferencing (VCoIP) and data transmission over Internet Protocols (DoIP).

Like Angel's Halo-Proteus, Helios would soar over a particular region to provide an array of telecommunication services. At the end of its mission, the aircraft would be glided back to an airport for maintenance and refitting. Hence, between landings, the platform operates for up to six months continuously in the stratosphere above the weather and commercial air traffic. The stratospheric platform carries a communications payload that communicates with different kind of users on the ground within a 30 to 600 mile footprint and connects them with a ground Gateway station tied directly to a fibre-optic backbone, or alternatively to GEO and non-GEO satellite constellations, therefore, minimizing the need for building any backhaul infrastructure. In this sense, the broadband capacity per aircraft platform is projected to be 5 Gb/s for the first generation system; moreover, multiple platforms can serve the same area and further reuse the same frequency spectrum. Due to their unique operating position in the sky, the platforms can also share a spectrum with terrestrial and satellite systems.

As was mentioned, the SkyTower platform connects users within its footprint of 30 to 600 miles in diameter to one or two Gateway stations on the ground that can be tied directly into a central switch and fibre-optic backbone of the telecommunications network. This means that for business and consumer subscribers service costs are dramatically lower. For instance, SkyTower, based on recent analyses completed with telecommunications service providers and system developers, projects that the capital cost per subscriber to deploy a fixed broadband system including redundant back-up platforms, is a fraction of the cost of other alternatives such as cable network and satellite infrastructure.

In addition, small, low-cost, stationary user antennas can be used, due to the unique tight turning radius of the aircraft, which makes it appear geostationary from the ground. The platform's closer proximity to the Earth surface enables much higher frequency reuse than satellites, resulting in more than 1,000 times the local access capacity compared to a GEO satellite. Multiple stratospheric platforms can serve the same area, further reusing the same frequency spectrum and multiplying system capacity.

In this way, SkyTower is now teaming with leading communications system operators and regional service providers to pursue global market opportunities and plans to launch the first telecommunications service within three years. Agreements have already been executed with multiple telecommunications service providers to jointly assess and pursue opportunities in their respective regional markets.

10.1.2.3. Broadcast and Broadband Battlefield Communications

AeroVironment and NASA experts have worked together for years in the development of the solar-electric aircraft technology. SkyTower and AeroVironment have been working closely with domestic and international regulatory bodies to obtain the approvals required to provide telecommunications services from a platform in the stratosphere. Authorization was obtained from the FCC to support the telecommunications testing and all flight tests in controlled US airspace are authorized by the FAA. At this point, SkyTower believes that the platform's extremely lightweight and environmentally benign characteristics, combined with its highly efficient use of frequency spectrum and energy, help to facilitate favorable regulatory support. Thus, this company plans to launch the first commercial service, fixed wireless broadband infrastructure and is in advanced discussions with multiple domestic and international service providers interested in SkyTower's revolutionary "last-mile" (local access) solution. Full service could begin in selected areas by sometime in 2004; parts of the system specifications are presented at a glance in **Table 10.2**.

Table 10.2. SkyTower System Specification at a Glance

Operating Altitude and Latitude	18.28 to 21.34 km on all Latitudes
Flight Duration & Max Speed	Up to six months between two landings with 150 m/h
Payload Weight & Power	Up to 227 kg and 5 kW
Flight Control	Redundant unmanned/autonomous system with manual override
Max Coverage	Typical 50 miles in diameter urban and 600 miles regional coverage
Emission & Look Angle	100% non-polluting and up to 90°
Frequency Spectrum	Current ITU allocation of 2; 27-32 and 47 GHz broad frequency range
Broadband Throughput	5 Gb/s per perform 1-G (multiple platform can serve same area)

Therefore, in collaboration with the CRL and TAO study divisions of the Japan Ministry of telecommunications and NASA, in June of 2002 SkyTower successfully organized the two following tests of the aircraft at altitudes of about 20 km:

1. Test No 1: Digital Broadcast High Definition Video (HDTV) – The SkyTower aircraft made 12 orbits on station position broadcasting transmissions with a Toshiba-built airborne HDTV transmitter via UHF channel (FCC Special Temporary Authorization). The transmission signal provided a picture-perfect video broadcast signal to a fixed receiver on the ground, at twice the resolution of conventional broadcast transmissions. Because of its much higher look angle, SkyTower platforms can fill in "urban canyons", coverage areas missed by terrestrial and satellite broadcast transmissions due to tall buildings, terrain and the like and can do so using a fraction of the power. During the tests, a 24 Mb/s data rate was achieved using only 1 W of Tx power, less than 1/10,000[th] of the power used by a typical terrestrial broadcast transmitter that has to overcome all transmissions blocked by tall buildings, trees and other obstructions to cover the same area. SkyTower's local footprint can also help satellite broadcasters to overcome capacity challenges that limit their ability to provide local channels within each market.

2. Test No 2 & 3: IMT-2000 (3-G) Mobile Application – The aircraft was flying over 7 hours on station for both tests with a NEC-built airborne transceiver for the transmission of video using off-the-shelf NTT DoCoMo video handsets (64 Kb/s) and Internet surfing up to 384 Kb/s using a laptop with a wireless modem on a frequency of 2 GHz user links (FCC Special Temporary Authorization). The SkyTower platform will provide wireless service operators with a low-cost infrastructure solution and a broad range of telecommunications applications (launch of a single aircraft provides instant city/region-wide coverage).

Future tests by SkyTower will demonstrate fixed wireless broadband capabilities such as ultra high-speed Internet access at user speeds in excess of 50 Mb/s.

10.1.2.4. Features of the SkyTower System

Unique few features of these SkyTower unmanned solar-electric aircraft that make them appealing platforms for telecommunications applications include:

1) Long flight duration of up to six months or more;

2) Minimal maintenance costs due to few moving parts (each motor has one moving part);

3) High redundancy levels (aircraft could lose multiple motors and still maintain stationary or land safely, most failures do not require an immediate response by a ground operator);

4) Highly autonomous mode enables only one ground operator to control multiple aircraft;

5) Use of solar energy to minimize fuel costs;

6) Tight turn radius which makes the platform appear geostationary from the ground equipment perspective (enables use of stationary user antennas) and enables multiple aircraft to serve the same area using the same frequency spectrum and

7) Flexible flight facility requirements (plane can take off from even a dirt field and in less distance than the length of its wingspan).

Compared with Terrestrial Systems, the SkyTower system offers lower cost in many cases, higher coverage angles, instantaneous coverage, faster and easier deployment, minimal backhaul requirements and the system is easy maintainable, upgradeable and relocatable.

Compared with Satellite Systems, the SkyTower system offers lower and scalable cost of service, higher coverage angles, higher service capacity (a single SkyTower platform can provide over 1,000 b/s per km^2 broadband capacity of a typical satellite), lower power requirements, cheaper launch on position with lower risk. The System is easy maintainable, upgradeable and relocatable and a local footprint with the possibility of regional and global coverage using many aircraft in constellation.

Compared with solar powered airships/blimps, the SkyTower system provides technical viability and station-keeping in winds, meanwhile, compared with jet/piston-powered aircraft, the SkyTower system provides economic viability and reliability/flight duration.

Early target market applications include fixed broadband and third-generation mobile service for high-speed Internet, voice, video and different data rate connectivity. At this point, other high-potential applications will be pursued on a global basis, including direct broadcasts and wireless telephony.

Therefore, SkyTower will be the lowest costing and most flexible wireless communication solution for connecting mobile users, offices and homes to the high-speed fibre backbone in many markets. Other competitive advantages include rapid deployment, easy technology upgrades and station-keeping, minimum infrastructure build out, flexible deployment and redeployment, scalability and the ability to match investment to market demand.

10.2. Airships GSPS

The new airship projects offer cost-effective systems for GSPS by using special unmanned and non-fuel solar-cell powered balloons with an estimated endurance of several months. In comparison with aircraft and airship systems it is difficult now to say which one will be better for the future reliable GSPS. There are several airships such as: SkyStation Global Network, SkyLARK Network, StratCon (StratoSat) Global Network, TAO Network, etc.

10.2.1. Sky Station Global Network

 The Sky Station stratospheric airship system is being built by Sky Station International Inc. (SSI) and an international consortium of companies including Finmeccanica S.p.A. Alenia Aerospace of Italy, Thomson-CSF Communications of France, Scaled Composites Inc. of the USA, Lindstrand Balloons Ltd. of the UK and Spar Aerospace of Canada.

The technology to provide wireless service from the stratosphere is called Stratospheric Telecommunications Service (STS), which will commence with the first Sky Station platform deployment in 2004. Sky Station platforms will be implemented in accordance with user demand, as expressed by responsible organizations in each country. A single Sky Station platform delivering 3G services will be able to support more than three million users in a metropolitan area. In addition, dynamically steerable antennas can automatically and instantaneously reallocate capacity as demand changes throughout the day. With this technology, commuter routes for example, will receive more capacity during rush hours, business districts during the business day, stadiums during games, etc. The system can direct 1,000 variable spot beams into its 400 km diameter service area footprint. There are no issues with tower placement, dead zones, the environment or bureaucratic impediments. Thus, the Sky Station system will offer better line-of-sight and fewer obstacles improve signal quality to the handset. Optically linked Sky Station platforms could transmit voice, data and video at speeds and volumes comparable to fibre-optic cables and provide a radio backbone link between metropolitan areas, countries and even continents, worldwide.

10.2.1.1. System Overview

The Sky Station company, an emerging provider of low-cost broadband Internet capacity via stratospheric platforms, announced on 28 June 1999 the completion of the Platform Conceptual Design Review by Airship Technologies Services, Ltd. Airship Technologies is the world's leading design and manufacturing company in the field of advanced technology airships, with a team who span 27 years of international airship experience and programs.

At this point, Sky Station is developing a technology that represents one of the most significant innovations since the satellite. With Sky Station's unique STS, a system of lighter-than-air platforms will be maintained in geostationary positions over metropolitan areas using advanced proprietary technologies. Worldwide regulatory approval for the use of stratospheric platforms was granted by the ITU in November 1997 and by the US FCC earlier that year. The Word Radio communications Conference (WRC) designated 600 MHz of spectrum in the 47 GHz band for use by Sky Station STS.

Before all, in August 1996 Finmeccanica S.p.A Alenia Aerospazio, Divisione Spazio from Italy and SSI signed an agreement intended to lead to the production of 250 stratospheric telecommunications payloads over the next seven years. The stratospheric payloads will operate in the 47 GHz frequency range and will serve as the backbone for the world's first globally available broadband wireless telecommunications system. Otherwise, the principal application of the new technology is expected to be wireless broadband Internet services. Sky Station broadband Internet user terminals are ultimately expected to cost under 100 US $ and to offer T1/E1 service for as little as a dollar a day and a PCMCIA card can be used to upgrade existing laptop computers to become high-speed Internet terminals.

In May 1997, the FCC approved the use of Sky Station technology in the 47 GHz frequency band. Stratospheric telecommunications technology uses the Sky Stations airship located 21 kilometres above the Earth to benefit American consumers and provide millions of wireless high-speed Internet links directly to consumer laptop and desktop computers. The FCC decision is the first governmental approval ever given for stratospheric telecommunication fixed and mobile services.

The Sky Station system will provide wireless T1/E1 (2 Mb/s uplink and 10 Mb/s downlink) links directly to laptop and personal computers. This kind of high-speed Internet service may be used for portable videophone and Web TV applications. At this point, the SSI company is able to accomplish its personal T1/E1 service because of the ultra-high channel capacity only available in metropolitan areas from a stratospheric altitude.

Then, in September 1999, Lockheed Martin Global Telecommunications (LMGT) received a contract to provide end-to-end systems integration for SSI Development Corporation's STS. LMGT will assist SSI in its plans to provide low-cost broadband Internet capacity via airships. Using advanced proprietary technologies; a system of lighter-than-air platforms will be maintained in geostationary positions over metropolitan areas. LMGT will lead a global team of companies, each of which is recognized as a leader in its field, to develop and deploy this turnkey system. The value of the contract was not disclosed.

10.2.1.2. Sky Station Airship Network

The industrial team building the Sky Station system includes LMGT as the end-to-end system integrator; Alenia as the primary payload developer; Astrium GmbH (formerly Dornier Satellitensysteme GmbH and now a corporate unit of DaimlerChrysler Aerospace of Germany), will be a developer for the power subsystem and the Thales Group (formerly Thomson) of France as a Gateway Earth station manufacturer and payload sub-contractor.

The Sky Station system is an airship network of lighter-than-air platforms, which are held in a geostationary position in the stratosphere approximately 21 km over a major metropolitan area, utilizing proprietary technologies. Each Sky Station stratospheric platform is equipped with a telecommunications payload, delivering a variety of wireless telecommunications services. Sky Station's first deployed service will be the Third Generation (3G) mobile solution, which brings multimedia to mobile phones. On the other hand, remote sensing and monitoring devices can also be installed on the stratospheric platform, providing invaluable continuous data collection.

Sky Station plans to deploy at least 250 Sky Station platforms, one about 21 km above every major city in the world. There may be more around large population centres, such as Tokyo or London and additional platforms can be added at any time to increase capacity over specific regions. Because each platform will utilize the same telecommunications payload and due to the mobile nature of Sky Station service, users will enjoy the 3G wireless services when they are at home, walking or driving.

Sky Station platforms can be of variable size, depending on market demand. The average platform will be approximately 200 m long and 60 m in diameter at the widest point. That makes a Sky Station platform about as wide as a football field and 2 times as long. Flight safety is a major consideration of the Sky Station stratospheric system development effort. Each platform will be constructed to the same exacting aerospace standards as commercial aircraft and satellites. Sky Station platforms will be FAA certified and each individual platform will be inspected and approved prior to launch.

Multiple safety features are being integrated into the entire system, the most important of which is the on-board monitoring system. Installed in each Sky Station platform, it reports back a steady stream of information to the Ground Control Station (GCC). In such a way, this early warning system enables SSI to anticipate problems and take the appropriate corrective actions before they actually occur. Hence, platforms can be recalled for repair if necessary. A new platform will be deployed in advance to replace the existing one, so there will be no interruption of service.

Accordingly, the use of state-of-the-art envelope materials and weaves virtually eliminates the probability of any catastrophic rupture of the main hull. In the unlikely event of a loss of buoyancy, the automated master control system is enabled to safely propel the platform to a water landing. In fact, safety is built into the Sky Station STS system with multiple redundancies to insure safe, continuous STS to users on a global basis.

10.2.1.3. Regional and Global Stratospheric Telecommunication Service

Located in the stratosphere 21 km above the Earth, each SSI platform acts as a very low orbiting satellite, providing high density, capacity and speed of service with low-power requirements and no latency to an entire metropolitan and suburban area-extending out into rural areas. No other existing or proposed technology offers this combination of very high density service and low cost. Spectrum in the 47 GHz band has already been designated globally by the ITU as well as the FCC for use by high-altitude stratospheric platforms, paving the way for planned commercial service to commence in the year 2004.

When deployed over a big city or metropolitan area, a Sky Station platform immediately provides the necessary communications infrastructure to combine voice, mobile data and broadband wireless service to hundreds of thousands of users. The system will cover both suburban and rural areas with the facilities of remote service. Subscribers can transmit directly to the platform, where on-board switching routes traffic directly to other Sky Station subscribers within the same platform coverage area. Traffic destined for subscribers outside the platform coverage area is routed through ground stations to the public networks, or in the next stage, to other platforms serving nearby cities.

All developing nations need low-cost access to high-density telecommunications links to support accelerated economic development and inclusion in the Information Revolution. Sky Station's platforms provide the fastest, easiest and least expensive way to bring advanced services to the developing world. One Sky Station platform alone provides voice service for millions of subscribers at a lower cost than any current or proposed system.

The Sky Station system is also providing different Mobile Solutions of services. The ability to communicate anywhere anytime is an integral part of today's global media and culture. At this point, a worldwide standard (3rd Generation Wireless or IMT-2000) is the next significant advance in a broadband wireless service for mobile, portable and fixed users. The STS system is the ideal means for low-cost rapid deployment of mobile services and the Sky Station system is participating in the development and delivery of 3rd Generation (3G) Cellular services.

Today's telecommunications networks have become stressed by the explosive growth of the Internet. As more users tie up lines for longer periods of time, the usage patterns for which the networks were originally designed have been fundamentally altered. Consumers have become increasingly dissatisfied with the slow speed of current dial-up access and are demanding higher speed solutions.

Figure 10.5. Sky Station Stratospheric Platform Configuration

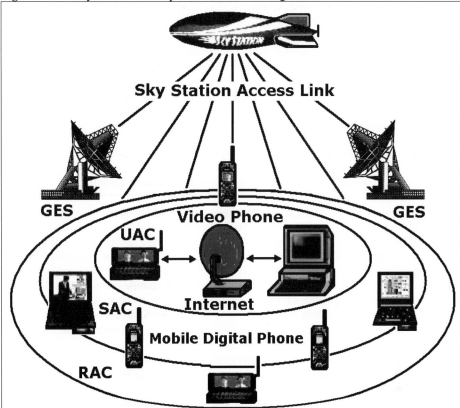

Courtesy of Webpage: "Sky Station Stratospheric Platforms Project" from Internet

Thus, Sky Station satisfies this demand by delivering personal T1/E1 broadband services to the mass market at a lower cost than existing or announced alternatives. With data rates bursting to 2 Mb/s uplink and 10 Mb/s downlink, subscribers will enjoy high speed Internet browsing and hosting, as well as other broadband services, such as video conferencing.

In **Figure 10.5.** is shown the Sky Station Space segment, which comprises one spaceship with access link for three coverage ranges: Urban Area Coverage (UAC), Suburban Area Coverage (SAC) and Rural Area Coverage (RAC). The ground segment includes several GES, mobile digital phone users, videophones, Internet access and E-mail service.

Sky Station customers will use standard, off-the-shelf 3G handsets and devices from any of the leading manufacturers around the world. These 3G handsets can be used for a multitude of services, from Internet browsing and E-mail, video chat/video conferencing, local and long distance telephony, on-line remote monitoring and security. In summary, any mobile multimedia service will be compatible with the Sky Station services.

The STS will provide both regional in the 2 GHz frequency band and global service, in the 47 GHz band. The telecommunications technical details and service characteristics for a typical 2 GHz band Broadband stratospheric platform are as follows:

1) One platform will be placed, with several GES for each metropolitan area.

2) The altitude of each airship will be around 20 to 21 km in the stratosphere.

3) Coverage area has to be about a 1,000 km diameter footprint with >1,000 spot beams per platform, with capacity of 1.77 Gb/s per 10 MHz allocation (5 MHz downlink and 5 MHz uplink) dynamically spread across the footprint and with the equivalent of 316,000 simultaneous 8 Kb/s telephone calls with 50% voice activity.

4) This system will employ frequency spectra in bands identified by the WRC for use with Third Generation Mobile Terrestrial Systems (1885 to 1980 MHz, 2010 to 2025 MHz and 2110 to 2170 MHz in Regions 1 and 3 and 1885 to 1980 MHz and 2110 to 2160 MHz in Region 2).

5) Signal protocols of this system will be QPSK modulation for the subscriber, WCDMA and CDMA2000 multiple access and multiple communications protocols will be supported at transport and network levels.

6) Subscriber information will be transferred at rates of 6.0 Kb/s for voice, 384 Kb/s to 2 Mb/s for data and with power requirements of 25 mW.

In addition, telecommunications technical details and service characteristics for typical 47 GHz band broadband stratospheric platforms are as follows:

a). About 250 airships will be placed worldwide, each operating independently and initially connected via ground Earth stations and existing public networks. Future platforms will be equipped with interplatform communications, namely with platform-to-platform links.

b). The altitude of each airship will be around 21 km in the stratosphere.

c). Coverage area has to be about 150 km in diameter or 19,000 km^2, including specified locations outside this zone with 700 spot beams per platform and angle of elevation >15°.

d). The system will employ a spectrum of 100 MHz in 47.9 to 48.2 GHz for the uplink and 100 MHz in 47.2 to 47.5 GHz for the downlink.

e). Signal protocols for this system will use modulation of QPSK for subscribers, 64QAM for Ground Earth Stations (GES) and the communications protocol will be FDMA/TDMA uplink and TDM downlink.

f). Subscriber information will use data rates at 2Mb/s uplink/10Mb/s downlink and power requirements of 100 to 250 mW.

h). Lifespan is 5 to 10 years for both types of airships.

10.2.1.4. Evaluation of the Sky Station System

There are several unique attributes that allow Sky Station to offer a broad array of services with low operating costs:

1) Sky Station platforms do not require a launch vehicle, they can move under their own power throughout the world or remain stationary and they can be brought down to Earth, refurbished and re-deployed without service interruption.

2) Each platform is independent for a dedicated area of coverage. Namely, once a platform is in position, it can immediately begin delivering service to its service area without the need to deploy a global infrastructure or a constellation of platforms to operate.

3) The altitude enables the Sky Station system to provide a higher frequency reuse and thus higher capacity than other wireless systems.

4) The inexpensive platforms and GES make it the lowest cost wireless infrastructure conceived to date per subscriber. The capabilities and low cost of the Sky Station system will revolutionize telecommunications.

5) Joint venture companies and government authorities located in each country will control the Sky Station platforms serving their region to ensure the best service offerings tailored to the local market. Offerings can change as a region develops.

6) Sky Station platforms are environmentally friendly, because they are powered by solar technology and non-polluting fuel cells.

7) The 21 km altitude provides subscribers with short paths through the atmosphere and unobstructed line-of-sight to the platform. Therefore, with small antennas and low power requirements, the Sky Station system allows for a wide variety of fixed and portable user terminals to meet almost any service need.

There is only one temporary disadvantage of this system. Namely, Sky Station has not yet designed interplatform optical links between nearby airships to provide better coverage and service. Meanwhile, according to the General project, the optical link is planned for the next stage. Thus, optically linked Sky Station platforms could transmit voice, data and video at speeds and volumes comparable to fibre-optic cable (Tb/s) and provide a wireless backbone telecommunications link between metropolitan areas, countries and continents. Sky Station is seeking global regulatory approval for this service on its airships.

10.2.2. TAO (SkyNet) Network

 A Research and Development program (R&D) on a Stratospheric Communications Platform (SCP) airship system is in progress since April 1998. The final goal of this project is to realize the SCP airship system, being capable of an acceptable long-duration station-keeping flight at a stratospheric altitude of about 20 km. The achievements will enable advanced wireless fixed and mobile communications, digital direct and relay broadcasting, modern broadband transmission and high-resolution observations and monitoring of the remote, rural and global environment. This advanced SCP program is promoted in collaboration with the Communications Research Laboratory of Japan (CRL), National Space Development Agency of Japan (NASDA), Japan Marine Science and Technology Centre (JAMSTEC) and the Telecommunications Advancement Organization of Japan (TAO).

10.2.2.1. Airship Platform System Description

The stratospheric platform is an unmanned airship kept at a stratospheric altitude of about 20 km for multimedia communications and Earth observation purposes. It is equipped with communications payload, observation sensors or other equipment. With the aim of quickly developing a stratospheric platform having great potential, the STA and MPT research groups began conducting the relevant research work in 1998. The SCP system is designed similar to a satellite space segment as a relay station to receive signals from ground stations using feeder links and to retransmit them to subscribers using service links. Therefore, an airship like a satellite is carrying a payload with corresponding transponders and antenna system. At any rate, the launch of SCP into position is much simpler than putting a satellite into any orbit. After careful preparation in the hanger space, the airship is launched in 4 Ascent phases through the troposphere and Interface location point in the stratosphere and finally, it shifts to the station-keeping position. The recovery phase goes in the opposite direction, namely, the airship is slowly moved from the station-keeping position towards the Interface point and from there descends down to the ground in 4 descent phases.

Figure 10.6. TAO Stratospheric Platform with Main Components

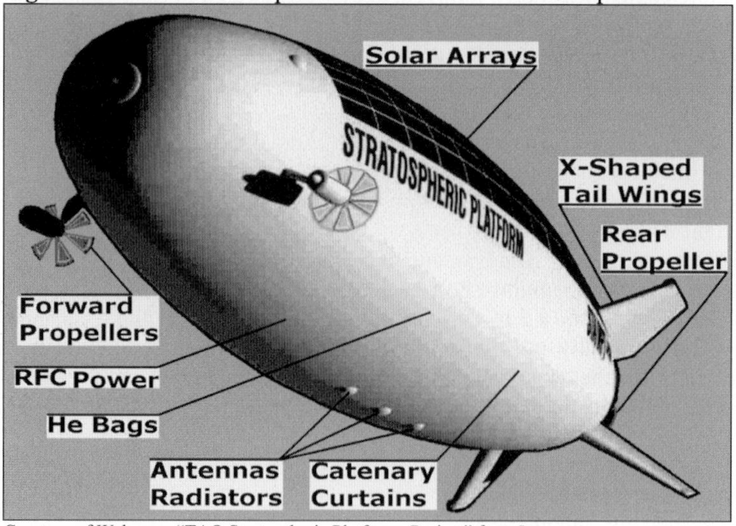

Courtesy of Webpage: "TAO Stratospheric Platforms Project" from Internet

The airship construction has a semi-rigid hull of ellipsoidal shape, with an overall length of about 200 m, see **Figure 10.6.** It is composed of an air-pressurized hull for maintaining a fixed contour and internal special bags filled with the buoyant helium gas.

Two air ballonets are installed inside the hull to keep the airship at a required attitude. For a load balance to the lifting force, catenary curtains are connected to the lower rigid keel, directly attached to the envelope. Propulsive propellers are mounted on both the stem and stern of the airship and tail fins are installed on the rear end of the hull. A solar photovoltaic power subsystem of solar cells and Regenerative Fuel Cells (RFC) is provided to supply a day/night cycle of electricity for airship propulsion.

The length of an airship in general is about 250 m and 60 m diameter. This is about 4 times as long as Jumbo jet passenger airplanes and so, its weight is about 32 tons. Thus, 50% of the weight corresponds to those of structures and membrane materials. Hence, solar arrays and fuel batteries, related to the electric power system, are also heavy. And the weight of mission equipment is supposed to be about 1 ton.

The necessary condition for an airship to float at a certain altitude is that the gravity and buoyancy forces, which are exerted on the airship, are in a state of equilibrium. When the shape and volume of the airship are supposed to be constant, unlike a balloon, the buoyant force at an altitude of 20 km becomes about 1/15 that at sea level. Accordingly, a buoyancy of 15 times as much is necessary for equilibrium. Therefore, in order to float a SCP in the stratosphere, it is necessary to make the weight of the airship light and to make the buoyancy as large as possible. Inside the airship there are several internal bags filled with He gas to obtain enough buoyancy.

10.2.2.2. Outline of the Tracking and Control System

In order to operate unmanned SCP safely, it is necessary to construct a tracking and control system and establish operational technique on board the platform and on the ground. Based on SCP technologies, continuous observation and prediction of weather conditions and monitoring of operational conditions of onboard equipment, appropriate countermeasures can be taken in time, even in case of dangerous atmospheric phenomena or abnormal performances of on-board equipment. At this point, the TAO airship system has to develop adequate TT&C solutions on board the platform and on the ground as well.

Airships can be affected strongly by wind, therefore, when the preliminary decision for launching or recovering of an airship is to be made, it is necessary to predict the weather, especially wind direction and speed, in advance and estimate whether:

1) The airship deviates from the area, within which the tracking and control system works effectively and

2) The launch or recovery can be conducted safely.

Based on this estimation, a final decision to launch or recover is made. After the last checks, the airship is released. It starts to ascend due to the effects of the buoyancy. Near the tropopause, which is the layer between the troposphere and stratosphere, it continues to ascend, being drifted by the jet stream. Finally, the airship arrives in position at an altitude of about 20 km. After this operation, the airship is moved to the station-keeping position and the mission operation is started.

Once an airship is launched, it can be used for a maximum of three years. Namely, an airship is periodically, about every three years, recovered and the He gas and onboard equipment condition is checked. After these routine checks, it is launched again. The life of an airship is supposed to be about ten years.

10.2.2.3. SCP Network Coverage

The main question of the future system is how many airships are necessary to cover all a particular territory or country and can this system can be global? A 15 Stratospheric Platform arrangement is necessary to cover all the territory of the Japanese islands for communications and broadcasting systems, under the condition of 22 km airship altitude with a minimum elevation angle of $10°$. A single airship can cover a certain service area independently, so that, for example, the service can be started from an area with a large population and the number gradually increased. This possibility of flexible business development is one of the merits of SCP systems. The service area that one airship can cover generally depends on certain available numbers of Tx, Rx, antennas, methods of modulation and transmission and many other factors. Otherwise, the final intention of this project is to offer service to other regions and certain countries and if economical and technical evaluations are correct, it will provide global coverage. The concept of the system is very advanced in comparison with similar projects and has almost no disadvantages.

10.2.2.4. Ground System Features

On board the airship there is mission equipment to provide Multimedia and Broadcasting for Fixed and Mobile communications services and an Earth observation and Disaster Monitoring System. Therefore, airships are expected to have the following mission features:

1) Broadband communications and broadcasting are possible with small-sized and very low-power terminals, because of much shorter propagation distances compared with existing satellite systems.

Figure 10.7. TAO Fixed Communication Network

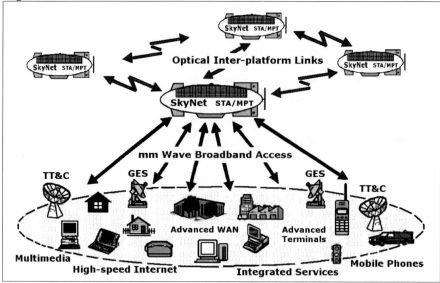

Courtesy of Webpage: "TAO Stratospheric Wireless Platform Project" from Internet

2) High-quality and cost-effective communications and broadcasting are possible with a smaller number of ground stations, due to significantly better line-of-sight conditions, less wave-blocking and multi-path effects compared with existing ground systems. Meanwhile, compared with satellite systems, the propagation distance is shorter by about 1/1800. Consequently, as EM signal propagation losses and delay distortions become much smaller, broadband communications and broadcasting are possible with smaller sized and lower power fixed and mobile terminals.

3) By establishing interplatform multimedia broadband links, high speed communications and broadcasting networks, comparable to optical fibre systems will be possible, which may enable the realization of novel communications and broadcasting systems.

4) Optimum system configurations are possible owing to the flexible operations of platform systems, which can enable expansion to a world communications system.

In reality, digital TV Broadcasting by SCP will use about 15 CSP to cover all of Japan, where over 10,000 stations are necessary at present. This system will be an advanced mobile communications complement to terrestrial systems at very low cost. Thus, access to for wireless communications systems will be more cost-effective than optical fibre systems. The TAO system will enable much better communications and broadcasting infrastructures for developing countries and will promote spreading (diffusion) in these countries. Emergency communications systems will retain communications links in an emergency by moving into position over the area. Various remote-sensing services will be available for radio wave observations, aerial photographs, meteorological observations, etc.

The CSP System is designed for Fixed and Mobile Multimedia two-way Communications. The ground segment consists of Ground Earth Stations (GES) or Gateways and fixed, semi fixed and/or mobile terminals, with corresponding auto tracking and focusing antenna systems for all applications, respectively.

Figure 10.8.   TAO Mobile Communication Network

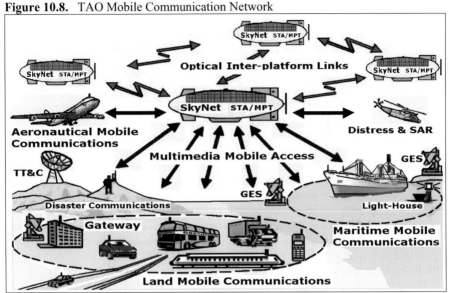

Courtesy of Webpage: "TAO Stratospheric Wireless Platform Project" from Internet

A complete fixed broadband multimedia service program with all applications is presented in Figure **10.7**. Fixed ground terminals can be a self-contained PC office configuration with modem, or as an integrated part of an advanced LAN or WAN, laptop, fixed telephone set in office or public and mobile or cellular phone equipment.

Mobile multimedia communications access will offer maritime, land, aeronautical and personal handheld applications and services, illustrated in **Figure 10.8**. At this point mobile ground terminals can be PC portable or laptop configurations interfaced to the transceiver with adequate antennas or self-contained mobile or portable/in vehicle transceiver units with mobile auto tracking antenna and personal handheld terminals with built-in antenna.

Maritime mobile communications will provide two-way commercial, distress and safety communications for ships, fishing fleets and other kinds of boats integrated with GEO and GNSS systems. In the framework of this service, there will be additional activities like buoy and lighthouse control, marine pollution, maritime investigation and SAR missions.

Land mobile communications will provide services for all kinds of vehicles like trains, buses, trucks and cars, including personal mobile terminal, cellular service and emergency communications for natural disasters, which can be enhanced with equipment for tracking and GNSS facilities. The SCP has to substitute or integrate current cellular systems.

Aeronautical mobile communications will provide commercial and distress service for all kinds of aircraft integrated with GEO, GPS and other GNSS to provide CNS service.

The broadcasting system using the SCP airship constellation will provide the following types of services: **1)** digital broadcasting; **2)** complementary terrestrial digital broadcasting to fixed and mobile stations; **3)** terrestrial rebroadcasting programs; **4)** relay broadcasting of HDTV and radio programs; **5)** movable broadcasting on demand, using mobile equipped stations; **6)** broadcasting for limited suburban regions, isolated islands, rural and remote places; **7)** field pickup from the SCP; and **8)** emergency news and observations.

10.2.2.5. Electric Power and Motion of Stratospheric Platform

Unmanned airships are maintained in the stratospheric zone constantly at an altitude of 20 to 22 km, where the winds and other meteorological conditions are calm. The mean temperature is -60 to $-50°$ C and the atmospheric pressure is about 50 hPa. The altitude of 20 km is about 60 times higher than the Tokyo TV Tower, so better lines-of-sight are obtained. Moreover, multi-path effects will be significantly reduced in most areas because of higher elevation angles. There are no clouds in the stratosphere, so perfectly clean solar energy can be used without atmospheric pollution.

The concept of airship launch and recovery features an upwind orientation of the airship heading, assuming limited wind speed at the ground surface for takeoff and landing. In additional, for ascents and descents, the airship heading should be kept always in an upwind direction, while non-powered flights should require only the buoyant lift into the jet stream. Stratospheric Platforms do not use any fuel for motion, so the only necessary energy is supplied by electric power, obtained from solar cells. The wide part in the graphic of an airship's topside is the solar cell array. Solar cells can supply clean energy without CO_2 gas generation, so that it is said to be kind to the Earth and is an ideal form of energy. Only electric power of about 200 kW is necessary for stratospheric platforms and it is used for the following reasons:

1) To rotate propulsive propellers and supply electric power to the various positioning and attitude controlling systems for station-keeping.

2) To operate the on-board payload, for the mission systems of communications, broadcasting and earth observation systems about 10 kW is necessary.

3) To charge the fuel battery in the daytime and to use it at night, when no solar electric power is obtained.

The position of an airship is as stationary as a GEO satellite. Namely, as the airship is used as a platform for communications and broadcasting, it is necessary for it to be stationary, like a broadcasting satellite. Meanwhile, the position of the airship cannot be permanently stationary because of different parameters, which have an influence on the moving range. At this point, it is necessary to provide permanent control of platform moving and when necessary to correct its position using station-keeping correction motors with forward and rear propellers. Consequently, as the maximum wind speed is over 20 m/sec at altitudes of 20 to 22 km, it will be very difficult for the airship to be controlled as strictly as geostationary satellites. The moving range of a stratospheric platform is very slow and in such a way, at present, requirements of station-keeping are within a radius of 1 km in the horizontal plane and +/–1km in the perpendicular direction and under this condition the design and fabrication of onboard equipment is considered.

A receiving antenna on the ground is set with the bore sight (the axis of the antenna beam) to the airship, so that the maximum receiving power can be obtained. If the position of the airship is changed, the receiving power decreases. Therefore, the airship is controlled so as to keep the position as stable as possible. For example, for a 20 cm diameter antenna, a position change of 600 m induces 3 dB or 50% power loss. This tendency depends upon the antenna size and operational frequency. In this instance, the larger an antenna radius or the higher the frequency, the receiving power becomes less and vice versa. Accordingly, to satisfactorily solve these problems, an automatically pointing corresponding adjustable antenna is being considered and in this way, it is necessary to develop a low-cost and small-sized antenna with automatic tracking system.

REFERENCES

1. Books

[01] Acerov A.M. & other, "Morskaya radiosvyaz i radionavigaciya", Transport, Moskva, 1987.
[02] Baylin F., "World satellite yearly", Baylin Publications, Boulder, 1998.
[03] Blonstein L., "Communications satellites, the technology of space communications", Heinemann, London, 1987.
[04] Calcutt D. & other, "Satellite communications, Principles and applications", Edward Arnolds, London, 1994.
[05] Campbell J., "Understanding GMDSS", Waterline, Shrewsbury, 1998.
[06] Chetty P.R.K., "Satellite technology and its applications", TAB, Blu Ridge Summit, 1993.
[07] Dalgleish D.I., "An introduction to satellite communications", IEE, London, 1989.
[08] Davidoff M.R., "The satellite experimenter's handbook", ARRL, Newington, 1984.
[09] Elbert B.R., "Ground segment and earth station handbook", Artech House, Boston – London, 2001.
[10] Elbert B.R., "Introduction to satellite communications", Artech House, London, 1987.
[11] Elbert B.R., "The satellite communication applications handbook", Artech House, London, 1997.
[12] El-Rabani A., "Introduction to GPS", Artech House, Boston-London, 2002.
[13] Evans B.G., "Satellite communication systems", IEE, London, 1991.
[14] Feher K., "Digital communications, Satellite Earth Station Engineering", Prentice-Hall, Englewood Cliffs, 1983.
[15] Freeman R.L., "Radio systems design for telecommunications (1-100 GHz)", John Wiley, Chichester, 1987.
[16] Fujimoto K. & other, "Mobile antenna systems handbook", Artech House, London, 1994.
[17] Gagliardi R.M., "Satellite communications", Van Nostrand Reinhold, New York, 1984.
[18] Gallagher B., "Never Beyond Reach", Inmarsat, London, 1989.
[19] Galic R., "Telekomunikacije satelitima", Skolska knjiga, Zagreb, 1983.
[20] Gordon G.D. & other, "Principles of communications satellites", John Wiley, Chichester, 1993.
[21] Grant A.E. & other, "Communication technology update", Focal Press, Boston, 2000.
[22] Group of authors, "Fifth international conference on satellite systems for mobile communications and navigation", IEE, London, 1996.
[23] Group of authors, "Fourth international conference on satellite systems for mobile communications and navigation", IEE, London, 1988.
[24] Group of authors, "Handbook - Mobile Satellite Service (MSS)", ITU, Geneva, 2002.
[25] Group of authors, "Handbook on Satellite Communications", ITU, Geneva, 2002.
[26] Group of authors, "Morskaya radiosyaz", Transport, Leningrad, 1985.
[27] Group of authors, "Pomorska enciklopedija – VI/VII volume", Jugoslavenski leksikografski zavod, Zagreb, 1983.
[28] Group of authors, "Radiowave propagation information for predictions for earth-to-space path communications", ITU, Geneva, 1996.
[29] Group of authors, "Utilisation des satellites pour les recherches et le sauvetage", Cepadues, Toulouse, 1984.
[30] Guinard F., "Maritime applications of Inmarsat-A and Inmarsat-C", Inmarsat, London, 1992.
[31] Ha T.T., "Digital satellite communications", Macmillan Publications, New York, 1986.
[32] Hadden A.D., "Personal communications networks, Practical implementation", Artech House, London, 1995.
[33] Heath S., "Multimedia and communications technology", Focal Press, Oxford, 1999.
[34] Higgins J., "Satellite Newsgathering", Focal Press, Oxford, 2002.
[35] Huurdeman A.A., "Guide to telecommunications transmission systems", Artech House, Boston-London, 1997.
[36] Jagoda A. & other, "Mobile communications", John Wiley, Chichester, 1995.
[37] Jamalipour A., "Low earth orbital satellites for personal communication networks", Artech House, London, 1998.
[38] Kadish J.E. & other, "Satellite communications fundamentals", Artech House, Boston-London, 2000.
[39] Kantor L.Y., "Sputnikovaya svyaz i veschanie", Radio i svyaz, Moskva, 1988.
[40] Kantor L.Y. & other, Sputnikovaya svyaz i problema geostacionarnoy orbiti", Radio i svyaz, Moskva, 1988.
[41] Kaplan D.E., "Understanding GPS principles and applications", Artech House, Boston-London, 1996.
[42] Law P.E., "Shipboard antennas", Artech House, Washington, 1983.
[43] Lee C.Y.W., "Mobile communications engineering", McGraw-Hill, London, 1982.
[44] Lees G.D. & other, "Handbook for marine radio communications", LLP, London, 1999.
[45] Long M., "World satellite almanac", Howard W. Sams, Indianopolis, 1987.
[46] Lukatela G. & other, "Digitalne telekomunikacije", Gradjevinska knjiga, Beograd, 1984.
[47] Maral G. & other, "Satellite communications systems", John Wiley, Chichester, 1994.
[48] Mladenov M.K. & other, "Periferni ustroystva za personaliny kompyutri", Tehnika, Sofia, 1987.
[49] Monroe J.W. & other, "Marine radionavigation and communications", Cornell Maritime Press, Centreville, 1998.
[50] Nenirovskiy A.S. & other, "Radio-releinie i sputnikovie sistemi peredachi", Radio i svyaz, Moskva, 1986.
[51] Noll E.M., Landmobile and marine radio technical handbook", Howard W. Sams, Indianapolis, 1985.

[52] Novik L.I. & other, "Sputnikovaya svyaz na more", Sudostroenie, Leningrad, 1987.
[53] Ohmori S. & other, "Mobile satellite communications", Artech House, Boston–London, 1998.
[54] Olsen J.M. & other, "An introduction to GMDSS", Nofus, Gravdal, 1994.
[55] Orehov A.A., "Radiopriemnie ustroystva morskogo sudna", Transport, Moskva, 1987.
[56] Pascall S.P. & other, "Commercial satellite communications", Focal Press, Oxford, 1997.
[57] Perishkin I.M. & other, "Sistemi podvizhnoy radiosvyazi", Radio I svyaz, Moskva, 1986.
[58] Pirumov V.S. & other, "Radio-elektronika v voyne na more", Voenoe izdatelstvo, Moskva, 1987.
[59] Pratt T. & other, "Satellite communications", TAB, Blue Ridge Summit, 1986.
[60] Prentiss S., "Satellite communications", TAB, Blue Ridge Summit, 1987.
[61] Radovanovic A., "PC modemske komunikacije", Tehnicka knjiga, Beograd, 1991.
[62] Ricci F.J. & other, "U.S. military communications", Computer Science Press, Rockville, 1986.
[63] Richharia M., "Mobile Satellite Communications – Principles and Trends", Addison-Wesley, Harlow, 2001.
[64] Richharia M., "Satellite communications system – Design principles", Macmillan, Basingstoke, 1995.
[65] Sheriff R.E. & other, "Mobile satellite communication networks", Wiley, Chichester, 2001.
[66] Smith P.C. & other, GMDSS for navigators", Butterworth Heinemann, Oxford, 1994.
[67] Solovev V.I. & other, "Svyaz na more", Sudostroenie, Leningrad, 1978.
[68] Stajic D. & other, "Racunarske telekomunikacije i mreze", Tehnicka knjiga, Beograd, 1991.
[69] Sternfeld A., "Vestacki sateliti", Tehnicka knjiga, Beograd, 1958.
[70] Tetley L. & other, "Electronic aids to navigation – Position fixing", Edward Arnold, London, 1991.
[71] Tetley L. & other, "Understanding GMDSS", Edward Arnold, London, 1994.
[72] Torrieri D.J., "Principles of military communication systems", Artech House, Dedham, 1982.
[73] Van Trees H.L., "Satellite communications", IEEE, New York, 1979.
[74] Venskauskas K.K. "Sistemi i sredstva radiosvyazi morskoy podvizhnoy sluzhbi", Sudostroenie, Leningrad, 1986.
[75] Viola B., "L'operatore di radiocomunicazioni del servizio mobile marittimo", Trevisini, Milano, 1987.
[76] Walker J., "Mobile information systems", Artech House, Boston – London, 1990.
[77] Waugh I., "The maritime radio and satellite communications manual", Waterline, Shrewsburg, 1994.
[78] Zovko-Cihlar B., "Sumovi u radiokomunikacijama", Skolska knjiga, Zagreb, 1987.
[79] Zhilin V.A., "Mezhdunarodnaya sputnikova sistema morskoy svyazi – Inmarsat", Sudostroenie, Leningrad, 1988.

2. Papers

[01] Agius A.A. & other, "Antenna Design for the ICO Handheld Terminal" 10th ICAP '97, Edinburgh, 1997.
[02] Agius A.A. & other, "Intelligent Handset Antenna Research within Mobile VCE", CCSR, Surrey, 1997.
[03] Agius A.A. & other, "QHS – Characteristics of a Proposed Wire Antenna for an SPCN Handheld Terminal", CCSR, Surrey, 1997.
[04] Agius A.A. & other, "The Design of Specifications for Satellite PCN Handheld Antennas", CCSR, Surrey, 1997.
[05] Fletcher G.D. & other, "The Silex optical interorbit link experiment", ECE Journal, Vol 3/ 6, IEE, Stevenage, 1991.
[06] Gallagher G.D., "Widening maritime satcom", Communication International", Journal No.11, Wealdstone, 1986.
[07] Gatenby P.V. & other, "Optical intersatellite links", ECE Journal, Vol. 3, No. 6, IEE, Stevenage, 1991.
[08] Hijiri Y., "The talking ship", Communications, International Radio Officer's Journal/2, INROC, Brussels, 1984.
[09] Ilčev D.St., "Acquirements of GMDSS certificates", Navigation, No. 2, Kotor, 1995.
[10] Ilčev D.St., "Characteristics and development of maritime satellite communications systems", Master work, Macedonian University, Faculty of Electrical Engineering, Skopie, 1994.
[11] Ilčev D.St., "Characteristics and development of global mobile satellite communications for maritime, land and aeronautical applications", Doctoral thesis, Serbian University, Faculty of Electrical Engineering, Beograd, 2000.
[12] Ilčev D.St., "COSPAS-SARSAT system", Nautical Courier, No. 4, Beograd, 1987.
[13] Ilčev D.St., "History of maritime radiocommunications", Our Sea, No. 5-6, Dubrovnik, 1987.
[14] Ilčev D.St., "Improving of radio and electronic equipment on board of Yugoslav merchant fleet after abolition and sanctions", Navigation, No. 3, Kotor, 1995.
[15] Ilčev D.St., Integration in communications and navigation", Nautical Courier, No. 2, Beograd, 1989.
[16] Ilčev D.St., "Integration in maritime communications", Collection of papers, No. 1, Maritime Faculty, Kotor, 1988.
[17] Ilčev D.St., "Inmarsat standard-A satellite terminals", Collection of papers, Maritime Faculty, Dubrovnik, 1989.
[18] Ilčev D.St., "Land earth station", Collection of Papers, Maritime Faculty, No. 16, Kotor, 1991.
[19] Ilčev D.St., "Maritime satellite communications", Nautical Courier, No.4, Beograd, 1987.
[20] Ilčev D.St., "Maritime information system", Navigation, No. 3, Kotor. 1995.
[21] Ilčev D.St., "Perspective of global mobile maritime satellite communications", Sixth International Conference on Electronic Engineering in Oceanography, Churchill College, Cambridge, 1993.
[22] Ilčev D.St., "Ship earth station", Radio Connection, No. 1, Beograd, 1989.
[23] Ivancic D.W. & others, "Application of Mobile-IP to Space and Aeronautical Networks", NASA, Cleveland, 2002.
[24] Kawashima R. & other, "On the see keeping qualities of fishing by field measurement - III", Japan Navigation Institute, Tokyo, 1976.
[25] Lundberg O., "Bringing modern communications to the communicators", CI Journal, No. 4, Wealdstone, 1985.

[26] Mochie G., "Future looks promising for the onboard maintainer", Safety at Sea, Journal No. 11, Redhill, 1988.
[27] Rohit G. & others, "Traffic Management for TCP/IP over Satellite ATM Networks" The Ohio State University, USA, 2000.
[28] Ruoss H.O. & other, "Optimized Slot Antenna for Handheld Mobile Telephone", 10th ICAP '97, Edinburgh, 1997.
[29] Sanford J., "The role of satellite communications in modern ship management", Communication, International Radio Officer's journal, No. 2, Brussels, 1983.
[30] Snowball A.E., "Global maritime mobile service via – The Inmarsat system now and in the future", Telecommunications Journal, No. 6, ITU, Geneva, 1986.
[31] Stevens W., "The Slow Start, Congestion Avoidance, Fast Retransmit and Fast Recovery Algorithms", NOAO, Tucson, AZ, 1997.
[32] Tuomas A., "Mobile IPv6 Security", Microsoft Research Ltd, Cambridge, 2002.
[33] Wilkinson C.F., "X.400 electronic mail", Electronics communication Engineering journal", Vol. 3, No. 6, IEE, Stevenage, 1991.
[34] Wright D., "Satcom development promote marine safety", Safety at Sea, Journal No. 6, London, 1983.
[35] Zaharov V. & other, "Smart Antenna Application for Satellite Communications with SDMA", Journal of Radio Electronics, Moscow, 2001.

3. Manuals

[01] "An introduction to the use of satcom for air traffic services and flight operations", Inmarsat, London, 1998.
[02] "Basic documents", Inmarsat, London, 1986.
[03] "Capsat mini-M mobile telephone users manual TT3060A", Thrane & Thrane, Soeborg, 1998.
[04] "CMA-2102 satcom High-Gain antenna", Canadian Marconi, Ville St-Laurent, 1995.
[05] "COSPAS-SARSAT 406 MHz distress beacon", COSPAS-SARSAT, London, 1996.
[06] "COSPAS-SARSAT LUT performance specification and design guidelines", COSPAS-SARSAT, London, 1995.
[07] "Description of the COSPAS-SARSAT space segment", COSPAS-SARSAT, London, 1996.
[08] "Description of the Globalstar system", Globalstar, San Jose, 1997.
[09] "Global maritime distress and safety system", IMO, London, 1987.
[10] "Globalstar backgrounder", Globalstar, San Jose, 1998.
[11] "Globalstar Telit Sat 550 Dual Mode GSM/Satellite Manual", Telit, Sgonico, 2000.
[12] "Globalstar Telit Sat 551 Car Kit Manual", Telit, Sgonico, 2000.
[13] "GMDSS" – Vol 5, Admiralty list of Radio Signals, Taunton, 1999.
[14] "GMDSS handbook", IMO, London, 1995.
[15] "GMDSS system operation manual", Raytheon, Kiel, 1998.
[16] "GMPCS Reference Book", ITU, Geneca, 2000.
[17] "Inmarsat aeronautical satcom services users conference", Inmarsat, Paris, 1999.
[18] "Inmarsat annual review", Inmarsat, London, 1989.
[19] "Inmarsat-3 ground control program", Inmarsat, London, 1993.
[20] "Inmarsat aeronautical services manual", Inmarsat, London, 1995.
[21] "Inmarsat-A maritime user's manual", Inmarsat, London, 1991.
[22] "Inmarsat-A mobile earth station", Users Guide, BT, London, 1990.
[23] "Inmarsat-B global phone 3000", GEC Marconi, Chelmsford, 1993.
[24] "Inmarsat-B high speed data service", Inmarsat, London, 1997.
[25] "Inmarsat-C Galaxy Sentinel user's guide", Trimble, Sunnyvale, 1997.
[26] "Inmarsat-C land mobile users handbook", Inmarsat, London, 1995.
[27] "Inmarsat-C maritime user's manual", Inmarsat, London, 1991.
[28] "Inmarsat data services user's guide", Inmarsat, London, 1996.
[29] "Inmarsat-M omniphone MTN 205", Marconi Marine, Chelmsford, 1996.
[30] "Inmarsat M & B technical summary", Inmarsat, London, 1992.
[31] "Inmarsat maritime communications handbook", Inmarsat, London, 1995.
[32] "Inmarsat Maritime Handbook", Inmarsat, London, 2002.
[33] "Inmarsat maritime operations", Inmarsat, London, 1996.
[34] "Inmarsat maritime services", Inmarsat, London, 1995.
[35] "Inmarsat maritime services information", Inmarsat, London, 1997.
[36] "Inmarsat maritime users manual", Inmarsat, London, 1984.
[37] "Inmarsat-P", Inmarsat, London, 1994.
[38] "Inmarsat-P system description", Inmarsat, London, 1995.
[39] "Inmarsat satellite communications services users handbook", Inmarsat, London, 1987.
[40] "Inmarsat standard-C technical notes for designer", Inmarsat, London, 1988.
[41] "International Inmarsat-E manual", Inmarsat, London 1993.
[42] "International satellite directory – Volume 1: The satellite industry", Design Publishers, Sonoma, 2000.
[43] "International satellite directory – Volume 2: Satellite systems and operators", Design Publishers, Sonoma, 2000.

[44] "Introduction to the COSPAS-SARSAT system", COSPAS-SARSAT, London, 1994.
[45] "IPSec Network Security", Documentation, Cisco Systems Inc., US, 2002.
[46] "JUE-310B Inmarsat-B SES Operation Manual", JRC, Tokyo, 2001.
[47] "JUE-75C Inmarsat-C MES Operation Manual", JRC, Tokyo, 2001.
[48] "KVH tracphone 50 mini-M manual", KVH Industries, Middlletown, 1997.
[49] "Landmobile Capsat-C transceiver", Thrane & Thrane, Soeborg, 1993.
[50] "Manual for use by the maritime mobile and maritime mobile-satellite services", ITU, Geneva, 1999.
[51] "Maritime applications of Inmarsat-A and Inmarsat-C", Inmarsat, London, 1992.
[52] "Maritime Capsat-C transceiver installation manual", Thrane & Thrane, Soeborg, 1997.
[53] "Maritime Safety Information Broadcasting" – Vol 3, Part 1, Admiralty list of Radio Signals, Taunton, 1999.
[54] "Oceanray 2C2 satellite communications terminal, Marconi Marine, Chelmsford, 1991.
[55] "Omniphone portable satellite communications terminal", Marconi Marine, Chelmsford, 1998.
[56] "Operator manual mini-M telephone system maritime SP4164A", S.P. Radio, Aalborg, 1998.
[57] "Operator's manual satcom 99-71896C", RDI, San Leandro, 1987.
[58] "Nera worldphone mini-M user's manual", Nera, Billingstad, 1992.
[59] "Radiowave propagation information for predictions for Earth-to-space path communications – Handbook", ITU, Radiocommunication Bureau, Geneva, 1996.
[60] "Safecom CM Inmarsat-C service manual", Philips, Copenhagen, 1993.
[61] "SafetyNET user's handbook", Inmarsat, London, 1995.
[62] "Sailor operator and technical manual for Inmarsat-C transceiver H2095B", S.P. Radio, Aalborg, 1995.
[63] "Sailor GMDSS console H2192", S.P. Radio, Aalborg, 1996.
[64] "Satellite communications system MX211A", Magnavox, Torrance, 1984.
[65] "Saturn 3 standard-A installation operator's and system description manuals", EB, Nesbru, 1986.
[66] "Saturn Bm C2 operator's manual", Nera, Billingstad, 1996.
[67] "Saturn C marine version operator's manual", Nera, Billingstad, 1995.
[68] "Single Mode Iridium Handset", Kyocera, Kanagawa, 1999.
[69] "Standard-C SES demodulator/decoder", Inmarsat, London, 1988.
[70] "Standard-C SES STR 1500C", Raytheon Anschutz, Kiel, 1998
[71] "System definition manual for the standard-C communication system", Inmarsat, London, 1988.
[72] "Technical requirements for Inmarsat standard-A coast earth station", Inmarsat, London, 1989.
[73] "Technical requirements for Inmarsat standard-A ship earth station", Inmarsat, London, 1983.
[74] "T-4000 high gain antenna system", Tecom, Chatsworth, 1995.
[75] "Technical reference guide", Inmarsat, London, 1995.
[76] "The Iridium system", Motorola, Libertyville, 2000.
[77] "Transportable Inmarsat-B earth station", California Microwave, New York, 1995.
[78] "Tron 30S/30S MkII Operation Manual", Jotron, Tjodalyng, 1999.
[79] "Tron 40S Operation Manual", Jotron, Tjodalyng, 1998.
[80] "Tron 45S/SX Operation Manual", Jotron, Tjodalyng, 1998.
[81] "Tron AIR Operation Manual", Jotron, Tjodalyng, 1999.
[82] "TT-3060A Capsat Mobile Telephone Users Manual", Thrane & Thrane, Soeborg, 1998.
[83] "User's Guide", Sea Launch Company, Long Beach, 1996.

4. Brochures

[01] "Aero industry directory", Inmarsat, London 1997.
[02] "Above and Beyond" (2 brochures), Globalstar, San Jose, 1999.
[03] "A compuship directory – Maritime satcom terminals", Inmarsat, London 1998.
[04] "A move-it directory – Mobile satcom terminals", Inmarsat, London 1998.
[05] "Big Dish", Thrane & Thrane, Soeborg, 1998.
[06] "Capsat GMDSS Dual Mode system", Thrane & Thrane, Soeborg, 1999.
[07] "Charges directory for satellite communications services", Inmarsat, London 1998.
[08] "Commitment, Creativity and Challenge", Thrane & Thrane, Soeborg, 1998.
[09] "Design and installation guidelines for Inmarsat-A SES", Inmarsat, London 1992.
[10] "Design and installation guidelines for Inmarsat-B SES", Inmarsat, London 1993.
[11] "Design and installation guidelines for Inmarsat-C SES", Inmarsat, London 1992.
[12] "Facts about Inmarsat", Inmarsat, London, 1999.
[13] "Freedom to communicate", Iridium Africa, Cape Town, 1998.
[14] "Get the picture", Sea Tel, Concord, 1999.
[15] "Global Aeronautical Communications Solutions", Inmarsat, London 2001.
[16] "Global Fleet Management", Thrane & Thrane, Soeborg, 1998.
[17] "Globalstar 1999 Review", Globalstar, San Jose, 1999.
[18] "Global Mobile Communications Solutions", Inmarsat, London 2001.
[19] "Globalstar-GSP 288/2900 Fixed Satellite Phone", Qualcomm, San Diego, 2000.

[20] "Globalstar EF-200 Satellite Phone", Globalstar, San Jose, 2001.
[21] "Globalstar ICS550 satellite telephone", ICS, Arundel, 1999.
[22] "Globalstar mobile phone R290 satellite", Ericsson, Stockholm, 1999.
[23] "Globalstar Telit Sat 550 Dual Mode Satellite/GSM 900", Telit, Sgonico, 2000.
[24] "Globalstar Telit Sat 551 Car Kit for Sat 550", Telit, Sgonico, 2000.
[25] "Globalstar Telit Sat 600 Dual Mode Satellite/GSM 900", Telit, Sgonico, 2000.
[26] "Globalstar Telit Sat 601 Car Kit for Sat 600", Telit, Sgonico, 2000.
[27] "Globalstar mobile phone R290 satellite", Ericsson, Stockholm, 1999.
[28] "Globalstar Qualcomm Marine Kit", Globalstar, San Jose, 1999.
[29] "Globalstar Vodacom's Satellite Services, Vodacom, Sandton, 1999.
[30] "ICS550 Globalstar satellite/GSM telephone system", ICS, Arundel, 1999.
[31] "Inmarsat Swift64 Aeronautical HSD Services – Facts", Inmarsat, London 2001.
[32] "Inmarsat-A", BT Inmarsat, London 1991.
[33] "Inmarsat Aeronautical Services – Facts", Inmarsat, London 2001.
[34] "Inmarsat directory", Lloyd's Press, London 1992.
[35] "Inmarsat phone mini-M", Inmarsat, London 1996.
[36] "Inmarsat maritime communications", Inmarsat, London 1999.
[37] "Inmarsat maritime operations", Inmarsat, London 1996.
[38] "Inmarsat maritime services information", Inmarsat, London 1997.
[39] "Inmarsat SES design and installation guidelines", Inmarsat, London 1996.
[40] "Iridium for maritime communication", Iridium Africa, Cape Town, 1998.
[41] "Iridium – Motorola wings" (3 brochures), Motorola, Libertyville, 1999.
[42] "Iridium – MultiExchange Unit MXU2000R", Motorola, Libertyville, 1999.
[43] "Iridium Technical Background", Washington, 1998.
[44] "Iridium – The world in your hands", Kyocera, Kanagawa, 1999.
[45] "Maritime services", Globalstar, San Jose, 1999.
[46] "Maritime Systems", Thrane & Thrane, Soeborg, 2000.
[47] "Marconi space systems", Marconi, Portsmouth, 1986.
[48] "Marine stabilized antenna systems", Sea Tel, Concord, 1999.
[49] "Maritime communications by satellite", Inmarsat, London, 1983.
[50] "Maritime Communication Systems", Thrane & Thrane, Soeborg, 1999.
[51] "Mission update", Sea Launch, Long Beach, 1999.
[52] "Mobile communications unlimited", Inmarsat, London, 1991.
[53] "Nera F33F55/F77", ", NERA, Billingstad, 2003.
[54] "Nera M2M Satellite Solutions" ", NERA, Billingstad, 2004.
[55] "Nera SatLink Two-way Satellite Broadband Solutions" ", NERA, Billingstad, 2004.
[56] "Nera WorldCommunicator", NERA, Billingstad, 1999.
[57] "Preview of Iridium products", Kyocera, Kanagawa, 1998.
[58] "Product information", Globalstar, San Jose, 1999.
[59] "Sailor Iridium Multi Channel MC4000", Sailor, Aalborg, 1999.
[60] "Sailor Iridium systems", Sailor, Aalborg, 1999.
[61] "Sailor Satellite Systems", Sailor, Aalborg, 1999.
[62] "Sailor SP4164A Mini-M", Sailor, Aalborg, 1999.
[63] "Samsung satellite/cellular phone", ICO, London, 1999.
[64] "Sat-906 System", Rockwell Collins, Reading, 1999.
[65] "Satcom-5000 System", Rockwell Collins, Reading, 2000.
[66] "Satcom-6000 System", Rockwell Collins, Reading, 2000.
[67] "Satellite communications on the move", Inmarsat, London, 1987.
[68] "Satellite Series - Iridium Accessory Brochure", Motorola, Libertyville, 1998.
[69] "Satellite Series 9501 Iridium Pager", Motorola, Libertyville, 1998.
[70] "Satellite Series 9500 Iridium Portable Telephone", Motorola, Libertyville, 1999.
[71] "Satellite technology, orbits and launches tutorial", Inmarsat, London, 1997.
[72] "Satellite teleport in the hearth of Europe", GT&T, Louvain-La-Neuve, 2000.
[73] "Scada", Eutelsat, Paris, 1999.
[74] "Scansat-M 9010 Satellite Telephone", Skanti, Vaerloese, 1994.
[75] "Sea Launch – At-A-Glance", Sea Launch Co LLC, Long Beach, 2000
[76] "Shipping Emergencies-SAR and GMDSS", Focus on IMO, IMO, London, 1999.
[77] "Shore-to-ship calling procedures", Inmarsat, London, 1984.
[78] "SITA Aircraft Communications", Set of pamphlets, SITA, Vienna, 2001.
[79] "Stay in Touch" (2 brochures), Thrane & Thrane, Soeborg, 1998.
[80] "Technical background", Iridium, Washington, 1999.
[81] "The Iridium System", Iridium, Washington, 2000.
[82] "The Microsat series", GT&T, Louvain-La-Neuve, 1998.

[83]   "The world in your hands", Kyocera, Kanagawa, 1998.
[84]   "Transportable Inmarsat-B Earth Station LYNXX", California Microwave, Hauppauge, 1995.
[85]   "Tron 30S MkII", Jotron, Tjodalyng, 1999.
[86]   "Tron 40S", Jotron, Tjodalyng, 1998.
[87]   "Tron 45S/SX", Jotron, Tjodalyng, 1998.
[88]   "Tron EPIRBs", Jotron, Tjodalyng, 1999.
[89]   "TT-3000M Aero-M System", Thrane & Thrane, Soeborg, 2000.
[90]   "TT-3002L Capsat MicroCap Antenna", Thrane & Thrane, Soeborg, 1998.
[91]   "TT-3005L Land Mobile Antenna", Thrane & Thrane, Soeborg, 1998.
[92]   "TT-3005M Maritime Antenna", Thrane & Thrane, Soeborg, 1998.
[93]   "TT-3008BBig Dish Antenna for TT-3080A Capsat Messenger", Thrane & Thrane, Soeborg, 1998.
[94]   "TT-3062A Capsat Rod Telephone", Thrane & Thrane, Soeborg, 1999.
[95]   "TT-3062D Capsat Compact Carphone", Thrane & Thrane, Soeborg, 1999.
[96]   "TT-3066A Capsat Big Dish Telephone", Thrane & Thrane, Soeborg, 1998.
[97]   "TT-3606E Message Terminal", Thrane & Thrane, Soeborg, 2000.
[98]   "TT-6000 Inmarsat-C LES Access, Control & Signalling Equipment", Thrane & Thrane, Soeborg, 2000.
[99]   "We've got the whole world talking" (3 brochures), Globalstar, San Jose, 1999.

5. Periodicals

[01]   "Aeronautical satellite news - ASN", Inmarsat, London, 1985-1995.
[02]   "Communications", INROC journal, Brussels, 1981-1985.
[03]   "Communications magazine", Globalstar, San Jose, 1997- 2000.
[04]   "Cospas-Sarsat Information Bulletin", Inmarsat, London, 1998-2000.
[05]   "Electronics Communication Engineering Journal", IEE, London, 1979-2001.
[06]   "Focus on", IMO journal No.1, London, 1992.
[07]   "Inmarsat Facts", Inmarsat, London, 1985-2000.
[08]   "Inmarsat News", Inmarsat, London, 1985-2000.
[09]   "Inside track", Inmarsat, London, 1995-2001.
[10]   "Ocean voice", Inmarsat, London, 1981- 2003.
[11]   "Marifacts", Comsat journal No. 1, Washington, 1991.
[12]   "Marinet systems news", Marinet Systems journal No. 4, Liverpool, 1991.
[13]   "Sea Launch Update", Sea Launch Co LLC, Long Beach, 1999-2001
[14]   "The Messenger", Thrane & Thrane, Soeborg, 2000.
[15]   "Via Inmarsat", Inmarsat, London, 1995-2003.
[16]   "Via Satellite", Phillips, Pittsfield, 1999-2003.

6. PC Mediums

[01]   "An International Partnership" (CD disc), Sea Launch Co LLC, Long Beach, 2000.
[02]   "COSPAS-SARSAT System Documentation", COSPAS-SARSAT, London, 2001.
[03]   "Globalstar – Above and Beyond" (CD disc), Globalstar, San Jose, 2001.
[04]   "Globalstar Papers/Manuals" (ZIP Iomega disc), Globalstar, San Jose, 2001.
[05]   "Globalstar – We've Got the whole World Talking" (CD disc), Globalstar, San Jose, 2001.
[06]   "Iridium Big LEO Satellite System (CD disc), Iridium LCC, Tempe, June 2002.
[07]   "Inmarsat Global Area Network" (CD), Inmarsat, London, 2001.
[08]   "Inmarsat Global Mobile Satellite Solutions" (CD), Inmarsat, London, 2001.
[09]   "Inmarsat Interactive Network" (CD), Inmarsat, London, 2001.
[10]   "Manuals and Pictures of products" (CD disc), Thrane & Thrane, Soeborg, 2001.
[11]   "SAT 906" (CD disc), Rockwell-Collins, Cedar Rapids, 2001.

ABOUT THE AUTHOR

The Author was born on 18 August 1944 in Jambol, Bulgaria. He received a B.Eng degree in Maritime Radio Engineering from the Faculty of Maritime Studies at Kotor of Podgorica University, Montenegro; received a BSc Eng (Hons) degree in Maritime Communications from the Faculty of Maritime and Transportation Studies of the University at Rijeka, Croatia and received an MSc academic degree in Electrical Engineering from the Faculty of Electrical Engineering, Telecommunication Department of the University at Skopie, Macedonia, in 1971, 1986 and 1994, respectively.

He also passed, in spring 1995, an on-site GMDSS training course on the Poseidon simulator at the Maritime Training Centre in Varna, Bulgaria.

Prof. Ilčev holds the certificates for Radio operator 1st class (Morse); for GMDSS 1st class Radio Electronic Operator and Maintainer and for Master Mariner without limitations.

Early in 2000, the author submitted his Doctoral dissertation for evaluation to the Faculty of Electrical Engineering of the University of Belgrade, Serbia, with the following theme: Characteristics and Development of Global Mobile Satellite Communications for Maritime, Land and Aeronautical Applications. The thesis has been positively evaluated for the PhD academic degree.

Between 1971 and 1987 he was engaged as a Radio and a Deck Officer onboard ships, as a Radio Electronic Engineer of a Coastal Radio Station and as an Electro-radio and Safety Superintendent for a shipping company.

Since 1987 he was a General Manager of IS Marine Radio in Kotor, Montenegro, for Sales, Installation, Service and Engineering of Radio, Satellite, GMDSS, Nautical and Electronic devices (E-mail: ismarineradio@hotmail.com). Since 1985 the Company was an Inmarsat partnership member (www.inmarsat.com).

At the same time, he has been a Nautical School teacher at Kotor and member of the Harbor Master's Board of Montenegro State, former Yugoslavia, for Radiotelegraphy and Radiotelephony training and examinations in accordance with SOLAS, IMO and ITU recommendations and regulations for Merchant Officer certifications. He has also been a Lecturer on the Faculty of Maritime Studies at Kotor, Podgorica University, Montenegro.

In the meantime, he was a Deck and Radio/GMDSS Officer on the ships Suisse Atlantique S.A. from Lausanne and of the Mediterranean Shipping Company (MSC) from Geneva.

Working for many years in the shipping industry, coastal radio stations and maritime education and engineering fields, he was involved in Maritime Transportation Technology, GMDSS, SAR and Mobile Radio/Satellite Communications, Navigation, Tracking and Surveillance for Maritime and later, for Land and Aeronautical applications.

He is the author of many papers and manuals about maritime Inmarsat, Cospas-Sarsat, Global Mobile Distress and Safety Systems (GMDSS), Search and Rescue (SAR) systems, maritime navigation, electronics, electric and information systems.

Currently, he is a citizen of South African and is nominated as a part-time professor of Satellite Communications in the postgraduate studies program of Westville Campus at Durban KwaZulu-Natal University.

Two years ago, Prof. Ilčev ceased to be an Associate Member of the IEE in London.

ACRONYMS

a	- Large Semi-major Axis of Elliptical Orbit
A	- Apogee
A	- Azimuth Angle
A/D	- Analog to digital signal conversion
AAC	- Aeronautical Administrative Communications
AAC-AOC	- Airline Administrative and Airline Operational Control
AAIC	- Accounting authority for satellite traffic invoicing
ABR	- Available Bit Rate
ACARS	- Aircraft Communications Addressing and Reporting System
ACARS	- Aircraft Communication and Reporting System
ACARS	- ARINC Communication and Reporting System
ACC	- Aeronautical Control Centre
ACK	- Acknowledgement
ACO	- Aeronautical Communications Organization
ACR	- Allowed Cell Rate
ACS	- Attitude Control System
ACS	- Assembly and Command Ship
ACSE	- Antenna Control and Signaling Equipment of LES
ACSSB	- Amplitude Companded SSB
ACU	- Above Cockpit Unit (Antenna Equipment of AES)
ACU	- Antenna Control Unit
ADE	- Above Deck Equipment (Antenna Equipment of SES)
ADM	- Adaptive Delta Modulation
ADPCM	- Adaptive Differential PCM
ADPS	- Aeronautical Data Processing System
ADREP	- Accident/Incident Data Reporting
ADSL	- Asymmetrical Digital Subscriber Line
ADSS	- Automatic Dependent Surveillance System
AEEC	- Airline Electronic Engineering Committee
AEIS	- Aeronautical Enroute Information Service
AES	- Aircraft (Aeronautical) Earth Station
AFC	- Automatic Frequency Control
AFIS	- Airborne Flight Information Service
AFTAX	- Aeronautical Fixed Telecommunication Automatic Exchange
AFTN	- Aeronautical Fixed Telecommunication Network
AGC	- Automatic Gain Control
AHD	- Above Haul Device (Antenna Equipment of VES)
AHNIS	- Aeronautical Highlights and Navigation Information Services
AIDC	- ATS Interfacility Data Communications
AIES	- Aeronautical Information Enroute Service
AIRCOM	- Aeronautical Communications
AKM	- Apogee Kick Motor
AL	- Accuracy Lateral
ALC	- Automatic Level Control
ALOHA	- A random multiple accesses devised at the University of Hawaii
AM	- Amplitude Modulation
AMBE	- Advanced Multi-band Excitation

AMES	- Aeronautical and Maritime Engineering Satellite (Japanese program)
AMSC	- Aeronautical MSC
AMSC	- American Mobile Satellite Consortium
AMSS	- Aeronautical Mobile Satellite Service (System)
AMVER	- Automated Mutual Assistance Vessel Rescue System
ANS	- Aeronautical Navigation System
ANSS	- Aeronautical Navigation Satellite System
AOC	- Aeronautical (Airline) Operational Control
AOC	- Advanced Operational Capability
AOC	- Attitude and Orbit Control
AOR	- Atlantic Ocean Region (Old system)
AORE	- Atlantic Ocean Region East (New system)
AORW	- Atlantic Ocean Region West (New system)
APC	- Adaptive Predictive Coding
APC	- Aeronautical Passenger Communications
APL	- Airport Pseudolites
Apogee	- More distant point of satellite from the Earth
APS	- Air Passenger Services
APT	- Automatic Picture Transmission
AR	- Axial Ratio
ARC	- Aeronautical Radio Communications
Arinc	- Aeronautical Radio Inc.
ARMMD	- Automatic Remote Monitoring and Messaging Data (M2M)
ARNSS	- Aeronautical Radionavigation Satellite Service
ARQ	- Automatic Repeat Request
ARTS	- Automated Radar Terminal System
ASCII	- American Standard Code for Information Exchange
ASI	- Aeronautical Safety Information
ASIU	- ATM Satellite Interworking Unit
ASK	- Amplitude Shift Keying
ASQF	- Application Specific Qualification Facility
ASR	- Airport Surveillance Radar
ASTP	- Aviation Security Training Packages
Astra	- Applications of Space Technology to the Requirements of Aviation
ATAS	- Aeronautical Transportation Augmentation System
ATC	- Air Traffic Control
ATCAS	- ATC Automation System
ATCDPS	- ATC Data Processing System
ATFM	- Air Traffic Flow Management
ATIS	- Automatic Terminal Information Service
ATM	- Air Traffic Management
ATM	- Asynchronous Transfer Mode
ATS	- Air Traffic Service
AUSREP	- Australian Report (as an AMVER)
AV	- Accuracy Vertical
AVCS	- Aerial Vehicle Communications System
AVICOM	- Avicom Japan Co. Ltd.
AvSat	- Aviation Satellite Program
AVHRR	- Advanced Very High Resolution Radiometer
AVSEC	- Aviation Security
AWGN	- Additive White Gaussian Noise

AWRS	- Aeronautical Weather Report Services
AWT	- Applied Weather Technology
b	- Small Semi-major Axis of Elliptical Orbit
b/s	- Bits per second (Baud)
BA	- British Airways
BACS	- Broadband Aeronautical Communications Service
Baud	- 1 b/s
BBMSS	- Broadband MSS
BC	- Before Christ
BCH	- Bose Chadhuri Hocquenghem Code
BCM	- Block Coded Modulation
BCMSS	- Broadcast MSS
BCU	- Below Cockpit Unit (AES Transceiver Unit)
BDE	- Below Deck Equipment (SES Transceiver Equipment)
Beacon	- All types of emergency satellite beacons used in COSPAS-SARSAT System
BER	- Bit Error Rate
BES	- Base Earth Station (Gateway)
BEST	- Bandwidth Efficient Satellite Transport
BGAN	- Broadband GAN
BHD	- Below Haul Device (VES Transceiver Equipment)
BISDN	- Broadband Integrated Service Digital Network
BMCS	- Broadband Maritime Communication Service
BOL	- Beginning Of Life
BPSK	- Binary PSK
BSU	- Beam Steering Unit
BT	- British Telecom
BVS	- Bonvoyage System (WX)
c	- Axis Between Centre of the Earth and Centre of Ellipse
C/M	- Carrier-to-Multipath Ratio
C/N	- Carrier to Noise Ratio
C/No	- Carrier to Noise Power Density Ratio
CAASD	- Corporation's Centre for Advanced Aviation System Development
CAB	- Civil Aviation Bureau
CAC	- Civil Aviation Community
CADIN	- Common Aeronautical Data Interchange Network
c-band	- Centimetre band
C-band	- Frequency band on 6/4 GHz (Tx/Rx) for Feeder Link (from 4 to 8 GHz)
CBR	- Constant Bit Rate
CCF	- Central Control Facility
CCIR	- International Radio Consultative Committee
CCIT	- International Telecommunications Consultative Committee
CCITT	- International Telegraph and Telephone Consultative Committee
CD	- Compact Disk
CDDA	- Crossed-Drooping Dipole Antenna
CDMA	- Code Division Multiple Access
CDR	- Critical Design Review
CELP	- Code Excited Linear Prediction
CES	- Coast Earth Station (Maritime and Land Mobile)
CIMS	- Customer Information Management System
CIS	- Commonwealth of Independent States (Former USSR)
CIS	- Communication and Information System

cm	- Centimetre Waves (SHF)
CM	- Coded Modulation
CN	- Correspondent Nodes
CNES	- Centre National d'Etudes Spatiales (France)
CNS	- Communications, Navigation and Surveillance
CNSO	- Civil Navigation Satellite Overlay (Inmarsat)
CoA	- Care-of Address
COE	- Committee on ECDIS
COMSAT	- Communications Satellite Corporation
Cospas	- Space System for Search of Distress Vessels and Airplanes - In Russian КОСПАС: Космическая Система Поиска Аварийных Судов и Самолетов
CPDLC	- Control Pilot Data Link Communications
CPF	- Central Processing Facility
CPFSK	- Continuous Phase Frequency Shift Keying
CPU	- Central Processor Unit
CQD	- Come Quick Distress (A UK Precursor to SOS)
CRL	- Communications Research Laboratory of Japan
CRS	- Coast Radio Station (Maritime)
CRT	- Cathode Ray Tube
CSC	- Common Signaling Channel
CSCF	- Called State Control Function
CSS	- Circuit Switched Service
CWnd	- Congestion Window
d	- Distance Between Satellite and the Earth's Surface
D&E	- Demonstration and Evaluation
D/A	- Digital to Analogue Signal Conversion
DAB	- Direct Audio Broadcasting
Dam	- Decametre waves (HF)
DAMA	- Demand Assigned Multiple Access
DASA	- Daimler Chrysler Aerospace AG
dB	- Decibel
DBMC	- Data Base Main Computer
DBPSK	- Differential Binary PSK
DBS	- Direct Broadcasting Satellite
DC	- Direct Current
DCCP	- Distribution and Concentration Communication Processor
DCM	- Digital Circuit Multiplication
DCPR	- Data Collection Platform Repeaters
DCS	- Data Collection System
DCTE	- Data-Circuit Terminating Equipment
DDR	- Data Distribution Region
deci mm	- Deci millimetre Waves (VEHF)
DECT	- Digital European Cordless Communication
DGPS	- Differential GPS
DIP/LNA	- Diplexer/Low Noise Amplifier
DIT	- Digital Image Transfer
DLP	- Data Link Processing
DM	- Delta Modulation
dm	- Decimetre Waves (UHF)
DME	- Distance Measuring Equipment
DMG	- Distress Message Generator

DND	- Department of National Defense (Canada)
DOD	- Tepartment of Defense
DOE	- Department of Energy
DOP	- Dilution of Precision
DPCM	- Differential PCM
DPSK	- Differential PSK
DRS	- Direct Readout Service
DS	- Direct Sequence
DSB-SC	- Double Side Band Suppressed Carrier
DSC	- Digital Selective Call
DSI	- Digital Speech Interpolation
DTAX	- Domestic Telecommunication Automatic Exchange
DTE	- Data Terminal Equipment
DVB	- Direct Video Broadcasting
DVB-S	- DVB-Satellite
DVB- T	- DVB-Terrestrial
DVB-RCS	- Digital Video Broadcasting-Return Channel
DVP	- Development Verification Platform
DVSI	- Digital Voice Systems Incorporated
E	- Eccentric Anomaly
EC	- European Commission
ECAC	- European Civil Aviation Conference
ECDIS	- Electronic Chart Display and Information Systems
ECS	- European Communications Satellite
EDCT	- Estimated Departure Clearance Time (Aviation)
EGC	- Enhanced Group Call
EGNOS	- European Geostationary Navigation Overlay Service
EHF	- Extremely High Frequency from 30 to 300 GHz (mm-band)
EIRP	- Effective Isotropic Radiated Power
ELMSS	- European Land Mobile Satellite System
ELT	- Emergency Locator Transmitter (Aeronautical application)
EM	- Electromagnetic
EMCA	- European Maritime Core Area
EME	- Externally Mounted Equipment
EMS	- European Mobile System
EMSS	- Experimental Mobile Satellite System (Japanese program)
EOL	- End of Life
EP	- Electric Power
EPIRB	- Emergency Position Indicating Radio Beacon (Maritime application)
ERAST	- Environmental Research Aircraft and Sensor Technology
ESA	- European Space Agency
ESNP	- European Satellite Navigation Programme
ESOC	- European Space Operation Centre
ESTB	- EGNOS System Test Bed
ETA	- Estimated Time of Arrival (Shipping and Airways)
ETD	- Estimated Time of Departure (Shipping)
FTP	- File Transport Protocol
ETS	- Engineering Test Satellite (Japanese program)
ETSI	- European Telecommunications Standard Institute
EU	- Electronics Unit
EUMETSAT	- Exploration of Meteorological Satellites

Eurocontrol	- European Organization for the Safety of Air Navigation
Eutelsat	- European Telecommunications Satellite Organization
EVGC	- Enhanced Voice Group Call
FAA	- Federal Aviation Administration of USA
FANS	- Future Air Navigation Systems
Fax	- Facsimile
FCC	- Federal Communications Commission (US)
FDM	- Frequency Division Multiplexing
FDMA	- Frequency Division Multiple Access
FDP	- Flight Data Processing
FEC	- Forward Error Correction
FES	- Fixed Earth Station
FFPP	- Flexible Flight Profile Planning
FH	- Frequency Hopping
FIR	- Flight Information Region
FleetNET	- Inmarsat EGC broadcast of ship-owner data to part of all fleet
FLS	- Forward Link Subsystem
FM	- Frequency Modulation
FoIP	- Fax over IP
FQR	- Factory Qualification Review
FRec	- Fast Recovery
FRet	- Fast Retransmit
FSK	- Frequency Shift Keying
FSS	- Fixed Satellite Service
FSS	- Flight Standards Service
FSSE	- Flight Safety Satellite Equipment
FTP	- File Transfer Protocol
G	- Universal Gravitational Constant
G/T	- Ratio of system gain to system noise Temperature
GABS	- Globalstar Accounting & Billing System
GACCS	- Global Aeronautical Corporate and Commercial System
GADSS	- Global Aeronautical Distress and Safety System
GAN	- Global Area Network
GASP	- Global Aviation Safety Plan (ICAO)
GASSC	- Global Aeronautical Safety Satellite Communications
Gateway	- GES (BES)
GBAS	- Ground Based Augmentation System
GBO	- Globalstar Business Office
Gb/s	- Gigabit per second
GCC	- Gateway Control Centre
GCC	- Ground Control Station
GDN	- Globalstar Data Network
GDSS	- Global Determination Satellite System
GEO	- Geostationary Earth Orbit
GEOLUT	- GEO Local User Terminal
GEOSAR	- GEO Search and Rescue
GES	- Ground Earth Station (Aeronautical Mobile)
GES	- Gateway Earth Stations
GFC	- Ground Forecasting Centre
GFR	- Guaranteed Frame Rate
GGS	- Gateway Ground Station

GHz	- Gigahertz is Radio Frequency Unit of 1000 MHz
GIC	- Ground Integrity Channel
GIO	- Geosynchronous Inclined Orbit
GLONASS	- Global Navigation Satellite System (In Russian: ГЛОНАСС - Глобальная Навигационнаяа Спутниковая Система)
GLS	- GNSS Landing System
GMBSS	- Global Maritime Broadcasting Satellite System
GMDSS	- Global Maritime Distress and Safety System
GMPSC	- Global Mobile Personal Satellite Communications
GMR	- Geostationary Mobile Radio (of Himaward Japanese Series)
GMS	- Geosynchronous Meteorological Satellite
GMS	- Ground Monitoring Stations
GMSC	- Global Mobile Satellite Communications
GMSS	- Global Mobile Satellite Systems
GMT	- Global Meteorological Technologies
GNSS	- Global Navigation Satellite System (ICAO Definition)
GNSS-P	- GNSS Panel
GOES	- Geostationary Operational Environmental Satellite
GOES	- Global Observation Environmental Satellite
GPRS	- General Packet Radio Service
GPS	- Global Positioning System
GR	- Guaranteed Rate
GRS	- Geostationary Ranging Station
GRS	- Ground Radio Station (Aeronautical)
GSAS	- Global Satellite Augmentation System
GSSII	- Global Sea State Information via Internet
GSM	- Global Service for Mobile communications or Group Special Mobile
GSPS	- Global Stratospheric Platform Systems
GTO	- Geostationary Transfer Orbit
GTS	- Global Telemaque Service
GW	- Gateway
h	- Altitude of Satellite Above the Earth's Surface
HAPS	- High Altitude Platform Station
HDTV	- High Definition TV
HEC	- Hybrid Error Correction
HEO	- Highly (Highinclined) Elliptical Orbit
HF	- High Frequency from 3 to 30 MHz (Dam-band)
HGA	- High Gain Antenna
HLP	- Horizontal Linear Polarization
Hm	- Hectometre Waves (MF)
HoA	- Home Address
HPA	- High Power Amplifier
HPBW	- Half Power Beamwidth
HSD	- High Speed Data over 9600 b/s, max 56/64 Kb/s
HSO	- Hybrid Satellite Orbits
HTTP	- Hyper-Text Transfer Protocol
Hz	- Hertz is Basic Radio Frequency Unit (1 Cycle/sec)
i	- Inclination Angle
IA	- Initial Approach
IAMSAR	- International Aeronautical and Maritime SAR Manual
IATA	- International Air Transport Association

IAWVG	- Instrumental Approach with Vertical Guidance
IBF	- Input Bandpass Filter
ICAO	- International Civil Aviation Organization
ICF	- Integrated Communications Facility
ICO	- Intermediate Circular Orbits
IDEC	- Integrated Development and Evaluation Centre
IDP	- Input Data Processing
IDU	- Indoor Unit
IEC	- International Electrotechnical Commission
IETF	- Internet Engineering Task Force
IF	- Intermediate Frequency
IFR	- Instrument Flight Rule
IFRB	- International Frequency Registration Board
IGA	- Intermediate Gain Antenna
IGN	- Inmarsat Ground Network
IGP	- Ionospheric Grid Points
IH	- Inmarsat Hemisphere
IHO	- International Hydrographic Organization
IKE	- Internet Key Exchange
ILS	- Instrument Landing System
IM	- Intermodulation
IMBE	- Improved Multi Band Excitation
IMCO	- Intergovernmental Maritime Consultative Organization (Former IMO)
IME	- Internally Mounted Equipment
IMO	- International Maritime Organization (New name)
IMOSAR	- IMO Search and Rescue Manual
INMARSAT	- INternational MARitime SATellite
INR	- Interference and Noise Ratio
INSAT	- Indian National Satellite System
Intelsat	- International Telecommunications Satellite organization
Intersputnik	- East European satellite communications organization formed by Russia
IOL	- Inter-Orbit Link
IOR	- Indian Ocean Region
IP	- Internet Protocol
IPSec	- IP Security Protocol
IPv4	- IP version 4
IPv6	- IP version 6
ISAS	- Inmarsat Satellite Augmentation System
ISDN	- Integrated Services Digital Network
ISL	- Inter-Satellite Link
ISP	- Internet Service Provider
ISPS	- International Ship and Port Security
ISTS	- Intelligent Satellite Transport System
ITS	- Intelligent Transportation System
ITU	- International Telecommunications Union
ITU-R	- ITU Radio
IWG	- Interoperability Working Group
JAL	- Japan Air Lines
JAMSTEC	- Japan Marine Science and Technology Centre
JAPREP	- Japan Report (like AMVER)
JMA	- Japan Meteorological Agency

JTIDS	- Joint Tactical Information Distribution System
K	- Boltzmann Constant
Ka-band	- Frequency band between 24 and 40 GHz
Kb/s	- Kilobit per second
K-band	- Frequency band between 18 and 24 GHz
KDD	- Kokusai Denshin Denwa (Japanese Telecom)
kHz	- Kilohertz is Radio Frequency Unit of 1000 Hz
km	- Kilometre
Km	- Kilometre Waves (LF)
Ku-band	- Frequency band between 12 and 18 GHz
kW	- Kilo Watt
LAA	- Local Augmentation Area
LAAS	- Local Area Augmentation System
LAN	- Local Area Network
L-band	- Frequency band on 1,6/1,5 GHz (Tx/Rx) for Service Link (from 1 to 2 GHz)
LCC	- Launch Control Centre
LDM	- Linear Delta Modulation
LEO	- Low Earth Orbit
LEOLUT	- LUT Local User Terminal
LEOSAR	- LEO Search and Rescue
LES	- Land Earth Station (All Mobile Applications)
LF	- Low Frequency from 30 to 300 kHz (Km-band)
LGA	- Low Gain Antenna
LGF	- LAAS Ground Facility
LHCP	- Left Hand Circular Polarization
LLM	- L-band Land Mobile
LMDS	- Local Multipoint Distribution Service
LMGT	- Lockheed Martin Global Telecommunications
LMSC	- Land MSC
LMSS	- Land Mobile Satellite Service (System)
LNA	- Low Noise Amplifier
LNSS	- Land Navigation Satellite System
LO	- Local Oscillator
LOA	- Length Over All
LOS	- Line-of-Sight
LP	- Launch Platform
LPC	- Linear Predictive Coding
LRNSS	- Land Radionavigation Satellite Service
LRPT	- Low Rate Picture Transmission
LSD	- Low Speed Data up to 2400 b/s
LTAS	- Land Transportation Augmentation System
LUT	- Local User Terminal (COSPAS-SARSAT receiving GES)
M	- Mass of the Earth Body
M	- Point of the Observer or Mobile
M2M	- Mashine-to-Mashine (SCADA)
MA	- Multiple Access
MAA	- Microstrip Array Antenna
Marecs	- Maritime European Communication Satellites
Marisat	- Maritime Satellite (system established 1976 by USA)
Marots	- Maritime Orbital Test Satellite
MAYDAY	- Distress Signal in Mobile Radiotelephony

Mb/s	- Megabit per second
m-band	- Metre band
MBCM	- Multiple BCM
MCC	- Mission Control Centre
MCC	- Master Control Centre
MCPC	- Multiple Channels per Carrier
MCR	- Minimum Cell Rate
MCS	- Master Control Station
MDCRS	- Meteorological Data Collection and Reporting System
MDSS	- Medium Data Rate Satellite System
MELPC	- Multipulse Excited LPC
MEO	- Medium Earth Orbit
MERSAR	- Merchant Ship Search and Rescue Manual
MES	- Mobile Earth Station
METAR	- Meteorological Aviation Reports
MF	- Medium Frequency from 300 to 3000 kHz (Hm-band)
MGA	- Medium Gain Antenna
MHz	- Megahertz is Radio Frequency Unit of 1000 kHz
MIFR	- Master International Frequency Register
MIPv6	- Mobile IP version 6
MIRP	- Manipulated Information Rate Processor
MLQ	- Maximum Likelihood Quantization
MLMSS	- Meteorological MSS
mm	- Millimetre Waves (EHF)
MM	- Motion Media
MMMSC	- Multimedia MSC
MMSC	- Maritime Mobile Satellite Communications
MMSS	- Maritime Mobile Satellite Service (System)
MMW	- Millimetre Wavelength
MN	- Mobile Node
MNSS	- Maritime Navigation Satellite System
MOBSAT	- Mobile Satellite Group
MORFLOT	- Ministry of Merchant Marine (former USSR)
Morya	- Russian (the former USSR) program for MMSS
MOU	- Memorandum of Understanding (COSPAS-SARSAT)
MPA	- Microstrip Patch Antenna
MPDS	- Mobile Packet Data Service
MPEG	- Moving Picture Expert Group
MPEG-2	- Moving Picture Expert Group 2
MRDU	- Multichannel Receiver Decoder Unit
MRNSS	- Maritime Radionavigation Satellite Service
MRS	- Monitoring and Ranging Station
ms	- Millisecond
MSA	- Mobile Satellite Antenna
MSAS	- MTSAT Satellite-based Augmentation System
MSAT	- Mobile Satellite System
MSB	- Mobile Satellite Broadcasting
MSC	- Mobile Satellite Communications
MSG	- Meteosat Second Generation
MSI	- Mobile Satellite Internet
MSI	- Maritime Safety Information

MSK	- Minimum Shift Keying
MSMC	- Message Switching Main Computer
MSS	- Mobile Satellite Service (System)
MRDU	- Multichannel Receiver Decoder Unit
MSUA	- Mobile Satellite Users Association
MTAS	- Maritime Transportation Augmentation System
MTB	- Mediterranean Test Bed
MTCM	- Multiple TCM
MTSAT	- Multi-functional Transport Satellite
MTU	- Mobile Terminal Unit
MUSAT	- Canadian program for LMSS
m/v	- Motor Vessel
MUT	- Mobile User Terminal
MUX	- Multiplexer
MWIF	- Mobile Wireless Internet Forum
NAS	- National Air (Airspace) System
NASA	- National Aeronautical and Space Administration (formed by US)
NASDA	- National Space Development Agency (formed by Japan)
NAVAREA	- IMO Global Navigational Areas
NAVTEX	- Narrow-band Direct-printing Telepgraphy System
NCC	- Network Control Centre
NCF	- Network Control Functions
NCS	- Network Coordination (Control) Station
NDWS	- Noble Denton Weather Services
NES	- Navigation Earth Station
NLES	- Navigation LES
NMA	- Norwegian Mapping Authority
NMF	- Network Management Functions
NMS	- Network Management Station
NOAA	- National Oceanic and Atmospheric Administration (USA)
NOC	- Network Operations Centre
Non-GEO	- Non GEO Satellite Configuration
NOTAM	- Notice to Airman
NPA	- Non-Precision Approach
NSTB	- National Satellite Test Bed
NVNG MSS	- Non-Voice Non-GEO MSS
NWS	- National Weather Service
NX	- Navigation Report (Warning)
OAM	- Operation and Maintenance
OBF	- Output Bandpass Filter
OCC	- Operation Control Centre
OCD	- Oceanic Clearance Delivery
ODL	- Oceanic Data Link
ODP	- Oceanic Air Traffic Control Data Processing
ODU	- Outdoor Unit
OMUX	- Output Multiplexer
O-QPSK	- Offset Quadrature Phase Shift Keying
ORR	- Operational Readiness Review
OSN	- Operational Support Network
OTG	- Ocean Traffic Generator
OTS	- Orbital Test Satellite

p	- Focal parameter
P	- Sub-satellite Point
PA	- Precision Approach
PABX	- Private Automatic Branch Exchange
PACF	- Performance Assessment and System Checkout Facility
PAMA	- Pre-Assigned Multiple Access
PANPAN	- Urgency Signal in Mobile Radiotelephony
P-band	- Frequency band between 200 and 400 MHz
PC	- Personal Computer
PCM	- Pulse Code Modulation
PCR	- Peak Cell Rate
PCS	- Personal Communications Service
PDC	- Personal Digital Cellular system designed in Japan for operation in either the 800 MHz and 1,50 GHz band
PDH	- Plesiochronous Digital Hierarchy
PDN	- Public Data Network
PDS	- Processed Data Stream
PE	- Propulsion Engine
PEO	- Polar Earth Orbit
Perigee	- More nearest point of satellite from the Earth
PES	- Portable (Personal) Earth Station
PFD	- Power Flux Density
PIB	- Pre-flight Information Bulletin
PID	- Pager Identity
PLB	- Personal Locator Beacon (Land or personal applications)
PLCP	- Physical Layer Convergence Protocol
PM	- Phase Modulation
PMSS	- Personal Mobile Satellite Service (System)
PN	- Pseudo (Pseudorandom) Noise
PNA	- Public Network Access
POR	- Pacific Ocean Region
PPP	- Point-to-Point Protocol
PPS	- Precise Positioning Service
PRMA	- Packed Reserved Multiple Access
PSDN	- Public Switched Data Network
PSK	- Phase Shift Keying
PSPDN	- Public Switched Private Data Network
PSTN	- Public Switched Telephone Network
PTT	- Post, Telegraph and Telephone
PVT	- Position, Velocity and Time
QAM	- Quadrature Amplitude Modulation
QHA	- Quadrifilar Helix Antenna
QoS	- Quality of Service
QPSK	- Quadrature PSK
R	- Equatorial Radius of the Earth
R&D	- Research and Development
R&R	- Rural and Remote
RAC	- Rural Area Coverage
RAIM	- Receiver Autonomous Integrity Monitoring
RC	- Requesting Channel
RCC	- Rescue Coordination Centre

RDDI	- Radio Direction Distance Information
RDI	- Radio Direction Information
RDMA	- Random (Packet) Division Multiple Access
RDP	- Radar Data Processing
RDSS	- Radio Determination Satellite Service
RF	- Radio Frequency
RFC	- Request for Comments
RFC	- Regenerative Fuel Cells
RFP	- Request for Proposal
RFU	- RF Unit
RHCP	- Right Hand Circular Polarization
RIMS	- Ranging and Integrity Monitoring Stations
RLS	- Return Link Subsystem
RNSS	- Radionavigation Satellite Service
RPY	- Roll, Pitch, Yaw
RR	- Radio Regulations
RS	- Reed-Solomon
RSC	- Rescue Sub Centre
RSS	- Reference and Synchronization Subsystem
RSM	- Reduction of Separation Minima
RTS	- Remote Troubleshooting System
RTT	- Round-Trip Time
RWnd	- Receiver Advertised Window
Rx	- Receiver
s	- Seconds
S	- Satellite
S/N	- Signal to Noise ratio
s/s	- Steam Ship
SA	- Service Area
SAC	- Special Access Code
SAC	- Suburban Area Coverage
SACK	- Selective Acknowledgment
SafetyNET	- Inmarsat EGC Based System for MSI Transmission
SAN	- Satellite Access Node
SAR	- Search and Rescue
SARP	- SAR Receiver-Processor and Memory
SARP	- Standard and Recommended Practices
SARP	- Standards and Recommended Practices
SARR	- SAR Repeater
Sarsat	- Search and Rescue Satellite Aided Tracking
SART	- SAR Radar Transponder
SATC	- Satellite Air Traffic Control
SAU	- Satellite Antenna Unit
S-band	- Frequency band between 2 and 4 GHz
SBAS	- Satellite Based Augmentation System
SBF	- Short Backfire
SBIR	- Small Business Innovative Research
SCADA	- Supervisory Control and Data Acquisition (M2M)
SCC	- Satellite Control Centre
SCORE	- Signal Communicating by Orbiting Relay Equipment
SCP	- Stratospheric Communications Platform

SCPC	- Signal Channel Per Carrier
SCS	- Satellite Control Station (System)
SCU	- Satellite (Subscriber) Communication Unit
SDH	- Synchronous Digital Hierarchy
SDK	- Software Developer's Kit
SDM	- System Definition Manual (of Inmarsat)
SDMA	- Space Division Multiple Access
SDU	- Satellite Data Unit
SECURITE	- Safety Signal in Mobile Radiotelephony
SEHA	- Safety Enhancement in High Altitudes
SELA	- Safety Enhancement in Low Altitudes
SES	- Ship Earth Station
SHF	- Super High Frequency between 3 and 30 GHz (cm-band)
SID	- Standard Interface Description
SIM	- Subscriber Identity Module
SINR	- Signal to Interference and Noise Ratio
SIR	- Signal to Interference Ratio
SIS	- Signal in Space
SITA	- Societé Internationale de Télécomommunications Aeronautique
SMAP	- Spot Management and Planning
SME	- Small to Medium-size Enterprise
SMGC	- Surface Movement Guidance and Control
SMS	- Short Message Service
SoHo	- Small office Home office
SOIT	- Satellite Operational Implementation Team
SOLAS	- Safety of Life at Sea Convention
SONET	- Synchronous Optical Network
SOS	- Distress Signal in Mobile Radiotelegraphy
SP	- Structure Platform
SPOC	- SAR Point of Contact
SPS	- Standard Positioning Service
SRS	- Ship Radio Station (MF/HF/VHF Conventional Mobile Radio)
SSB	- Single Side Band
SSI	- Sky Station International Inc.
SSMA	- Spread Spectrum Multiple Access
SSPA	- Solid State Power Amplifier
SSThres	- Slow Start Threshold
SSTV	- Slow Scan Television
STE	- Secure Telephone Equipment
STM	- Synchronous Transfer Mode
STS	- Stratospheric Telecommunications Service
STS	- Space Transportation System
SUDS	- Secondary Users Data Station
SUT	- SatLink User Terminal
SWAT	- SBAS Worldwide Availability Tool
Swift64	- Aeronautical HSD Solution
TAF	- Terminal Area Forecasts
TAO	- Telecommunications Advancement Organization of Japan
TAV	- TransAtmospheric Vehicle
Tb/s	- Terabit per second
TBUE	- Test Bed User Equipment

TC	- Thermal Control
TCC	- Traffic Control Centre
TCE	- Traffic Channel Equipment
TCM	- Trellis Coded Modulation
TCP	- Transmission Control Protocol
TCS	- Terrestrial Communications Subsystem
TDM	- Time Division Multiplexing
TDMA	- Time Division Multiple Access
TDMoIP	- TDM over IP
TDRSS	- Tracking and Data Relay Satellite System
TEC	- Total Electron Content
Tel	- Telephone
TES	- Transportable Earth Station
Tlg	- Telegraphy
Tlx	- Telex
TMI	- Telesat Mobile Inc.
TMS	- Test Master Station
TRAD	- Terminal Radar Alphanumerical Display
TT&C	- Tracking, Telemetry, and Command
TTN	- Terrestrial Telecommunications Network
TTT	- Safety Signal in Mobile Radiotelegraphy
Tty	- Teletypewriter (Teleprinter)
TV	- Television
TVRO	- TV Receive Only
TWIP	- Terminal Weather Information for Pilots
TWT	- Traveling Wave Tube
TWTA	- Traveling Wave Tube Amplifier
Tx	- Transmitter
UAC	- Urban Area Coverage
UAV	- Unmanned Aerial Vehicles
UBR	- Undefined (Unspecified) Bit Rate
UDI	- Unrestricted Digital Information
UHF	- Ultra High Frequency between 300 MHz and 3 GHz (dm-band)
UN	- United Nations
USB	- Unified S-band
USB	- Universal Serial Bus
USNS	- US Navy Ship
USS	- US Ship
UST	- User Service Terminal
UT	- User Terminal
VARB	- Van Allen Radiation Belts
VAS	- Value Added Service
VBR	- Variable Bit Rate
VC	- Video Conferencing
VC	- Virtual Channels
VCoIP	- VideoConference over IP
VDB	- VHF Data Broadcast
VDL	- VHF Digital Link
VDU	- Video Display Unit
VEHF	- Very Extremely High Frequency from 300 GHz to 3 THz (deci mm-band)
VES	- Vehicle Earth Station

VFR	- Visual Flight Rules
Version-2	- Two GAN units linked together via Bonding cable double throughput to 128 Kb/s
VHF	- Very High Frequency between 30 and 300 MHz (m-band)
VLP	- Vertical Linear Polarization
VoIP	- Voice over IP
Volna	- Russian (Ex-USSR) program for MMSS and AMSS
VOR	- VHF Omnidirectional Ranging
VPN	- Virtual Private Networks
VPoIP	- VideoPhone over IP
VS/VD	- Virtual Source/Virtual Destination
VSAT	- Very Small Aperture Terminal
VSWR	- Voltage Standing Wave Ratio
VTS	- Vessels Traffic Services
WAA	- Wide Augmentation Area
WAAS	- Wide Area Augmentation System
WAD	- Wide Area Differential
WADGNSS	- Wade Area Differential GNSS
WAN	- Wide Area Network
WANAV	- Wide Area Navigation (RNAV)
WARC	- World Administrative Radio Conference
WARC-ST	- WARC-Satellite
W-CDMA	- Wireless-CDMA
WDCP	- Weather Data Collection Platforms
WEFAX	- WX Fax
WIPP	- WAAS Integrity Performance Panel
WMO	- World Meteorological Organization
WMS	- Wide-Area Master Stations
WQHA	- Wire Quadrifilar Helix Antenna
WRC	- World Radio (Radiocommunications) Conference
WRS	- Wide-Area Reference Stations
WWW	- World-Wide Web (Internet)
WX	- Weather Report (Warning)
X25/X70/X400	- Standards for messaging and data systems specified by the CCITT
X-band	- Frequency band between 8 and 12 GHz
XXX	- Urgency Signal in Maritime Mobile Radiotelegraphy
XPD	- Cross-Polar Discrimination
XPI	- Cross-Polar Isolation
XSA	- Cross-Slot Array Antennas
δ	- Angle of Antenna Radiation
ε	- Elevation Angle
ϕ	- Phase Angle
Θ	- True Anomaly
Π	- Perigee
ν	- Angular Speed of the Earth's Rotation
λ	- Geographical Longitude
φ	- Geographical Latitude
Ψ	- Central or Sub-satellite Angle
ω	- Frequency Angle
ω	- Argument of the Perigee
Ω	- Right Ascension of an Ascending Node Angle

INDEX